physica status solidi (a)

VOLUME 12 · NUMBER 2 · 1972

KADEMIE-VERLAG · BERLIN

Classification Scheme

1. Structure of Crystalline Solids
   1.1 Perfectly Periodic Structure
   1.2 Solid-State Phase Transformations
   1.3 Alloys. Metallurgy
   1.4 Microstructure (Magnetic Domains See 18; Ferroelectric Domains See 14.4.1)
   1.5 Films
   1.6 Surfaces
2. Non-Crystalline State
3. Crystal Growth
4. Bonding Properties
5. Mössbauer Spectroscopy
6. Lattice Dynamics. Phonons
7. Acoustic Properties
8. Thermal Properties
9. Diffusion
10. Defect Properties (Irradiation Defects See 11)
    10.1 Metals
    10.2 Non-Metals
11. Irradiation Effects (X-Ray Diffraction Investigations See 1 and 10)
12. Mechanical Properties (Plastic Deformations See 10)
    12.1 Metals
    12.2 Non-Metals
13. Electron States
    13.1 Band Structure
    13.2 Fermi Surfaces
    13.3 Surface and Interface States
    13.4 Impurity and Defect States
    13.5 Elementary Excitations (Phonons See 6)
        13.5.1 Excitons
        13.5.2 Plasmons
        13.5.3 Polarons
        13.5.4 Magnons
14. Electrical Properties. Transport Phenomena
    14.1 Metals. Semi-Metals
    14.2 Superconductivity. Superconducting Materials and Devices
    14.3 Semiconductors
        14.3.1 Films
        14.3.2 Surfaces and Interfaces
        14.3.3 Devices. Junctions (Contact Problems See 14.3.4)
        14.3.4 High-Field Phenomena, Space-Charge Effects, Inhomogeneities, Injected Carriers (Electroluminescence See 20.3; Junctions See 14.3.3)
    14.4 Dielectrics
        14.4.1 Ferroelectrics
15. Thermoelectric and Thermomagnetic Properties
16. Photoconductivity. Photovoltaic Effects
17. Emission of Electrons and Ions
    17.1 Field Emission Microscope Investigations
18. Magnetic Properties
    18.1 Paramagnetic Properties
    18.2 Ferromagnetic Properties
        18.2.1 Ferromagnetic Films
    18.3 Ferrimagnetic Properties
    18.4 Antiferromagnetic Properties

(Continued on cover three)

physica status solidi (a)

applied research

Volume 12 · Number 2 · Pages 317 to 686, K65 to K124, and A9 to A16

August 16, 1972

PSSA 12(2) 317—686, K65—K124, A9—A16 (1972)

AKADEMIE-VERLAG · BERLIN

Subscriptions and orders for single copies should be addressed to

AKADEMIE-VERLAG GmbH, 108 Berlin, Leipziger Straße 3 – 4 or to Buchhandlung KUNST UND WISSEN, Erich Bieber, 7 Stuttgart 1, Wilhelmstr. 4 – 6 or to Deutsche Buch-Export und -Import GmbH, 701 Leipzig, Postschließfach 160

Editorial Note:

"physica status solidi (a)" undertakes that an original paper *accepted for publication* before the 23rd of any month will be published within 50 days of this date unless the author requests a postponement. In special cases there may be some delay between *receipt* and *acceptance* of a paper due to the review and, if necessary, revision of the paper.

Schriftleiter und verantwortlich für den Inhalt: Professor Dr. Dr. h. c. P. Görlich, 102 Berlin, Neue Schönhauser Str. 20 bzw. 69 Jena, Humboldtstr. 26. Chefredakteur: Dr. H.-J. Hänsch. Redaktionskollegium: Prof. Dr. E. Gutsche, Dr. H.-J. Hänsch, Dr. H. Lange, Dr. S. Oberländer. Anschrift der Schriftleitung: 102 Berlin, Neue Schönhauser Str. 20, Fernruf: 4 22 33 80. Verlag: Akademie-Verlag GmbH, 108 Berlin, Leipziger Str. 3–4, Fernruf: 22 04 41, Telex-Nr. 112 020, Postscheckkonto: Berlin 350 21. – Die Zeitschrift „physica status solidi (a)" erscheint jeweils am 16. des Monats. Bezugspreis eines Bandes 120,– M. Bestellnummer dieses Bandes: 1085/12. Jeder Band enthält zwei Hefte. Gesamtherstellung: VEB Druckerei „Thomas Müntzer" Bad Langensalza. – Veröffentlicht unter der Lizenznummer 1310 a des Presseamtes beim Vorsitzenden des Ministerrates der Deutschen Demokratischen Republik.

Contents

Short Notes

Pre-printed Titles

Systematic List

The Author Index of Volume 12 Begins on Page 687

(It will be delivered together with Volume 13, Number 1.)

Review Article

phys. stat. sol. (a) **12**, 325 (1972)

Subject classification: 19; 13.4; 22; 22.4; 22.4.2; 22.5; 22.5.3; 22.6; 22.8; 22.8.1; 22.8.2; 22.9

Department of Physics, University of Calgary, Calgary, Alberta Canada

A Survey of the EPR Spectra of Gd^{3+} in Single Crystals

By

H. A. Buckmaster and Y. H. Shing

Contents

1. Introduction

The paramagnetic properties of the gadolinium ion Gd^{3+} have had a long and interesting history. In 1923 Woltjer and Kamerlingh Onnes [1] discovered that gadolinium sulphate behaved in accordance with Langevin's theory at liquid helium temperatures. Subsequently Debye [2] and Giauque [3] proposed independently in 1926 and 1927 that this behaviour could be utilized to produce magnetic cooling. This proposal was realized by Giauque and MacDougall [4] in 1933 using this salt. Other Gd^{3+} compounds have contributed to the development of the adiabatic demagnetization technique. It was natural, when the technique of electron paramagnetic resonance (EPR) was developed in 1945, that Gd^{3+} compounds would be studied extensively. Interest in the paramagnetic properties of this ion was renewed in 1956 after it was successfully utilized by Scovil et al. [5] as the active material in the first operational solid-state maser. Consequently the ground state splittings of Gd^{3+} in crystalline electric fields with various symmetries have been measured extensively using EPR as a function of temperature. These measurements were facilitated by its unusually long relaxation time which enables its EPR spectrum to be observed at temperatures as high as ≈ 290 °K. Gadolinium occupies a unique position in the lanthanide series of elements since the $4f^n$ electronic configuration is half-filled. It is well known that the 4f electrons in this series exhibit, to first approximation, a spin–orbit coupling of the Russell-Saunders type. The $4f^7$ configuration gives rise to an $^8S_{7/2}$ free-ion ground state in this approximation according to Hund's rules. Experimentally it is known that this approximation is inadequate to explain the EPR and optical spectra of Gd^{3+} since the g-value of the ground state is less than the expected free electron value of 2.0023. Moreover, contrary to the predictions based on this approximation, the eightfold degeneracy of this state is partially removed by the crystalline electric potential to which it is exposed when it is substituted into a diamagnetic host lattice. The group-theoretical considerations of Bethe [6] show that a crystalline electric potential with cubic symmetry splits an eightfold degenerate free-ion ground state into two doublets and one quartet and that axial symmetry splits this state into four doublets. In general, the observed ground state splittings of Gd^{3+} in crystalline electric potentials of various symmetries are $\approx 0.2\ cm^{-1}$. Consequently they are readily accessible to direct measurement using EPR techniques at microwave frequencies with commonly available laboratory magnetic fields. It is challenging to study the ground state splittings of an S-state ion like Gd^{3+} because no satisfactory theoretical explanation has been found for the magnitude of these splittings. Lacroix [7] calculated the g-value for the ground state $^8S_{7/2}$ of Gd^{3+} by considering perturbation due to the excited states through the spin–orbit coupling $\zeta \sum_i \boldsymbol{l}_i \cdot \boldsymbol{S}_i$. He obtained a g-value of 1.9930 by taking account of the $^6P_{7/2}$ excited state assuming that the value of the spin–orbit coupling constant ζ was $1540\ cm^{-1}$. Wybourne [8] determined that the best-fit value of ζ for Gd^{3+} was $1480\ cm^{-1}$ by using an intermediate coupling scheme to fit the experimental energy levels. He obtained a g-value of 1.99454 for the ground state using this best-fit value of ζ. These theoretical g-values are larger than the experimental g-value of (1.9918 ± 0.001). Wybourne [8] also reviewed the various perturbation mechanisms that may be responsible for the ground state splittings of Gd^{3+} in a crystalline electric potential of axial symmetry and calculated the magnitude of the splitting produced by each mechanism. He concluded that no single mechanism or combination of these mechanisms could explain the observed splittings. Sharma [9] concluded from a limited study of the EPR spectrum of Gd^{3+}, when it is substituted into the C_{2v} symmetry of the host lattice LnF_3 where Ln $\equiv$ La, Ce, Pr, Nd, that good agreement could be obtained using an ionic model of this system. Buckmaster et al. [10] reexamined these perturbation mechanisms and compared the predicted splittings with the experimentally determined fine structure splitting parameters describing the EPR spectrum of the Gd^{3+} ion when it is substituted systematically into the C_{3h} symmetry of the host lattice $Ln(C_2H_5SO_4)_3 \cdot 9\,H_2O$ where Ln $\equiv$ La, Ce, Pr, Nd, Sm, Eu, Tb, Ho, Er, Tm, Yb, Y. They concluded from these measurements that no success can be expected from any mechanism which is based on an ionic model. These contradictory conclusions appear to imply that the nature of the interaction is more complicated than had been anticipated previously. The temperature dependence study of the spin Hamiltonian parameters of Gd^{3+} in $CaWO_4$ by Harvey and Kiefte [11] tends to confirm the conclusion of Buckmaster et al. [10]. Consequently a theoretical explanation of the splitting of the ground state of the Gd^{3+} ion in a crystalline electric potential is not yet possible despite the extensive literature available describing the results of experimental studies on this ion. It is appropriate that these results should be collected together in a systematic fashion since they are scattered throughout the literature using a variety of nomenclatures. It is hoped that the availability of this literature review will encourage renewed theoretical effort to uncover the underlying physical explanation for the existence and magnitude of the ground state splittings in Gd^{3+}.

It is conventional to fit the experimental EPR spectra for the Gd^{3+} ion using the spin Hamiltonian formalism first proposed by Abragam and Pryce [12]. This formalism is a zero-order approximation to the general spin Hamiltonian describing the ground state behaviour of the electrons in a paramagnetic ion under the influence of a crystalline electric potential, an external magnetic field, and the magnetic field due to its nucleus. Buckmaster and Chatterjee [13] have discussed the merit of using a generalized spin Hamiltonian to analyse the EPR and ENDOR spectra of the Eu^{2+} ($4f^7$, $^8S_{7/2}$) ion. It is possible that this approach will find favour in the future when magnetic field measurement techniques are refined to yield greater accuracy. Presently the available accuracy is not sufficient to justify the inclusion of any of the additional terms appearing in the generalized spin Hamiltonian but not appearing in the conventional spin Hamiltonian when analysing EPR data although their inclusion is justified when ENDOR data is analysed. The hyperfine structure in the EPR spectra of the Gd^{3+} ion arises from the isotopes Gd^{155} and Gd^{157} which both have a nuclear spin of 3/2. This structure is seldom resolved because the magnitude of the hyperfine splitting term is almost always small compared with the observed linewidth of the EPR transitions for Gd^{3+} which is, typically, 1 to 4 mT. Relatively little systematic work has been done on the linewidths nor on relaxation times, so this information is omitted from this summary.

2. The Crystalline Electric Field Potential at the Site of a Paramagnetic Ion

The crystalline electric potential at the site of a paramagnetic ion can be expressed in terms of spherical harmonics (Hutchings [14])

$$V(r, \theta, \varphi) = \sum_{l=0}^{\infty} \sum_{m=-l}^{+l} r^l \gamma_l^m Y_l^m(\theta, \varphi), \tag{1}$$

where

$$\gamma_l^m = \sum_j \left(\frac{4\pi}{2l+1}\right) \frac{q_j}{R_j^{(l+1)}} (-1)^m Y_l^{-m}(\theta_j, \varphi_j),$$

$Y_l^m(\theta, \varphi)$ is a spherical harmonic,

(r, θ, φ) are the spherical coordinates of the unpaired electron of the paramagnetic ion,

and

$(R_j, \theta_j, \varphi_j)$ are the spherical coordinates of the j-th ligand ion with charge q_j.

The crystalline electric potential $V(r, \theta, \varphi)$ in equation (1) can be rewritten following Prather [15] in terms of tesseral harmonics to avoid imaginary quantities:

$$V(r, \theta, \varphi) = \sum_{l=0}^{\infty} \sum_{m=0}^{l} r^l \left[C_l^m Z_{lm}^c(\theta, \varphi) + S_l^m Z_{lm}^s(\theta, \varphi)\right], \tag{2}$$

where the tesseral harmonics are defined by

$$\left.\begin{aligned} Z^{c}_{lm} &\equiv \frac{1}{\sqrt{2}}\left[Y^m_l + (-1)^m Y^{-m}_l\right] = \left[\frac{2l+1}{2}\frac{(l-m)!}{(l+m)!}\right]^{1/2} P^m_l(\cos\theta)\frac{\cos m\varphi}{\sqrt{\pi}},\\ Z^{s}_{lm} &\equiv \frac{i}{\sqrt{2}}\left[Y^m_l - (-1)^m Y^{-m}_l\right] = \left[\frac{2l+1}{2}\frac{(l-m)!}{(l+m)!}\right]^{1/2} P^m_l(\cos\theta)\frac{\sin m\varphi}{\sqrt{\pi}},\\ Z_{l0} &= Y^0_l, \end{aligned}\right\} \tag{3}$$

$P^m_l(\cos\theta)$ is an associated Legendre polynomial,

$$\left.\begin{aligned} C^m_l &= \sum_j \left(\frac{4\pi}{2l+1}\right)\frac{q_j}{R_j^{(l+1)}} Z^c_{lm}(\theta_j, \varphi_j)\\ &\text{and}\\ S^m_l &= \sum_j \left(\frac{4\pi}{2l+1}\right)\frac{q_j}{R_j^{(l+1)}} Z^s_{lm}(\theta_j, \varphi_j)\,. \end{aligned}\right\} \tag{4}$$

The explicit expansion of equation (1) or (2) is subject to the symmetry invariant of the crystalline electric potential at the site of the paramagnetic ion.

Hutchings [14] applied equations (2), (3), and (4) to the cubic symmetry and derived the crystalline electric potential with respect to the fourfold, sixfold, and eightfold cubic coordination. The crystalline electric potential has the same form for each of these cubic coordinations and is given by

$$V_{\text{cubic}} = D_4\left\{Y^0_4(\theta,\varphi) + \sqrt{\frac{5}{14}}\left[Y^4_4(\theta,\varphi) + Y^{-4}_4(\theta,\varphi)\right]\right\} + + D_6\left\{Y^0_6(\theta,\varphi) - \sqrt{\frac{7}{2}}\left[Y^4_6(\theta,\varphi) + Y^{-4}_6(\theta,\varphi)\right]\right\}, \tag{5}$$

where D_4 and D_6 are the cubic crystal field parameters. The calculation of the matrix elements of V_{cubic} can be facilitated using (5) expressed in the operator form of Stevens's [16] operator equivalents O^m_l. These operators have the same transformation properties as the corresponding tesseral harmonics Z^c_{lm}. Equation (5) becomes

$$V_{\text{cubic}} = d_4\left[O^0_4 + 5\,O^4_4\right] + d_6\left[O^0_6 - 21\,O^4_6\right], \tag{6}$$

where

$$d_4 = \frac{3}{16\sqrt{\pi}}\frac{D_4}{r^4}\beta_J\langle r^4\rangle\,,$$

$$d_6 = \frac{1}{32}\left(\frac{13}{\pi}\right)^{1/2}\frac{D_6}{r^6}\cdot\gamma_J\langle r^6\rangle\,,$$

and the multiplicative factors β_J and γ_J have been tabulated for iron group and lanthanide ions by Bleaney and Stevens [17] and Elliott and Stevens [18]. Alternatively the matrix elements of the crystalline electric potential can be calculated using the tensor operator technique of Racah [19, 20]. He introduced a tensor operator $C^{(l)}_m$ which is related to the spherical harmonic Y^m_l by

$$C^{(l)}_m = \left(\frac{4\pi}{2l+1}\right)^{1/2} Y^m_l\,. \tag{7}$$

The matrix elements of tensor operators $C_m^{(l)}$ can be calculated systematically using the Wigner-Eckart theorem and the concept of fractional parentage elaborated by Racah [20]. The formula for the calculation of these matrix elements in an l^n configuration is given by Wybourne [21]:

$$\langle l^n \alpha\, S\, L\, J\, M \left| C_q^{(k)} \right| l^n \alpha\, S'\, L'\, J'\, M' \rangle =$$

$$= (-1)^{J-M} \begin{pmatrix} J & k & J' \\ -M & q & M' \end{pmatrix} \langle l^n \alpha\, S\, L\, J\, || C^{(k)} || l^n\, \alpha'\, S'\, L'\, J' \rangle =$$

$$= (-1)^{J-M} \begin{pmatrix} J & k & J' \\ -M & q & M' \end{pmatrix} \langle l\, || C^{(k)} ||\, l \rangle \, \langle l^n \alpha\, S\, L\, J\, || U^{(k)} ||\, l^n \alpha'\, S'\, L'\, J' \rangle \,, \qquad (8)$$

where

$$\langle l^n \alpha\, S\, L\, J\, || U^{(k)} ||\, l^n \alpha'\, S'\, L'\, J' \rangle = (-1)^{\,S+L'+J+k}\, [(2J+1)\,(2\,J'+1)]^{1/2} \times$$

$$\times \begin{Bmatrix} J & J' & k \\ L' & L & S \end{Bmatrix} \langle l^n \alpha\, S\, L || U^{(k)} ||\, l^n \alpha' S' L' \rangle \,, \qquad (9)$$

and $\begin{pmatrix} J & k & J' \\ -M & q & M' \end{pmatrix}$ and $\begin{pmatrix} J & J' & k \\ L' & L & S \end{pmatrix}$ are the 3 j and 6 j symbols, respectively. The doubly reduced matrix elements of the unit tensor operator $U^{(k)}$, $\langle l^n \alpha\, S\, L |\, | U^{(k)} |\, | l^n \alpha'\, S'\, L' \rangle$, have been tabulated by Nielson and Koster [22] for p^n-, d^n-, and f^n-configurations and $k = 2(1)\,6$, or may be calculated using the formula (2-92) of Wybourne [21].

Substituting equations (7) and (8) into equation (5) and comparing with equation (6), it can be found that the multiplicative factors, α_J, β_J, and γ_J in the Stevens operator equivalent calculations correspond to the reduced matrix elements $\langle l^n \alpha\, S\, L\, J\, || C^{(k)} ||\, l^n \alpha'\, S'\, L'\, J' \rangle$ for $k = 2$, 4, and 6, respectively. The calculations of the reduced matrix elements of $C^{(k)}$ given by (8) and (9) have been well documented. The evaluations of the multiplicative factors α_J, β_J, and γ_J in the literature is obscured although limited tabulations of these factors are available. It is concluded that the tensor operator technique is more convenient to employ in these calculations than the Stevens operator equivalent method.

For noncubic point groups the explicit form of the crystalline electric potential can be obtained by considering the invariant conditions for the tesseral harmonics under the various symmetry operations. The conditions are listed below.

C_n (n-fold rotation): The invariant condition of C_n requires $V(\theta, \varphi) = V\,[\theta, \varphi + (2\,\pi/n)]$. Since the tesseral harmonics are functions of $\cos m\,\varphi$ or $\sin m\,\varphi$ (equation (3)), m can be only an integral multiple of n, i.e., $m = k\,n$, $k = 1, 2, 3, \ldots$.

C_2' (twofold rotation about the y-axis): The coordinates (r, θ, φ) under the transformation of C_2' become $(r, \pi - \theta, \pi - \varphi)$. This transformation limits the order of the tesseral harmonics l to be even for Z_{lm}^{c} and to be odd for Z_{lm}^{s}.

I (inversion): Under the operation of I, the coordinatinates (x, y, z) become $(-x, -y, -z)$. The invariant condition requires that the order of the tesseral harmonics is even.

σ_h (horizontal reflection plane): The operation of σ_h results in the change of (x, y, z) to $(x, y, -z)$ or the change of (r, θ, φ) to $(r, \pi - \theta, \varphi)$. The tesseral harmonic indices l and m both must be either even or odd for it to be invariant under the operation σ_h.

σ_v (vertical reflection plane): A vertical plane of symmetry exists at $\varphi = \beta$ if $V(\theta, \beta + \alpha) = V(\theta, \beta - \alpha)$ for all values of α. According to this condition the vertical reflection planes for Z^c_{lm} are located at $\beta = 0$, π/m, $2\pi/m, \ldots, (m-1)\pi/m$, and the vertical reflection planes for Z^s_{lm} are located at $\beta = \pi/2m$, $3\pi/2m, \ldots, (2m-1)\pi/2m$. It can be seen that the xz plane $\beta = 0$ is a symmetry plane for all Z^c_{lm} and the yz plane ($\beta = \pi/2$) is a symmetry plane for Z^c_{lm} when m is even and for Z^s_{lm} when m is odd.

$S_4(\sigma_h C_4)$: The combined operation of σ_h and C_4 transforms (r, θ, φ) into $(r, \pi - \theta, \varphi + \pi/2)$. The invariant condition requires that both m and $(2l - m)/2$ must be even.

The nonvanishing terms in the expansion given by (2) can be easily deduced with this set of invariant conditions of various symmetry operations. Prather [15] listed the nonvanishing coefficients in the crystalline electric potential for noncubic point groups. This list is reproduced in Table 1. Two points should be noted when using Table 1: The odd parity terms (l odd) in the crystalline electric potential can be neglected when the crystalline electric potential operates within only one electronic configuration. The maximum l-value in the expansion is given by $2l$ where l is the orbital quantum number of the unpaired electron. When both C and S are shown, the first appearing S-coefficient is shown in parenthesis. This represents the fact that although both C and S are consistent with the required symmetry, the S-term may be suppressed by a suitable rotation about the z-axis. This may be done for anyone (C, S) pair in a point group, but only for one.

The crystalline electric potential of a paramagnetic ion with unpaired f-orbital electrons and C_{3h} site symmetry can be obtained using Table 1:

$$V_{C_{3h}} = r^2 C_2^0 Z_{20}(\theta, \varphi) + r^4 C_4^0 Z_{40}(\theta, \varphi) + r^6 C_6^0 Z_{60}(\theta, \varphi) + r^6 C_6^6 Z^c_{66}(\theta, \varphi) \equiv$$
$$\equiv C_{20} Y_2^0(\theta, \varphi) + C_{40} Y_4^0(\theta, \varphi) + C_{60} Y_6^0(\theta, \varphi) + C_{66}[Y_6^6(\theta, \varphi) + Y_6^{-6}(\theta, \varphi)] \equiv$$
$$\equiv B_0^2 C_0^{(2)} + B_0^4 C_0^{(4)} + B_0^6 C_0^{(6)} + B_6^6 [C_6^{(6)} + C_6^{(6)}], \tag{10}$$

where the odd parity terms have been neglected, the term $Z^s_{66}(\theta, \varphi)$ has been suppressed, and C_l^m, C_{lm}, and B_m^l are the crystal field parameters which are associated to the tesseral harmonics Z_{lm}, the spherical harmonics Y_l^m, and the tensor operators $C_m^{(l)}$, respectively.

3. The Spin Hamiltonian for Gd^{3+} in Single Crystals

EPR data for a paramagnetic ion in a crystal is traditionally described using a spin Hamiltonian. A general description of each term in a spin Hamiltonian is given by Bowers and Owen [23]. The spin Hamiltonian terms pertinent to the description of the EPR data for Gd^{3+} are now discussed briefly.

3.1 Zero-magnetic-field dependent terms

3.1.1 Fine-structure terms

The fine-structure terms in a spin Hamiltonian describe the crystalline electric field splitting of the ground state of a paramagnetic ion. The fine-

Table 1

Nonvanishing coefficients in the crystalline electric potential for noncubic point groups

M	N	$D_{\infty h}$	$C_{\infty v}$	D_{6h}	D_6	C_{6h}	C_{6v}	D_{3h}	D_{3d}	D_{4h}	D_{2h}	D_{2d}	C_{4h}	D_4	C_{4v}	C_6	S_6	C_{3h}	D_3	C_{3v}	C_4	C_{2h}	S_4	D_2	C_{2v}	C_3	C_2	$C_s=\sigma_h$	$S_2=I$
0	1		C				C								C	C				C	C				C	C	C		
	2	C	C	C	C	C	C	C	C	C	C	C	C	C	C	C	C	C	C	C	C	C	C	C	C	C	C	C	C
	3		C				C								C	C				C	C				C	C	C		
	4	C	C	C	C	C	C	C	C	C	C	C	C	C	C	C	C	C	C	C	C	C	C	C	C	C	C	C	C
	5		C				C								C	C				C	C				C	C	C		
	6	C	C	C	C	C	C	C	C	C	C	C	C	C	C	C	C	C	C	C	C	C	C	C	C	C	C	C	C
1	1																											C(S)	
	2																												C(S)
	3																											C S	
	4																												C S
	5																											C S	
	6																												C S
2	2										C											C(S)		C	C		C(S)	C S	C S
	3											S											C(S)	S	C		C S		
	4										C											C S		C	C		C S	C S	C S
	5											S											C S	S	C		C S		
	6										C											C S		C	C		C S	C S	C S
3	3							S										C(S)	S	C						C(S)		C S	
	4								C								C(S)		C	C						C S			C S
	5							S										C S	S	C						C S		C S	
	6								C								C S		C	C						C S			C S
4	4									C	C	C	C(S)	C	C						C(S)	C S	C S	C	C		C S	C S	C S
	5													S	C						C S			S	C		C S		
	6									C	C	C	C S	C	C						C S	C S	C S	C	C		C S	C S	C S
5	5																											C S	
	6																												C S
6	6			C	C	C(S)	C	C	C		C					C(S)	C S	C S	C	C		C S		C	C	C S	C S	C S	C S

structure terms have the same symmetry properties as the crystalline electric potential at the site of the paramagnetic ion. Table 1 can be used to obtain the fine-structure terms in the spin Hamiltonian for a paramagnetic ion with noncubic site symmetry since they are subject to the same symmetry invariance. Conventionally the fine-structure terms for Gd^{3+} are expressed in terms of Stevens [16] operator equivalents. For example, the fine-structure terms for Gd^{3+} with cubic site symmetry are given by (cf. equation (6)),

$$B_4\,(O_4^0 + 5\,O_4^4) + B_6\,(O_6^0 - 21\,O_6^4)\,, \tag{11}$$

where O_l^m are Stevens operator equivalents and B_4 and B_6 are adjustable parameters which cannot be related simply to the crystal field parameters d_4 and d_6 since the splitting mechanisms for the gorund $^8S_{7/2}$ state of Gd^{3+} are not yet understood (cf. Wybourne [8]).

Alternatively the fine-structure terms may be expressed .in terms of the tensor operators $T_{lm}(\boldsymbol{S})$ which will be discussed in Section 4.

3.1.2 Hyperfine-structure terms

The hyperfine structure in the EPR spectrum of Gd^{3+} is not normally observed since the linewidth usually obscures this structure. The hyperfine structure for the system of Gd^{3+} in ThO_2 has been resolved by Low and Shaltiel [24] and Abraham et al. [25] since the EPR linewidths are ≈ 0.1 mT. The hyperfine-structure term in a spin Hamiltonian is given by $\boldsymbol{S} \cdot \mathbf{A} \cdot \boldsymbol{I}$ where $\mathbf{A}$ is the hyperfine splitting tensor.

3.2 Linear-magnetic-field dependent term

3.2.1 The first-order Zeeman term

In general the first-order Zeeman term is given by $\beta\,\boldsymbol{B} \cdot \mathrm{g} \cdot \boldsymbol{S}$, where β is the Bohr magneton, $\boldsymbol{B}$ is the magnetic field flux density, g is the spectroscopic splitting factor which is a second-rank symmetric tensor, and $\boldsymbol{S}$ is the electron spin. The g-tensor can be represented by its three principal values g_x, g_y, and g_z if the coordinate system is appropriately chosen. The g-factor is very nearly isotropic for Gd^{3+} in most crystals since the ground state of Gd^{3+} is an orbital S-state to good approximation.

3.2.2 Higher-order Zeeman terms

The linear-magnetic-field dependent terms associated with third- and fifth-order electron spin operators can be written as

$$\sum_{i=x,y,z} (g'\,\beta\,B_i\,S_i^3 + g''\,\beta\,B_i\,S_i^5)\,.$$

These higher-order Zeeman terms have been experimentally determined for Gd^{3+} in ThO_2 and $La(C_2H_5SO_4)_3 \cdot 9\,H_2O$. The upper bound of the value of g' was found to be 2×10^{-7} for Gd^{3+} in ThO_2 by Marshall and Tilton [26], and to be 2×10^{-6} for Gd^{3+} in $La(C_2H_5SO_4)_3 \cdot 9\,H_2O$ by Buckmaster et al. [27].

4. The Tensor Operator Formulation of the Spin Hamiltonian

Racah [19] defined the "irreducible tensor operator of rank l" as the operator T_l whose $(2\,l + 1)$ components T_{lm} $(m = -l, -l + 1, \ldots, +l)$ satisfy the same commutation rules with respect to total angular momentum $\boldsymbol{J}$ as the spherical harmonic operators $Y_l^m(\theta, \varphi)$ and the same transformation properties as the spherical harmonics $Y_l^m(\theta, \varphi)$. These commutation rules are

$$\left.\begin{aligned} &[J_\pm, T_{lm}] = [(l \mp m)(l \pm m + 1)]^{1/2} \, T_{l,m\pm1} \\ \text{and} \quad & \\ &[J_z, T_{lm}] = m \, T_{lm} , \end{aligned}\right\} \tag{12}$$

where $J_\pm = J_x \pm i \, J_y$.

The tensor operator[1]) T_{lm} can be conveniently transformed between two reference frames using the relation

$$T_{lm}(\boldsymbol{J}') \equiv R(\alpha\,\beta\,\gamma) \, T_{lm}(\boldsymbol{J}) \, R^{-1}(\alpha\,\beta\,\gamma) = \sum_{m=-l}^{+l} D_{m'm}^{(l)} \, T_{lm'}(\boldsymbol{J}) , \tag{13}$$

where $R(\alpha\,\beta\,\gamma)$ is the rotation operator, (α, β, γ) are the Euler angles describing the rotation, $D_{m'm}^{(l)} = \langle m' | \, R(\alpha\,\beta\,\gamma) \, | l \, m \rangle$, and $\boldsymbol{J}$ and $\boldsymbol{J}'$ are the angular momenta in the initial and final reference frames, respectively. On the other hand the Stevens operator equivalents O_l^m cannot be transferred using equation (13) since they do not satisfy the definition of the irreducible tensor operator. The transformation of a spin Hamiltonian expressed in tensor operator formalism has been discussed in detail by Buckmaster et al. [28].

The explicit form of $T_{lm}(\boldsymbol{J})$ expressed in terms of the angular momentum $\boldsymbol{J}$ can be deduced from the first-rank operator $T_{lm}(\boldsymbol{J})$. The first-rank tensor operators are equivalent to the spherical components of a vector:

$$\left.\begin{aligned} T_{1\,\pm1}(\boldsymbol{J}) &= \mp \frac{1}{\sqrt{2}} J_\pm , \\ T_{10}(\boldsymbol{J}) &= J_z , \end{aligned}\right\} \tag{14}$$

where $J_\pm = J_x \pm i \, J_y$. Tensor operators of higher rank can be generated using the normalized first-rank tensor operator defined in (14), the relation $T_{ll}(\boldsymbol{S}) = [T_{11}(\boldsymbol{S})]^l$ which was given by Rose [29], and the ladder operation given in (12). The tensor operators $T_{lm}(\boldsymbol{J})$ have been tabulated for $l = 0(1)7$ by Buckmaster et al. [28]. The relationship between T_{lm} and the Stevens operator equivalent O_l^m is given by

$$O_l^m = \text{const} \, [T_{lm} + (-1)^m \, T_{l\,-m}] , \tag{15}$$

where the constants are given in the later section when the spin Hamiltonian is expressed using either the O_l^m or T_{lm} operators. The matrix elements of the tensor operators $T_{lm}(\boldsymbol{J})$ can be calculated systematically using the Wigner-Eckart theorem and are given by

$$\langle \alpha \, J \, M | \, T_{lm}(\boldsymbol{J}) \, | \alpha' \, J' \, M' \rangle = (-1)^{J-M} \langle \alpha \, J || \, T_l(\boldsymbol{J}) \, || \alpha' \, J' \rangle \begin{pmatrix} J & l & J' \\ -M & m & M' \end{pmatrix} , \tag{16}$$

[1]) The term "irreducible" is omitted for simplicity. In what follows the "tensor operator" implies irreducible unless otherwise specified.

where

$$\langle \alpha\, J \| \, T_l(\boldsymbol{J}) \, \| \alpha'\, J' \rangle = \left[\frac{l!\, l!\, (2\,J + l + 1)!}{2^l (2\,l)!\, (2\,J - l)!}\right]^{1/2} \delta(\alpha, \alpha')\, \delta(J, J') \qquad (17)$$

and α, α' are additional quantum numbers used to specify the eigenstates.

The numerical evaluation of these matrix elements can be easily performed using (16) and (17) and the numerical table of 3 j symbols by Rotenberg et al. [30].

5. Tabulation of the Data Describing the EPR Spectra of Gd^{3+}

The data describing the EPR spectra of the Gd^{3+} ion have been grouped for convenience according to the site symmetry of the crystalline electric potential. Each table is introduced by a presentation of the appropriate spin Hamiltonian for each symmetry in both the operator equivalent and tensor operator formalism. The adjustable parameters in the spin Hamiltonian can be determined experimentally by fitting the measured EPR spectra to the spin Hamiltonian. The fine-structure parameters b_l^m may be expressed in the units of wave number (cm^{-1}), of frequency (GHz, MHz), or of magnetic field (T, G). The conversion relations are

$$\left.\begin{aligned} b_l^m\,(\mathrm{GHz}) &= 10^{-3} \times b_l^m\,(\mathrm{MHz}) = 29.97925 \times b_l^m\,(\mathrm{cm}^{-1}) \\ \text{and} \qquad & \\ b_l^m\,(\mathrm{GHz}) &= g \times 13.99600 \times b_l^m\,(\mathrm{T}) \\ &= g \times 13.99600 \times 10^{-4} \times b_l^m\,(\mathrm{G})\,, \end{aligned}\right\} \qquad (18)$$

where g is the spectroscopic splitting factor. The data in the tables are grouped according to the host lattice. The information includes the dilution of the Gd^{3+} ion with respect to the host ion, the measurement temperature and the measurement frequency, the fine- and hyperfine-structure constants, the g-factor, and the source of information. The fine- and hyperfine-structure constants are expressed in units of frequency (GHz, MHz), since this is the most convenient unit to use in the EPR spectroscopy.

Explicit expressions can be derived for the energy of the eight states when the magnetic field is applied parallel to the symmetry axis of the crystalline electric potential using second-order perturbation methods. This approach is justified when the electronic Zeeman term is large compared to the fine-structure terms. The hyperfine-structure term is always neglected in such calculations. The appropriate expressions for the energy differences for ($\Delta M = \pm 1$) transitions are included in the materials prefacing each table.

5.1 Cubic symmetry

The analysis given in Sections 2 and 3 can be employed to derive the operator form of the spin Hamiltonian describing EPR spectra of Gd^{3+} in crystalline electric fields with various symmetries. The spin Hamiltonian for the Gd^{3+} ion in crystals with cubic site symmetry is

$$\mathscr{H}_{\mathrm{cubic}} = \beta\, \boldsymbol{B} \cdot \mathrm{g} \cdot \boldsymbol{S} + \left(\frac{1}{60}\right) b_4\, (O_4^0 + 5\, O_4^4) + \left(\frac{1}{1260}\right) b_6\, (O_6^0 - 21\, O_6^4) + \boldsymbol{S} \cdot \mathbf{A} \cdot \boldsymbol{I}, \qquad (19)$$

where the spectroscopic splitting factor g, the fine-structure constants b_4 and b_6, and the hyperfine-structure constant **A** are treated as the adjustable parameters. The remaining symbols are defined in Section 3 and the factors (1/60) and (1/1260) are introduced with the fine-structure constants b_4 and b_6 respectively for numerical convenience.

The spin Hamiltonian given in equation (19) is the most commonly used form in the literature because the Stevens operator equivalents O_l^m were first introduced into the EPR spectrocopy. Section 4 showed that the Stevens operator equivalent O_l^m is related to the tensor operators $T_{l\pm m}$ by (15) but does not transform as conveniently as T_{lm}. Thus it is preferable to express (19) in terms of the tensor operators T_{lm}, particularly if the Zeeman magnetic field is not directed parallel to the symmetry axis, since then it can be conveniently transformed using equation (13):

$$\mathcal{H}_{\text{cubic}} = \beta \boldsymbol{B} \cdot \mathbf{g} \cdot \boldsymbol{S} + B_4 \left\{ T_{40}(\boldsymbol{S}) + \sqrt{\frac{5}{14}}\,[T_{44}(\boldsymbol{S}) + T_{4-4}(\boldsymbol{S})] \right\} + $$
$$+ B_6 \left\{ T_{60}(\boldsymbol{S}) - \sqrt{\frac{7}{2}}\,[T_{66}(\boldsymbol{S}) + T_{6-6}(\boldsymbol{S})] \right\} + \boldsymbol{S} \cdot \mathbf{A} \cdot \boldsymbol{I}\,, \qquad (20)$$

where

$$B_4 = \left(\frac{\sqrt{70}}{30}\right) b_4 \quad \text{and} \quad B_6 = \left(\frac{\sqrt{231}}{315}\right) b_6\,. \qquad (21)$$

The secular matrix of $\mathcal{H}_{\text{cubic}}$ given in (19) factors into four 2×2 matrices when the applied magnetic field is parallel to the [1, 0, 0] direction if the hyperfine interaction term is neglected. Abraham et al. [25] have shown that the eigenvalues for the eight $|M_s\rangle$ states may be obtained exactly for this direction:

$$\left.\begin{aligned}
E_{\pm 7/2} &= \left(\frac{13\,c + 5\,d \pm 3\,G}{2}\right) \pm \frac{1}{2}\left[\left(3\,c - 9\,d \mp \frac{2}{3}\,G\right)^2 + \frac{140\,G^2}{9}\right]^{1/2},\\
E_{\pm 5/2} &= \left(\frac{5\,c + 7\,d \pm G}{2}\right) \pm \frac{1}{2}\,[(5\,c + 7\,d \mp 2\,G)^2 + 12\,G^2]^{1/2}\,,\\
E_{\pm 3/2} &= \left(\frac{5\,c + 7\,d \mp G}{2}\right) \pm \frac{1}{2}\,[(5\,c + 7\,d \pm 2\,G)^2 + 12\,G^2]^{1/2}\,,\\
E_{\pm 1/2} &= \left(\frac{13\,c + 5\,d \mp 3\,G}{2}\right) \pm \frac{1}{2}\left[\left(3\,c - 9\,d \pm \frac{2}{3}\,G\right)^2 + \frac{140\,G^2}{9}\right]^{1/2},
\end{aligned}\right\} \qquad (22)$$

where $G = g_{\parallel}\,\beta\,B$, $c = 4\,b_4$, and $d = 4\,b_6$.

EPR data for the Gd^{3+} ion in crystals with cubic site symmetry are listed in Table 2.

Table 2

Spin Hamiltonian parameters of Gd^{3+} in single crystals with cubic site symmetry

host lattice	dilution	temperature (°K)	frequency (GHz)	g	b_4 (MHz)	b_6 (MHz)	A (MHz) and remarks	ref.
CaF_2		room	9.2	1.9918 ±0.001	(±)*)139.70 ±0.9	(±) 0.21 ±0.9		[34]
	0.01%	290	9.2	1.993 ±0.001	(±) 140.08 ±0.4	(±) 0.04 ±0.04	temperature dependence of b_4 between 290 and 1.4 °K	[35]
	0.05 wt%	290	24	1.991 ±0.002	−138.65 ±3.7	+3.0 ±1.5	sign corrected in the later paper	[36], [37]
	0.15 at %	283	9.2	1.9926 ±0.0010	(±) 140.60 ± 1.5	(±) 0.03 ±0.9	temperature dependence of b_4 between 950 and 77 °K	[38]
CdF_2	1 to 0.03%	room	12.4	1.992 ±0.002	(±) 142.10 ± 0.9	(±) 0.0 ±0.9		[39]
	0.01%	room	13	1.992 ±0.002	(±) 129.66 ± 3.7	(±) 0.0 ±1.5		[40]
	0.01%	77	13	1.992 ±0.001	(±) 152.14 ± 3.7	(±) 0.0 ±1.5		[40]
	0.1 at %	77	10	1.9917 ±0.0002	−150.600 ± 0.018	−0.3969 ±0.0038	$g' = 10^{-5}(3.4 \pm 0.5)$	[58]
	0.01%	4.2	9.5	1.99149 ±0.00002	−151.236 ±0.012	−0.3937 ±0.001	$g' = 10^{-5}(-3.2 \pm 0.1)$ $g'' = 10^{-6}(1.6 \pm 0.1)$	[41]
	0.01%	1.5	13	1.992 ±0.001	(±) 153.64 ± 3.7	(±) 0.0 ±1.5		[40]
BaF_2	0.02%	room	9.2	1.9921 ±0.001	(±) 108.52 ±0.9	(±) 0.6 ±0.9		[42]
	—**)	room	9	1.984 ±0.020	106.50 ± 1.50	(∓) 3.0 ±0.75		[59]
	—	295	10	1.9912 ±0.0005	−107.63 ± 0.22	−0.40 ±0.10		[60]

*) (±) indicates that only the relative sign of the parameter was determined.
**) — indicates not available.

Table 2 (continued)

host lattice	dilution	temperature (°K)	frequency (GHz)	g	b_4 (MHz)	b_6 (MHz)	A (MHz) and remarks	ref.
BaF_2	0.15 at%	283	9.2	1.9921 ±0.001	(±) 109.12 ± 1.5	(±) 0.54 ±0.9	temperature dependence of b_4 between 950 and 77 °K	[38]
	–	77	9	1.982 ±0.02	(±) 110.70 ± 1.50	(∓) 3.82 ±0.75		[59]
	–	77	10	1.9917 ±0.0005	−112.65 ± 0.22	−0.352 ±0.075		[60]
	–	4.2	10	1.9916 ±0.0005	−113.17 ± 0.22	−0.427 ±0.075		[60]
SrF_2	0.02%	room	9.2	1.9923 ±0.001	(±) 122.91 ±1.5	(±) 1.80 ±1.2		[42]
	0.15 at%	283	9.2	1.9934 ±0.001	(±) 122.02 ± 1.5	(±) 0.75 ±0.9	temperature dependence of b_4 between 950 and 77 °K	[38]
$SrCl_2$	0.5 wt%	290	9.0	1.9906 ±0.0010	(±) 29.68 ± 0.075	(∓) 0.15 ±0.075		[43]
	0.5 wt%	294.5	–	1.9925 ±0.0005	− 44.65 ± 0.05	−0.0925 ±0.0125	the discrepancy between this work and previous work is explained	[44]
	0.5 wt%	77.0	–	1.9925 ±0.0005	49.30 ± 0.05	−0.09 ±0.0125		[44]
	0.5 wt%	4.2	–	1.9925 ±0.0005	− 49.825 ± 0.05	−0.1275 ±0.0125		[44]
	0.5 wt%	≈2.0	–	1.9922 ±0.0005	− 49.825 ± 0.05	−0.135 ±0.0125		[44]
ThO_2	0.2 wt%	297	10.0	1.9917 ±0.0002	−165.475 ± 0.05	−1.1775 ±0.0375		[45]
	–	290	10.0	1.9913 ±0.0005	(±) 164.78 ± 0.21	(±) 0.787 ±0.21	$A_{155} = 11.84 \pm 0.42$; $A_{157} = 15.97 \pm 0.33$	[24]
	–	90	10.0	1.991 ±0.001	(±) 168.59 ± 0.56	(±) 1.26 ±0.56		[24]
	<1:10^5	77	10.0	1.99161 ±0.00003	−169.394 ± 0.03	−1.313 ±0.003	$A_{155} = -12.061 \pm 0.008$; $A_{157} = -15.813 \pm 0.008$	[46]
	0.2 wt%	77	10.0	1.9917 ±0.0002	−169.375 ± 0.05	−1.2375 ±0.0375		[25]
	0.2 wt%	20	10.0	1.9917 ±0.0002	−169.775 ± 0.05	−1.235 ±0.05		[25]
	0.2 wt%	4	10.0	1.9917 ±0.0002	−169.775 ± 0.05	−1.255 ±0.05		[25]
	0.2 wt%	≈2	10.0	1.9917 ±0.0002	−169.625 ± 0.05	−1.225 ±0.1	$A_{155} = 12.04 \pm 0.08$; $A_{157} = 15.7 \pm 0.11$	[25]

Table 2 (continued)

host lattice	dilution	temperature (°K)	frequency (GHz)	g	b_4 (MHz)	b_6 (MHz)	A (MHz) and remarks	ref.
CeO_2	0.05%	300	9.4	1.991 ±0.001	(±) 161.14 ± 0.15	(±) 0.525 ±0.525	A_{155} = 11.62 ± 0.28; A_{157} = 15.21 ± 0.28	[47]
	<0.01%	290	10.0	1.991 ±0.001	(−) 161.70 ± 0.375	(−) 0.618 ±0.187		[48]
	0.02%	297	10.0	1.9921 ±0.0007	−161.90 ± 0.15	−1.125 ±0.25		[45]
	0.02%	77	10.0	1.9918 ±0.0005	−161.175 ± 0.15	−1.05 ±0.25		[45]
	0.02%	20	10.0	1.9921 ±0.0005	−166.875 ± 0.15	−1.15 ±0.25		[45]
	0.02%	≈10	10.0	1.9912 ±0.0005	−166.775 ± 0.15	−1.275 ±0.25		[45]
	0.02%	4.2	10.0	1.9907 ±0.0005	−166.525 ± 0.15	−1.125 ±0.25		[45]
	0.02%	≈2	10.0	1.9908 ±0.0005	−166.525 ± 0.15	−1.025 ±0.25		[45]
CdTe		77	—	1.9869 ±0.0003	− 72.16 ± 0.18	0.78 ±0.18		[49]
BaO	$2.5:10^5$	77	9.0	1.9918 ±0.0002	− 5.715 ± 0.14	+2.20 ±0.14	A_{155} = 14.38±0.56; A_{157} = 10.96±0.56	[50]
CaO	0.05%	290	9.5	1.9913 ±0.0005	(−) 34.78 ± 0.3	(+) 3.45 ±0.3	sign corrected in the later paper	[51], [52]
	—	290	10.0	1.9922 ±0.0005	− 34.33 ± 0.3	+3.22 ±0.3		[37], [53]
	—	77	10.0	1.9922 ±0.0005	− 36.50 ± 0.15	+3.52 ±0.15	sign corrected in the later paper	[37], [53]
	0.05%	77	9.5	1.9908 ±0.0005	− 36.27 ± 0.3	+3.48 ±0.3		[51]
	0.05%	4.2	9.5	1.9925 ±0.0005	− 36.57 ± 0.3	+3.57 ±0.3		[51]
	—	4.2	10.0	1.9922 ±0.0005	− 36.57 ± 0.15	+3.60 ±0.15		[53]
SrO	—	290	9.0	1.9912 ±0.0005	≈ 18	—		[52]
	—	290	—	1.991 ±0.001	(−) 17.69 ± 0.9	—		[54]
	—	77	—	1.991 ±0.001	(−) 19.49 ± 0.9	—		[54]
	—	4.2	—	1.989 ±0.001	(−) 17.39 ± 1.5	—		[54]

Table 2 (continued)

host lattice	dilution	temperature (°K)	frequency (GHz)	g	b_4 (MHz)	b_6 (MHz)	A (MHz)	ref.
MgO	–	room	9	1.9920 ±0.0005	−100.81 ± 0.15	+6.37 ±0.08	A_{155} = 11.54±0.15; A_{157} = 15.17±0.15	[111]
	–	77	9	1.9920 ±0.0005	−104.78 ± 0.15	+6.60 ±0.08		[111]
	–	4.2	9	1.9920 ±0.0005	−104.93 ± 0.15	+6.60 ±0.08		[111]
$BaTiO_3$	Ba^{2+} site 0.2 to 0.01 at%	425	16.0	1.995 ±0.003	≈ 18			[55]
	Ti^{4+} site 0.1 to 0.2 at%	436	9.0	1.992 ±0.004	(±) 69.85 ± 1.5	(∓) 5.4 ±1.5		[56]
	Ti^{4+} site 0.1 to 0.2 at%	300	9.0	1.992 ±0.004	(±) 71.95 ± 3.0	(∓) 11.99 ± 9.0		[56]
	Ti^{4+} site 0.1 to 0.2 at%	245	9.0	1.992 ±0.005	(±) 68.95 ± 3.0	(∓) 6.296 ± 9.0		[56]
	Ti^{4+} site 0.1 to 0.2 at%	77	9.0	1.992 ±0.007	(±) 71.95 ± 6.0	(±) 18.0 ±12.0		[56]
$SrTiO_3$	0.1 at%	300	16.0	1.992 ±0.002	− 17.09 ± 0.6	+ 1.5 ± 0.9		[55]
$KTaO_3$	0.1 to 0.03 at%	77	9.0	1.990 ±0.002	− 20.98 ± 0.9	+ 1.5 ± 1.5		[57]
	0.1 to 0.03 at%	4.2	9.0	1.990 ±0.002	− 24.40 ± 0.81	+ 1.41 ± 0.9		[57]
Cs_2ZrCl_6	–	room	10	1.9923 ±0.0002	(±) 55.67 ± 0.06	(∓) 1.95 ± 0.06		[61]
	–	room	35	1.9921 ±0.0003	(±) 55.52 ± 0.06	(∓) 1.71 ± 0.06		[61]

5.2 Tetragonal symmetry

The system of tetragonal symmetry includes the seven point groups C_4, C_{4v}, C_{4h}, S_4, D_4, D_{4h}, and D_{2d}. Using Table 1 the crystalline electric potential for the Gd^{3+} ion in crystals with tetragonal site symmetry is

$$V_{\text{tetra}} = C_{20}\, Y_2^0(\theta, \varphi) + C_{40}\, Y_4^0(\theta, \varphi) + C_{60}\, Y_6^0(\theta, \varphi) + C_{44}\, [Y_4^4(\theta, \varphi) + Y_4^{-4}(\theta, \varphi)] + $$
$$+ C_{64}\, [Y_6^4(\theta, \varphi) + Y_6^{-4}(\theta, \varphi)] + i\, S_{64}\, [Y_6^4(\theta, \varphi) - Y_6^{-4}(\theta, \varphi)]\,, \qquad (23)$$

where the odd parity terms have been omitted.

Consequently the fine-structure terms in the spin Hamiltonian for the Gd^{3+} ion with tetragonal site symmetry are identical in form with equation (23). The imaginary term in (23) vanishes for point groups C_{4v}, D_4, D_{4h}, and D_{2d}. The imaginary term is usually omitted in the fine-structure terms since the contribution to the fine-structure splittings due to this term is generally small. The spin Hamiltonian for the Gd^{3+} ion in crystals with tetragonal site symmetry is

$$\mathcal{H}_{\text{tetra}} = \beta\, \boldsymbol{B} \cdot \mathbf{g} \cdot \boldsymbol{S} + \left(\frac{1}{3}\right) b_2^0\, O_2^0 + \left(\frac{1}{60}\right) b_4^0\, O_4^0 + \left(\frac{1}{1260}\right) b_6^0\, O_6^0 + \left(\frac{1}{60}\right) b_4^4\, O_4^4 + $$
$$+ \left(\frac{1}{1260}\right) b_6^4\, O_6^4 + \boldsymbol{S} \cdot \mathbf{A} \cdot \boldsymbol{I}\,, \qquad (24)$$

where b_l^m are the fine-structure constants for tetragonal symmetry and the remaining symbols have been defined previously. Equation (24) can be expressed using tensor operators:

$$\mathcal{H}_{\text{tetra}} = \beta\, \boldsymbol{B} \cdot \mathbf{g} \cdot \boldsymbol{S} + B_{20}\, T_{20}(\mathbf{S}) + B_{40}\, T_{40}(\mathbf{S}) + B_{60}\, T_{60}(\mathbf{S}) + $$
$$+ B_{44}\, [T_{44}(\mathbf{S}) + T_{4-4}(\mathbf{S})] + B_{64}\, [T_{64}(\mathbf{S}) + T_{6-4}(\mathbf{S})] + \boldsymbol{S} \cdot \mathbf{A} \cdot \boldsymbol{I}\,, \quad (25)$$

where

$$\left.\begin{aligned} B_{20} &= \left(\frac{2}{\sqrt{6}}\right) b_2^0\,, \quad B_{40} = \left(\frac{\sqrt{70}}{30}\right) b_4^0\,, \quad B_{60} = \left(\frac{\sqrt{231}}{315}\right) b_6^0\,, \\ B_{44} &= \left(\frac{1}{30}\right) b_4^4\,, \quad \text{and} \quad B_{64} = \left(\frac{1}{315}\sqrt{\frac{11}{6}}\right) b_6^4\,. \end{aligned}\right\} \qquad (26)$$

When the applied magnetic field is parallel to the z-axis, which is the fourfold symmetry axis, the off-diagonal matrix elements of the spin Hamiltonian may be considerably smaller than the diagonal elements. A second-order perturbation calculation is then adequate to obtain eigenvalues of the spin Hamiltonian. Rosenthal [31] has shown that the resonant magnetic fields of the seven

($\Delta M = \pm 1$) transitions resulting from a perturbation calculation are

$$\left.\begin{aligned}
&|\pm\tfrac{7}{2}\rangle \leftrightarrow |\pm\tfrac{5}{2}\rangle;\\
&g_{\|}\,\beta\,B = h\,\nu \mp (6\,b_2^0 + 20\,b_4^0 + 6\,b_6^0) -\\
&\quad - P\left\{\frac{7\,[1 + (\frac{5}{7})\,\varrho]^2}{1 \pm 3\,Q \mp R \pm 3\,S} - \frac{15\,[1 - (\frac{1}{3})\,\varrho]^2}{1 \pm Q \mp 5\,R \mp 7\,S}\right\},\\
&|\pm\tfrac{5}{2}\rangle \leftrightarrow |\pm\tfrac{3}{2}\rangle;\\
&g_{\|}\,\beta\,B = h\,\nu \mp (4\,b_2^0 - 10\,b_4^0 - 14\,b_6^0) - P\left\{\frac{15\,[1 - (\frac{1}{3})\,\varrho]^2}{1 \pm Q \mp 5\,R \mp 7\,S}\right\},\\
&|\pm\tfrac{3}{2}\rangle \leftrightarrow |\pm\tfrac{1}{2}\rangle;\\
&g_{\|}\,\beta\,B = h\,\nu \mp (2\,b_2^0 - 12\,b_4^0 + 14\,b_6^0) -\\
&\quad - P\left\{\frac{15\,[1 - (\frac{1}{3})\,\varrho]^2}{1 \mp Q \pm 5\,R \pm 7\,S} - \frac{7\,[1 + (\frac{5}{7})\,\varrho]^2}{1 \mp 3\,Q \pm R \mp 3\,S}\right\},\\
&|-\tfrac{1}{2}\rangle \leftrightarrow |+\tfrac{1}{2}\rangle;\\
&g_{\|}\,\beta\,B = h\,\nu - P\,\{7\,[1 + (\tfrac{5}{7})\,\varrho]^2\} \times\\
&\quad \times \left\{\frac{1}{1 + 3\,Q - R + 3\,S} + \frac{1}{1 - 3\,Q + R - 3\,S}\right\},
\end{aligned}\right\} \qquad (27)$$

where h is Planck's constant, ν is the measurement frequency,

$$\varrho = \frac{b_6^4}{b_4^4}, \quad P = \frac{(b_4^4)^2}{20\,g_{\|}\,\beta\,B}, \quad Q = \frac{b_2^0}{g_{\|}\,\beta\,B}, \quad R = \frac{b_4^0}{2\,g_{\|}\beta\,B}, \quad \text{and} \quad S = \frac{b_6^0}{2\,g_{\|}\beta\,B}.$$

EPR data for the Gd^{3+} ion in crystals with tetragonal site symmetry are listed in Table 3.

5.3 Hexagonal symmetry

The system of hexagonal symmetry includes the seven point groups C_6, C_{6v}, C_{6h}, C_{3h}, D_6, D_{6h}, and D_{3h}. The crystalline electric potential of the Gd^{3+} ion at a site with hexagonal symmetry was given by equation (10). Using (10) the spin Hamiltonian for the Gd^{3+} ion in crystals with hexagonal site symmetry is

$$\begin{aligned}\mathcal{H}_{\text{hexa}} = \beta\,\boldsymbol{B}\cdot\mathbf{g}\cdot\boldsymbol{S} + \left(\frac{1}{3}\right) b_2^0\,O_2^0 + \left(\frac{1}{60}\right) b_4^0\,O_4^0 + \left(\frac{1}{1260}\right) b_6^0\,O_6^0 +\\
+ \left(\frac{1}{1260}\right) b_6^6\,O_6^6 + \boldsymbol{S}\cdot\mathbf{A}\cdot\boldsymbol{I}\,,\end{aligned} \qquad (28)$$

where b_l^m are the fine-structure constants for hexagonal symmetry. Equation (28) can be also expressed using tensor operators:

$$\begin{aligned}\mathcal{H}_{\text{hexa}} = \beta\,\boldsymbol{B}\cdot\mathbf{g}\cdot\boldsymbol{S} + B_{20}\,T_{20}(\boldsymbol{S}) + B_{40}\,T_{40}(\boldsymbol{S}) + B_{60}\,T_{60}(\boldsymbol{S}) +\\
+ B_{66}\,[T_{66}(\boldsymbol{S}) + T_{6-6}(\boldsymbol{S})] + \boldsymbol{S}\cdot\mathbf{A}\cdot\boldsymbol{I}\,,\end{aligned} \qquad (29)$$

Table 3

Spin Hamiltonian parameters of Gd^{3+} in single crystals with tetragonal site symmetry

host lattice	dilution	temperature (°K)	frequency (GHz)	$g_{\parallel}$	$g_{\perp}$	b_2^0 (GHz)	b_4^0 (GHz)	b_6^0 (GHz)	b_4^4 (GHz)	b_6^4 (GHz)	ref.
CaF_2	0.01%	room	9.2	1.993 ±0.003	–*)	(±) 4.4616**) ±0.014	(±) 0.0688 ±0.0048	(±) 0.00183 ±0.0037	(±) 0.455 ±0.02	0.0 ±0.21	[62]
	–	room	8 to 51	1.992 ±0.001	–	(±) 4.449 ±0.01	(±) 0.069 ±0.002	(±) 0.002 ±0.001	(±) 0.437 ±0.005	0.0	[63]
	0.05 at%	room	35.0	1.992 ±0.001	1.992 ±0.001	−4.4519 ±0.003	−0.0693 ±0.0003	−0.0021 ±0.0003	−0.4407 ±0.003	+0.0159 ±0.0015	[64]
CdF_2	1 to 0.03%	room	21.4	1.992 ±0.002	–	(±) 0.1259 ±0.009	(±) 0.0045 ±0.0009	(±) 0.0006 ±0.0015	(±) 0.1355 ±0.003	0.0 ±0.0009	[39]
SrF_2	0.02%	room	9.2	1.992 ±0.003	–	(±) 3.3676 ±0.023	(±) 0.0592 ±0.0096	(±) 0.00183 ±0.011	(±) 0.172 ±0.06	(±) 0.021 ±0.124	[42]
$BaTiO_3$	0.2 to 0.01 at%	300	16.0	1.995 ±0.003	–	−0.8802 ±0.003	+0.012 ±0.003	+0.0048 ±0.003	−0.006 ±0.003	0.0	[55]
$SrTiO_3$	0.1 at%	77	16.0	1.992 ±0.002	–	−0.7003 ±0.0015	−0.0144 ±0.0015	−0.00075 ±0.0015	−0.0075 ±0.0009	+0.0003 ±0.0015	[55]
	0.1 at%	4.2	16.0	1.992 ±0.002	–	−1.0867 ±0.0015	−0.0097 ±0.0015	+0.0042 ±0.0015	−0.0126 ±0.0015	−0.00207 ±0.0015	[55]
$CaWO_4$	<1:10³	77	23.0	1.9915 ±0.0004	1.9916 ±0.0004	−2.7482 ±0.003	−0.07195 ±0.0006	−0.0018 ±0.0009	−0.4350 ±0.003	0.0 ±0.0009	[65]a)
	–	290	37.0	–	–	(−) 2.6753	–	–	–	–	[66]
	0.5 at%	293	10	1.99135	1.99155	−2.68003	−0.06841	+0.00075	−0.4212	+0.01259	[11]
	0.5 at%	155	10	1.99120	1.99157	−2.72715	−0.07063	+0.00078	−0.4322	+0.01133	[11]
	0.5 at%	78	10	1.99130	1.99188	−2.74847	−0.07135	+0.00075	−0.4356	+0.01196	[11]
	0.5 at%	4.2	10	1.99160	1.99108	−2.75755	−0.07171	+0.00072	−0.4377	+0.00881	[11]
$BaWO_4$	0.1 at%	room	37.5	1.991	–	(±) 2.4327	(±) 0.0362	0.0	(∓) 0.3149	0.0	[67]
	–	290	37.0	–	–	(−) 2.4433	–	–	–	–	[66]
	–	77	37.0	–	–	(−) 2.6502	–	–	–	–	[66]
$PbWO_4$	–	room	9.4	1.9923 ±0.0005	1.9922 ±0.0005	−2.4607 ±0.003	−0.0511 ±0.00015	≦0.0003	−0.3970 ±0.003	+0.00840 ±0.0015	[68]
$SrWO_4$	0.1 at%	room	37.5	1.991 ±0.002	–	(−) 2.6022 ±0.006	(−) 0.0511 ±0.003	0.0 ±0.003	−0.3568 ±0.015	0.0 ±0.009	[69]
	–	77	37.0	–	–	(−) 2.6972	–	–	–	–	[66]
$CaMoO_4$	0.02 to 1 at%	room	9.4	1.9916 ±0.0003	1.9920 ±0.0003	−2.5638 ±0.003	−0.0505 ±0.00009	≦0.00009	−0.2767 ±0.003	+0.00960 ±0.0006	[70]
	–	77	37.0	–	–	(−) 2.6297	–	–	–	–	[66]

*) Not available.
**) (±) indicates that only the relative sign of the parameter was determined.
a) The hyperfine-structure constants were found to be $A_{155} = (12.40 \pm 0.055)$ MHz; $A_{157} = (16.28 \pm 0.055)$ MHz.

Table 3 (continued)

host lattice	dilution	temperature (°K)	frequency (GHz)	$g_{\parallel}$	$g_{\perp}$	b_2^0 (GHz)	b_4^0 (GHz)	b_6^0 (GHz)	b_4^4 (GHz)	b_6^4 (GHz)	ref.
$CdMoO_4$	–	room	37.0	1.991 ±0.001	–	(−) 2.5482 ±0.006	(−) 0.0570 ±0.006	0.0 ±0.006	(−) 0.2938 ±0.009	0.0 ±0.009	[71]
	–	77	37.0	–	–	(−) 2.6469	–	–	–	–	[66]
$PbMoO_4$	0.5%	290	10 to 30	1.992 ±0.001	1.992 ±0.001	(−) 2.3983 ±0.006	(−) 0.0360 ±0.006	0.0 ±0.006	(−) 0.2728 ±0.030	0.0 ±0.030	[72]
$BaMoO_4$	0.1%	room	9.4	1.992 ±0.0005	1.992 ±0.0005	(−) 2.1840 ±0.003	(−) 0.02590 ±0.0003	≦0.00009	(−) 0.2111 ±0.0003	(+) 0.00540 ±0.0015	[73]
	–	290	37.0	–	–	(−) 2.1783	–	–	–	–	[66]
	–	77	37.0	–	–	(−) 2.3081	–	–	–	–	[66]
$SrMoO_4$	–	290	37.0	–	–	(−) 2.4199	–	–	–	–	[66]
	–	77	–	1.992 ±0.001	1.992 ±0.001	(−) 2.5003 ±0.006	(−) 0.0390 ±0.003	0.0 ±0.003	(−) 0.2009 ±0.045	0.0 ±0.045	[74]
$ZrSiO_4$	0.02 to 0.66%	300	9.0	1.9917 ±0.0005	1.9912 ±0.0010	−1.0870 ±0.0003	−0.02278 ±0.00009	+0.00183 ±0.00009	+0.1925 ±0.0021	0.003 ±0.003	[75]b)
	0.002%	290	9.0 and 37.5	1.9918 ±0.0001	1.9915 ±0.0005	(±) 1.057 ±0.001	(±) 0.023 ±0.0005	(∓) 0.0020 ±0.0003	(∓) 0.193 ±0.002	0.0	[76]c)
	0.002%	77	9.0 and 37.5	1.9918 ±0.0001	1.9915 ±0.0005	(±) 1.184 ±0.001	(±) 0.022 ±0.0005	(∓) 0.0022 ±0.0003	(∓) 0.20 ±0.01	0.0	[76]
	0.02 to 0.66%	77	9.0	1.9914 ±0.0005	1.9907 ±0.0010	−1.1914 ±0.0006	−0.02389 ±0.00018	+0.00183 ±0.00018	+0.1991 ±0.0045	0.0 ±0.009	[75]
	0.02 to 0.66%	4.2	9.0	1.9911 ±0.0005	1.9904 ±0.0010	−1.2004 ±0.0003	−0.02389 ±0.00009	+0.00183 ±0.00009	+0.1994 ±0.0036	0.0 ±0.006	[75]
	0.02 to 0.66%	1.5	9.0	1.9915 ±0.0005	1.9908 ±0.0010	−1.2010 ±0.0003	−0.02398 ±0.00009	+0.00186 ±0.00009	+0.2003 ±0.0045	0.0 ±0.006	[75]
$HfSiO_4$	0.02 to 0.66%	300	9.0	1.9912 ±0.0005	1.9914 ±0.0010	−0.9213 ±0.0003	−0.02740 ±0.00009	+0.00189 ±0.00009	+0.1904 ±0.0036	0.0 ±0.003	[75]d)
	0.02 to 0.66%	77	9.0	1.9914 ±0.0005	1.9910 ±0.0010	−1.0610 ±0.0003	−0.02854 ±0.00009	+0.00204 ±0.00009	+0.1934 ±0.0018	0.0 ±0.003	[75]
	0.02 to 0.66%	4.2	9.0	1.9910 ±0.0005	1.9903 ±0.0010	−1.0721 ±0.0006	−0.02872 ±0.00018	+0.00204 ±0.00018	+0.1979 ±0.0045	0.0 ±0.009	[75]
$ThSiO_4$	0.02 to 0.66%	300	9.0	1.9918 ±0.0005	1.9915 ±0.0010	−0.8904 ±0.0003	−0.02033 ±0.00009	+0.00129 ±0.00009	+0.0914 ±0.0018	0.0 ±0.003	[75]e)
	0.02 to 0.66%	77	9.0	1.9913 ±0.0005	1.9913 ±0.0010	−0.9315 ±0.0003	−0.02114 ±0.00009	+0.00138 ±0.00009	+0.0887 ±0.0048	0.0 ±0.006	[75]
	0.02 to 0.66%	4.2	9.0	1.9913 ±0.0005	1.9910 ±0.0010	−0.9357 ±0.0003	−0.02120 ±0.00009	+0.00132 ±0.00009	+0.0938 ±0.0042	0.0 ±0.006	[75]
	0.02 to 0.66%	≈1.5	9.0	1.9914 ±0.0005	1.9909 ±0.0010	−0.9360 ±0.0003	−0.02120 ±0.00009	+0.00135 ±0.00009	+0.0938 ±0.0024	0.0 ±0.003	[75]

b) The hyperfine-structure constants were found to be $A_{155} = (12.91 \pm 0.055)$ MHz; $A_{157} = (16.92 \pm 0.084)$ MHz.
c) The hyperfine-structure constants were found to be $A_{155} = (9.67 \pm 0.027)$ MHz; $A_{157} = (12.63 \pm 0.027)$ MHz.
d) The hyperfine-structure constants were found to be $A_{155} = (12.96 \pm 0.055)$ MHz; $A_{157} = (17.00 \pm 0.055)$ MHz.
e) The hyperfine-structure constants were found to be $A_{155} = (12.66 \pm 0.055)$ MHz; $A_{157} = (16.64 \pm 0.055)$ MHz.

Table 3 (continued)

host lattice	dilution	temperature (°K)	frequency (GHz)	$g_{\parallel}$	$g_{\perp}$	b_2^0 (GHz)	b_4^0 (GHz)	b_6^0 (GHz)	b_4^4 (GHz)	b_6^4 (GHz)	ref.
YVO_4	0.5 at%	300	24.0	1.9915 ±0.0002	1.9915 ±0.0002	−1.3239 ±0.0006	−0.00450 ±0.0003	+0.00240 ±0.0018	+0.1244 ±0.0012	0.0 ±0.0006	[31], [77]
	0.2%	298	9.0	1.9920 ±0.0008	1.9916 ±0.0008	−1.3191 ±0.0024	−0.00420 ±0.0006	+0.00150 ±0.0006	+0.1223 ±0.0045	0.0018 ±0.0045	[78]
	0.2%	78	9.0	1.9916 ±0.0008	–	−1.4240 ±0.0024	−0.00450 ±0.0006	+0.00150 ±0.0006	–	–	[78]
	0.2%	4.2	9.0	1.9900 ±0.001	–	−1.4360 ±0.0045	−0.00480 ±0.0012	+0.00150 ±0.0006	–	–	[78]
	0.5 at%	4.2	10 and 34	1.991 ±0.001	1.991 ±0.001	−1.4390 ±0.003	−0.00480 ±0.0006	+0.00138 ±0.0003	+0.1019 ±0.003	0.00048 ±0.00042	[77]
YPO_4	–	300	24.0	1.9915 ±0.0002	1.9915 ±0.0002	−2.1822 ±0.0003	−0.01229 ±0.0003	+0.00120 ±0.0003	+0.0636 ±0.0012	0.0 ±0.0006	[77]
	0.2%	298	9.0	1.9915 ±0.0005	1.9914 ±0.0005	−2.1810 ±0.0021	−0.01200 ±0.0003	+0.00120 ±0.0003	+0.0639 ±0.0045	0.0045 ±0.0045	[78]
	0.2%	78	9.0	1.9911 ±0.0005	–	−2.2200 ±0.0021	−0.0126 ±0.0003	+0.00120 ±0.0003	–	–	[78]
	0.2%	4.2	9.0	1.9892 ±0.0008	–	−2.2185 ±0.006	−0.0129 ±0.0009	+0.00150 ±0.0003	–	–	[78]
	–	4.2	10 and 34	1.991 ±0.001	1.991 ±0.001	−2.2254 ±0.0018	−0.01244 ±0.00018	+0.00147 ±0.00015	+0.0684 ±0.0015	0.0 ±0.0015	[77]
$YAsO_4$	–	300	24.0	1.9917 ±0.0002	1.9917 ±0.0002	−0.9488 ±0.0006	−0.0126 ±0.0003	+0.00120 ±0.0015	+0.1304 ±0.0006	0.0 ±0.0006	[77]
	0.2%	298	9.0	1.9926 ±0.0005	1.9926 ±0.0005	−0.9545 ±0.0012	−0.0120 ±0.0003	+0.0009 ±0.0003	+0.1280 ±0.003	0.0012 ±0.003	[78]
	0.2%	78	9.0	1.9920 ±0.0005	–	−0.9926 ±0.0012	−0.0126 ±0.0003	+0.0012 ±0.0003	–	–	[78]
	0.2%	4.2	9.0	1.9915 ±0.0005	–	−0.9977 ±0.0018	−0.0126 ±0.0003	+0.0012 ±0.0003	–	–	[78]
$ScVO_4$	0.2%	300	variable frequency	1.9917 ±0.0002	– –	−1.1213 ±0.0005	−0.0076 ±0.0002	+0.0025 ±0.0003	0.149 ±0.001	−0.0009 ±0.0004	[79]

where

$$B_{20} = \left(\frac{2}{\sqrt{6}}\right) b_2^0\,, \quad B_{40} = \left(\frac{\sqrt{70}}{30}\right) b_4^0\,, \quad B_{60} = \left(\frac{\sqrt{231}}{315}\right) b_6^0\,, \quad \text{and} \quad B_{66} = \left(\frac{1}{315}\right) b_6^6\,.$$

In general the magnitude of the term $b_6^6\,O_6^6$ is small compared with that of the other fine-structure terms. It is a fairly good approximation to neglect the term $b_6^6\,O_6^6$ since the matrix elements of this term are far-off diagonal and small in magnitude. The resonant magnetic fields of ($\Delta M = \pm 1$) transitions when the magnetic field $\boldsymbol{B}$ is directed parallel to the z-axis, are

$$\left.\begin{aligned}
|\pm\tfrac{7}{2}\rangle \leftrightarrow |\pm\tfrac{5}{2}\rangle\,; &\quad g_{\parallel}\,\beta\,B = h\,\nu \mp (6\,b_2^0 + 20\,b_4^0 + 6\,b_6^0)\,,\\
|\pm\tfrac{5}{2}\rangle \leftrightarrow |\pm\tfrac{3}{2}\rangle\,; &\quad g_{\parallel}\,\beta\,B = h\,\nu \mp (4\,b_2^0 - 10\,b_4^0 - 14\,b_6^0)\,,\\
|\pm\tfrac{3}{2}\rangle \leftrightarrow |\pm\tfrac{1}{2}\rangle\,; &\quad g_{\parallel}\,\beta\,B = h\,\nu \mp (2\,b_2^0 - 12\,b_4^0 + 14\,b_6^0)\,,\\
|+\tfrac{1}{2}\rangle \leftrightarrow |-\tfrac{1}{2}\rangle\,; &\quad g_{\parallel}\,\beta\,B = h\,\nu\,,
\end{aligned}\right\} \tag{30}$$

where h is Planck's constant and ν is the measurement frequency.

EPR data for the Gd^{3+} ion in crystals with hexagonal site symmetry are listed in Table 4. The hyperfine constant **A** is omitted in this table since the hyperfine structure has not been resolved for this symmetry.

5.4 *Trigonal symmetry*

The system of trigonal symmetry includes five point groups: C_3, $C_{3i}(S_6)$, C_{3v}, D_3, and D_{3d}. The spin Hamiltonian for the Gd^{3+} ion in crystals with trigonal site symmetry is

$$\begin{aligned}
\mathcal{H}_{\text{tri}} = \beta\,\boldsymbol{B}\cdot\mathrm{g}\cdot\boldsymbol{S} &+ \left(\frac{1}{3}\right) b_2^0\,O_2^0 + \left(\frac{1}{60}\right) b_4^0\,O_4^0 + \left(\frac{1}{1260}\right) b_6^0\,O_6^0 + \\
&+ \left(\frac{1}{3}\right) b_4^3\,O_4^3 + \left(\frac{1}{36}\right) b_6^3\,O_6^3 + \left(\frac{1}{1260}\right) b_6^6\,O_6^6 + \boldsymbol{S}\cdot\mathbf{A}\cdot\boldsymbol{I}\,,
\end{aligned} \tag{31}$$

where the symbols have been defined previously. Equation (31) can be expressed using tensor operators:

$$\begin{aligned}
\mathcal{H}_{\text{tri}} = \beta\,\boldsymbol{B}\cdot\mathbf{g}\cdot\boldsymbol{S} &+ B_{20}\,T_{20}(\boldsymbol{S}) + B_{40}\,T_{40}(\boldsymbol{S}) + B_{60}\,T_{60}(\boldsymbol{S}) + \\
&+ B_{43}\,[T_{4-3}(\boldsymbol{S}) - T_{43}(\boldsymbol{S})] + B_{63}\,[T_{6-3}(\boldsymbol{S}) - T_{63}(\boldsymbol{S})] + \\
&+ B_{66}\,[T_{66}(\boldsymbol{S}) + T_{6-6}(\boldsymbol{S})] + \boldsymbol{S}\cdot\mathbf{A}\cdot\boldsymbol{I}\,,
\end{aligned} \tag{32}$$

where

$$B_{20} = \left(\frac{2}{\sqrt{6}}\right) b_2^0\,, \quad B_{40} = \left(\frac{\sqrt{70}}{30}\right) b_4^0\,, \quad B_{60} = \left(\frac{\sqrt{231}}{315}\right) b_6^0\,,$$

$$B_{43} = \left(\frac{\sqrt{2}}{6}\right) b_4^3\,, \quad B_{63} = \left(\frac{1}{18}\sqrt{\frac{11}{5}}\right) b_6^3\,, \quad \text{and} \quad B_{66} = \left(\frac{1}{315}\right) b_6^6\,.$$

Strictly speaking (see Table 1), the imaginary terms such as

$$B'_{63}\,i\,[T_{6-3}(\boldsymbol{S}) + T_{63}(\boldsymbol{S})] \quad \text{and} \quad B'_{66}\,[T_{66}(\boldsymbol{S}) - T_{6-6}(\boldsymbol{S})]$$

Table 4
Spin Hamiltonian parameters of Gd^{3+} in single crystals with hexagonal site symmetry

host lattice	dilution	temperature (°K)	frequency (GHz)	$g_{\parallel}$	$g_{\perp}$	b_2^0 (GHz)	b_4^0 (GHz)	b_6^0 (GHz)	b_6^6 (GHz)	ref.
$La(C_2H_5SO_4)_3 \cdot 9\,H_2O$	–	room	–	1.992 ±0.001	–	+0.5729 ±0.003	−0.01133 ±0.0006	+0.00141 ±0.00045	–	[25]
	–	room	10.0	–	–	+0.5600 ±0.014	+0.0099 ±0.0003	−0.0009 ±0.0003	–	[80]
	0.1 at%	290	35.0	1.99180 ±0.0001	1.991835 ±0.00011	+0.582642 ±0.0004	−0.011514 ±0.00007	+0.001389 ±0.00008	−0.01071 ±0.0019	[81]
	0.1 at%	290	10.0	1.991922 ±0.00006	1.991964 ±0.00008	+0.583724 ±0.00065	−0.011542 ±0.00006	+0.001389 ±0.00008	−0.01103 ±0.0022	[81]
	0.1 at%	290	5.0	1.992307 ±0.00032	1.992465 ±0.00018	+0.584093 ±0.0004	−0.011567 ±0.00010	+0.001409 ±0.00012	−0.01229 ±0.0032	[81]
	–	288	9.0	1.99201 ±0.0002	–	+0.585314 ±0.00018	−0.011590 ±0.000015	+0.001373 ±0.00024	+0.01229 ±0.0006	[82]
	–	280	9.0	1.99206 ±0.0002	–	+0.588792 ±0.00018	−0.011665 ±0.000015	+0.001385 ±0.00024	−0.01229 ±0.0006	[82]
	–	271	9.0	1.99200 ±0.0002	–	+0.592210 ±0.00018	−0.011710 ±0.000015	+0.001394 ±0.00024	−0.01229 ±0.0006	[82]
	–	242	9.0	1.99196 ±0.0002	–	+0.601294 ±0.00018	−0.011839 ±0.000015	+0.001406 ±0.00024	−0.01259 ±0.0006	[82]
	–	232.4	9.0	1.992 ±0.001	–	+0.59629 ±0.003	−0.01139 ±0.0006	+0.001289 ±0.00045	–	[25]
	–	190	9.0	1.99192 ±0.0002	–	+0.612176 ±0.00018	−0.011998 ±0.000015	+0.001457 ±0.00024	−0.01229 ±0.0006	[82]
	–	165	9.0	1.99196 ±0.0002	–	+0.614994 ±0.00018	−0.012028 ±0.000015	+0.001466 ±0.00024	−0.01274 ±0.0006	[82]
	–	131	9.0	1.99188 ±0.0002	–	+0.616343 ±0.00018	−0.012067 ±0.000015	+0.001496 ±0.00024	−0.01304 ±0.0006	[82]
	–	90	9.0	1.9915 ±0.0002	–	+0.613975 ±0.00018	−0.012010 ±0.000015	+0.001517 ±0.00024	–	[82]
	0.5%	90	10.0	1.988 ±0.002	1.986 ±0.003	+0.61368 ±0.006	−0.01187 ±0.0009	+0.001889 ±0.0003	+0.01049 ±0.0015	[83]

Table 4 (continued)

host lattice	dilution	temperature (°K)	frequency (GHz)	$g_{\parallel}$	$g_{\perp}$	b_2^0 (GHz)	b_4^0 (GHz)	b_6^0 (GHz)	b_6^6 (GHz)	ref.
$La(C_2H_5SO_4)_3 \cdot 9H_2O$	0.5	90	24.0	1.9880 ±0.001	1.9901 ±0.001	+0.61577 ±0.006	−0.01082 ±0.0018	+0.002938 ±0.0015	+0.0006 ±0.003	[84]
	0.5%	90	24.0	1.9902 ±0.002	1.9901 ±0.001	+0.61817 ±0.006	−0.01193 ±0.0009	+0.001229 ±0.0009	+0.01764 ±0.0015	[84]*
	0.1% and 1.0%	77	9.0	1.991 ±0.002	–	+0.61158 ±0.006	−0.01109 ±0.0009	+0.001199 ±0.0006	–	[85]
	0.01%	77	35.0	1.991 ±0.001	–	+0.61158 ±0.006	−0.01115 ±0.0009	+0.001289 ±0.0006	–	[86]
	–	77	–	1.992 ±0.001	–	+0.60678 ±0.003	−0.01160 ±0.0006	+0.001139 ±0.00045	–	[25]
	–	77	9.0	1.99200 ±0.0002	–	+0.61257 ±0.00018	−0.01203 ±0.000015	+0.001538 ±0.00024	−0.01274 ±0.0006	[82]
	0.5%	20	10.0	1.989 ±0.001	1.989 ±0.001	+0.59899 ±0.003	−0.01172 ±0.00045	+0.001589 ±0.00015	+0.01199 ±0.0015	[83]
	–	4.2	9.0	1.9915 ±0.0002	–	+0.60138 ±0.00018	−0.01182 ±0.000015	+0.001532 ±0.00024	–	[82]
	0.01%	4	35.0	1.991 ±0.001	–	+0.59059 ±0.006	−0.01154 ±0.0009	+0.001439 ±0.0006	–	[86]
	–	4	–	1.992 ±0.001	–	+0.59329 ±0.003	−0.01151 ±0.0006	+0.001439 ±0.00045	–	[25]
$Y(C_2H_5SO_4)_3 \cdot 9H_2O$	0.5%	293	–	1.987 ±0.010	–	+0.46318 ±0.0045	−0.01078 ±0.0012	+0.002099 ±0.0012	–	[87]
	0.01%	290	35.0	1.991 ±0.001	–	+0.46468 ±0.006	−0.01133 ±0.0009	+0.001439 ±0.0006	–	[86]
	0.5%	77	–	1.987 ±0.002	–	+0.46918 ±0.003	−0.01139 ±0.0009	+0.001499 ±0.0009	–	[87]
	0.01%	77	35.0	1.991 ±0.001	–	+0.47067 ±0.006	−0.01181 ±0.0009	+0.001619 ±0.0006	–	[86]
	0.5%	4.2	–	1.993 ±0.002	–	+0.45449 ±0.003	−0.01139 ±0.0009	+0.001499 ±0.0009	–	[87]
	0.01%	4	35.0	1.991 ±0.001	–	+0.45568 ±0.006	−0.01172 ±0.0009	+0.001799 ±0.0006	–	[86]

*) Parameters obtained from ($\Delta M = \pm 2$) transitions.

Table 4 (continued)

host lattice	dilution	temperature (°K)	frequency (GHz)	$g_{\parallel}$	$g_{\perp}$	b_2^0 (GHz)	b_4^0 (GHz)	b_6^0 (GHz)	b_6^6 (GHz)	ref.
$Ce(C_2H_5SO_4)_3 \cdot 9\,H_2O$	0.01%	77	35.0	–	–	+0.58160 ±0.006	−0.01181 ±0.0009	+0.001529 ±0.0006	–	[86]
	0.01%	4	35.0	–	–	+0.56061 ±0.006	−0.01193 ±0.0009	+0.001679 ±0.0006	–	[86]
$Pr(C_2H_5SO_4)_3 \cdot 9\,H_2O$	–	room	10.0	–	–	+0.54502 ±0.0045	+0.01139 ±0.0003	−0.001199 ±0.0003	–	[80]
$Nd(C_2H_5SO_4)_3 \cdot 9\,H_2O$	–	room	10.0	–	–	+0.52674 ±0.009	+0.01139 ±0.0006	−0.001199 ±0.0003	–	[80]
$Sm(C_2H_5SO_4)_3 \cdot 9\,H_2O$	–	room	10.0	–	–	+0.50935 ±0.0057	+0.01139 ±0.0006	−0.00090 ±0.0003	–	[80]
$Eu(C_2H_5SO_4)_3 \cdot 9\,H_2O$	–	room	10.0	–	–	+0.49945 ±0.0036	+0.01139 ±0.003	−0.001199 ±0.0003	–	[80]
$Tb(C_2H_5SO_4)_3 \cdot 9\,H_2O$	–	room	10.0	–	–	+0.48027 ±0.0069	+0.01139 ±0.0006	−0.001199 ±0.0003	–	[80]
$Dy(C_2H_5SO_4)_3 \cdot 9\,H_2O$	–	room	10.0	–	–	+0.47427 ±0.0015	+0.001139 ±0.0003	−0.001499 ±0.0003	–	[80]
	0.01%	1.2	35.0	–	–	+0.46768 ±0.006	−0.01178 ±0.0009	+0.001649 ±0.0006	–	[86]
$Ho(C_2H_5SO_4)_3 \cdot 9\,H_2O$	–	room	10.0	–	–	+0.48087 ±0.0033	+0.01169 ±0.0003	−0.00090 ±0.0003	–	[80]

Table 4 (continued)

host lattice	dilution	temperature (°K)	frequency (GHz)	$g_{\parallel}$	$g_{\perp}$	b_2^0 (GHz)	b_4^0 (GHz)	b_6^0 (GHz)	b_6^6 (GHz)	ref.
$Er(C_2H_5SO_4)_3 \cdot 9\,H_2O$	–	room	10.0	–	–	+0.45119 ±0.0015	+0.01109 ±0.0003	−0.001199 ±0.0003	–	[80]
$Tm(C_2H_5SO_4)_3 \cdot 9\,H_2O$	–	room	10.0	–	–	+0.44219 ±0.0048	+0.01109 ±0.0003	−0.001199 ±0.0003	–	[80]
$Yb(C_2H_5SO_4)_3 \cdot 9\,H_2O$	–	room	10.0	–	–	+0.43410 ±0.0024	+0.01139 ±0.0006	−0.001199 ±0.0003	–	[80]
$LaCl_3$	undoped	293.5	9.0	1.9917 ±0.0005	–	+0.02218 ±0.00048	+0.006326 ±0.00012	+0.000630 ±0.00012	+0.002638 ±0.0003	[88]
	0.5 wt%	293.5	9.0	1.9923 ±0.0005	–	+0.02443 ±0.0003	+0.005156 ±0.00006	+0.002129 ±0.00006	–	[89]
	0.01 at%	≈290.0	10.0	1.991 ±0.001	–	+0.02506 ±0.0003	+0.005037 ±0.00012	+0.001919 ±0.00045	–	[90]
	0.01 at%	90.0	10.0	1.991 ±0.001	–	+0.04797 ±0.00006	+0.006386 ±0.00015	+0.000750 ±0.00015	+0.004197 ±0.0009	[90]
	0.5 wt%	77.0	9.0	1.9919 ±0.0005	–	+0.04905 ±0.0003	+0.006416 ±0.00006	+0.000630 ±0.00006	–	[89]
	0.5 wt%	20.0	9.0	1.9924 ±0.0005	–	+0.05339 ±0.0003	+0.006386 ±0.00006	+0.000630 ±0.00006	–	[89]
	0.5 wt%	≈10.0	9.0	1.9917 ±0.0005	–	+0.05339 ±0.0003	+0.006386 ±0.00006	+0.000630 ±0.00006	–	[89]
	0.5 wt%	4.0	9.0	1.9927 ±0.0005	–	+0.05300 ±0.0003	+0.006476 ±0.00006	+0.000690 ±0.00006	–	[89]
	0.5 wt%	≈ 2.0	9.0	1.9926 ±0.0005	–	+0.05282 ±0.0003	+0.006535 ±0.00006	+0.000720 ±0.00006	–	[89]

Table 4 (continued)

host lattice	dilution	temperature (°K)	frequency (GHz)	$g_\parallel$	$g_\perp$	b_2^0 (GHz)	b_4^0 (GHz)	b_6^0 (GHz)	b_6^6 (GHz)	ref.
$LaBr_3$	1.0%	room	9.0	1.99	1.99	+0.15859	+0.02671	+0.001793	+0.02245	[91]
	0.5 wt%	294.0	9.0	1.9920 ±0.0005	–	+0.13722 ±0.0003	+0.04173 ±0.00006	−0.001007 ±0.00006	–	[89]
	1.0%	78.0	9.0	1.99	1.99	+0.19223	+0.03094	+0.02057	+0.02341	[91]
	0.5 wt%	77.0	9.0	1.9922 ±0.0005	–	+0.17796 ±0.0003	+0.04248 ±0.00006	−0.001034 ±0.00006	–	[89]
	0.5 wt%	4.0	9.0	1.9922 ±0.0005	–	+0.18515 ±0.0003	+0.04263 ±0.00006	−0.001016 ±0.00006	–	[89]
	1.0%	4.2	9.0	1.99	1.99	+0.19454	+0.03157	+0.02266	–	[91]
	0.5 wt%	2.0	9.0	1.9922 ±0.0005	–	+0.18353 ±0.0003	+0.04032 ±0.00006	−0.00974 ±0.00006	–	[89]
$La(SO_4)_3 \cdot 9\,H_2O$	1.0%	room	9.0	1.99	1.99	−1.91777	−0.007794	+0.0006	≈0.0006	[91]
	1.0%	78.0	9.0	1.99	1.99	−1.93216	−0.009893	+0.0012	≈0.006	[91]
	1.0%	4.2	9.0	1.99	1.99	−1.93456	−0.009593	+0.0018	–	[91]
$La(OH)_3$	1.0%	77.0	24.0	1.992 ±0.001	1.992 ±0.001	−0.8214 ±0.006	−0.006895 ±0.0015	+0.001499 ±0.0015	–	[92]
$Y(OH)_3$	1.0%	300.0	10 and 24	1.992 ±0.001	1.992 ±0.001	−0.40832 ±0.0018	−0.005996 ±0.0009	+0.001499 ±0.0015	–	[92]
	1.0%	77.0	10 and 24	1.992 ±0.001	1.992 ±0.001	−0.39093 ±0.0018	−0.006595 ±0.0006	+0.001799 ±0.0006	+0.01589 ±0.0003	[92]

should be included in (31) or (32) for point groups C_3 and $C_{3i}(S_6)$. In general these imaginary terms are neglected in the spin Hamiltonian because its effects on the fitting of EPR spectra are small. Marshall and Serway [32] have shown that the resonant magnetic fields of the seven ($\Delta M = \pm 1$) transitions can be calculated from (30) or (31) using a perturbation method when the applied magnetic field $\boldsymbol{B}$ is directed parallel to the z-axis (the threefold symmetry axis):

$$\left.\begin{aligned}
&|\pm\tfrac{7}{2}\rangle \leftrightarrow |\pm\tfrac{5}{2}\rangle;\\
&g_{\parallel}\beta B = h\nu \mp (6\,b_2^0 + 20\,b_4^0 + 6\,b_6^0) -\\
&\qquad - P\left\{\frac{7\,(1+\varrho)^2}{1 \pm 4\,Q \mp 2\,R \pm 2\,S} - \frac{4\,[1-(\frac{7}{4})\,\varrho]^2}{1 \pm 2\,Q \mp 22\,R}\right\} -\\
&\qquad - T\,\{(1 \pm Q \pm 10\,R \pm S)^{-1} - (1 \mp Q \mp 10\,R \mp S)^{-1}\},\\
&|\pm\tfrac{5}{2}\rangle \leftrightarrow |\pm\tfrac{3}{2}\rangle;\\
&g_{\parallel}\beta B = h\nu \mp (4\,b_2^0 - 10\,b_4^0 - 14\,b_6^0) -\\
&\qquad - P\left\{\frac{4\,[1-(\frac{7}{4})\,\varrho]^2}{1 \pm 2\,Q \mp 22\,R}\right\} - T\,\{(1 \mp Q \mp 10\,R \mp S)\}^{-1},\\
&|\pm\tfrac{3}{2}\rangle \leftrightarrow |\pm\tfrac{1}{2}\rangle;\\
&g_{\parallel}\beta B = h\nu \mp (2\,b_2^0 - 12\,b_4^0 + 14\,b_6^0) -\\
&\qquad - P\left\{\frac{7\,(1+\varrho)^2}{1 \pm 4\,Q \mp 2\,R \pm 2\,S} - \frac{4\,[1-(\frac{7}{4})\,\varrho]^2}{1 \mp 2\,Q \pm 22\,R}\right\},\\
&|+\tfrac{1}{2}\rangle \leftrightarrow |-\tfrac{1}{2}\rangle;\\
&g_{\parallel}\beta B = h\nu + 2\,P\left\{\frac{7\,(1+\varrho)^2}{[1-(4\,Q-2\,R+2\,S)^2]} - \frac{4\,[1-(\frac{7}{4})\,\varrho]^2}{[1-(2\,Q-22\,R)^2]}\right\},
\end{aligned}\right\} \quad (33)$$

where

$$\varrho = \frac{b_6^3}{b_4^3}, \quad P = \frac{20\,(b_4^3)^2}{3\,g_{\parallel}\beta B}, \quad Q = \frac{b_2^0}{g_{\parallel}\beta B}, \quad R = \frac{b_4^0}{3\,g_{\parallel}\beta B},$$

$$S = \frac{b_6^0}{g_{\parallel}\beta B}, \quad \text{and} \quad T = \frac{7\,(b_6^6)^2}{6\,g_{\parallel}\beta B}.$$

EPR data of the Gd^{3+} ion in crystals with trigonal site symmetry are listed in Table 5. There are six fine-structure constants in the spin Hamiltonian to be experimentally determined. In some cases the effect of b_4^3, b_6^3, and b_6^6 are not resolved and are represented in a linear combination form.

5.5 Orthorhombic or lower symmetry

The minimum symmetry possessed by a point group belonging to the orthorhombic system is two orthogonal twofold rotation axes or two orthogonal twofold improper rotation axes. The three point groups belonging to this system are C_{2v}, D_z, and D_{2h}. The spin Hamiltonian for the Gd^{3+} ion in crystals with

Table 5

Spin Hamiltonian parameters of Gd^{3+} in single crystals with trigonal site symmetry

host lattice	dilution	temperature (°K)	frequency (GHz)	$g_{\parallel}$	$g_{\perp}$	b_2^0 (GHz)	b_4^0 (GHz)	b_6^0 (HGz)	b_4^3 (GHz)	b_6^3 (GHz)	b_6^6 (GHz)	ref.
BaF_2	0.02%	room	9.2	1.992 ±0.001	–	(±) 0.4086 ±0.0024	(∓) 0.07315 ±0.0018	–	(∓) 0.1005 ±0.0022	–	–	[42]
	–	room	10 and 36	1.991 ±0.001	–	(±) 0.412 ±0.005	(∓) 0.072 ±0.005	(±)0.002 ±0.005	(±) 0.1045 ±0.0003	(±) 0.00017 ±0.00015	(±) 0.0 ±0.1	[93]
	0.1 wt%	300	10	1.9923 ±0.0005	–	−0.4095 ±0.0003	+0.07357 ±0.00009	−0.000839 ±0.00009	−0.10391 ±0.00015	−0.000248 ±0.00026	−0.0018 ±0.0066	[60]
	0.1 wt%	77	10	1.9924 ±0.0005	–	−0.45248 ±0.00018	+0.07732 ±0.00006	−0.000749 ±0.00006	−0.10905 ±0.00010	−0.000111 ±0.00014	−0.0066 ±0.0036	[60]
	0.1 wt%	4.2	10	1.9921 ±0.0005	–	−0.4608 ±0.0006	+0.07729 ±0.00015	−0.000779 ±0.00015	−0.10941 ±0.00022	−0.000103 ±0.00031	−0.0060 ±0.0081	[60]
$CaF_2(OH^-)$	0.1 to 0.001%	room	9.2	1.990 ±0.002	–	+1.2212 ±0.0023	+0.04228 ±0.0012	+0.0 ±0.0012	+0.03573 ±0.007	+0.00216 ±0.0054	+0.02804 ±0.0028	[42]
$CaF_2(O^{2-})$	0.01 to 0.001%	room	9.2	1.990 ±0.003	–	+4.9960 ±0.0046	−0.07851 ±0.0024	+0.00554 ±0.0036	+0.09290 ±0.021	+0.0 ±0.0054	+0.02804 ±0.056	[42]
	1:10⁴	room	10 to 50	1.992 ±0.001	–	−5.00 ±0.01	+0.078 ±0.002	+0.0 ±0.001	−2.12 ±0.005	+0.05 ±0.005	+0.0 ±0.05	[94], [95]
$CdF_2(O^{2-})$	10^{-4} to 5×10^{-3}	room	9	1.9919 ±0.0003	1.992 ±0.001	+2.2984 ±0.0023	−0.08612 ±0.0012	+0.00092 ±0.0018	+0.1186 ±0.0014	−0.000216 ±0.0011	+0.0028 ±0.028	[96]
$SrF_2(O^{2-})$	0.02%	room	9.2	1.992 ±0.003	–	+5.4447 ±0.023	−0.06039 ±0.012	+0.0092 ±0.018	+0.04145 ±0.014	+0.0 ±0.011	+0.0 ±0.28	[42]
ZnO	–	room	35	1.987 ±0.002	1.978 ±0.002	+2.5493 ±0.0014	+0.04728 ±0.0008	+0.00139 ±0.00028	–	–	–	[97]
CdSe	–	77	–	1.985 ±0.002	–	+2.4148 ±0.003	+0.02072 ±0.0012	+0.00108 ±0.0009	+0.05246 ±0.0045	–	–	[49]

Table 5 (continued)

host lattice	dilution	temperature (°K)	frequency (GHz)	$g_{\parallel}$	$g_{\perp}$	b_2^0 (GHz)	b_4^0 (GHz)	b_6^0 (GHz)	b_4^3 (GHz)	b_6^3 (GHz)	b_6^6 (GHz)	ref.
Al_2O_3	$<10^{-3}$	room	24	1.9912 ±0.0005	–	+3.0966 ±0.006	+0.07795 ±0.003	+0.0030 ±0.0015	$\|b_4^3\| =$ =0.05486± ±0.003	$\|b_6^3\| \leq$ ≤0.003	$\|b_6^6\| =$ =0.01499± ±0.0015	[98]
$CaCO_3$	–	room	10	1.992 ±0.001	1.993 ±0.001	−1.368 ±0.002	−0.0033 ±0.001	+0.0042 ±0.001	$\|b_4^3\| \approx$ ≈0.020	–	$\|b_6^6\| =$ =0.031± ±0.005	[99]
	parts per 10^6	77	10 and 37.5	1.9918 ±0.0001	1.9918 ±0.0001	−1.7005 ±0.0002	+0.00761 ±0.00014	+0.00516 ±0.00014	~0.01126	+0.00229	0.0027	[32]*)
$LaAlO_3$	–	689	9	1.991 ±0.0005	–	+0.2009 ±0.003	–	–	–	–	–	[100]
	–	635	9	1.991 ±0.0005	–	+0.2788 ±0.003	–	–	–	–	–	[100]
	–	575	9	1.991 ±0.0005	–	+0.4515 ±0.003	+0.01679 ±0.0009	+0.0027 ±0.0006	–	–	–	[100]
	–	505	9	1.9910 ±0.0005	–	+0.6245 ±0.003	+0.01739 ±0.0009	+0.0030 ±0.0006	–	–	–	[100]
	–	415	9	1.9910 ±0.0005	–	+0.9204 ±0.003	+0.01796 ±0.0009	+0.0030 ±0.0006	–	–	–	[100]
	–	358	9	1.9915 ±0.0005	–	+1.0103 ±0.003	+0.01838 ±0.0009	+0.0030 ±0.0006	–	–	+0.02968 ±0.0036	[100]
	–	293	9	1.9908 ±0.0005	1.986 ±0.010	+1.1128 ±0.003	+0.01850 ±0.0009	+0.0030 ±0.0006		–	+0.02278 ±0.0036	[100]
	–	273	9	1.9911 ±0.0005	1.984 ±0.010	+1.1506 ±0.003	+0.01868 ±0.0009	+0.0027 ±0.0006		–	+0.02818 ±0.0036	[100]
	–	203	9	1.9904 ±0.0005	1.978 ±0.010	+1.2792 ±0.003	+0.01925 ±0.0009	+0.0027 ±0.0006	–	–	+0.02159 ±0.0036	[100]
	–	80	10	1.9911 ±0.0005	1.978 ±0.010	+1.4366 ±0.003	+0.01925 ±0.0009	+0.0027 ±0.0006	–	–	+0.03058 ±0.0036	[100]
	–	20	10	1.9909 ±0.0005	1.979 ±0.010	+1.4711 ±0.003	+0.01937 ±0.0009	+0.0027 ±0.0006	–	–	+0.02398 ±0.0036	[100]

*) The hyperfine-structure constants A were determined with $A_{155} = (12.12 \pm 0.011)$ MHz; $A_{157} = (15.90 \pm 0.014)$ MHz.

Table 5 (continued)

host lattice	dilution	temperature (°K)	frequency (GHz)	$g_{\parallel}$	$g_{\perp}$	b_2^0 (GHz)	b_4^0 (GHz)	b_6^0 (GHz)	b_4^3 (GHz)	b_6^3 (GHz)	b_6^6 (GHz)	ref.
$Bi_2Mg_3(NO_3)_{12} \cdot 24\ H_2O$	–	290	10	1.9917 ±0.0001	1.9983 ±0.0001	+0.35346 ±0.00012	+0.00115 ±0.00007	+0.001264 ±0.000002	–	–	+0.016065 ±0.00032	[101]
	1%	90	10	1.990 ±0.004	1.992 ±0.004	+0.3717 ±0.003	+0.0027 ±0.0003	+0.0018 ±0.0003	–	–	+0.03598 ±0.003	[102]
	–	77	10	1.9926 ±0.0001	1.9919 ±0.0001	+0.38775 ±0.00012	+0.00355 ±0.00004	+0.001762 ±0.00004	–	–	+0.01638 ±0.00064	[101]
	1%	20	10	1.992 ±0.003	1.992 ±0.003	+0.3717 ±0.003	+0.0027 ±0.0003	+0.0018 ±0.0003	–	–	+0.03598 ±0.003	[102]
$Bi_2Zn_3(NO_3)_{12} \cdot 24\ H_2O$	–	290	10	1.9914 ±0.0001	1.9988 ±0.0001	+0.34660 ±0.00012	+0.00158 ±0.00004	+0.000902 ±0.000002	–	–	+0.01449 ±0.00064	[101]
	–	77	10	1.9912 ±0.0001	1.9886 ±0.0001	+0.37465 ±0.00012	+0.00337 ±0.00007	+0.000962 ±0.000002	–	–	+0.01386 ±0.00064	[101]
$La_2Mg_3(NO_3)_{12} \cdot 24\ H_2O$	–	290	10	1.9917 ±0.0001	1.9922 ±0.0001	+0.33117 ±0.00012	+0.00104 ±0.00004	+0.00135 ±0.00004	–	–	+0.011655 ±0.00032	[101]
	–	77	10	1.9917 ±0.0001	1.9930 ±0.0001	+0.34624 ±0.00012	+0.00273 ±0.00004	+0.00131 ±0.00006	–	–	+0.1449 ±0.0032	[101]
$La_2Zn_3(NO_3)_{12} \cdot 24\ H_2O$	–	290	10	1.9918 ±0.0001	1.9924 ±0.0001	+0.34856 ±0.0024	+0.00187 ±0.00007	+0.001465 ±0.000002	–	–	+0.000032 ±0.000032	[101]
	–	77	10	1.9928 ±0.0001	1.9914 ±0.0001	+0.36497 ±0.00012	+0.00357 ±0.0004	+0.001880 ±0.000002	–	–	+0.014175 ±0.00032	[101]
LaHAPI	0.3%	300	23	1.991 ±0.003	1.989 ±0.003	+2.1066 +0.0075 -0.0006	+0.0039 ±0.0048	−0.0024 ±0.0030	–	–	+0.02188 ±0.0027	[39]
	0.3%	90	23	1.9931 ±0.0005	–	+2.4793 +0.0075 −0.0006	+0.00839 ±0.0048	−0.0033 ±0.003	+0.02398 ±0.0054	+0.0024 ±0.0042	+0.02069 ±0.0015	[39]
	undiluted	300	23	1.990 ±0.002	–	+2.1315 ±0.015	+0.00516 ±0.0011	−0.0024 ±0.0015	–	–	–	[39]

orthorhombic site symmetry is

$$\mathcal{H}_{\text{rhombic}} = \beta \boldsymbol{B} \cdot \text{g} \cdot \boldsymbol{S} + \left(\frac{1}{3}\right) b_2^0 O_2^0 + \left(\frac{1}{60}\right) b_4^0 O_4^0 + \left(\frac{1}{1260}\right) b_6^0 O_6^0 + \left(\frac{1}{3}\right) b_2^2 O_2^2 + \\ + \left(\frac{1}{60}\right) b_4^2 O_4^2 + \left(\frac{1}{1260}\right) b_6^2 O_6^2 + \left(\frac{1}{60}\right) b_4^4 O_4^4 + \left(\frac{1}{1260}\right) b_6^4 O_6^4 + \\ + \left(\frac{1}{1260}\right) b_6^6 O_6^6 + \boldsymbol{S} \cdot \boldsymbol{A} \cdot \boldsymbol{I} . \tag{34}$$

Equation (34) can be reexpressed using tensor operators:

$$\mathcal{H}_{\text{rhombic}} = \beta \boldsymbol{B} \cdot \text{g} \cdot \boldsymbol{S} + B_{20} T_{20}(\boldsymbol{S}) + B_{40} T_{40}(\boldsymbol{S}) + B_{60} T_{60}(\boldsymbol{S}) + \\ + B_{22} [T_{22}(\boldsymbol{S}) + T_{2-2}(\boldsymbol{S})] + B_{42} [T_{42}(\boldsymbol{S}) + T_{4-2}(\boldsymbol{S})] + \\ + B_{62} [T_{62}(\boldsymbol{S}) + T_{6-2}(\boldsymbol{S})] + B_{44} [T_{44}(\boldsymbol{S}) + T_{4-4}(\boldsymbol{S})] + \\ + B_{64} T_{64}(\boldsymbol{S}) + T_{6-4}(\boldsymbol{S})] + B_{66} [T_{66}(\boldsymbol{S}) + T_{6-6}(\boldsymbol{S})] , \tag{35}$$

where the relationships between B_{20} and b_2^0, B_{40} and b_4^0, B_{60} and b_6^0, B_{44} and b_4^4, B_{64} and b_6^4, B_{66} and b_6^6 have been given in Sections 5.2 and 5.3, and where

$$B_{22} = \left(\frac{1}{3}\right) b_2^2 , \qquad B_{42} = \left(\frac{\sqrt{7}}{60}\right) b_4^2 , \qquad B_{62} = \left(\frac{1}{315} \sqrt{\frac{11}{5}}\right) b_6^2 .$$

For points groups belonging to the monoclinic system the imaginary terms such as $B'_{42}\, i\, [T_{42}(\boldsymbol{S}) - T_{4-2}(\boldsymbol{S})]$, $B'_{62}\, i\, [T_{62}(\boldsymbol{S}) - T_{6-2}(\boldsymbol{S})]$, etc., should be added to equation (35). For point groups belonging to the triclinic system, the tensor operators with odd m-numbers such as $B_{21} [T_{21}(\boldsymbol{S}) - T_{2-1}(\boldsymbol{S})]$, etc. should also be included in the spin Hamiltonian. The point groups C_2, C_s, and C_{2h} belong to the monoclinic system, and the point groups C_1 and $C_i(I)$ belong to the triclinic system.

Singh et al. [33] have shown that the resonant magnetic fields of the seven ($\Delta M = \pm 1$) transitions calculated when $\boldsymbol{B} \parallel z$ from (34) to the second order are given by

$$\left.\begin{aligned}
&|\pm\tfrac{7}{2}\rangle \leftrightarrow |\pm\tfrac{5}{2}\rangle ; \quad g_{\parallel} \beta B = h\nu \mp (6\, b_2^0 + 20\, b_4^0 + 6\, b_6^0) + \\
&\qquad + E\left(\frac{45}{1 \pm 3F} - \frac{21}{1 \pm 5F}\right), \\
&|\pm\tfrac{5}{2}\rangle \leftrightarrow |\pm\tfrac{3}{2}\rangle ; \quad g_{\parallel} \beta B = h\nu \mp (4\, b_2^0 - 12\, b_4^0 - 14\, b_6^0) + \\
&\qquad + E\left(\frac{60}{1 \pm F} - \frac{21}{1 \pm 5F} - \frac{45}{1 \pm 3F}\right), \\
&|\pm\tfrac{3}{2}\rangle \leftrightarrow |\pm\tfrac{1}{2}\rangle ; \quad g_{\parallel} \beta B = h\nu \mp (2\, b_2^0 - 10\, b_4^0 + 14\, b_6^0) + \\
&\qquad + E\left(\frac{21}{1 + 5F} - \frac{45}{1 \pm 3F} - \frac{120\, F}{1 - F^2}\right), \\
&|+\tfrac{1}{2}\rangle \leftrightarrow |-\tfrac{1}{2}\rangle ; \quad g_{\parallel} \beta B = h\nu + E\left(\frac{90}{1 - 9F^2} - \frac{120}{1 - F^2}\right),
\end{aligned}\right\} \tag{36}$$

Table 6

Spin Hamiltonian parameters of Gd^{3+} in single crystals with orthorhombic or lower site symmetry

host lattice	dilution	temperature (°K)	frequency (GHz)	g	b_2^0 (GHz)	b_4^0 (GHz)	b_6^0 (GHz)	b_2^2 (GHz)	other fine-structure constants (GHz)	ref.
LaF_3	0.01 at%	77	10	1.991 ±0.001	+0.71153 ±0.003	+0.01685 ±0.0006	−0.00051 ±0.0006	+0.01676 ±0.006	b_4^2 = −0.07825 ±0.009; b_4^4 = +0.1316 ±0.009; b_6^2 = −0.01829; b_6^4 = −0.02188 ±0.015; b_6^6 = −0.01799	[9]
	0.01 at%	90	10	1.990 ±0.001	(+) 0.7165 ±0.003	(−) 0.01679 ±0.0006	(+) 0.00042 ±0.0006	(−) 0.01499 ±0.006	b_4^2 = (+)0.0809 ±0.009; b_4^4 = (−)0.1289 ±0.009; b_6^4 = (−)0.003 ±0.015; $b_6^2+b_6^6$ = (+)0.0255 ±0.009	[103]
	0.01 at%	room	10	1.9934 ±0.0015	+0.70247 ±0.003	+0.01532 ±0.0006	−0.00063 ±0.0006	+0.05765 ±0.006	b_4^2 = −0.07285 ±0.009; b_4^4 = +0.1271 ±0.009; b_6^2 = −0.03028; b_6^4 = −0.02188 ±0.015; b_6^6 = −0.03118	[9]
PrF_3	0.01 at%	room	10	1.9941 ±0.003	+0.77742 ±0.003	+0.01496 ±0.0006	−0.00051 ±0.0006	+0.08010 ±0.006	b_4^2 = −0.08334 ±0.009; b_4^4 = +0.1283 ±0.009; b_6^2 = −0.02159; b_6^4 = −0.0030 ±0.015; b_6^6 = −0.01979	[9]
CeF_3	0.01 at%	room	10	1.9938 ±0.002	+0.74019 ±0.003	+0.01655 ±0.0006	−0.00051 ±0.0006	+0.05582 ±0.006	b_4^2 = −0.07555 ±0.009; b_4^4 = +0.1370 ±0.009; b_6^2 = −0.04107; b_6^4 = −0.03687 ±0.015; b_6^6 = −0.05066	[9]
NdF_3	0.01 at%	room	10	1.9914 ±0.003	+0.79562 ±0.003	+0.01868 ±0.0006	−0.00390 ±0.0006	+0.16905 ±0.006	b_4^2 = −0.06955 ±0.009; b_4^4 = +0.1124 ±0.009; b_6^2 = −0.04377; b_6^4 = +0.04767 ±0.015; b_6^6 = −0.00690	[9]

Table 6 (continued)

host lattice	dilution	temperature (°K)	frequency (GHz)	g	b_2^0(GHz)	b_4^0(GHz)	b_6^0(GHz)	b_2^2(GHz)	other fine-structure constants (GHz)	ref.
$LaCl_3 \cdot 7\,H_2O$	–	room	10	$g_x = 1.998 \pm 0.003$ $g_y = 2.000 \pm 0.003$ $g_z = 1.989 \pm 0.003$	$(\pm)$ 0.39363 ± 0.00015	$(\pm)$ 0.0006 ± 0.0003	–	$(\mp)$ 0.2254 ± 0.0006	$\frac{3}{8} b_4^0 - \frac{5}{2} b_4^2 + \frac{35}{8} b_4^4 = 0.0(\pm)0.0003$ $\frac{3}{8} b_4^0 + \frac{5}{2} b_4^2 + \frac{35}{8} b_4^4 = (\pm)0.0006(\pm)0.0003$	[104]
$NdCl_3 \cdot 6\,H_2O$	0.05 at%	room	9	$g_\parallel = 1.987 \pm 0.002$ $g_\perp = 1.986 \pm 0.002$	$(\pm)$ 1.8348	$(-)$ 0.03298	$(-)$ 0.00125	$(-)$ 1.1520	$b_4^4 - b_4^2 = (+)0.1359$ $b_6^6 - b_6^4 + b_6^2 = (-)0.02406$	[33]
$YCl_3 \cdot 6\,H_2O$	0.16%	293	9	$g_x = 1.988 \pm 0.004$ $g_y = 1.993 \pm 0.004$ $g_z = 1.999 \pm 0.004$	$+1.6009 \pm 0.015$	-0.009 ± 0.003	–	$+2.3024 \pm 0.045$	–	[105]
$PrCl_3 \cdot 7\,H_2O$	0.05 at%	room	9	$g_\parallel = 1.976 \pm 0.002$ $g_\perp = 1.960 \pm 0.002$	$(+)$ 1.1604	$(+)$ 0.00644	$(+)$ 0.00205	$(-)$ 0.7285	$b_4^4 - b_4^2 = (+)0.02138$ $b_6^6 - b_6^4 + b_6^2 = (-)0.04148$	[33]
$La(NO_3)_3 \cdot 6\,H_2O$	0.05 at%	room	9	$g_\parallel = 1.995 \pm 0.002$ $g_\perp = 1.986 \pm 0.002$	$(+)$ 0.6885	$(+)$ 0.002597	$(+)$ 0.00028	$(-)$ 0.6210	$b_4^4 - b_4^2 = (+)0.01812$ $b_6^6 - b_6^4 + b_6^2 = (-)0.00933$	[106]
$Nd(NO_3)_3 \cdot 6\,H_2O$	0.05 at%	room	9	$g_\parallel = 1.984 \pm 0.002$ $g_\perp = 1.992 \pm 0.002$	$(+)$ 1.3329	$(+)$ 0.00203	$(+)$ 0.00150	$(-)$ 1.0885	$b_4^4 - b_4^2 = (+)0.03583$ $b_6^6 - b_6^4 + b_6^2 = (+)0.01105$	[106]

Table 6 (continued)

host lattice	dilution	temperature (°K)	frequency (GHz)	g	b_2^0(GHz)	b_4^0(GHz)	b_6^0(GHz)	b_2^2(GHz)	other fine-structure constants (GHz)	ref.
TiO_2	0.1 at%	295	9.4	1.9930	(+) 0.3994	(+) 0.05718	(−) 0.00279	–	–	[107]
	0.1 at%	77	9.4	1.9941	(+) 0.3374	(+) 0.05638	(−) 0.0028	(+) 0.06363	$b_4^2 = (-)0.0067$ $b_6^4 = 0.00899$ $b_4^4 = (-)0.05456$ $b_6^6 + b_3^2 = +0.0134$	[107]
	0.1 at%	1.8	9.4	1.9986	(+) 0.3015	(+) 0.06014	(−) 0.0028	–	–	[107]
$BaTiO_3$	Ba^{2+} site 0.2 at%	250	9	1.992 ±0.005	+0.8013 ±0.003	+0.00839 ±0.003	−0.00120 ±0.0015	+0.01679 ±0.006	$b_4^2 = +0.0336 \pm 0.009$ $b_4^4 = -0.0306 \pm 0.009$ $b_6^4 = +0.0057 \pm 0.006$ $b_6^2 + b_6^6 = 0.0198 \pm 0.009$	[56]
YGaG	–	300	36	1.992 ±0.003	+1.3224 ±0.0015	−0.1265 ±0.0015	+0.0018 ±0.0015	−0.8085 ±0.0015	$b_4^4 = +0.06326 \pm 0.0015$ $b_6^4 = +0.0003 \pm 0.0015$	[108]
	–	4.2	36	1.992 ±0.003	+1.3446 ±0.0015	−0.1355 ±0.0015	+0.0124 ±0.0015	−0.8508 ±0.0015	$b_4^4 = +0.06625 \pm 0.0015$ $b_6^4 = +0.00388 \pm 0.0015$	[108]
YAlG	–	300	36	1.990 ±0.003	+2.3315 ±0.0015	−0.1406 ±0.0015	+0.0113 ±0.0015	−0.2551 ±0.0015	$b_4^4 = +0.06955 \pm 0.0015$ $b_6^4 = +0.00728 \pm 0.0015$	[108]
	–	4.2	36	1.990 ±0.003	+2.3276 ±0.0015	−0.1457 ±0.0015	+0.01385 ±0.0015	−0.2905 ±0.0015	$b_4^4 = +0.07045 \pm 0.0015$ $b_6^4 = +0.01286 \pm 0.0015$	[108]
$CdWO_4$	0.01%	300	24	1.99 ±0.01	(∓) 2.2275 ±0.027	(∓) 0.009 ±0.009	–	(±) 0.7195 ±0.135	$b_2^1 = (\mp)1.0493 \pm 1.109$ $b_4^3 = (\mp)1.02 \pm 1.109$ $b_4^1 = (\mp)0.420 \pm 0.540$ $b_4^4 = (\pm)0.300 \pm 0.120$ $b_4^2 = (\pm)0.150 \pm 0.105$	[109]
$KY(MoO_4)_2$	0.1at%	room	37	$g_x = g_y = g_z = 1.992 \pm 0.002$	+2.472 ±0.005	−0.064 ±0.005	–	+0.258 ±0.005	$b_4^4 = 0.145 \pm 0.005$ $b_4^2 = 0.020 \pm 0.005$	[110]

where

$$F = \frac{b_2^0}{g_{\parallel} \beta B} \quad \text{and} \quad E = \frac{(b_2^2)^2}{18 g_{\parallel} \beta B}.$$

EPR data for the Gd^{3+} ion in crystals with orthorhombic or lower site symmetry are listed in Table 6. The small fine-structure constants are given in the linear combination form. The hyperfine constant is not listed since the hyperfine structure is not resolved for this symmetry.

References

[1] H. R. Woltjer and H. Kamerlingh Onnes, Kon. Acad. Wetensch. Proc., Amsterdam **32**, 772 (1923).
[2] P. Debye, Ann. Phys. (Germany) **81**, 1154 (1926).
[3] W. F. Giauque, J. Amer. Chem. Soc. **49**, 1864 (1927).
[4] W. F. Giauque and D. P. MacDougall, Phys. Rev. **43**, 768 (1933).
[5] H. E. D. Scovil, J. Feher, and H. Seidel, Phys. Rev. **105**, 762 (1957).
[6] H. A. Bethe, Ann. Phys. (Germany) **3**, 133 (1929).
[7] R. Lacroix, Helv. phys. Acta **30**, 374 (1957).
[8] B. G. Wybourne, Phys. Rev. **148**, 317 (1966).
[9] V. K. Sharma, J. chem. Phys. **54**, 496 (1971).
[10] H. A. Buckmaster, R. Chatterjee, and Y. H. Shing, Canad. J. Phys. **50**, 991 (1972).
[11] J. S. M. Harvey and H. Kiefte, Canad. J. Phys. **49**, 995 (1971).
[12] A. Abragam and M. H. L. Pryce, Proc. Roy. Soc. **A205**, 135 (1951).
[13] H. A. Buckmaster and R. Chatterjee, J. Magnetic Resonance **5**, 1 (1971).
[14] M. T. Hutchings, Solid State Phys. **16**, 227 (1964).
[15] J. L. Prather, Atomic Energy Levels in Crystals, U.S. Nat. Bur. Standards, Monograph 19, Washington 1961.
[16] K. W. H. Stevens, Proc. Phys. Soc. **A65**, 209 (1952).
[17] B. Bleaney and K. W. H. Stevens, Rep. Progr. Phys. **16**, 108 (1953).
[18] R. J. Elliott and K. H. W. Stevens, Proc. Roy. Soc. **A218**, 553 (1953).
[19] G. Racah, Phys. Rev. **62**, 438 (1942).
[20] G. Racah, Phys. Rev. **63**, 367 (1943).
[21] B. G. Wybourne, Spectroscopic Properties of Rare Earths, Interscience Publishers, New York 1965.
[22] C. W. Nielson and G. F. Koster, Spectroscopic Coefficients for the p^n, d^n, and f^n Configurations, M.I.T. Press, Cambridge (Mass.) 1963.
[23] K. D. Bowers and J. Owen, Rep. Progr. Phys. **18**, 304 (1955).
[24] W. Low and D. Shaltiel, J. Phys. Chem. Solids **6**, 315 (1958).
[25] M. M. Abraham, E. J. Lee, and R. A. Weeks, J. Phys. Chem. Solids **26**, 1249 (1965).
[26] S. A. Marshall and R. A. Tilton, Phys. Letters (Netherlands) **30A**, 216 (1969).
[27] H. A. Buckmaster, R. Chatterjee, and Y. H. Shing, J. Magnetic Resonance **4**, 85 (1971).
[28] H. A. Buckmaster, R. Chatterjee, and Y. H. Shing, phys. stat. sol. (a), **13**, 9 (1972).
[29] M. E. Rose, Elementary Theory of Angular Momentum, Wiley, New York 1957 (p. 146).
[30] M. Rotenberg, R. Bivins, N. Metropolis, and J. K. Wooten, Jr., The $3j$ and $6j$ Symbols, M.I.T. Press, Cambridge (Mass.) 1959.
[31] J. Rosenthal, Phys. Rev. **164**, 363 (1967).
[32] S. A. Marshall and R. A. Serway, Phys. Rev. **171**, 345 (1968).
[33] G. B. Singh, G. C. Upreti, and P. Venkateswarlu, J. chem. Phys. **46**, 2885 (1967).
[34] C. Ryter, Helv. phys. Acta **30**, 353 (1957).
[35] K. Horai, J. Phys. Soc. Japan **19**, 2241 (1964).
[36] W. Low, Phys. Rev. **109**, 265 (1958).

[37] W. Low and R. S. RUBINS, Phys. Letters (Netherlands) **1**, 316 (1962).
[38] T. REVAI, Soviet Phys. — Solid State **10**, 1014 (1968).
[39] J. M. BAKER and F. I. B. WILLIAMS, Proc. Phys. Soc. **78**, 1340 (1961).
[40] P. EISENBERGER and P. S. PERSHAN, Phys. Rev. **167**, 292 (1968).
[41] R. H. BORCHERTS, T. COLE, and T. HORN, J. chem. Phys. **49**, 4880 (1968).
[42] J. SIERRO, Helv. phys. Acta **36**, 505 (1963).
[43] W. Low and V. ROSENBERGER, Phys. Rev. **116**, 621 (1959).
[44] M. M. ABRAHAM, L. A. BOATNER, C. B. FINCH, and E. J. LEE, Phys. Letters (Netherlands) **25A**, 230 (1967).
[45] M. M. ABRAHAM, L. A. BOATNER, C. F. FINCH, E. J. LEE, and R. A. WEEKS, J. Phys. Chem. Solids **28**, 81 (1967).
[46] S. A. MARSHALL, Phys. Rev. **159**, 191 (1967).
[47] I. V. VINOKUROV, Z. N. ZONN, and V. A. IOFFE, Soviet Phys. — Solid State **7**, 814 (1965).
[48] K. KOMET, W. LAW, and R. C. LINARES, Phys. Letters (Netherlands) **19**, 473 (1965).
[49] R. S. TITLE Phys. Rev. **138**, A631 (1965).
[50] K. E. MANN and L. V. HOLROYD, phys. stat. sol. **28**, K27 (1968).
[51] A. J. SHUSKUS, Phys. Rev. **127**, 2022 (1962).
[52] A. J. SHUSKUS, J. chem. Phys. **41**, 1885 (1964).
[53] W. Low and R. S. RUBINS, Paramagnetic Resonance **1**, 79 (1963).
[54] W. Low and J. T. SUSS, Phys. Letters (Netherlands) **11**, 115 (1964).
[55] L. RIMAI and G. A. DE MARS, Phys. Rev. **127**, 702 (1962).
[56] T. TAKEDA, J. Phys. Soc. Japan **24**, 533 (1968).
[57] H. UNOKI and T. SAKUDO, J. Phys. Soc. Japan **21**, 1730 (1966).
[58] S. R. D. SMITH and T. COLE, J. chem. Phys. **52**, 1286 (1970).
[59] J. E. DRUMHELLER, J. chem. Phys. **38**, 970 (1963).
[60] L. A. BOATNER, R. W. REYNOLDS, and M. M. ABRAHAM, J. chem. Phys. **52**, 1248 (1970).
[61] S. MANIV and A. GABAY, J. Phys. Chem. Solids **31**, 2579 (1970).
[62] J. SIERRO and R. LACROIX, C. R. Acad. Sci. (France) **250**, 2686 (1960).
[63] I. V. VINOKUROV, M. M. ZARIPOV, YU. E. POLSHII, V. G. STEPANOV, G. K. CHIRKIN, and L. YA. SHEKUN, Soviet Phys. — Solid State **4**, 1637 (1963); Fiz. tverd. Tela **4**, 2238 (1962).
[64] K. ZDANSKY and A. EDGAR, Phys. Rev. **B3**, 2133 (1971).
[65] C. F. HEMPSTEAD and K. D. BOWERS, Phys. Rev. **118**, 131 (1960).
[66] V. M. VINOKUROV and V. G. STEPANOV, Soviet Phys. — Solid State **9**, 844 (1967).
[67] M. M. ZARIPOV, L. I. POTKIN, M. I. SAMOILOVICH, and V. G. STEPANOV, Soviet Phys. — Solid State **8**, 2766 (1967).
[68] M. L. MEILMAN, Soviet Phys. — Solid State **8**, 2924 (1967).
[69] M. M. ZARIPOV, A. A. MANENKOV, and G. K. CHIRKIN, Soviet Phys. — Solid State **8**, 212 (1966).
[70] M. L. MEILMAN and N. V. SOLOVEV, Soviet Phys. — Solid State **7**, 2512 (1965).
[71] V. F. DERNOV-PEGAREV, M. M. ZARIPOV, M. I. SAMOILOVICH, and V. G. STEPANOV, Soviet Phys. — Solid State **8**, 197 (1966).
[72] I. N. KURKIN and L. YA. SHEKUN, Soviet Phys. — Solid State **6**, 1560 (1965).
[73] M. L. MEILMAN, M. I. SAMOILOVICH, L. I. POTKIN, and N. I. SERGEEVA, Soviet Phys. — Solid State **8**, 1864 (1967).
[74] I. N. KURKIN and V. G. STEPANOV, Soviet Phys. — Solid State **7**, 223 (1965).
[75] M. M. ABRAHAM, G. W. CLARK, C. B. FINCH, R. W. REYNOLDS, and H. ZELDES, J. chem. Phys. **50**, 2057 (1969).
[76] D. R. HUTTON and G. J. TROUP, Brit. J. appl. Phys. **15**, 405 (1964).
[77] J. ROSENTHAL, R. RILEY, and V. RANON, Phys. Rev. **177**, 625 (1969).
[78] G. KUHL, Z. Phys. **225**, 277 (1969).
[79] W. URBAN, phys. stat. sol. **37**, 181 (1970).
[80] C. R. VISWANATHAN and E. Y. WONG, J. chem. Phys. **49**, 966 (1968).

[81] H. A. Buckmaster, R. Chatterjee, and Y. H. Shing, J. Phys. C **4**, 832 (1971).
[82] I. R. Dagg, R. C. Kemp, and H. F. Summons, J. Phys. C **2**, 1098 (1969).
[83] B. Bleaney, H. E. D. Scovil, and R. S. Trenam, Proc. Roy. Soc. **A223**, 15 (1954).
[84] H. A. Buckmaster, Canad. J. Phys. **34**, 150 (1956).
[85] R. J. Richardson and S. Lee, Phys. Rev. **B1**, 108 (1970).
[86] J. Dweck and G. Seidel, Phys. Rev. **151**, 289 (1966).
[87] P. Wysling, Solid State Commun. **7**, 873 (1969).
[88] R. W. Reynolds and L. A. Boatner, J. chem. Phys. **54**, 1827 (1971).
[89] L. A. Boatner and M. M. Abraham, Phys. Rev. **163**, 213 (1967).
[90] C. A. Hutchison, B. R. Judd, and D. F. D. Pope, Proc. Phys. Soc. **B70**, 514 (1957).
[91] D. R. Johnston, E. Y. Wong, and O. M. Stafsudd, J. chem. Phys. **44**, 2693 (1966).
[92] P. D. Scott, J. chem. Phys. **54**, 5384 (1971).
[93] A. A. Antipin and I. N. Kurkin, Soviet Phys. — Solid State **6**, 730 (1964).
[94] V. M. Vinokurov, M. M. Zaripov, Yu. E. Polskii, V. G. Stepanov, G. K. Chirkin, and L. Ya. Shekun, Soviet Phys. — Solid State **5**, 436 (1963); Fiz. tverd. Tela **5**, 599 (1963).
[95] V. M. Vinokurov, Soviet Phys. — Solid State **5**, 2126 (1964).
[96] J. M. Moret, J. Webber, and R. Lacroix, Helv. phys. Acta **41**, 243 (1968).
[97] A. Hausmann, Solid State Commun. **7**, 579 (1969).
[98] S. Geschwind and J. P. Remeika, Phys. Rev. **122**, 757 (1961).
[99] V. A. Atsarkin, V. G. Lushnikov, and L. P. Sorokina, Soviet Phys. — Solid State **7**, 1912 (1966).
[100] W. Low and A. Zusman, Phys. Rev. **130**, 144 (1963).
[101] H. A. Buckmaster, J. C. Dering, and D. J. I. Fry, J. Phys. C **1**, 599 (1968).
[102] R. S. Trenam, Proc. Phys. Soc. **A66**, 118 (1953).
[103] D. A. Jones, J. M. Baker, and D. F. D. Pope, Proc. Phys. Soc. **74**, 249 (1959).
[104] M. Weger and W. Low, Phys. Rev. **111**, 1526 (1958).
[105] D. Meierling and W. Ullmann, Z. Naturf. **23a**, 1971 (1968).
[106] G. B. Singh and P. Venkateswarlu, J. chem. Phys. **46**, 4765 (1967).
[107] S. Yamaka, J. Phys. Soc. Japan **18**, 1557 (1963).
[108] L. Rimai and G. A. de Mars, J. appl. Phys. Suppl. **33**, 1254 (1962).
[109] L. Kratěna, K. Ždánský, and V. Šik, phys. stat. sol. **28**, 175 (1968).
[110] M. M. Zaripov, L. D. Livanova, V. G. Stepanov, and M. L. Falin, Soviet Phys. — Solid State **8**, 188 (1966).
[111] M. M. Abraham, L. A. Boatner, Y. Chen, J. L. Kolopus, and R. W. Reynolds, Phys. Rev. **B4**, 2853 (1971).

(Received February 29, 1972)

Original Papers

phys. stat. sol. (a) **12**, 363 (1972)

Subject classification: 18.2.1

Central Institute for Cybernetics and Information Processes, Berlin (a), and Institute for Low Temperature and Structure Research, Wrocław (b)

Theory of the Transversal Shifting of Interacting Bloch Walls in Magnetic Films

By

R. Straubel (a) and W. J. Ziętek (b)

The transversal motion of a finite sequence of interacting parallel 180° Bloch walls in a magnetic film is theoretically studied, taking into account the exchange and anisotropy energy of the system, the stray-field interaction between neighbouring walls, and the magnetic "friction resistance" of the film. If an external magnetic field moves the first wall (driving wall) of the sequence with constant velocity toward the remaining walls, the equations of motion of the system are shown in the asymptotic limit to lead to a dynamically stable Bloch wall array which is being shifted uniformly under the mere influence of the driving wall. The theory is exemplified on a sequence of ten Bloch walls for which numerical asymptotic solutions are given and discussed.

Die transversale Verschiebung einer endlichen Folge von wechselwirkenden parallelen 180° Blochwänden in einer dünnen magnetischen Schicht wird theoretisch untersucht, bei Berücksichtigung der Austausch- und Anisotropieenergie des Systems, der Streufeldwechselwirkung zwischen benachbarten Wänden und des magnetischen „Reibungswiderstandes" der Schicht. Wenn ein äußeres Magnetfeld die erste Blochwand der Folge (Triebwand) mit konstanter Geschwindigkeit in Richtung der anderen Wände bewegt, führen die Bewegungsgleichungen des Systems im asymptotischen Grenzfall zu einer dynamisch stabilen Blochwandanordnung, die allein unter dem Einfluß der Triebwand gleichförmig verschoben wird. Als Beispiel werden numerische asymptotische Lösungen für eine Folge von zehn Blochwänden angegeben und diskutiert.

1. Introduction

While examining experimentally the domain structure of thin uniaxial magnetic films with the direction of easiest magnetization lying in the film plane, Williams and Sherwood [1] discovered the existence of stable double 180° Bloch walls. This was inasmuch surprising as domains in bulk uniaxial magnetic materials were known to be always separated by single 180° Bloch walls [2, 3]. A simple theoretical explanation of this phenomenon was given by Kaczér [4] who applied, somewhat inconsistently, the conventional phenomenological formalism [3, 5 to 7] in the asymptotic limit (single-wall case; see [7 to 11]) to a pair of close-spaced parallel 180° Bloch walls in a thin magnetic film. Besides the exchange and anisotropy energy of the walls Kaczér took also into account the interaction due to their stray fields originating from the magnetic poles of the wall on the upper and lower film surfaces (cf. Fig. 1). This has been accomplished by simply assigning to each wall an effective magnetic moment M and taking the interaction energy of those moments over the distance between the wall centres. Under these assumptions the minimization of the system energy leads, according to [4], to a stable configuration of the coupled walls but, surprisingly, the equilibrium distance between the walls appears to be the same for the cases A and B in Fig. 1.

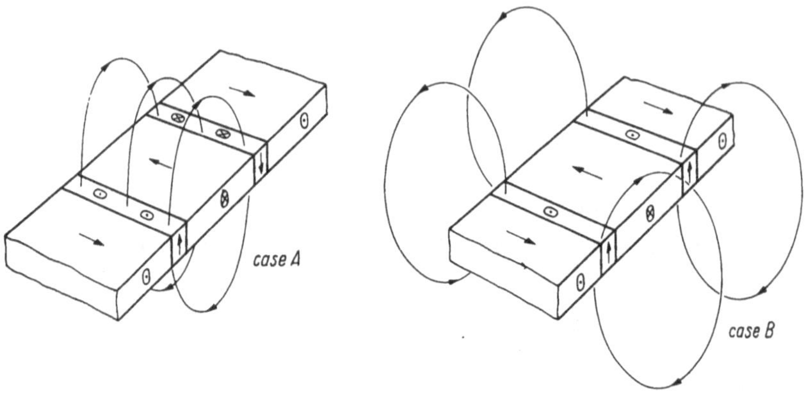

Fig. 1. Model of Bloch walls with antiparallel (A) and parallel (B) magnetic moments

This result is obviously wrong in the latter case (B), as the repulsion between walls with parallel magnetic moments should lead in a frictionless film to an infinite equilibrium distance. It is clear that this unphysical result is a consequence of applying the single-wall theory to a double-wall system, and that a finite equilibrium distance between equally polarized walls (case B in Fig. 1) can only result when taking into consideration the magnetic "friction resistance" of the film which counteracts the repulsion between those walls.

For this reason we re-examine this problem in the present paper, by employing the theory of periodic domain structures [12 to 15] to the more general case of k parallel Bloch walls with an arbitrary sequence of their magnetic polarization (i.e., parallel or antiparallel magnetic moments of neighbouring walls). Beyond that, upon introducing a phenomenological "friction force" we consider the transversal motion (i.e., perpendicular to the walls and hence to the anisotropy axis) of the Bloch wall sequence when the first (or driving) wall is being moved with constant velocity by an external magnetic field toward the remaining walls. It is shown that the Lagrange equations of the system lead in the asymptotic limit (infinite time) to a dynamically stable array of the Bloch walls which, in a train-like fashion, are being pushed along with constant velocity by the driving wall. The theory is illustrated by applying it to a sequence of ten 180° Bloch walls and presenting numerical solutions for this case.

2. The Model

We assume the magnetic film to lie in the xy co-ordinate plane, to be infinite in the x-direction, and to have the thickness d in the z-direction. The direction of easiest magnetization of the film (anisotropy axis) is assumed to be parallel to the y-axis, and its dimension in this direction finite but large compared with d. Accordingly, the 180° Bloch walls lie in the yz co-ordinate plane, their thickness being δ and their positions being described by the co-ordinate x_n, $n = 0, 1, 2, \ldots, k$, as shown in Fig. 2. Because of the finite thickness of the Bloch walls we have the following conditions for the distances r_n between neighbouring walls:

$$r_n \equiv x_n - x_{n-1} > \delta , \qquad n = 1, 2, \ldots, k . \tag{1}$$

Fig. 2. Model of magnetic film with 180° Bloch walls

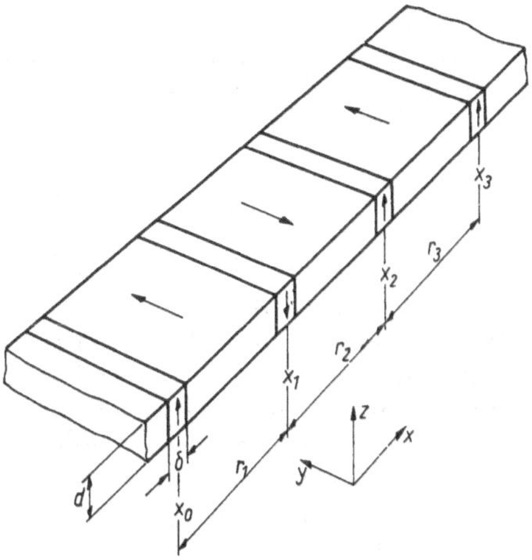

In a periodic domain structure, the Bloch wall distance r_n is equal to the domain width Δ. It has been shown in [9 to 11] for a variety of Bloch wall types that the conventional asymptotic single-wall approach (i.e., $\Delta \to \infty$) is inadequate if $\Delta/\delta < 2$. As this may well be the case in the problem under consideration, we must apply the multi-domain formalism. The dependence of the energy density σ (per unit area of the wall) of a 180° Bloch wall on the domain width Δ has been derived for uniaxial ferromagnets [13, 15]. Except for a factor 2, this dependence happens to be the same as that for the (90° [100] 90°) Bloch wall in b.c.c. ferromagnets derived in [8] and numerically examined in [10] (cf. equation (71) in [13] and equation (21) in [10]). As the analytical formula of this dependence is pretty complicated (it involves elliptic integrals of the first and second kind), we replace it here with the simpler expression

$$\sigma = \sigma_0 \left[1 + C_1 \frac{\delta_0}{\Delta} \exp\left(-\frac{C_2 \Delta}{\delta_0}\right)\right], \tag{2}$$

which, with the fitting constants

$$C_1 = 3.1 \quad \text{and} \quad C_2 = 0.5 \tag{3}$$

approximates quite satisfactorily (dash-dotted line in Fig. 3) the numerical curve σ/σ_∞ from Fig. 1 in [10] (bold line in Fig. 3), for $\Delta/\delta_0 > 1$. It is seen that

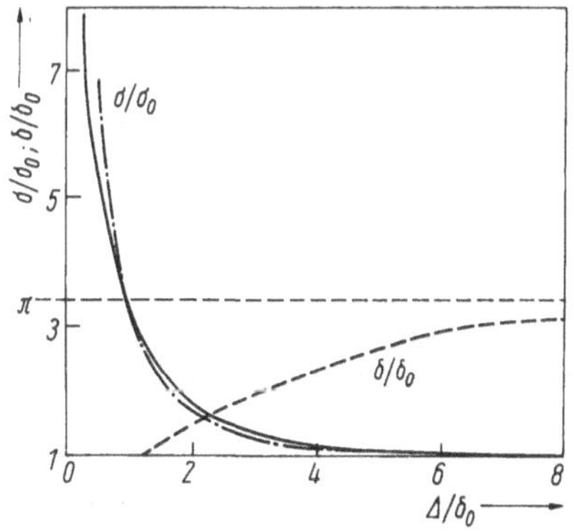

Fig. 3. Exact (solid curve) and approximate (dash-dotted curve) dependence of Bloch wall energy σ on domain width Δ and Bloch wall thickness δ as function of Δ, according to [10]

σ_0 in (2) coincides with the Bloch wall energy density σ_∞ in the asymptotic limit $\Delta \to \infty$ (single-wall case), while, according to [10, 13], the corresponding asymptotic thickness of the wall is $\delta_\infty = \pi\,\delta_0$. For a 180° Bloch wall in a uniaxial ferromagnet we have (see equations (32), (37), and (41) in [15])

$$\sigma_0 = 4\,(\tilde{A}\;K_1)^{1/2}\,, \qquad \delta_0 = \left(\frac{\tilde{A}}{K_1}\right)^{1/2}, \tag{4}$$

where $\tilde{A}$ and K_1 are the phenomenological exchange and (first) anisotropy constant, respectively [3, 6, 16].

In the theory of periodic domain structures, the Bloch wall energy σ may as well be interpreted as the "excess" exchange and anisotropy energy of a domain (deviation from complete magnetic saturation) due to the existence of Bloch walls. This is so because the integrals over Δ in calculating σ are clearly invariant under translations along the direction normal to the walls [8, 10, 11, 13, 15]. Thus, we substitute $\Delta \to r_n$ in (2) and, upon introducing the reduced (dimensionless) distances

$$\bar{r}_n = \frac{r_n}{\delta_0}\,, \tag{5}$$

we obtain the energy

$$\sigma_n = \sigma_0 \left[1 + \frac{C_1}{\bar{r}_n} \exp\left(-C_2\,\bar{r}_n\right)\right], \qquad n = 1, 2, \ldots, k\,, \tag{6}$$

which can be assigned to the n-th domain. The corresponding energy per unit length in the y-direction has the form

$$E_n^{\mathrm{d}} = \sigma_n\,d\,; \qquad n = 1, 2, \ldots, k\,. \tag{7}$$

Note that this energy decreases with increasing Bloch wall distance r_n ("repulsion"), and that it remains a good approximation of the Bloch wall energy as long as condition (1) is satisfied and neighbouring wall distances do not differ too much from each other ($r_n \approx r_{n-1}$; quasi-periodic domain structure).

The advantage of using the domain energy (7), (6) instead of the conventional Bloch wall energy (2) resides in the fact that in this way we exclude automatically the energy of the outer halves of the first ($n = 0$) and last ($n = k$) Bloch wall of the sequence (i.e., for $x < x_0$ and $x > x_k$). Clearly, the latter energy does not depend on the wall distances r_n and thus is immaterial for our considerations.

The next energy to be accounted for is due to the magnetic stray-field interaction between neighbouring Bloch walls. We shall take it into account by following the linear-dipole approach used in [4] and assign to each wall the effective magnetic moment $M = 2\,I_\mathrm{s}\,\delta\,d/\pi$ per unit length in the y-direction where I_s is the saturation magnetization of the film. Hence, the corresponding interaction energy of two neighbouring walls reads

$$E_n^{\mathrm{i}} = \mp\,\frac{M^2}{\pi\,\mu_0\,r_n^2} = \mp\,\frac{4\,I_\mathrm{s}^2\,\delta^2\,d^2}{\pi\,\mu_0\,\delta_0^2\,\bar{r}_n^2}\,, \quad n = 1, 2, \ldots, k \tag{8}$$

per unit length along the y-axis where μ_0 is the vacuum permeability. The upper and lower signs correspond to the case A and B in Fig. 1, respectively.

Note that, in contrast to the domain energy (7), with increasing Bloch wall distance r_n the stray-field interaction energy (8) increases in the case A (attraction) and decreases in the case B (repulsion). It should also be noted that E_n^{i}

depends additionally on r_n through the Bloch wall thickness δ which itself is a function of Δ [8 to 11]. For a 180° Bloch wall in a uniaxial ferromagnet this dependence is given by the dashed curve δ/δ_0 in Fig. 3 (see equation (60) in [13] and equations (20) and (23) and Fig. 1 in [10]). For simplicity, however, we shall in the present paper disregard this dependence and assume the effective magnetic moments M of the walls to be constant, like in [4]. This is a reasonable approximation for, say, $r_n > 5\,\delta_0$ (cf. Fig. 3) which is actually equivalent to assuming

$$M \approx 2\, I_s\, \delta_0\, d\,, \quad \delta \approx \delta_\infty = \pi\, \delta_0\,, \quad E_n^{\mathrm{i}} \approx \mp \frac{4\, I_s^2\, d^2}{\pi\, \mu_0\, \bar{r}_n^2}\,. \tag{8'}$$

3. The Static Case of Two Bloch Walls

To demonstrate the advantage of our model over that assumed in [4], we consider briefly the simple case of a single pair of Bloch walls and, with the notation $E_n^{\mathrm{d}} + E_n^{\mathrm{i}} \equiv E_n$, easily obtain from the condition

$$\frac{\partial E_n}{\partial r_n} = \frac{\partial E_n}{\partial \bar{r}_n} = 0 \tag{9}$$

the equation

$$\frac{\mp\, C_3 + \bar{r}_n\,(1 + C_2\, \bar{r}_n)\, \exp\,(-\,C_2\, \bar{r}_n)}{\bar{r}_n^3} = 0 \tag{10}$$

or, if $\bar{r}_n \neq \infty$,

$$\bar{r}_n\,(1 + C_2\, \bar{r}_n)\, \exp\,(-\,C_2\, \bar{r}_n) = \pm C_3\,, \tag{11}$$

where

$$C_3 = \frac{8\, I_s^2\, d}{\mu_0\, \pi\, \sigma_0\, C_1} \approx \frac{I_s^2\, d}{\mu_0\, \sigma_0}, \tag{12}$$

in the approximation (8′) and due to (3). The upper and lower signs in equation (10) and (11) correspond to the case A and B in Fig. 1, respectively.

One easily proves that the left-hand side L of (11) has a maximum, and that $L \to 0$ if $r_n \to 0$ or $r_n \to \infty$. This is shown schematically in Fig. 4. As expected, in the case B the only solution of equation (10) is $r_n = \infty$ which is physically correct, in contrast to the result obtained in [4].

In the case A it is readily seen from Fig. 4 that, depending on the value of C_3, equation (11) has two, one, or no solutions. The critical value C_3^{crit} above which no solutions of equation (11) exist is easily calculated by determining the maximum of the function L, which is found to be at

$$\bar{r}_{\max} = \frac{1 + \sqrt{5}}{2\, C_2} \approx 3.2 \tag{13}$$

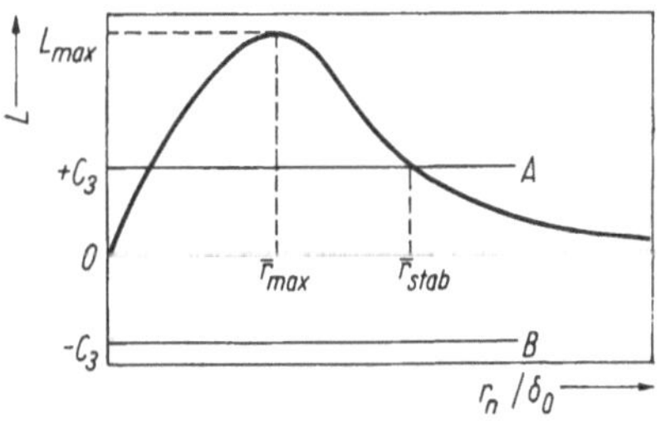

Fig. 4. Solution of equation (11) for a single Bloch wall pair

and leads to the result

$$C_3^{\mathrm{crit}} = C_2^2\, \bar{r}_{\max}^3 \exp\left(-C_2\, \bar{r}_{\max}\right) \approx 1.6\,. \tag{14}$$

Of the two solutions for $C_3 < C_3^{\mathrm{crit}}$ it is always the larger Bloch wall distance $\bar{r}_{\mathrm{stab}} > \bar{r}_{\max}$ which corresponds to an energy minimum. Thus, for the stable solutions we have $\bar{r}_{\mathrm{stab}} \to \infty$ as $C_3 \to 0$ and $\bar{r}_{\mathrm{stab}} \to \bar{r}_{\max}$ as $C_3 \to C_3^{\mathrm{crit}}$. Note that for $C_3 = 0$ equation (10) also leads to the solution $\bar{r}_n = \infty$. Moreover, it should be noticed that the inequality $r_{\mathrm{stab}} > 3.2\,\delta_0$, which holds for the stable solutions, justifies the approximation (8′) and, even in the extreme case $\delta = \delta_\infty = \pi\,\delta_0$, ensures condition (1) to be met.

4. The Equations of Motion

According to [16, 17], to the moving Bloch wall an effective and velocity-independent mass m (per unit length in the y-direction, in our case) and hence a kinetic energy can be ascribed. Let us assume that the first Bloch wall ($n = 0$) of the sequence is moved by an external force (e.g., magnetic field) with constant velocity $\dot{x}_0 = x_0/t$ along the x-axis, and that H_0 is the threshold field needed to get a wall moving in the film (magnetic friction resistance). Then, the Hamiltonian of the system reads

$$\mathcal{H} = \sum_{n=1}^{k} \left(\frac{1}{2}\, m\, \dot{x}_n^2 + E_n\right), \qquad E_n \equiv E_n^{\mathrm{d}} + E_n^{\mathrm{i}}\,. \tag{15}$$

(The kinetic energy of the driving wall is constant and thus immaterial.) Hence, with $\mathcal{L} = m \sum \dot{x}_n^2 - \mathcal{H}$ the Lagrange equations of the system have the form

$$m\, \ddot{x}_n + \frac{\partial\,(E_n + E_{n+1})}{\partial x_n} = f_n(\dot{x}_n)\,, \qquad n = 1, 2, \ldots, k-1\,; \tag{16}$$

$$m\, \ddot{x}_k + \frac{\partial E_k}{\partial x_k} = f_k(\dot{x}_k)\,,$$

where $f_n(\dot{x}_n)$ and $f_k(\dot{x}_k)$ are non-conservative friction forces related to H_0. For later purposes, we note the useful identities

$$\left.\begin{aligned}
&\frac{\partial E_n}{\partial x_n} = \frac{\partial E_n}{\partial r_n} = \frac{1}{\delta_0}\frac{\partial E_n}{\partial \bar{r}_n} = -\frac{\partial E_n}{\partial x_{n-1}}\,,\\
&\frac{\partial E_n^{\mathrm{d}}}{\partial x_n} = -\frac{\sigma_0\, C_1\, d}{\delta_0\, \bar{r}_n^2}\left(1 + C_2\, \bar{r}_n\right) \exp\left(-\,C_2\, \bar{r}_n\right),\\
&\frac{\partial E_n^{\mathrm{i}}}{\partial x_n} = \pm\, \frac{2\, M^2}{\pi\, \mu_0\, \delta_0^3\, \bar{r}_n^3}
\end{aligned}\right\} \tag{17}$$

that follow for $n = 1, 2, \ldots, k$ from equations (1) and (5) to (8).

In specifying $f_n(\dot{x}_n)$ we utilize the experimental result

$$\dot{x} = \beta\,(H - H_0) \tag{18}$$

obtained in [18] for a 180° Bloch wall moving with constant velocity $\dot{x}$ under the influence of an external magnetic field $H > H_0$ (dynamic equilibrium; for details and other specifications of the friction force, see e.g., [3, 16]). Here, β and the threshold field H_0 are material constants characteristic of the magnetic film. Now, assume the magnetic film in Fig. 2 to be finite (but still very long) in the x-direction and divided at its centre $x = 0$ by a single 180° Bloch wall into two domains. Then, if the wall shifts from its equilibrium position at $x = 0$ to the point x under the influence of a homogeneous magnetic field H directed along the y-axis, its "potential" energy (reversible process; frictionless case) changes by the amount $-2 I_s H x d$ (per unit length in the y-direction). Hence, the Hamiltonian of the wall can be written as

$$\mathcal{H} = \frac{1}{2} m \dot{x}^2 - 2 I_s H d x . \tag{19}$$

If, in addition, there is a friction force (irreversible process) that leads to the relation (18) then, in accordance with the potential energy in (19) it should have the form

$$f(\dot{x}) = - K \dot{x} - 2 I_s H_0 d . \tag{20}$$

Indeed, equations (19) and (20) lead to the Lagrange equation

$$m \ddot{x} - 2 I_s H d = - K \dot{x} - 2 I_s H_0 d \tag{21}$$

which in the asymptotic limit $t \to \infty$, $\ddot{x} = 0$ (dynamic equilibrium) has the solution

$$\dot{x} = \frac{2 I_s d (H - H_0)}{K} \equiv \beta (H - H_0) \tag{22}$$

that coincides with (18) and relates the friction force constant K to the coefficient β.

5. The Asymptotic Solutions

The result (22) enables us to solve equation (16) in the asymptotic limit $t \to \infty$, $\ddot{x}_n = 0$, $\dot{x}_n = \dot{x}_0$ (dynamic equilibrium). One easily verifies that, by utilizing the formulae (17) and adding pairwise equations (16) in reversed order (i.e., for x_k and x_{k-1}, x_{k-1} and x_{k-2} etc.), the coupled equation system (16) separates in the asymptotic case into a set of k independent equations,

$$\frac{\partial E_n}{\partial \bar{r}_n} = - (k + 1 - n) 2 I_s d \delta_0 (H_0 + \dot{x}_0/\beta) , \tag{23}$$

as $f_n(\dot{x}_n) = - 2 I_s d (H_0 + \dot{x}_0/\beta)$ for all n, according to equations (20) and (22). With the parameter

$$C_4 = \frac{2 I_s \delta_0 (H_0 + \dot{x}_0/\beta)}{\sigma_0 C_1} = \frac{2 I_s \delta_0 H}{\sigma_0 C_1} \tag{24}$$

that depends on the velocity $\dot{x}_0$ of the driving wall and thus on the external magnetic field H (acting only on this wall), the equations (23) assume eventually the form

$$\frac{\mp C_3 + \bar{r}_n (1 + C_2 \bar{r}_n) \exp(- C_2 \bar{r}_n)}{\bar{r}_n^3} = (k + 1 - n) C_4 \tag{25}$$

or

$$\bar{r}_n (1 + C_2 \bar{r}_n) \exp(- C_2 \bar{r}_n) - (k + 1 - n) C_4 \bar{r}_n^3 = \pm C_3 , \quad n = 1, 2, \ldots, k \tag{26}$$

when making use of equations (15) and (6) to (8). The upper and lower signs of C_3 in the above equations corresponds to the case A and B in Fig. 1 repectively for each pair of neighbouring Bloch walls in the sequence.

Equations (25) and (26) constitute each a set of transcendental equations that must be solved numerically. They represent generalizations of (10) and (11), respectively, to the dynamic case of k Bloch walls moving with constant velocity in a film with friction resistance. In the frictionless static case we have $C_4 = 0$, as $\dot{x}_0 = H_0 = 0$, in which case (25) and (26) coincide with (10) and (11) respectively. This implies that the system is at static equilibrium only in the extreme case of the "monotonous" A-type sequence (upper sign of C_3 for all n), in which case we have a stable equidistant Bloch wall array.

The apparent advantage of equations (25) and (26) resides in the fact that we have a separate equation for each pair of neighbouring Bloch walls. Thus, upon solving for a given k the equations in the two extreme cases of the monotonous A- and B-type Bloch wall sequence (fixed sign of C_3 for all n), the problem is automatically solved for any of the 2^k different combinations in which A- and B-type Bloch wall pairs can be arranged in the sequence. Of course, this simplification is a consequence of restricting the stray-field interaction (8) to neighbouring walls and constitutes a fair approximation as long as condition (1) is not violated.

Let us denote by L^{A} and L^{B} the left-hand side of (25) for the case A and B, respectively, and by $L^{(n)}$ the left-hand side of (26). One easily proves that

$$L^{\mathrm{A}} \to 0-\,, \qquad L^{\mathrm{B}} \to 0+\,, \qquad L^{(n)} \to -\infty \tag{27}$$

as $\bar{r}_n \to \infty$ and

$$L^{\mathrm{A}} \to -\infty\,, \qquad L^{\mathrm{B}} \to +\infty\,, \qquad L^{(n)} \to 0+ \tag{28}$$

as $\bar{r}_n \to 0$. A closer inspection shows that, for positive and finite $\bar{r}_n$, only the functions L^{A} and $L^{(n)}$ may have extrema, namely, a maximum in the case of $L^{(n)}$ and a maximum and a minimum in the case of L^{A}. A further analysis leads to the conclusion that

$$L_{\max}^{(k)} > \cdots > L_{\max}^{(2)} > L_{\max}^{(1)} > 0\,, \qquad L_{\max}^{\mathrm{A}} > -L_{\min}^{\mathrm{A}} > 0\,. \tag{29}$$

Since

$$L^{\mathrm{B}}(\bar{r}_n) > L^{\mathrm{A}}(\bar{r}_n)\,, \qquad L^{(n+1)}(\bar{r}_n) > L^{(n)}(\bar{r}_n) \tag{30}$$

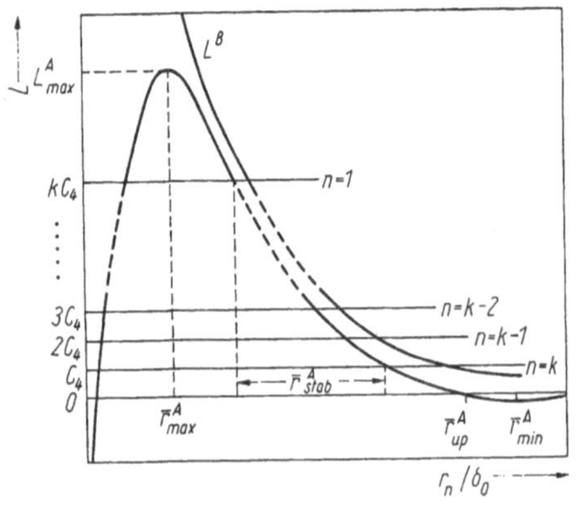

Fig. 5. Solutions of equations (25) for a sequence of k Bloch walls

Fig. 6. Solutions of equations (26) for a sequence of k Bloch walls

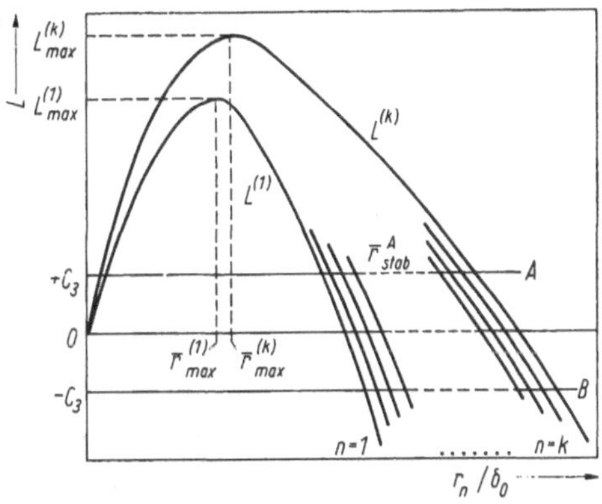

in the whole interval $0 < \bar{r}_n < \infty$, we eventually arrive at the schematic curves presented in Fig. 5 and 6 which correspond to the equation systems (25) and (26), respectively.

6. Discussion of Results

From Fig. 5 and 6 it is seen that for $C_4 > 0$ there exist always finite solutions in the case B, in contrast to the frictionless static case (10) where the only solutions are $\bar{r}_n = \infty$. It can be verified that the finite solutions are stable ones.

In the case A the situation is much the same as in the frictionless static case (10), namely, depending on the values of C_3, C_4, and k there exist two, one, or no solutions for each n. This is also evident from Fig. 5 and 6. In the case of two solutions, it is always the larger Bloch wall distance $\bar{r}^{A}_{stab}$ that corresponds to the energy minimum. Let us denote by $\bar{r}^{A}_{max}$ and $\bar{r}^{(n)}_{max}$ the distances that correspond to the maxima of the functions L^{A} and $L^{(n)}$, respectively, and by $\bar{r}^{A}_{up}$ and $\bar{r}^{A}_{min}$ the distances corresponding to $L^{A} = 0$ and to the minimum of L^{A}, respectively. One easily proves that

$$\bar{r}^{A}_{min} > \bar{r}^{A}_{up} \geqq \bar{r}^{A}_{stab} \geqq \bar{r}^{A}_{max}\,,\ \bar{r}^{(1)}_{max}\,. \tag{31}$$

Hence, $\bar{r}^{A}_{max}$ and $\bar{r}^{A}_{up}$ are the lower and upper limits of the stable solutions $\bar{r}^{A}_{stab}$, respectively, and the upper limit corresponds precisely to the frictionless static case $C_4 = 0$ (equation (10)). However, in contrast to equation (13) the analytical determination of $\bar{r}^{A}_{max}$ (or $\bar{r}^{(1)}_{max}$) is impossible, as $\bar{r}^{A}_{max}$ is the smaller of the two solutions of the transcendental equation

$$x\,(2 + 2\,x + x^2)\,e^{-x} = C_2\,C_3\,, \qquad x \equiv C_2\,\bar{r} \tag{32}$$

which must be solved numerically. Still, one can derive analytically the condition for the existence of real and positive solutions of equations (32) which, in a very good approximation, can be written in the form

$$C_2\,C_3 < 0.89\,. \tag{33}$$

This condition actually widens the interval (14) for C_3 to 1.78. If (33) is not fulfilled, the function L^{A} in Fig. 5 has no maximum. However, it is readily seen from Fig. 5 that the mere existence of the maximum does not yet ensure equation (25) to have solutions in the case A, as this requires the condition

$$C_3(r^{A}_{max})^{-3} - C^2 \exp\left(-\,C_2\,r^{A}_{max}\right) > 2\,(k + 1 - n)\,C_4 \tag{34}$$

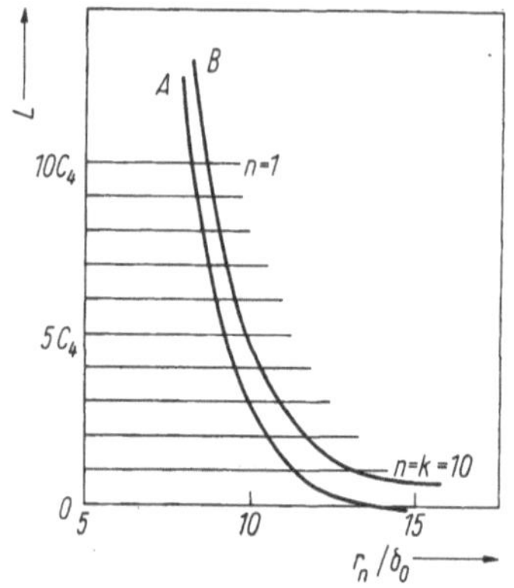

Fig. 7. Numerical solutions of equations (25) for a sequence of ten 180° Bloch walls with parameters $C_2 = 0.5$; $C_3 = 10^{-1}$; $C_4 = 10^{-4}$; $\bar{r}^A_{max} \approx 0.15$; $L^A_{max} \approx 14.81$; $\bar{r}^A_{min} \approx 17.2$; $L^A_{min} \approx 1.3 \times 10^{-5}$; $\bar{r}^A_{up} \approx 14.1$

to be satisfied which in the frictionless static case $C_4 = 0$ corresponds to equation (14). Thus, the inequality (34) constitutes the sufficient condition for the existence of stable solutions in the case A, with $\bar{r}^A_{max}$ determined from (32).

Note that (34) can also be utilized in determining the maximum number k_{max} of Bloch walls with A-type coupling which under specific conditions (wall type, film thickness, shifting velocity, etc.) can "safely" be shifted along the film without destroying the domains separating them. If $k > k_{max}$, then the $(k - k_{max} + 1)$ "tail" walls of the sequence $(k_{max} \leqq n \leqq k)$ move along with stable "bumper" domains between them (provided that the corresponding distances $\bar{r}^A_{stab}$ all satisfy condition (1)), while the $(k_{max} - 1)$ "head" walls $(1 \leqq n < k_{max})$ and the driving wall form apparently a screw-like entity with no domains in between.

We note also the general rules

$$\bar{r}^A_{n+1} > \bar{r}^A_n\,, \qquad \bar{r}^B_{n+1} > \bar{r}^B_n\,, \qquad \bar{r}^B_n > \bar{r}^A_n \tag{35}$$

which hold for the stable solutions and follow directly from Fig. 5 and 6. They show that, for a fixed k, out of all possible dynamically stable Bloch wall arrays the monotonous A- and B-type sequences are the shortest and the longest ones, respectively.

It is clear that the equation set (25) is easier to handle numerically than (26), as it requires the determination of only two numerical curves, L^A and L^B. This is illustrated in Fig. 7 which presents numerical results for a sequence of ten 180° Bloch walls, with the constants C_1, C_2 as in equation (3) and $C_3 = 10^{-1}$, $C_4 = 10^{-4}$. (According to (12) and (24), this corresponds, e.g., to $I_s^2 = 10^5\,\mathrm{erg\,cm^{-3}}$, $\sigma_0 = 10\,\mathrm{erg\,cm^{-2}}$, $\delta_0 = 10^{-6}\,\mathrm{cm}$, $d = 10^{-5}\,\mathrm{cm}$, $\mu_0 = 1$, $H = 5\,\mathrm{Oe}$.) The monotonous Bloch wall sequences corresponding to those solutions are shown in Fig. 8a and b, together with a mixed sequence (Fig. 8c) composed of alternating A- and B-type Bloch wall pairs. As $L^A_{max} \approx 15$ (cf. Fig. 8), we have in the case A stable solutions for sequences composed of up to 15×10^4 Bloch walls. However, $L^A \approx 0.01$ for $\bar{r}_n = 5$. Hence, only about 10^2 of those solutions satisfy the condition $\bar{r}^A_{stab} > 5$ which restricts the applicability of the approximation (8′).

It should be noted that even for $C_3 = 0$ (neglect of stray-field interactions) equations (25) and (26) have stable solutions, and then for arbitrary $k > 0$ and $C_4 > 0$ (cf. Fig. 6). Moreover, if $\dot{x}_0 \to 0$ (i.e., $H \to H_0$) the sequence comes to a halt, the stable Bloch wall distances being in either case (A and B) larger than during the motion (smaller C_4; cf. Fig. 5). Similarly, the sequence while in motion contracts with increasing $\dot{x}_0$ and/or H_0. It should also be stressed that, by suitably specifying the constants C_1 and C_2 according to the curves given in [10, 11], and upon modifying respectively the effective magnetic moments of the walls, equations (25) and (26) can be applied to any type of Bloch walls.

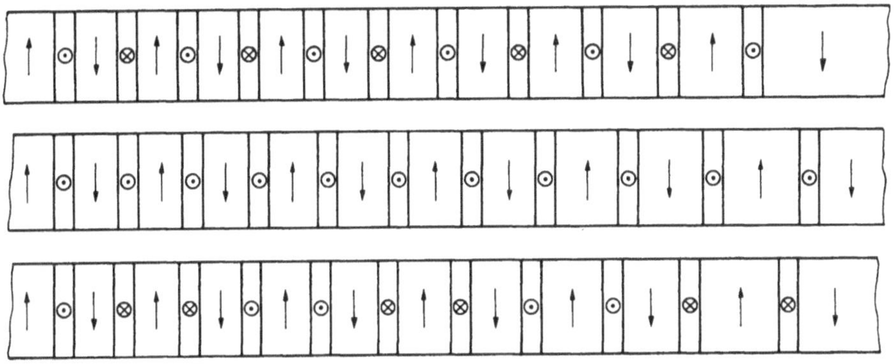

Fig. 8. Monotonous a) A-type and b) B-type Bloch wall sequence, and c) mixed sequence corresponding to the solutions from Fig. 7

To refine the theory as presented here, one would have to take into account the dependence of the Bloch wall thickness δ on r_n in equation (8), the stray-field interaction between neighbouring domains due to the magnetic poles on the lateral film surface (xz plane; cf. Fig. 2), and, perhaps, the stray-field interaction between next-nearest Bloch walls.

References

[1] H. J. Williams and R. C. Sherwood, J. appl. Phys. **28**, 548 (1957).
[2] C. Kittel and J. Galt, Solid State Phys. **3**, 437 (1956).
[3] A. Seeger (Ed.), Chemische Bindung in Kristallen und Ferromagnetismus, Springer-Verlag, Berlin/New York 1966.
[4] J. Kaczér, J. appl. Phys. **29**, 569 (1958).
[5] L. Landau and E. Lifshits, Phys. Z. SU **8**, 153 (1935).
[6] B. A. Lilley, Phil. Mag. **41**, 792 (1950).
[7] A. Wachniewski, Acta phys. Polon. **29**, 437 (1966); **30**, 647 (1966).
[8] A. Wachniewski and W. J. Zię̨tek, Acta phys. Polon. **32**, 21 (1967); **32**, 93 (1967); **33**, 581 (1968).
[9] W. J. Zię̨tek, Acta phys. Polon. **32**, 385 (1967).
[10] J. Ulner and W. J. Zię̨tek, Acta phys. Polon. **35**, 127 (1969).
[11] K. Durczewski and W. J. Zię̨tek, Acta phys. Polon. **35**, 307 (1969).
[12] M. J. Shirobokov, Dokl. Akad. Nauk SSSR **24**, 426 (1939).
[13] J. Klamut and W. J. Zię̨tek, Proc. Phys. Soc. **82**, 264 (1963).
[14] W. J. Zię̨tek, phys. stat. sol. **8**, 65 (1965).
[15] W. Wasilewski and W. J. Zię̨tek, Acta phys. Polon. **A42** (1972) (in the press).
[16] E. Kneller, Ferromagnetismus, Springer-Verlag, Berlin/Heidelberg/Göttingen 1962.
[17] U. Enz, Helv. phys. Acta **37**, 245 (1964).
[18] K. J. Sixtus and L. Tonks, Phys. Rev. **37**, 930 (1931).

(Received May 18, 1972)

phys. stat. sol. (a) **12**, 375 (1972)

Subject classification: 10.2; 22.5.2

A. F. Ioffe Physico-Technical Institute, Academy of Sciences of the USSR, Leningrad

Amplitude Dependent Internal Friction in NaF Crystals in the Temperature Range 295 to 4.2 °K

By

B. K. KARDASHEV, S. P. NIKANOROV, and O. A. VOINOVA

The internal friction is measured in NaF crystals in the temperature range 295 to 4.2 °K using the composite oscillator technique. A square proportionality of the amplitude dependent decrement on the vibrational strain amplitude has been found. The dislocation loop length distribution and the law for the dislocation–point defect pinning force are obtained on the basis of experimental data combined with the results of the theoretical work by Indenbom and Chernov. The dislocation–point defect binding energy is estimated.

Измерения внутреннего трения в кристалле NaF показали, что амплитудно-зависимый декремент пропорционален квадрату амплитуды колебаний в диапазоне температур от 295 до 4,2 °K. Сопоставление полученных данных с результатами теоретической работы Инденбома и Чернова позволили получить функцию распределения петель по длинам, определить характер силового закона взаимодействия дислокаций с точечным дефектом, а также оценить величину энергии связи дефекта с дислокацией.

1. Introduction

Internal friction measurement is a very sensitive method in studying elementary events of plastic deformation [1].

As a dislocation glides it should overcome, due to external stress, different obstacles. If the process takes place at a certain temperature different from absolute zero thermal agitation may contribute to it appreciably. Both the lattice and its defects can serve as obstacles. In alkali halide crystals the Peierls energy barrier is low as compared to those due to the lattice defects, such as point defects, dislocations, and so on. This is confirmed by the strong influence of the impurity content as well as the thermal and mechanical treatment on the mechanical parameters of these crystals [2]. Thus a dislocation in alkali halides can be regarded as a string pinned at some points (Koehler's model [3]). Such a model was helpful in building the theory of dislocation internal friction [4 to 6].

After Granato and Lücke [4] the continuous increase of internal friction is explained by the fact that the dislocation segments of different length break away at different stresses. This theory, however, does not take into account thermal agitation. Nevertheless, the experimental results obtained even at room temperature were in rather good agreement with the predictions.

Friedel [5] made an attempt to introduce a finite temperature into the amplitude dependent damping theory. Here the length distribution effect is not considered. The thermal fluctuation only leads to the continuous increase of internal friction in spite of equal dislocation segments.

The most common theoretical treatment of dislocation hysteresis was developed by Indenbom and Chernov [6]. They developed the amplitude dependent damping theory taking into account thermally activated dislocation depinning as well as the character of the segment length distribution.

Below the experimental data are given for the amplitude dependent internal friction in NaF crystals from 295 down to 4.2 °K and the analysis of the above-mentioned theories on the basis of these results.

2. Experimental Results and Discussion

The internal friction measurements were made on NaF single crystal specimens cleaved along {100} planes. The specimens had the shape of bars with about 30 mm length and cross section $3.0 \times 2.5\ mm^2$. Spectral analysis showed the impurity content to be less than 30 ppm for Mg and less than 10 ppm for Fe, Al, Cu, and Ca in total.

The two-component oscillator (sample and piezoquartz), having an eigenfrequency of about 100 kHz was excited on longitudinal waves. The experimental set-up has been described elsewhere [7].

Dislocations were introduced into the specimen at room temperature by means of four-point circular plastic bending. The dislocation density was fixed by etch pit technique. $FeCl_3$ in distilled water was used as etchant. The prestrain etch pit density was about $10^5\ cm^{-2}$ and after bending about $10^7\ cm^{-2}$.

Fig. 1 shows the amplitude dependent part of internal friction at different temperatures. The amplitude dependent decrement δ_H was determined as the difference between the total and amplitude independent decrement of the sample. The data are plotted in coordinates ln δ_H vs. vibrational strain amplitude ε_0.

Friedel [5] has obtained an expression for the decrement where the depinning activation energy is a linear function of applied stress. Such an approximation must result in straight lines for the ln δ_H–ε_0 dependence, their slope being proportional to the activation volume. Evidently, the dependence obtained is no straight line. The reason of this discrepancy is, perhaps, a more complicated dependence of the activation energy on external stress.

The authors of [6] deal with a more general case: alongside with thermal agitation they consider the effect of dislocation segment length distribution on hysteresis damping. If the average delay time of a dislocation at the energy barrier is approximately equal to the period of the external stress oscillation, the magnitude of the activation energy H is estimated:

$$H(L, \tau) = kT \ln\left(\frac{\nu_0}{\nu}\right) \quad (1)$$

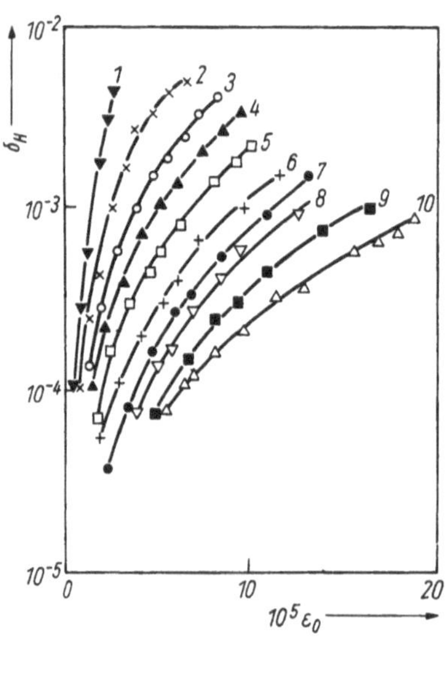

Fig. 1. Amplitude dependence of internal friction at different temperatures: (1) 295 °K, (2) 194 °K, (3) 142 °K, (4) 111 °K, (5) 80 °K, (6) 46 °K, (7) 35 °K, (8) 26 °K, (9) 17 °K, (10) 4.2 °K

and the following expression for δ_H is given as a function of temperature and applied stress amplitude τ_0 [6]:

$$\delta_H = \int_{L_{\min}}^{\infty} L^3 \left(\frac{\tau_H}{\tau_0}\right)^2 N(L)\,\mathrm{d}L\,, \tag{2}$$

where $2L$ dislocation segment length, τ external stress, k Boltzmann constant, T absolute temperature, ν_0 first mode of the dislocation segment oscillation frequency, ν frequency of the stress oscillation, $N(L)$ segment length distribution function, $\tau_H(L, T)$ stress at which the thermally activated breakaway of the segment of a certain length at a certain temperature occurs according to (1).

This equation includes also the approximation that the elastic field of the point defect acts on dislocations as a concentrated force equal to the total dislocation–point defect interaction force [8].

Thus, knowing $H(L, T)$ and $N(L)$ one can get the amplitude and temperature dependence of the hysteresis damping. But as far as these functions are not established the authors of [6] suggest to solve the inverse problem: to obtain the binding energy, the force interaction law, and the distribution function of the segment length just from internal friction experiments.

Neglecting the contribution of the dislocation energy to the energy barrier height resulting from the change of the dislocation line length under external stress, one can obtain $H(L, \tau)$ as a function of one parameter only [8]:

$$F(T) = b\,L\,\tau\,, \tag{3}$$

where b is the absolute value of the Burgers vector. $F(T)$ has the meaning of the total dislocation–point defect interaction force. Under these conditions and if $N(L)$ can be represented as a power function the effect of thermal activation and dislocation segment length distribution on the amplitude dependent internal friction can be specified:

$$\delta_H = f_1(T)\, f_2(\tau_0)\,.$$

It should be mentioned here that the distribution law determines in this case the dependence of the decrement upon the stress amplitude $f_2(\tau_0)$.

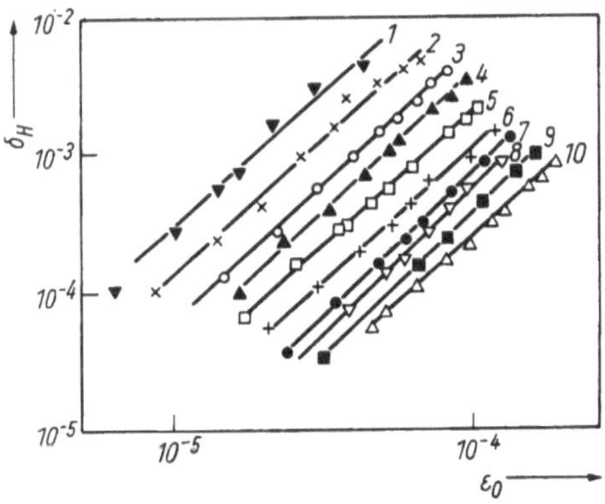

Fig. 2. Amplitude dependence of internal friction at different temperatures in logarithmic scale (see Fig. 1)

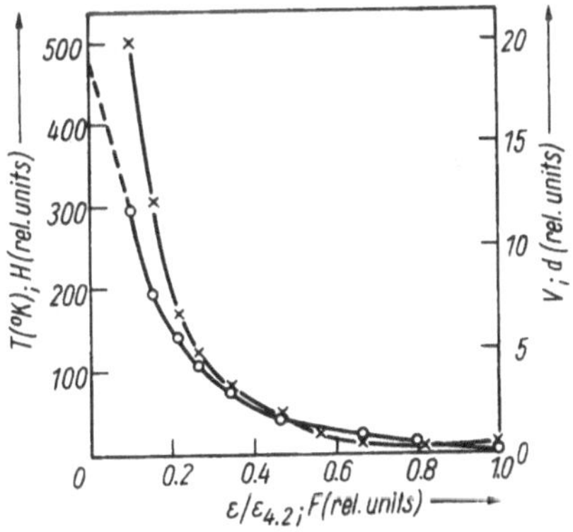

Fig. 3. Activation energy and activation volume versus resolved vibrational strain amplitude: ○ activation energy, × activation volume

Really, in Fig. 2 where the amplitude dependence of the decrement is plotted in a logarithmic scale, one can see a set of parallel lines which can be brought to one line by simple translation along the vibrational strain amplitude axis. (For $\{110\}$ $\langle 1\bar{1}0 \rangle$ slip systems in alkali halide crystals $\tau_0 = \frac{1}{2} E_{100}\, \varepsilon_0$, E_{100} being the Young modulus in $\langle 100 \rangle$ direction.)

The slope of these lines corresponds to a quadratic $\delta_H - \tau_0$ dependence. This leads to a power law of the segment length distribution $N(L) \sim L^{-6}$. As follows from Fig. 2, the distribution law does not change with temperature. Then it becomes clear why the final formulae of the theory of Granato and Lücke are in sufficiently good qualitative agreement with experiment for temperatures above 0 °K.

Using (1), (2), and (3) it has been shown in [6] that $\tau_0(T)$ at $\delta_H = \text{const}$ is proportional to $F(T)$, so the activation energy can be plotted as a function of F from experimental data, what has been done for the NaF crystal. Fig. 3 shows the temperature, or according to (1), the activation energy in terms of $k \ln (\nu_0/\nu)$ versus the resolved strain amplitude, the latter being proportional to the defect–dislocation interaction force $F(T)$. Extrapolating to zero force, we have estimated the binding energy between dislocation and pinning defect which amounts to 0.48 eV. (ν_0 is taken as 10^{10} s^{-1}.)

The derivative $\partial H/\partial F = d$ as a function of F defines the variation of the activation length d with the force F. The inverse function $F(d)$ can be interpreted as the dislocation–point defect force interaction law. Fig. 4 shows $F(d)$ in a logarithmic scale. It should be noted here that the deviation of the ln F dependence from the smooth curve can be caused by a rather approximate calculation of the derivative $\partial H/\partial F$. However, in passing to high stresses the independence of the activation distance upon the external stress is highly probable, as one can see on the $H(F)$ plot (Fig. 3).

Thus, the interaction law derived from experimental data can be compared with the theoretical calculations available, that makes it possible to specify the type of point defect and dislocation, which determine mainly the plastic properties of the investigated crystal.

3. Conclusions

1. The internal friction measurements within a wide temperature range reveal the quadratic amplitude dependence of hysteresis damping from room to liquid helium temperatures.

2. The comparison of the data obtained with the results of Friedel's theoretical treatment [5] shows the model suggested there to be inapplicable to our case.

3. Indenbom and Chernov's theoretical studies [6, 8] enable us to estimate the binding energy (0.48 eV) and to determine the character of the dislocation–point defect force interaction law for NaF crystal. The distribution law is shown not to change over the temperature range from 295 down to 4.2 °K.

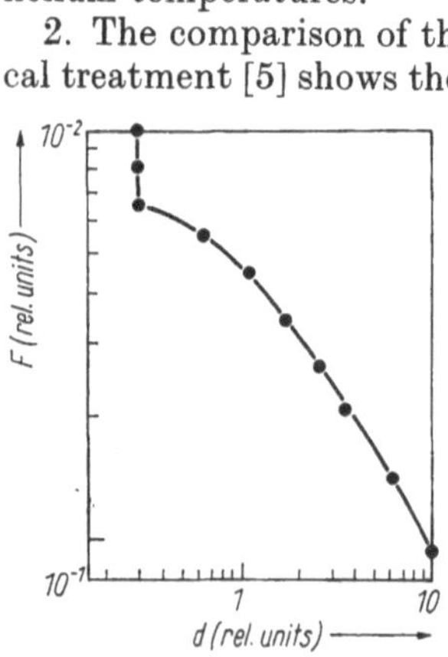

Fig. 4. Dislocation–point defect force interaction law

Acknowledgements

The work has been carried out in the laboratory guided by Prof. A. V. Stepanov, to whom the authors are indebted for continuous support and valuable discussion. We are also grateful O. M. Karlova for skilful technical assistance.

References

[1] W. P. MASON (Ed.), Physical Acoustics, Vol. III A: The Effect of Imperfections, Academic Press, 1966.
[2] E. YU. GUTMANAS and E. M. NADGORNYI, phys. stat. sol. **23**, 571 (1967).
[3] J. S. KOEHLER, Imperfections in Nearly Perfect Crystals, Wiley, 1952.
[4] A. V. GRANATO and K. LÜCKE, J. appl. Phys. **27**, 583 (1956).
[5] J. FRIEDEL, Dislocations, Pergamon Press, 1964.
[6] V. L. INDENBOM and V. M. CHERNOV, Coll. Mechanisms of Relaxation Phenomena in Solids, Obninsk 1970, preprint F.E.I. 215, Nauka, Moscow 1972.
[7] B. K. KARDASHEV and S. P. NIKANOROV, Fiz. tverd. Tela **13**, 160 (1971).
[8] V. M. CHERNOV and V. L. INDENBOM, Fiz. tverd. Tela **10**, 3331 (1968).

(Received May 17, 1972)

phys. stat. sol. (a) **12**, 381 (1972)

Subject classification: 10.2; 1.4; 22.1.2

A. V. Shubnikov Institute of Crystallography, Academy of Sciences of the USSR, Moscow

Scanning Electron Microscopy of Defects in Dielectric Films

By

V. G. Galstyan, S. V. Nosikov, F. P. Press, and V. V. Pastushkov

A technique of the local analysis of defects in silicon oxide thin films and the expansion of the defects in semiconducting crystalline substrates is given. The technique is based on the use of a scanning electron microscope (electron beam induced current and secondary electron emission modes of operation). It is shown that local junctions are formed at temperatures in the interval 450 to 560 °C as a result of interaction of Al and SiO_2 defects. It is suggested that the degradation of electrical properties of planar structures is due to the high density of defects arising near steps formed at the boundary between the original oxide film and the thin oxide layer thermally re-grown in the photolithographic openings.

Предлагается методика локального изучения дефектов тонких пленок двуокиси кремния и их передачи на подложку полупроводникового кристалла. Методика основана на использовании растрового электронного микроскопа. Показано, что образование локальных p–n переходов происходит при температурах 450 до 560 °C в результате взаимодействия алюминия с дефектом двуокиси кремния. Обнаруженная повышенная плотность дефектов на границе старого окисла (толстый слой) и вновь выращенного (тонкий окисел) является, вероятно, причиной ухудшения характеристик планарных структур.

1. Introduction

Dielectric films (such as SiO_2 and Si_3N_4 films) represent one of the basic elements of modern semiconductor structures. They serve as diffusion masks, interlayer isolation, and passivating layers. The structure of the films is usually not perfect and shows both defects of atomic size (vacancies, foreign ions) and microdefects (pinholes, inclusions, etc.) [1]. Although the microdefects may greatly influence the degradation of semiconductor structures [2], they have not yet been studied thoroughly up to now.

The presence of micro-inhomogeneities results, as a rule, in two undesirable effects: deterioration of dielectric film properties and local change in the physical properties of the semiconductor substrate due to the impurity penetration through the defects. The film defects seem to be "transferred" to the substrate leading to an undesirable change in the structure properties.

The present work deals with the scanning electron microscopy (SEM) of local p–n junctions formed by "transferring" microdefects to a semiconductor substrate.

2. Experimental Technique

The investigations were carried out with the JSM-2 scanning electron microscope operated in induced current mode [3] and secondary electron emission mode [4]. Local p–n junctions formed on transferring the defects to the substrate were found in the induced current mode at low magnification; the topography of revealed defects was examined in the secondary emission mode at high magnification. The experiments were performed on SiO_2 thermal films grown on silicon substrates. The oxide films were subjected to the following treatment: a) high-temperature donor or acceptor diffusion; the glasses formed were etched away according to the standard technique [1]; b) Al metallization followed by heat treatment in an inert gas atmosphere; the Al film was then removed by hot sulphuric acid. After that a Pt layer 300 Å thick was deposited on the SiO_2 surface by the help of an electron beam [5]. This film transparent to the electron beam acted as contact to the local areas formed and eliminated the dielectric surface polarization under electron probe radiation.

It should be noted that the minority-carrier diffusion length in the specimens studied was large enough to detect local p–n junctions in the induced current mode at low (200 to 300×) magnification. The experimental technique used permitted to scan individual defects on the dielectric film, to study their topography and their transference to the substrate.

3. Experimental Results and Discussion

3.1 The nature of defects in SiO_2 and the transference of the defects to the substrate during B and P diffusions

A typical picture of the local p–n junctions scanned in the induced current mode and formed by boron diffusion at 1000 °C for an hour through the defects in SiO_2 0.4 μm thick is shown in Fig. 1a. (The source is B_2O_3; the borosilicate glass is removed.) The detailed contrast analysis of about 200 local p–n junctions and associated defects permitted to reveal some regularities:

a) The local p–n junctions are found in the induced current mode as circular dark or bright spots depending upon the area from which the signal is taken (Fig. 1a).

b) The spot size and their contrast increase with the accelerating voltage and current of the electron beam. Besides additional spots can be found.

a 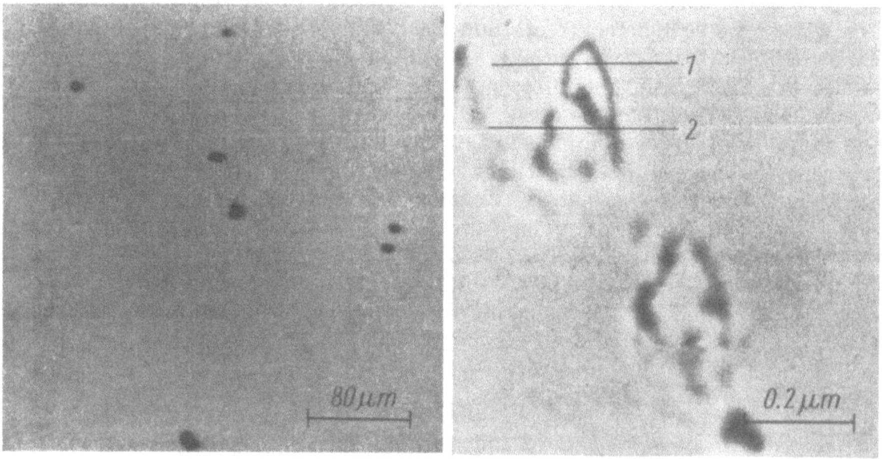 b

Fig. 1. Microphotographs of defects in SiO_2 and of local p–n junctions.
a) Local p–n junctions (electron-beam-induced current mode operation);
b) topography of a "small" damage in the SiO_2 film (secondary electron emission mode operation)

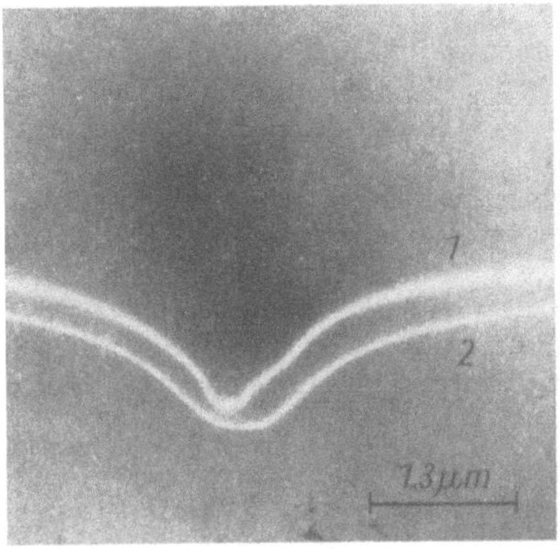

c

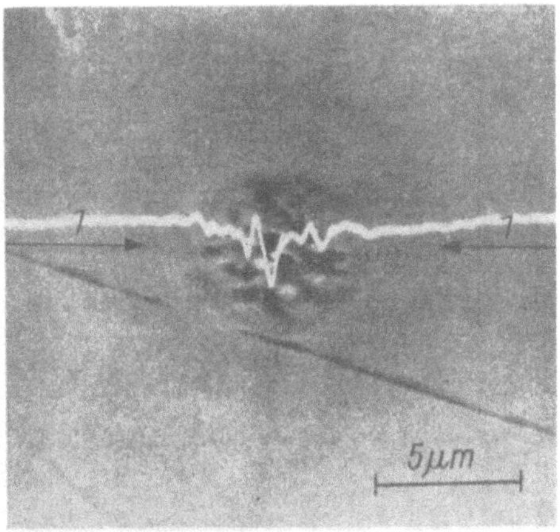

d

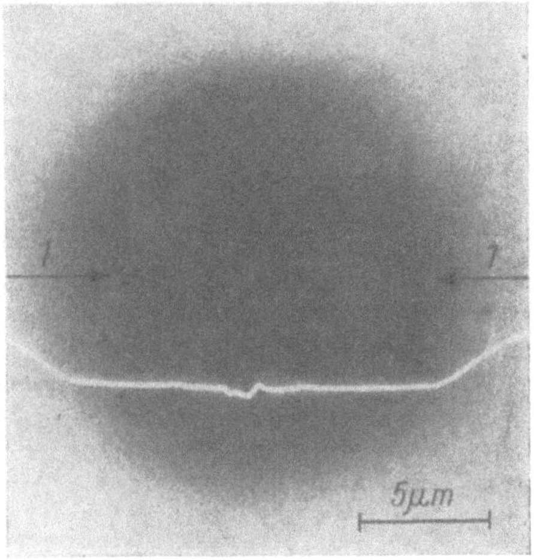

e

Fig. 1.
c) Local p–n junction in induced current mode corresponding to the defects shown in Fig. 1 b (oscillograms of induced current signal along lines 1 and 2);
d) topography of a "large" damage in the SiO_2 film in secondary emission mode (oscillogram of secondary electron signal taken along line 1–1);
e) local p–n junction in induced current mode corresponding to the defect shown in Fig. 1 d (oscillogram of induced current signal taken along line 1–1)

c) As a rule, a spot of larger size is associated with defects of larger size (cf. Fig. 1b and c with Fig. 1d and e).

d) The spot intensity increases towards the centre (Fig. 1c). On the oscillogram a flat extremum may be observed.

e) The defects in dielectric films may be of different shapes and dimensions ranging, as a rule, from 0.1 to 1.0 μm.

The value of the induced current from a p–n junction is, in general, a complicated function of electron beam parameters (the accelerating voltage and current), the diffusion length of minority carriers of a semiconductor material, and the properties of a p–n junction. In the fixed modes of the SEM operation and with constant diffusion length for a given specimen the difference in the contrast of local p–n junctions seems to be related only to differences in their properties.

A local p–n junction of small size may be represented as a semispherical collector of minority carriers. The value of the induced current signal is determined mainly by the area of the p–n junction, which is verified by the relationship observed between the geometrical dimensions of the defects and spots. Characteristic flat extrema observed on some spot oscillograms perhaps correspond to p–n junctions of significant dimensions. On scanning such a p–n junction the induced current signal is constant all over the junction area. To stabilize the semiconductor structure parameters one usually uses phosphorosilicate glass. The process of phosphorus doping may be accompanied by high-temperature annealing in an inert gas atmosphere [6].

To study the defect transference during the formation of the phosphorosilicate glass, P diffusion was carried out at 1100 °C for 1 min (a $POCl_3$ source was used), followed by an annealing process at 1200 °C for 10 min in argon. In the case discussed the phosphorosilicate glass was not removed prior to Pt contact deposition; therefore the local p–n junction formed in a defect must be isolated from the contact by the glass layer. However, we have observed local p–n

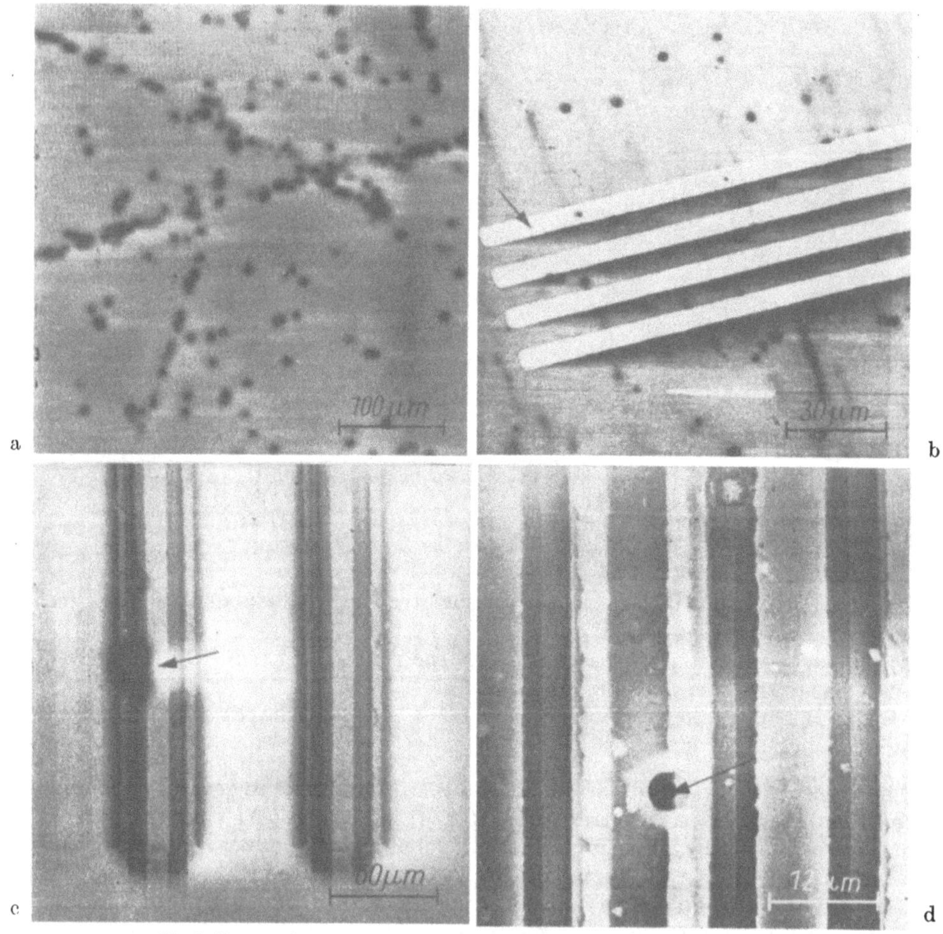

Fig. 2. Propagation of SiO_2 defects at the stabilizing action of phosphorus diffusion.
a) Local p–n junctions formed as a result of phosphorus penetration through the SiO_2 thin film (induced current mode);
b) high concentration of local p–n junctions near the interface between oxide layers of different thickness (bright bars correspond to areas of thin oxide; one of them is shown by the arrow);
c) local damage of the p–n junction in planar structure near the interface (shown by the arrow) (induced current mode of operation);
d) topography of the breakdown site corresponding to the damage in Fig. 2c (shown by the arrow) (secondary electron emission image)

junctions (Fig. 2a). The figure shows that in addition to the high density of local p–n junctions (2×10^4 cm^{-2}) an inhomogeneous distribution takes place. The distinct typical p–n junction lines were attributed to single-crystal substrate surface defects.

Since visualization of a p–n junction is possible when an electric contact to the local area exists, it is reasonable to assume that the conductivity of a phosphorosilicate tube, formed at the defect site, significantly exceeds the bulk conductivity of the defect. This effect may be caused by the increased concentration of impurities at the defect site [7]. A large number of defects are known to appear at the thin oxide–thick oxide boundary of planar structures [2]. This fact was verified by using a model for the thin oxide (0.2 μm)–thick oxide (0.7 μm) boundary with the estimation of the probability for forming local p–n junctions on the oxide step in the p-type substrate in the case of stabilizing phosphorus diffusion under the above conditions. The picture of the p–n junctions observed is illustrated in Fig. 2b. One can see that the density of the p–n junctions localized at the boundary is an order of magnitude higher than that in the thin and thick oxide.

We consider the contrast between the regions of thin and thick oxides observed in the induced current mode to be caused by a change in the Me–SiO_2–Si system capacitance due to the effect of the scanning electron beam. Such a contrast appears in the presence of a positive potential at the metal electrode above 10 V (in the presence of a negative potential the contrast is reversed). The regions of different brightness correspond to the regions of different thickness; the brighter ones are the strips of the oxide of smaller thickness. A similar situation takes place if the signal is taken from the substrate. The type of the contrast is determined only by the polarity of the region from which the signal is taken, which principally differs from the p–n junction contrast.

The mechanism of the phenomenon is rather complicated and is due to a change in the charge at the outer surface of the dielectric film, to the generation of the charge carrier pairs in the bulk of the dielectric layer, to a change in the properties of the interface, and to the generation of excess charge carriers in the underlying semiconductor layer. One can, in general, imagine that when a bias voltage is applied to the gate (or substrate), the charge on the capacitance is increased. This results in an increase in the signal appearing on the capacitance due to its change with electron beam action.

In planar structures the p–n junctions, as a rule, extend to the surface near the oxide steps. The formation of local diffusion regions, where the p–n junctions extend to the surface, perhaps alters the properties and can lead to the degradation of the characteristics. This is verified by the results obtained investigating p–n junctions of induced p-channel MOS structures having high leakage current.

A comparative analysis of the contrasts in the induced current mode of the p–n junctions with high leakage currents and normal ones permitted to detect local anomalies in defective junctions. A typical example is shown in Fig. 2c. The location of the anomalies approximately coincides with the region where the p–n junction extends to the oxide step. The relationship between the p–n junction anomalies and the defects on the SiO_2 film boundary was established during the investigation of the gate dielectric breakdown; the breakdowns were observed during secondary emission. The locations of the breakdowns, as a rule, coincided with the anomalous areas of the p–n junction. Fig. 2d shows the result of the damage of the boundary after the breakdown at the location of the anomaly shown in Fig. 2c. The centre of the cylindrical hole (shown by the arrow) through the SiO_2 film takes place exactly at the oxide step.

3.2 Investigation of the transference of the defects to SiO_2 during Al-metallization heat treatment

An Al film 0.3 μm thick was deposited on SiO_2 0.2 μm thick thermally grown on an n-type silicon substrate ($\varrho = 2\ \Omega$ cm). The heat treatment was done at different temperatures in the range 350 to 560 °C. Local p–n junctions were observed in the induced current mode of operation on specimens subjected to heat treatment for 10 to 20 min in argon at temperatures above 450 °C. The measurement of the Al–SiO_2–Si I–U characteristics verified the presence of p–n junctions.

After the heat treatment at 500 °C regions of adequate crystal orientation (triangles in the (111) plane; cf. Fig. 3a) were formed around the defects under the dielectric film. Such regions were observed by using an electron microscope and replica techniques described in [8]. The authors relate the formation of

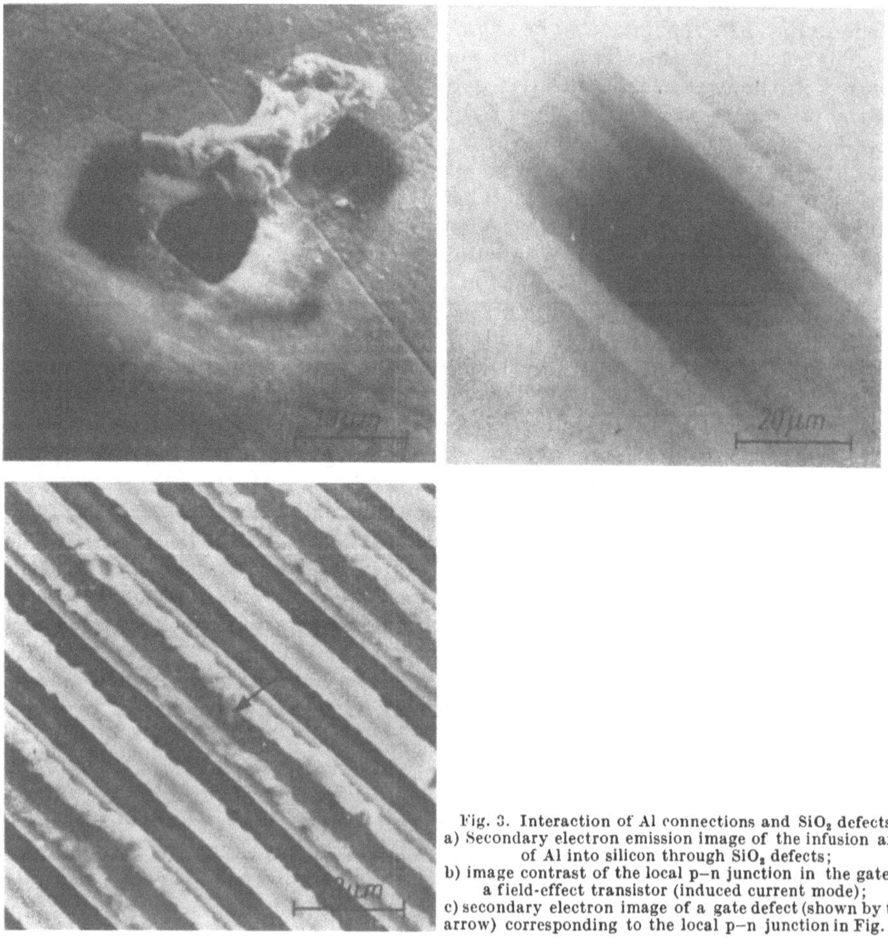

a b c

Fig. 3. Interaction of Al connections and SiO_2 defects.
a) Secondary electron emission image of the infusion area of Al into silicon through SiO_2 defects;
b) image contrast of the local p–n junction in the gate of a field-effect transistor (induced current mode);
c) secondary electron image of a gate defect (shown by the arrow) corresponding to the local p–n junction in Fig. 3b

triangles to the interaction between Al and SiO_2 at the defect sites and presume that these regions consist of Al_2O_3.

In [9] the formation of alloy regions is thought to be due to the exothermal reaction

$$4\,Al + 3\,SiO_2 \rightarrow 3\,Al_2O_3 + 2\,Si\,.$$

The authors of [10] consider the "exothermal hypothesis" to be groundless and ascribe the formation of the local areas to Al diffusion into silicon.

The formation of alloy p–n junctions of considerable size when using heat treatment of short duration (10 min), from our point of view, proves the validity of the exothermal hypothesis. Local p–n junctions formed during Al-metallization heat treatment at temperatures below the temperature of the Al–Si eutectic were observed on planar structures with high leakage current flowing through SiO_2. In Fig. 3b an example is given of a local p–n junction (dark spot) in the gate of a field-effect transistor isolated in the induced current mode. The secondary electron emission picture of the corresponding portion of the gate shows that the SiO_2 film is defective (Fig. 3c).

4. Conclusions

By using a scanning electron microscope we performed an investigation of microdefects in thermally grown SiO_2 and their transference in the process of B and P diffusion and during Al-metallization heat treatment. We studied the influence of the local regions formed during the transference of the defects on the semiconductor structure properties.

The following has been shown:

1. The formation of local p–n junctions takes place under conditions that are not too severe for B and P diffusion into the SiO_2 defects of 0.1 μm size.
2. High-temperature heat treatment of phosphorosilicate glass can lead to the formation of local diffusion regions with a density of 2×10^4 cm^{-2}.
3. Local p–n junctions in the p-type substrate are formed in the range 450 to 560 °C due to the interaction between Al and SiO_2 defects.
4. The probability of the formation of local p–n junctions along the boundary of oxide layers of different thickness is, at least, an order of magnitude higher than in a uniform layer. This may be one of the main causes of the degradation of the characteristics of the planar structure because their p–n junctions extend to the surface, as a rule, near the oxide steps.

References

[1] R. H. Burger and R. P. Donovan (Eds.), Fundamentals of Silicon Integrated Devices Technology, Vol. 1, Prentice-Hall, Inc., Englewood Cliffs (N.J.) 1967.
[2] L. C. Hamitter, Electronics **42**, 107 (1969).
[3] T. E. Everhart, O. C. Wells, and R. K. Matta, J. Electrochem. Soc. **111**, 920 (1964).
[4] C. W. Oatley and T. E. Everhart, J. Electronics **2**, 568 (1957).
[5] M. Kamosida and O. Takacu, Trans. Inst. Electron. and Commun. Eng. (Japan) **52C**, 490 (1969).
[6] B. E. Deal, M. Sklar, A. S. Growe, and E. N. Snow, J. Electrochem. Soc. **114**, 266 (1967).
[7] F. J. Cocca and K. G. Carroll, IEEE Trans. Electron Devices **15**, 962 (1969).
[8] N. J. Chou and G. M. Eldridge, J. Electrochem. Soc. **117**, 1287 (1970).
[9] J. McCarthy, Microelectronics and Reliability **9**, 187 (1970).
[10] J. A. Cunningham and R. H. Wackfield, Microelectronics and Reliability **9**, 519 (1970).

(Received June 21, 1972)

phys. stat. sol. (a) **12**, 389 (1972)

Subject classification: 10.2; 22.5.2

Département de Physique des Matériaux, associé au C.N.R.S., Université Claude Bernard de Lyon, Villeurbanne

Dislocation-Induced Precipitation of Colloids in Additively Coloured KCl

By

D. DURAND, G. CHASSAGNE, and J. SERUGHETTI

The rate of thermal transformation of F centres into colloids in additively coloured KCl increases with the dislocation density created by thermal quenching or by plastic deformation. A colloid formation process is proposed. The dislocation stress field acts on F centre diffusion towards nucleation sites, as impurity–vacancy aggregates, created by plastic deformation. The colloid growth rate is thus enhanced. The activation energy of the precipitation process is the same (0.57 eV) for plastically deformed samples or only quenched ones.

La vitesse de transformation des centres F en colloïdes dans les cristaux de KCl colorés additivement augmente avec la densité de dislocations créés par trempe ou par déformation plastique. On propose un processus de formation des colloïdes. Le champ de contrainte des dislocations agit sur la diffusion des centres F vers les sites de nucleation, tels que les agregats d'impuretés et de lacunes, créés par déformation plastique. La vitesse de croissance des colloïdes en est ainsi augmentée. L'énergie d'activation du processus de précipitation est la même (0,57 eV) que le cristal soit déformé plastiquement ou seulement trempé.

1. Introduction

In alkali halides coloured by stoichiometric excess of metal, as has been shown by several authors [1], F centres can coagulate to colloids of metal under thermal annealing in the dark. Amelinckx [2] showed the importance of dislocations in the formation of colloids which decorate them. Penley and Witte [3] studying the rate of creation of colloids in undeformed KCl crystals found that the precipitation process is best described as a dislocation-induced precipitation at low temperature annealing (200 °C). Sastry [4] in a discussion on the proposed mechanisms [2, 3] concluded that the diffusion of neutral F centres via the vacancy mechanisms results in the clustering of F centres in the bulk (or lattice) of the crystal more than around dislocations. It appears that the colloid precipitation mechanism in alkali halides has not yet been entirely established.

In an attempt to know the influence of dislocations on the rate of colloid nucleation we study the behaviour at 200 °C of F centres in additively coloured KCl crystals in which the dislocation density is varied by plastic deformation.

2. Experimental Procedure

KCl crystal purchased from Quartz and Silice Co. (positive divalent impurities about 3 ppm Ca^{2+} and 3 ppm Mg^{2+}) are cleaved along {100} ($20 \times 10 \times 5$ mm^3) from the same crystal; then they are additively coloured by a stoichiometric excess of potassium employing Van Doorn's technique [5] at 50 Torr and about 700 °C for 24 h, in order to ensure uniformity of the F centre concentration. Crystals are rapidly quenched after coloration in an argon jet at 25 °C. For this study the specimens are cleaved ($6 \times 6 \times 1$ mm^3) removing initial crystal faces. Crystals are always kept in a vacuum and in the dark. They are deformed at room temperature by compression along the ⟨001⟩ direction, parallel to their larger dimension, using an Inströn testing machine at a speed of 50 μm/min. The plastic strain is always lower than 4%. It must be noticed that the yield point variations of crystals either uncoloured or additively coloured and quenched are very small according to Suzuki's observations [6]. Samples are isothermally annealed at 200 °C for a maximum time of 130 h in a vacuum (10^{-3} Torr) in an attempt to minimize the interaction of the stoichiometric excess of potassium with the layer of water adsorbed on the surface [7, 8]. After annealing, the specimens are quenched in an argon jet. The annealing temperature of 200 °C is chosen to eliminate the risk of potassium colloid evaporation. The spectra are recorded at room temperature using a Beckman DK 2A recording type spectrophotometer. The total concentration of F centres or of the potassium excess is calculated employing Smakula's relation taking into account the F-aggregate centres.

3. Experimental Results

After additive coloration and after deformation the samples show only F, M, and R bands. No band is observed in the UV region of the spectrum. For deformed samples the decrease of the F band due to the direct F centre ionization by moving dislocations is negligible for the plastic strains used. In Table 1 columns 3 and 4 the initial F centre concentration and plastic strains of samples A and B are given. Fig. 1 and 2 show typical optical absorption curves before and after annealing (about 31 h) at a temperature of 200 °C for deformed and undeformed samples A and B, respectively. Before annealing (Fig. 1 curves 1 and 3) samples A show M and R bands which disappear after annealing for nearly 5 h at 200 °C. Samples B were annealed at 400 °C for a very short time after coloration and deformation in an attempt to destroy the F-aggregate centres (Fig. 2, curves 1 and 3).

Table 1

Experimental values of n and τ as determined from a fit of the equation $\alpha(t)/\alpha_0 = \exp(-(t/\tau)^n)$ and maximum of the colloid band after an annealing time of 30 h for samples A and B

crystal	sample	N_f (10^{17} cm^{-3})	plastic strain (%)	n	τ (h)	λ (nm)
A	1	2.8	0	0.59	28	741
	2	2.7	1.3	0.61	19.5	738
	3	2.9	2	0.62	18	732
	4	2.9	2.8	0.60	16.5	730
	5	2.4	3	0.60	16	725
	6	2.3	4	0.63	14	717
B	1	2.3	0	0.88	46	750
	2	2.3	0	0.86	39	750
	3	2.1	1	0.85	35	751
	4	2.3	2.1	0.81	22	750
	5	2.5	3	0.80	18	749
	6	2.4	3.2	0.89	18	747

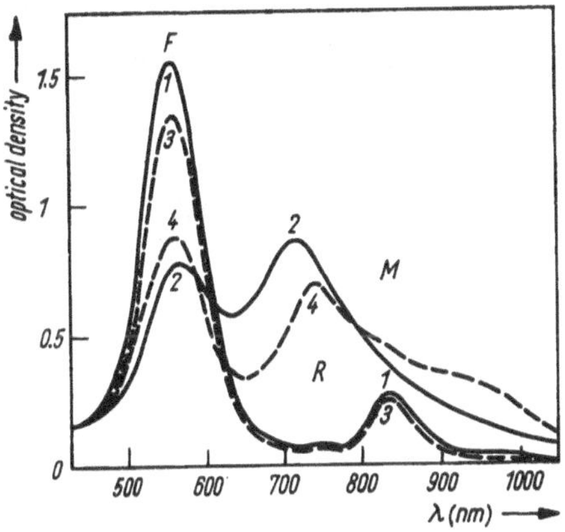

Fig. 1. Absorption spectra of samples A. – – – – – undeformed sample 1A; ——— deformed (4%) sample 6A; curves 1 and 3 before annealing, curves 2 and 4 after 31 h of annealing at 200 °C

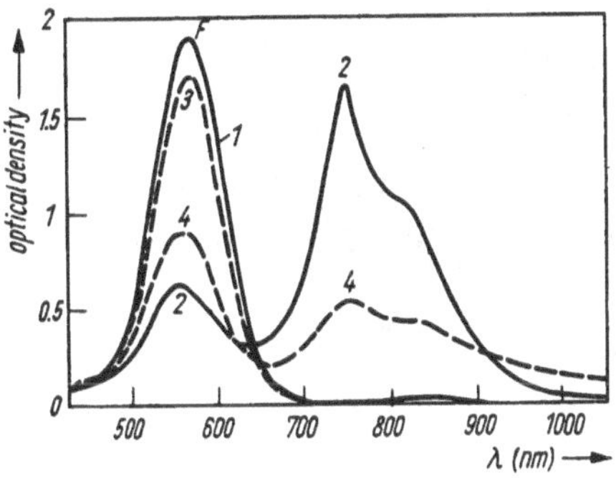

Fig. 2. Absorption spectra of samples B. – – – – – undeformed sample 1B; ——— deformed (3.2%) sample 6B; curves 1 and 3 before annealing, curves 2 and 4 after 32 h of annealing at 200 °C

In both cases, after annealing, the absorption bands peaked at 741 nm (Fig. 1, curve 4, undeformed sample A), 717 nm (Fig. 1, curve 2, deformed sample A), and 750 nm (Fig. 2, curves 4 and 2, undeformed and deformed samples B) are identified as colloidal bands. The criterion [1] that the line shape of the colloidal band is insensitive to the change of the crystal temperature from RT to LNT has been used to identify these bands as colloidal bands. It is of interest to note that an isobestic point appears after an annealing time of about 15 h. Such a point of constant extinction is the well recognized criterion for the presence of a two component system [9].

Fig. 1 curve 4 shows for the undeformed sample A two composite infrared bands which appear after an annealing time of about 4 h. These unidentified bands are not observed in deformed sample A even for a very long annealing time (130 h at 200 °C). The origin of these may be due either to colloids of a larger size or to impurities. The spectra of Fig. 2 (curves 2 and 4) show a band

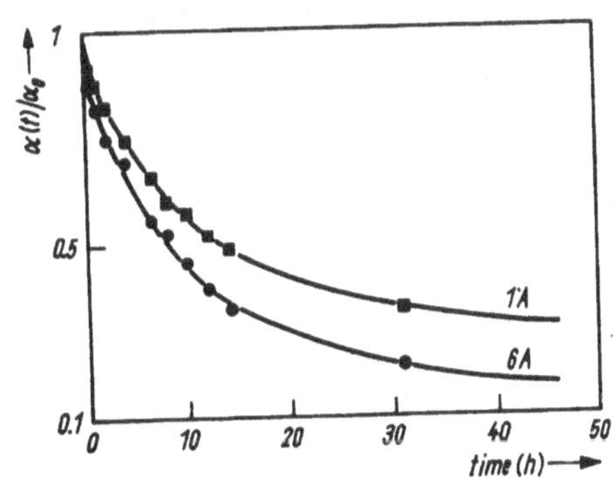

Fig. 3. Unprecipitated fraction of F centres vs. annealing time at 200 °C for undeformed sample 1A and deformed (4%) sample 6A

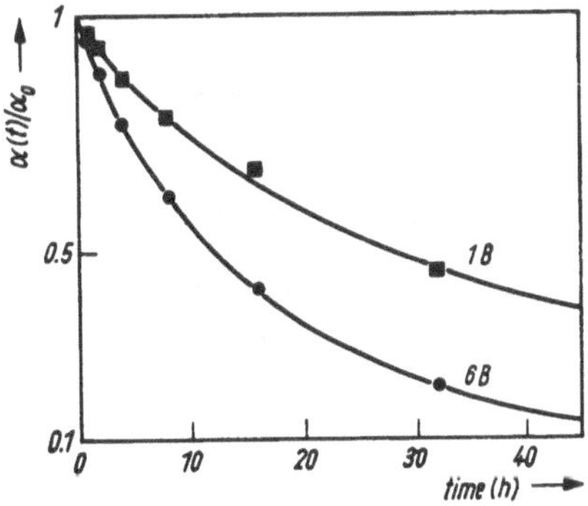

Fig. 4. Unprecipitated fraction of F centres vs. annealing time at 200 °C for undeformed sample 1B and deformed (3.2%) sample 6B

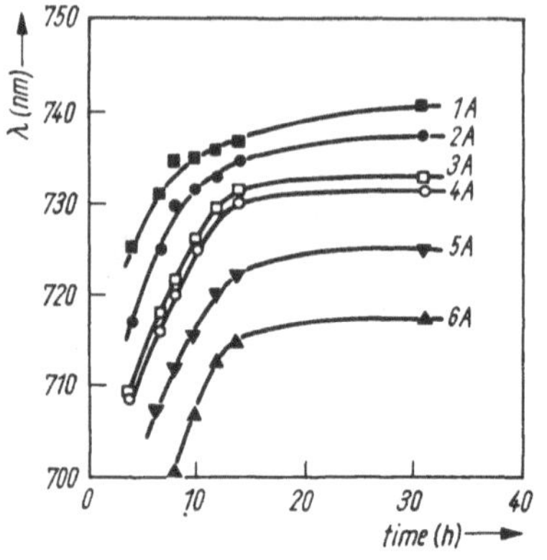

Fig. 5. Peak position of colloid band vs. annealing time for samples A

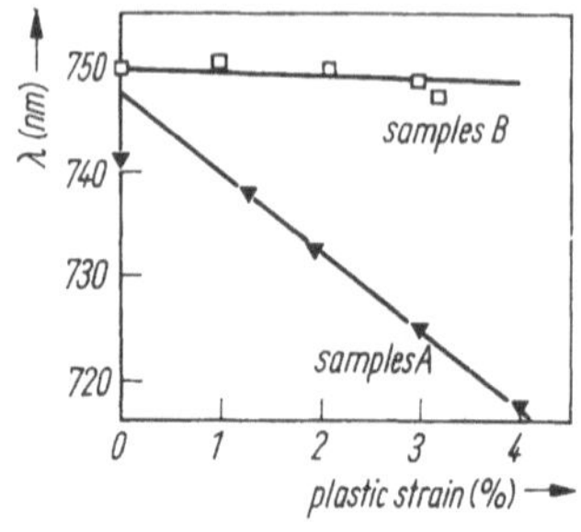

Fig. 6. Peak position of colloid band vs. plastic strain in samples A and B after about 30 h of annealing

peaked at nearly 830 nm for deformed and undeformed samples B. These bands, observed by Penley and Witte [3] disappear after an annealing time of about 100 h at 200 °C.

In all samples A and B thermal bleaching of the F band is faster for a deformed specimen than for an undeformed one (Fig. 3 and 4) and the bleaching rate of F centres is proportional to the plastic strain. Fig. 5 shows the variation of the maximum of the colloidal band versus annealing time for different plastic strains in samples A. In each case this maximum has a final value corresponding to the equilibrium between F centres and colloids. Fig. 6 curve A shows that for sample A this maximum decreases with the plastic strain at a given time.

For samples B the maximum of the colloidal band increases with annealing time, as for samples A, but this maximum does not vary with the plastic strain of the crystal at a given time (Fig. 6, curve B).

4. Discussion

According to the present results the following facts have to be interpreted:

1. the rate of thermal bleaching of F centres is increased in plastically deformed crystals. The F band decrease is accompanied by the increase of a colloidal band.

2. the wavelength of the colloidal band maximum evolves with the annealing time and tends towards a final value for all samples A and B. This final value is a function of the plastic strain only for samples A.

Item 1. shows the importance of dislocations in the colloid formation process. Dislocations can generate nucleation sites and act through their stress field. It is of interest to note that, even for undeformed crystals, dislocations have the same importance. After additive coloration at 700 °C, because of the rapid quenching, a stress gradient appears between the external and the internal parts of the crystal. The first part being in compression, the latter in tension a plastic deformation thus appears. The residual stresses due to this plastic deformation are observable under the microscope between crossed polarizers. The dislocation density after additive coloration and quenching is found by etching technique to be nearly 10^6 cm^{-2}. By measuring the rate of thermal bleaching of F centres as a function of temperature it could be shown that the effective activation energy (about 0.57 eV) is the same for deformed and undeformed specimens.

The nucleation sites can be produced during plastic deformation. Indeed it is admitted that dislocation interactions [10] produce many defects, as isolated vacancies, aggregate vacancies, dislocation dipoles. These defects being more numerous when the number of active slip systems is larger and the plastic strain higher. Dislocation interactions should generate an equal number of cation and anion vacancies and it is expected that vacancy pairs are formed because of electrostatic interactions between them. These vacancy pairs which are highly mobile, can form vacancy clusters which remain in registry with the lattice. This is confirmed by the study of the irradiation of deformed crystals [11]. Some information on the concentration and the size of these clusters can be obtained from measurements of density varations after plastic deformation [12] and from investigations of electron traps for colour-centre electrons produced by plastic deformation [13]. In our crystals it is reasonable from [12, 13] to admit that the single vacancy concentration is about 10^{16} cm^{-3} in undeformed quenched crystals and 5×10^{17} cm^{-3} in 4% deformed ones. The clusters are formed probably of a few anion and cation vacancies.

On the other hand in our crystals the positive divalent impurities (6 ppm) can screen the aged dislocations. During the plastic deformation it has been shown [14, 15] by electrical charge measurements that 10^{12} to 10^{14} cm^{-3} impurities are screened out. These impurities can form defects as impurity–vacancy pairs but also larger defects with vacancy clusters described above. These aggregates can be trapped in the tensile region of the dislocation stress field, and act as nucleation sites in the colloid formation. It has, indeed, been observed that in bending deformed samples, colloids are principally formed in the tensile region of the crystal.

Let us look now at F centres. With Chang [16] and Serughetti and Saucier [17] we can suppose that F centres can be trapped in the stress field of dislocations or dislocation dipoles created during plastic deformation. These centres, perturbed by this stress field, could diffuse towards dislocations. During their motion they may meet nucleation sites created near dislocations and may form complex centres like Scott R′ centres [18], the initial stage of the formation of potassium colloids. Thus, in the first stage of the precipitation process we do not observe an isobestic point, and the position of the colloidal band (item 2.) varies. All this seems to indicate that complex centres, other than colloids, are formed at this stage. This hypothesis is in good agreement with the observations of Jain and Sootha [19] who indicate that divalent impurities (about 10 ppm) increase the colloid precipitation rate by the initial formation of Scott R′ centres. Thus, by increasing the dislocation density, the number of nucleation sites is increaseed and the path-length of F centre diffusion towards dislocations is reduced. The thermal bleaching resulting from the F centre coagulation onto nucleation sites is increased likewise the K colloid precipitation in plastically deformed crystals.

An other process of nucleation site formation can be considered [2]. An F centre diffusing towards dislocations and arriving on a charged jog, moves this jog two ións further and transforms a potassium ion into a potassium atom. The process can then start over again, and finally the dislocation would climb leaving a row of potassium atoms. These atoms can diffuse along the dislocation and agglomerate to form the first colloid nuclei. However, it seems that in our case the first process described above is preponderant.

On the other hand, the study of the growth rate of the colloid particles should furnish information on the growth mechanism. Of the various precipitation mechanisms based on the assumption that the precipitation rate is determined by a diffusion process, the one that seems to be most applicable to the present case is a stress assisted precipitation mechanism. Ham [20] calculated the growth rate of nuclei in strained materials. He showed that for a regular array of dislocations replaced by a single dislocation at the centre of a cylinder of radius r_s (volume around the dislocation in which particles precipitate onto it), the concentration $\alpha(t)$ of unprecipitated imperfections at the time t is given by

$$\alpha(t) = \alpha_0 \exp\left(-\lambda_0^2 D t\right), \tag{1}$$

in which α_0 is the initial concentration, D the diffusion coefficient and

$$\lambda_0^2 \approx \frac{2}{r_s^2}\left[\ln\left(\frac{r_s}{R}\right) - \frac{3}{5}\right]^{-1}.$$

R is the radius of a cylindrical sink surrounded by the cylinder of radius r_s. The original length L of dislocation lines per unit volume is defined by

$$L = (\pi r_s^2)^{-1}.$$

In general the evolution of the unprecipitated fraction of imperfections versus annealing time may be given by

$$\frac{\alpha(t)}{\alpha_0} = \exp\left(-\left(\frac{t}{\tau}\right)^n\right), \tag{2}$$

where n may have the value 1 in the later stage of precipitation

$$\left(\frac{\alpha(t)}{\alpha_0} \leq 0.9\right).$$

When the precipitation mechanism is a limited diffusion precipitation mechanism, Ham [21] showed that the equation (2) is still valid. The exponent n is 3/2 as long as $\alpha(t)/\alpha_0 > 1/2$, it changes to $n = 1$ for $\alpha(t)/\alpha_0 < 1/2$ (long time). Thus the two mechanisms can only be distinguished for $0.5 < \alpha(t)/\alpha_0 < 0.9$, $n = 1$ in this range pointing to precipitation on dislocations. In this case

$$\tau^{-1} = \beta L D \tag{3}$$

with

$$\beta = 2\pi\left(\ln\left(\frac{r_s}{R}\right) - \frac{3}{5}\right)^{-1}.$$

In all cases equation (2) is not valid when there is a diffusion current from the precipitate to the solvent.

Table 1 summarizes the results in an attempt to fit the rate of bleaching of F centres in samples A and B using an equation of the form (2). Fig. 7 shows typical data for several samples A and B. It appears in Table 1 that the expo-

Fig. 7. Logarithm of the reciprocal of the unprecipitated fraction of F centres vs. annealing time at200 °C
--■---■-- deformed (4%) sample 6A;
--×---×-- undeformed sample 1A;
—△——△— deformed (3.2%) sample 6B;
—●——●— deformed (1%) sample 3B;
—○——○— undeformed sample 1B

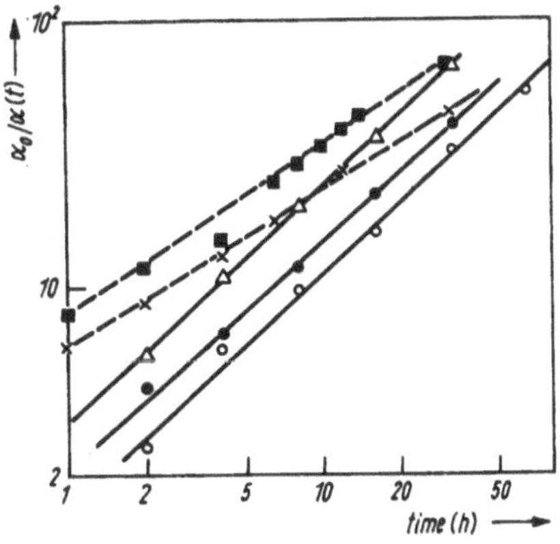

nent n is independent of the plastic strains and that the value of n is found smaller than 1. Then using [20] and [21] these values eliminate the hypothesis of a diffusion-limited precipitation mechanism proposed by [4] even for undeformed crystals. Fig. 8 shows that the relaxation time τ varies linearly with plastic strain and therefore with dislocation density L as shown in equation (3). The diffusion coefficient calculated from the equation (3) using $\beta = 1$ is found to be about 10^{-12} cm^2/s. This value does not allow us to determine the diffusing entity of either neutral or ionized F centres. So, at an annealing temperature of 200 °C for which there are very few ionized F centres, we shall admit, like [4] that it is the neutral F centres which diffuse by a vacancy mechanism.

Item 2. must now be considered. The peak position of the colloid band in experimental absorption spectra, after an annealing time of about 30 h, when compared with absorption spectra of K in KCl calculated [2] from Mie's theory indicates that the particle radius is: 27 nm in undeformed sample 1A ($\lambda =$ $= 741$ nm), 12 nm in deformed sample 6A ($\lambda = 717$ nm), and about 30 nm in deformed and undeformed samples 1B and 5B ($\lambda = 750$ nm). If we consider that all potassium colloid particles precipitate along dislocations with a spherical shape and a body centered structure, it is possible to estimate the colloid size when the dislocation density is known (Table 2). The agreement between

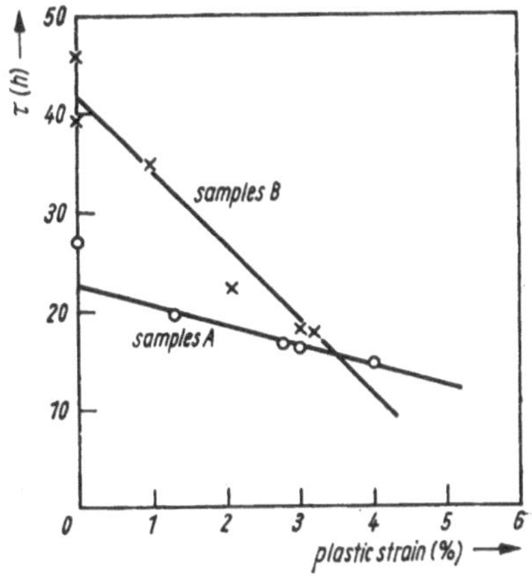

Fig. 8. Relaxation time τ vs. plastic strain for samples A and B

Table 2

Radius of the colloid particles estimated from the maximum of the colloid band using Mie's theory and estimated by us from the number of precipitated potassium atoms and from the dislocation density

sample	precipitated potassium atoms (10^{17} cm^{-3})	dislocation density (cm^{-2})	wavelength of the colloidal band (nm)	radius estimated from Mie theory (nm)	radius estimated by us (nm)
1 A	1.8	10^6	741	27	24
6 A	1.8	5×10^7	717	12	3.8
1 B	1.3	10^6	750	30	20
5 B	1.9	5×10^7	750	30	4.5

the radii calculated from Mie's theory and the value estimated by us is rather good for samples A. Thus the peak position variation of the colloid band may indicate a decrease in the colloid size as a function of the plastic strain. This can easily be explained by the previous assumptions that the plastic deformation creates more numerous nucleation sites. Likewise Fig. 5 shows an increase in the particle size with the annealing time.

However, the halfwidths of the colloid bands (about 0.4 eV) in samples A are not in agreement with those calculation from Mie's theory (about 0.16 eV). It appears that the greater symmetry and narrower band width of the calculated absorption curves as compared with the experimental curves indicate probably a distribution in the size and shape of the colloid particles in our samples.

In samples B the colloid precipitation rate is slower than in samples A (Fig. 3 and 4), Table 1 column 6. This could explain a more homogeneous colloid size distribution in samples B. The initial annealing at 400 °C of these samples may have made the number of nucleation sites uniform. Thus our calculations give a particle size which is too small in deformed samples B. The halfwidth of the experimental colloid bands in this case (about 0.28 eV) is larger than the value given by the Mie's theory for this particle size [2] but is not very different from that obtained by Jain and Sootha [19] in KCl crystals containing about 10 ppm divalent impurities.

Finally, in these experiments, the colloid precipitation in deformed and undeformed crystals requires the existence of clusters of defects. These defects created by plastic deformation favour the creation of complex centres, the initial stage of the colloid formation. The neutral F centre diffusion towards these nucleation sites assisted by the dislocation stress field leads to the development of these centres into colloids and their growth. It seems that it is not useful to invoke another process such as the clustering of F centres in the bulk [4] even for undeformed crystals.

References

[1] J. H. SCHULMAN and W. D. COMPTON, Color Centers in Solids, Pergamon Press, London 1962.

[2] S. AMELINCKX, Phil. Mag. **1**, 269 (1956).

[3] J. C. PENLEY and R. S. WITTE, J. appl. Phys. **33**, 2875 (1962).

[4] P. V. SASTRY, J. Phys. Soc. Japan **26**, 73 (1969).

[5] C. Z. van Doorn, Rev. sci. Instrum. **32**, 755 (1961).
[6] T. Suzuki and M. Doyama, J. Phys. Chem. Solids **10**, 77 (1959).
[7] D. Durand and J. Serughetti, C. R. Acad. Sci. (France) **272**, 1224 (1970).
[8] J. C. Penley and R. S. Witte, J. chem. Phys. **40**, 1550 (1964).
[9] W. T. Doyle, Phys. Rev. **111**, 1067 (1958).
[10] R. W. Davidge and P. L. Pratt, phys. stat. sol. **3**, 665 (1963).
[11] W. A. Sibley, C. M. Nelson, and J. H. Crawford, Jr., Phys. Rev. **139**, A1328 (1965).
[12] W. H. Vaughan, W. J. Leno, and R. Smoluchowski, Phys. Rev. **110**, 652 (1958).
[13] M. Ueta and W. Kanzig, Phys. Rev. **97**, 1591 (1955).
[14] F. Rueda and W. Dekeyser, Acta metall. **11**, 35 (1963).
[15] J. Serughetti, B. Schaeffer, C. H. S. Dupuy, and H. Saucier, Bull. Soc. Sav. **2**. 251 (1967).
[16] Chang, Phys. Rev. **138**, A839 (1965).
[17] J. Serughetti and H. Saucier, phys. stat. sol. **37**, 381 (1970).
[18] A. B. Scott and L. P. Bupp, Phys. Rev. **79**, 341 (1950).
[19] S. C. Jain and G. D. Sootha, Phys. Rev. **171**, 1075 (1968).
[20] F. S. Ham, J. appl. Phys. **30**, 915 (1958).
[21] F. S. Ham, J. Phys. Chem. Solids **6**, 335 (1958).

(Received April 7, 1972)

phys. stat. sol. (a) 12, 399 (1972)

Subject classification: 3; 22.3

Chemical Physics Division, Department of Physics, Eindhoven University of Technology

Growth and Habit of GaS Single Crystals Obtained from the Vapour Phase

By

R. M. A. Lieth

A comparative study of the sublimation and chemical transport of the compound GaS is reported. The parameters investigated in the closed system are the iodine concentration, the tube geometry, and undercooling. The rate of transport in case of sublimation is diffusion-controlled, in the iodine-assisted transport process at concentrations higher than 1 mg/cm^3, convection governs the process, while at lower concentrations the diffusion-controlled sublimation becomes important. In general, the lateral growth is fast when the chemical transport is used; the habit of the crystals is plate-like, the *c*-axis being normal to the plane of the platelet. A large amount of screw dislocations can be seen on the (0001) faces and the crystals exhibit in general n-type conductivity. If the sublimation technique is used, the crystals are not exclusively plate-like, the majority of the crystals exhibit needle-like and ribbon-like habit. With growth in the length direction, the growth axis is parallel to the *c*-axis, it takes place around screw dislocations and two crystallization mechanisms are assumed to take place simultaneously. The formation of a large number of two-dimensional nuclei between the arms of the screw dislocation seems to be responsible for a decrease of the mean surface migration distance X_s on the basal plane.

Es wird über eine vergleichende Untersuchung der Sublimation und des chemischen Transports der Verbindung GaS berichtet. Die im abgeschlossenen System untersuchten Parameter waren: Die Jodkonzentration, die Gefäßgeometrie und die Unterkühlung. Im Falle der Sublimation ist die Transportrate diffusionsgesteuert, im Jod-assistierten Transportprozeß bei Konzentrationen über 1 mg/cm^3 steuert Konvektion den Prozeß, während bei niedrigeren Konzentrationen diffusionsgesteuerte Sublimation bedeutsam wird. Im allgemeinen ist das laterale Wachstum schnell, wenn chemischer Transport benutzt wird; der Habitus des Kristalls ist plättchenförmig, wobei die *c*-Achse normal zur Plättchenebene ist. Ein großer Anteil von Schraubenversetzungen wird auf den (0001)-Ebenen beobachtet und die Kristalle zeigen im allgemeinen n-Leitfähigkeit. Wenn die Sublimationstechnik benutzt wird, sind die Kristalle nicht ausschließlich plättchenförmig, die Mehrzahl der Kristalle zeigen nadelförmigen und Streifenhabitus. Bei Wachstum in der Längsrichtung ist die Wachstumsachse parallel zur *c*-Achse; es findet um eine Schraubenversetzung herum statt und es wird angenommen, daß zwei Kristallisationsmechanismen simultan wirken. Die Bildung einer großen Zahl zweidimensionaler Kerne zwischen den Zweigen der Schraubenversetzung scheint für den Abfall des mittleren Oberflächenwanderungsabstands X_s auf der Basisebene verantwortlich zu sein.

1. Introduction

Single crystals of the semiconducting compound GaS can be grown from the vapour phase in closed systems by employing either the iodine-assisted transport technique [1, 2] or the sublimation method [2, 3]. In either case the closed system consists of an evacuated quartz ampoule, placed in a suitable temperature gradient in a horizontal furnace.

In previous work [2, 3] the differences observed in the obtained crystals have been mentioned; most interesting are the differences in habit and in the type of conductivity. Products obtained by chemical transport (C.T.) have in general a conductivity which is n-type and they show a plate-like habit. Those grown in sublimation (S.G.) exhibit a p-type conductivity and are — dependent on the growth temperature — needle-like, ribbon-like, or in the shape of small plates (Table 1).

In this work the two techniques are compared; the composition of the vapour, the transport rate, and the habit as a function of crystallization temperature are discussed. Furthermore an attempt is made to explain the differences in the growth mechanisms and the morphology.

2. Experimental Procedure

All quartz ampoules employed are 180 mm in length with inside diameters ranging from 10 to 35 mm, dependent on the sort of experiment. They were cleaned, filled, evacuated (10^{-5} Torr), and sealed-off according to a procedure described elsewhere [2], then placed in a horizontal dual-zone furnace with independently heated, automatically controlled zones. The two zones can maintain their temperature within ± 2 °C; Pt–Pt 10% Rh thermocouples placed near both ends of the ampoules are used to control the zone temperatures. In all experiments the source region (T_H) was kept at 930 °C; for C.T. work 800 °C $<$ $< T_L <$ 870 °C, for S.G. work 750 °C $< T_L <$ 910 °C. In all experiments high-quality GaS powder (cf. [2] and [3]) and for C.T. doubly sublimated iodine was used. The amount of powder was 2 g per 56.5 cm³ tube volume, the time 16 to 17 h for C.T. experiments and 100 to 105 h for S.G.

3. Composition of the Vapour

The composition of the vapour phase is different for the two processes. In the sublimation process the vapour consists of GaS molecules [4, 5] and dissociation of GaS into its components is negligible as long as $T_L >$ 830 °C. At lower temperatures an appreciable amount of free gallium is deposited in the coolest part of the tube and Ga_2S_3 is formed in that part of the ampoule where the temperature is higher than $\approx$ 870 °C. In the iodine process on the other hand the vapour consists of several species taking part in the transport process. Previously [3] it has been assumed that in the reaction of solid GaS and the halogen, followed by condensation of GaS in the cold region, the intermediate gaseous products consist of sulphur and a mixture of gallium mono-, di-, and tri-iodide. The steps in the reaction consisted of the formation of S_2 and the gallium iodides at T_H, followed by transport of these vapour species to T_L where subsequently the exotherm disproportionation reaction occurs and GaI_3 is formed while GaS is deposited. Transport of GaI_3 back to the T_H-region completes the cycle. This assumption was confirmed by Rustamov and Geidarov's investigations [6]; according to them iodine starts to react with the source material at about 300 °C; above 700 °C the vapour consists of the species GaI and S_2 which are transported to the cold zone where in the disproportionation reaction GaS is deposited. An over-all reaction of the type

$$GaS_{(s)} + \frac{1}{2} GaI_{3\,(v)} \leftrightarrows \frac{3}{2} GaI_{(v)} + \frac{1}{2} S_{2\,(v)} , \qquad (1)$$

which can be proposed to be predominant, analogous to the case $GaAs:I_2$ [7], can however be rejected on account of the low value of the equilibrium constant K of this reaction (see Fig. 1).

Reactions in which gallium iodides are formed such as

$$GaS_{(s)} + \frac{3}{2} I_{2(v)} \leftrightharpoons GaI_{3(v)} + \frac{1}{2} S_{2(v)} , \quad (2)$$

$$GaS_{(s)} + \frac{1}{2} I_{2(v)} \leftrightharpoons GaI_{(v)} + \frac{1}{2} S_{2(v)} , \quad (3)$$

and

$$GaS_{(s)} + I_{(v)} \leftrightharpoons GaI_{(v)} + \frac{1}{2} S_{2(v)} , \quad (4)$$

have more favourable K-values (see Appendix) which make them sufficiently important to be considered.

Another important reaction to be considered, especially at low halogen concentrations, is the sublimation reaction

$$GaS_{(s)} \leftrightharpoons GaS_{(v)} . \quad (5)$$

The results of the work of Jona and Mandel on ZnS:HCl and $ZnS:I_2$ [8, 9] and the calculations of Arizumi and Nishinaga [10] have emphasized the fact that all possible reactions have to be considered when describing such chemical transport processes.

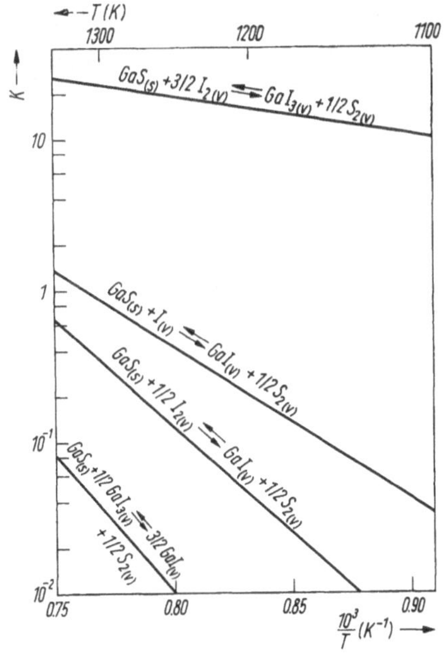

Fig. 1. Equilibrium constants K vs. $10^3/T$ calculated from thermodynamical data for the reactions (1), (2), (3), and (4). (See Appendix for calculations)

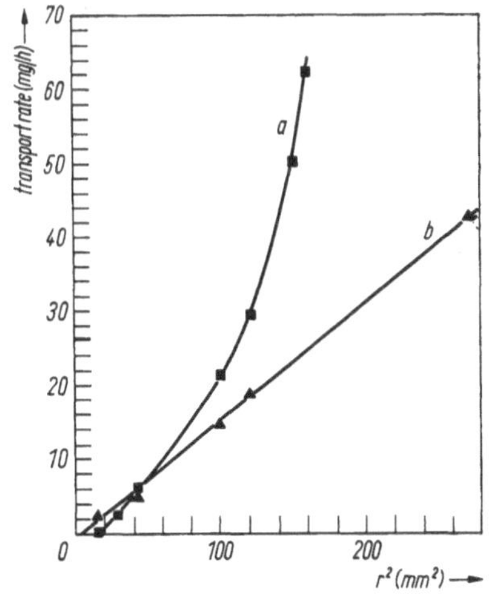

Fig. 2. Rate of transported GaS as a function of the square radius of the reaction ampoule. Tube length = = 180 mm; T_H = 930 °C; T_L = 850 °C. ■ iodine-assisted process; ▲ sublimation process

4. Transport Rate

The rate of transport was studied for both C.T. and S.G. as a function of ampoule geometry (Fig. 2), for C.T. as a function of the iodine concentration (Fig. 3), and for both processes as a function of undercooling, ΔT (Fig. 4). The dependence of the transport rate in mg/h on the square radius of the ampoule, as depicted in Fig. 2, shows a linear increase of curve (b) which is an indication that the rate is diffusion-controlled in the range 930 to 850 °C ($\Delta T = 80$ °C), while the strong increase of curve (a) indicates the increasing influence of convection.

No efforts were made to control nucleation; in the larger tubes (> 20 mm i.d.) the deposited material in C.T. was rather intergrown, with a large amount of the yield in microcrystalline form. Material transported in S.G. is always for more than 90% microcrystalline, condensed in the form of a crust on the tube wall. In 20 mm tubes the needles, ribbons, and plates grow on this crust, in wider tubes only an entangled mass of twinned ribbons is obtained. As a consequence of these results we employed in all later C.T. and S.G. work 20 mm tubes only.

In Fig. 3 the influence of the iodine content on the rate of transport is shown in a plot of the flux (mol/cm² s) of solid GaS collected in the growth region versus the amount of initially added iodine. It shows a minimum at 1 mg/cm³ for both types of tubes. The increase of the rate towards the left suggests that in the range of low concentrations diffusion is rate-determining, while at higher concentrations convection predominantes, although both transport mechanisms are operative. It seems that two different reactions occur: To the left of the minimum, reaction (5) and probably (4) seem to govern the rate of transport. As the formation of GaI_3 in the disproportionation reaction probably affects the equilibrium of (2), at the right of the minimum, reactions (3) and (4) govern the rate of transport. The observed increase in resistivity with decreasing iodine content [3, 11] and the change of conductivity type from n- to p-type when the amount of halogen is decreased [2], are in agreement with the idea that two

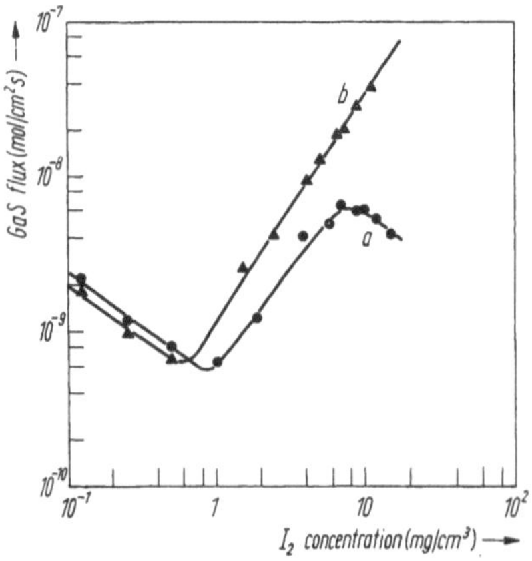

Fig. 3. Rate of transported GaS versus the iodine concentration for two sorts of ampoules. Curve (a) is for tubes with 11 mm i.d., curve (b) for tubes with 20 mm i.d. Tube length = 180 mm; $T_H = 930$ °C; $T_L = 850$ °C

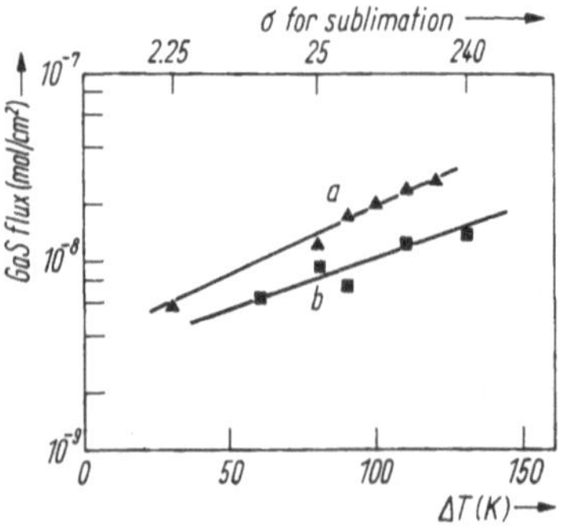

Fig. 4. Rate of transported GaS as a function of undercooling $\Delta T = T_H - T_L$; $T_H = 930$ °C. (a) Sublimation growth; (b) chemical transport. (GaS flux in mol/cm² s)

different reaction types are operative. The onset of convection gives the sharpest increase for (b), the decrease of (a) after a maximum at 6 mg/cm³ is not quite understood. A probable explanation is that the equilibrium of reactions (such as (3) and (4)) are increasingly displaced to the left with increasing pressures. This leaves reaction (2) to govern the rate. However the characteristic violet colour of free iodine has never been observed in our experiments, which makes the role of (2) doubtful. The reason that such an effect has not been observed in the case of curve (b), is probably due to a stronger convection in the 20 mm tubes than in the 11 mm tubes.

In Fig. 4 the transport rate is presented as a function of undercooling, ΔT. The flux is for S.G. material twice that of C.T. for an equal undercooling ΔT of 80 °C. The difference increases slightly with increasing ΔT. In C.T. for the growth of good quality crystals, ΔT never exceeded 60 °C. Higher ΔT values resulted in much intergrown material. However, the S.G. rate at $\Delta T = 30$ °C is equal to the C.T. rate at an undercooling $\Delta T = 60$ °C, whereas the yield of microcrystalline deposited material in S.G. is still strikingly larger than in C.T. As to the supersaturation σ in sublimation of GaS, an indication can be obtained from the slope of p_{GaS} versus $1/T$ [3, 5].[1]) No temperature-dependent vapour pressure can be calculated for the C.T. case owing to the lack of reliable information about the kind of processes taking place. The slope of sublimation, however, can be taken to be steepest; the vapour is saturated in the temperature range used, whereas in the halogen-assisted process there is no condensed iodide phase so that a much lower supersaturation is assumed in C.T. than in S.G.

Apart from such factors like transport rate and saturation the whole process of nucleation and growth probably affects the quality of the deposited GaS, and the steps considered to be of importance are:

a) the disproportionation of GaI into GaI_3 and gallium (as taking place in reactions (1), (3), and (4));

b) the formation of nuclei;

c) the subsequent diffusion and trapping of building units (Ga and S or GaS) on the growing surface.

For S.G., step a) is not relevant, step b) and c) are probably affected by the composition of the vapour.

The results discussed so-far show that with equal transport rate for both processes, the formation of only a few nuclei and the subsequent lateral outgrow into large single crystals is far more favoured in the halogen-assisted process than in the sublimation process.

5. Crystal Habit

The crystal habit of C.T. and S.G. material has been briefly described in a previous paper [2]. In C.T. experiments a decrease of the iodine concentration below ≈ 1 mg/cm³ results in an increasing amount of needle- and ribbon-like crystals; the conductivity is p-type analogous to that of sublimation-grown crystals.

Employing more than 1 mg/cm³, the first deposited plate crystals in C.T. grow flat along the tube wall, in a later stage of the process crystals grow out in all directions, not exclusively radially.

---

[1]) $\sigma = \alpha - 1 = (p - p_0)/p_0$ with p being the pressure at the evaporation temperature and p_0 the equilibrium pressure at the growth temperature.

S.G. crystals on the other hand assume various habits as a response to temperatures in the growth region and are not exclusively platelike, most of the material (> 90%) being microcrystalline like a crust on the tube wall. In a later stage a few of the crystallites grow out in the form of thin narrow ribbons, thin needles, small plates or rods, all dependent on T_L (see Table 1);

Table 1
Crystal habit as a function of temperature

	crystal growth forms in S.G.	growth temperature (°C)
thin small needles (straight and branched)	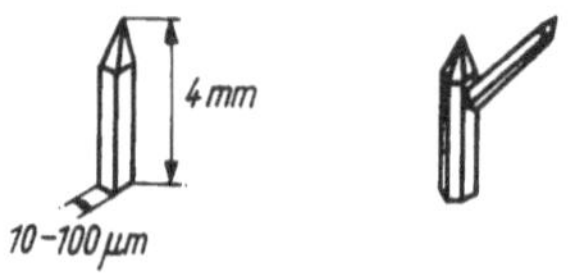	750 to 850
ribbons	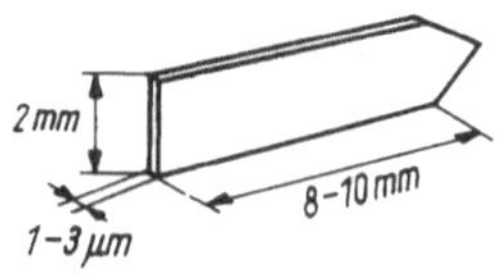	750 to 900
and		
twinned ribbons	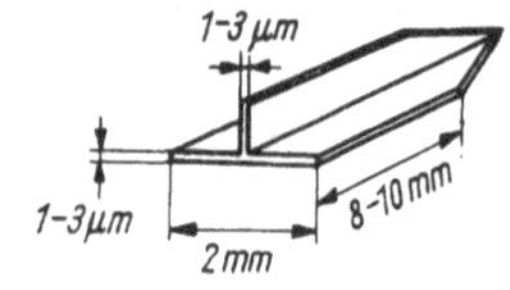	750 to 850 mostly in tubes >20 mm ∅
rods (straight and branched)	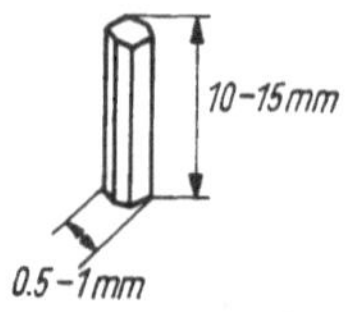	850 to 900
platelets	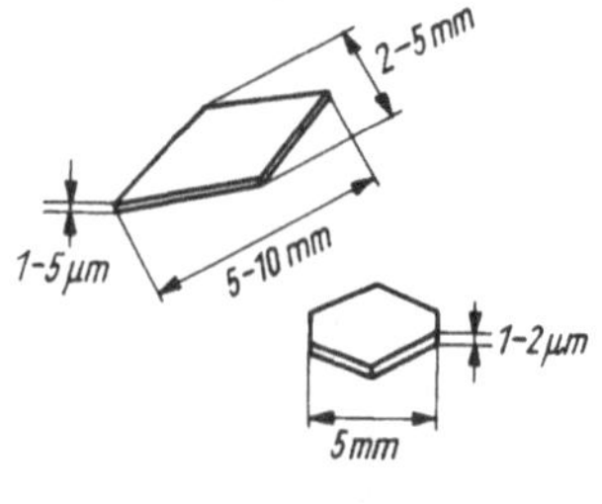	850 to 900
prisms and twinned prisms	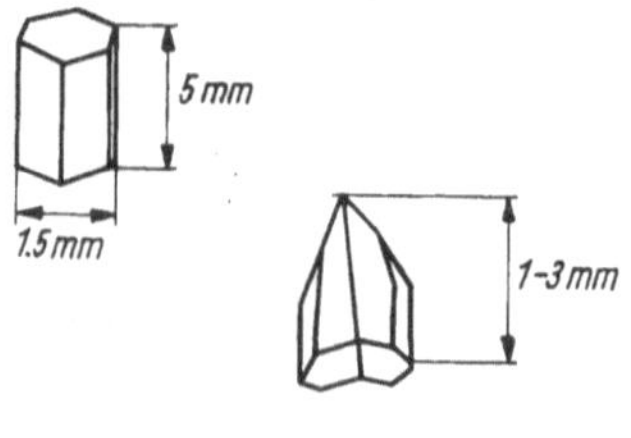	>900

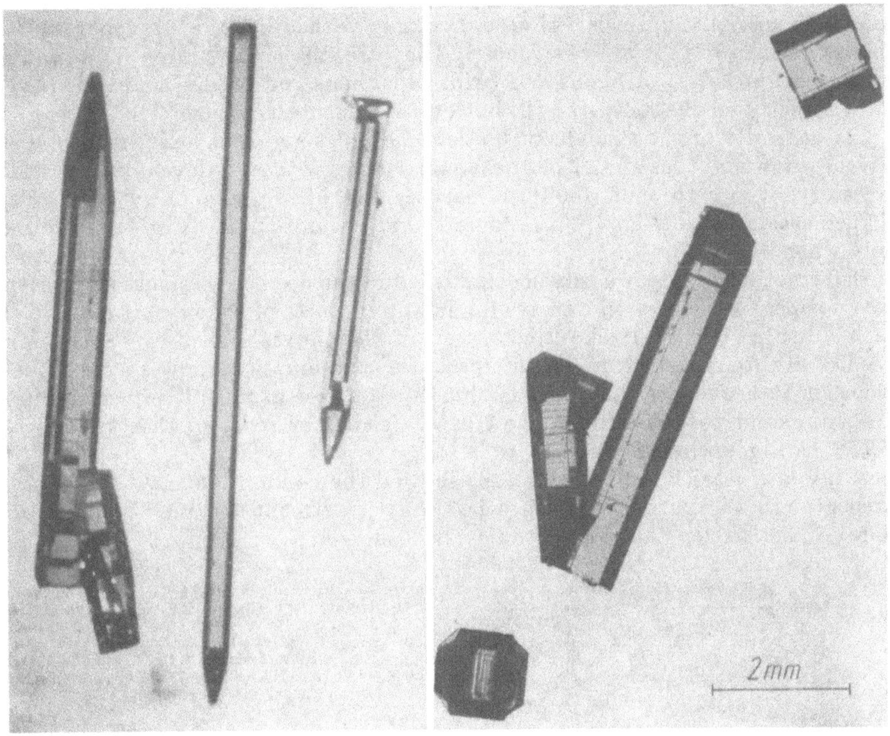

a b

Fig. 5. a) GaS needles with sharp tips grown in sublimation process. b) Branched GaS needle from sublimation process

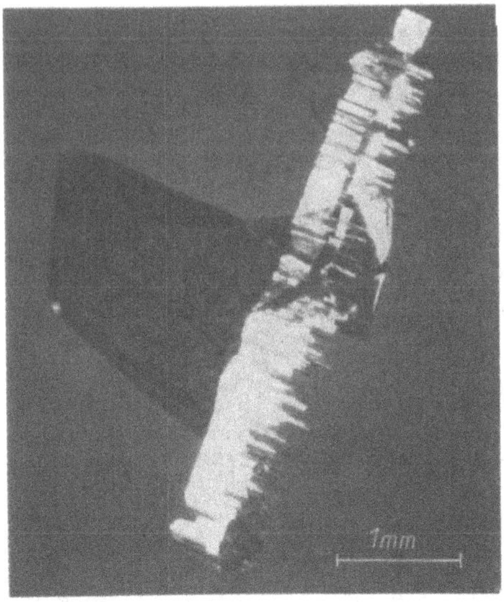

Fig. 6. Sublimation-grown rod with irregularly formed prism faces. Rod and platelet are intergrown

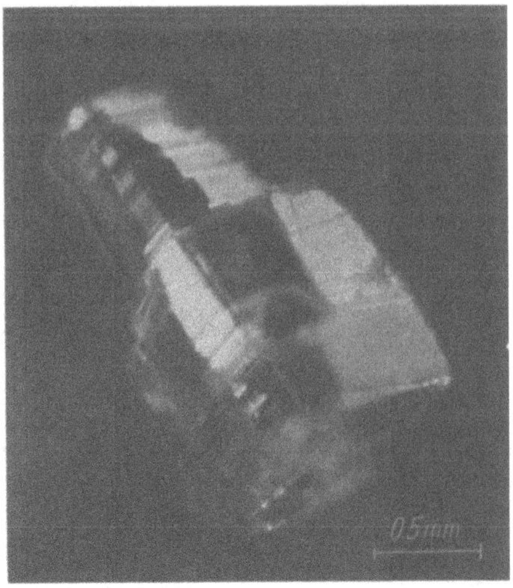

Fig. 7. Twinned GaS rod from sublimation

growth is radially inwards. However, plates with dimensions comparable to those obtained in C.T. are rare; mostly they are smaller and often found intergrown with needles. Ribbons and twinned ribbons and needles and rods never grow directly on the tube wall [2] but on the first formed substrate.

The majority of the straight or branched needles and rods have smooth, well developed prism faces and either pencil-sharp tips with well developed pyramid faces or flat-growth ends ((0001) faces) (see Fig. 5). Specimens with irregular shaped prism faces (Fig. 6) are seldom found, twinned rods (Fig. 7) are found only when $T_L > 900$ °C.

Deformation by slip is easily accomplished in the needles; pressure on a prism face normal to the growth axis results in slip of parts of the crystal relative to each other more or less like a collection of thin hexagonal discs. Sometimes needles are found which under microscopical examination show a "tube-like" inclusion in the length direction (cf. Fig. 8).[2]) Never in S.G., however, gallium droplets could be detected on the tips of needles or rods, so that growth by V.L.S. can be excluded.

X-ray analyses (Laue and rotation) showed the column-like crystals to have their growth axis parallel to their c-axis; heat treatment and high-temperature X-ray investigations have not been carried out yet.

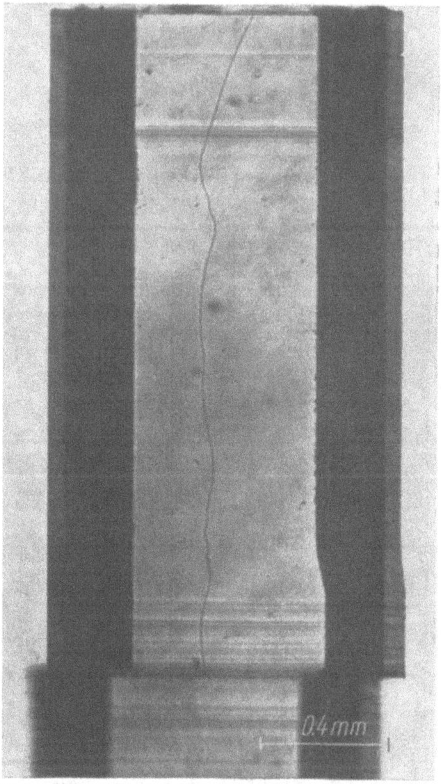

a

Fig. 8. a) GaS needle having a "tube-like" inclusion b) Microphotograph (interf. contrast) of top surface (0001) of the needle crystal of Fig. 8a. The cleavage step starts at the point where the "tube-like" inclusion emerges (marked ×) and runs to the edge of the crystal. The black "circle" is the "tube-like" inclusion spiraling into the crystal and emerging at ×

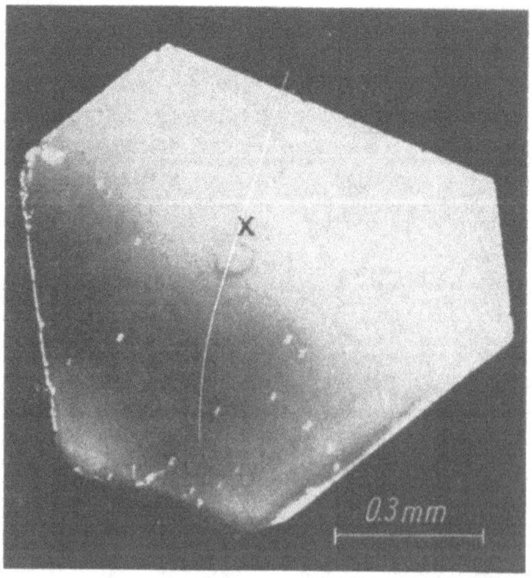

²) Precipitation of an unknown impurity probably decorates the screw dislocation and gives as a result the "tube-like" inclusion.

6. Growth Mechanism and Morphology

In this part an attempt is made to explain the habit differences mentioned. Using the Hartman-Perdok theory [12] the layer-like growth of GaS as obtained in C.T. is understandable. These authors show — on an energy basis — the relation between crystal structure and morphology. Crystal morphology is governed by chains of strong bonds running through the structure; such a chain of strong bonds is called a periodic bond chain (P.B.C.). This theory allows the classification of crystal forms in three classes with decreasing morphological importance: F-faces — flat faces, containing two or more coplanar P.B. chains in a "slice" of thickness d_{hkl}, S-faces — stepped faces, containing one P.B.C. per "slice", and K-faces — kinked faces, containing no P.B.C. The thickness d_{hkl} of such a "slice" is the minimum thickness of a growth layer and is determined by the periodicity of the surface layer; such a "slice" has the same composition as the crystal.

Environmental factors may have a special effect on this structure–morphology relationship; according to Sunagawa [13] the nature of F-, S-, and K-faces can be varied in response to those environmental factors. Impurities, for example, may act like kink poisons; a change in temperature may alter the relative growth rate of F-faces without inducing S- or K-faces [14]. The GaS structure can be described as a stacking of slices one on top of the other where the (0001) face, parallel to three bond chains, is an F-face. The projection on the (0001) face (Fig. 9a) shows those three chains, rotated over 60° relative to each other. The bond chains manifest themselves as straight parts of growth-layer edges, which is demonstrated in Fig. 10. Those edges have been determined by single-crystal X-ray analysis to be parallel to the $\{10\bar{1}0\}$ faces.

A projection of the GaS structure on a $(11\bar{2}0)$ face or on a $(10\bar{1}0)$ face, however, reveals no continuous bonds in the [0001] direction (see Fig. 9b) owing to the fact that the compound has a layered structure. Therefore the prism $\{10\bar{1}0\}$ faces are S-faces.

Lateral growth during C.T. is thus understandable; the building units are absorbed, they migrate over the surface and are trapped at kinks. In this way the S-faces grow rapidly normal to themselves which results in the growth of large plate-like crystals with smooth (0001) faces.

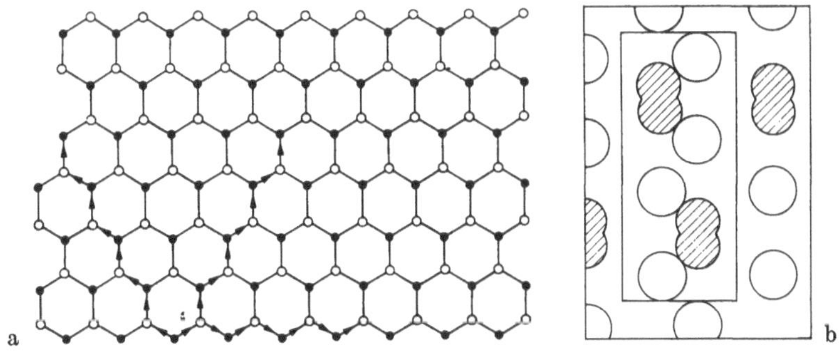

Fig. 9. a) Schematic illustration of the GaS structure projected on (0001). The direction of the three bond chains are indicated by arrows. ● S, ○ Ga. b) Schematic illustration of the GaS structure. Projection of two fourfold slabs on $(11\bar{2}0)$. Empty circles: S, hatched circles: Ga

Fig. 10. A GaS platelet grown in the sublimation process showing straight growth layer edges. The hexagonal terraces are the onset of needle growth

According to theory initiation of new growth layers is difficult on the smooth F-faces and screw dislocations presenting permanently available steps remove this difficulty. Growth normal to the basal plane is thus predicted to take place on screw dislocations.[3])

C.T.-grown flakes indeed show an abundance of hexagonal and rounded-off growth spirals; step heights in the order of 500 to 1500 Å have been found for such spirals [15].

Compared with the experiments of Kaldis on CdS [16] we would say that the growth of large plate-like GaS crystals in C.T. is not the result of a poisoning effect of iodine. Neutron activation data [4]) revealed a rather low iodine content in C.T.-grown crystals (< 10 ppm) [17].

In S.G., however, the growth of the (basal) F-faces normal to themselves exceed the growth rate of the (prism) S-faces normal to themselves. The pronounced growth in the length direction may in principle be explained on the basis of environmental influences like temperature changes or the presence of impurities [13, 14], or by factors which have a stimulating effect on growth at the tip, such as the presence of defects like screw dislocations or a decrease of the mean surface migration distance X_s of the building units. In considering this, the following experimental facts are of importance.

a) Interference contrast microscopy has revealed steps on the top faces of all needles and rods which were cleaved parallel to the (0001) face. On the other hand, small S.G. platelets have been found on which, even under the largest possible magnification, no screw dislocation could be observed on the basal (0001) face.

b) Although the initial high supersaturation has decreased after the deposition of the microcrystalline crust in the cold region, the value of σ is still high.

c) The role played by both heat conductivity and impurities is incompletely understood. A low heat conductivity favours the growth of needle-like crystals. This was observed by Kremheller [18] in experiments on ZnS. For the hexagonal GaS crystals the growth axis of the needles and rods is parallel to the

[3]) In Jackson's theory the atomically smooth interface is explained with the help of a high entropy of fusion, ΔS, which gives $\alpha > 2$.

[4]) Carried out at the R.C.N. Laboratories at Petten, Netherlands.

c-axis, which is the direction of highest heat dissipation. It is generally accepted that impurities can initiate screw dislocations which eventually can cause elongated growth. In our case this means in fact a variation in the nature of the so-called F- and S-faces [13].

Although one only can speculate as to what happens, it seems reasonable to suggest that the observed growth in the length direction may be explained in terms of a difference in the mean displacement X_s over the basal plane, whereas in C.T., growth units obviously migrate over considerable distances before they are built in at crystal edges, resulting in lateral growth of large flat plate crystals, and that growth normal to the basal plane needs a screw dislocation, whereas in S.G. the building units do not migrate over the (0001) plane over distances comparable to those in C.T. during needle growth. We assume that in S.G. two crystallization mechanisms can occur simultaneously; these are the crystallization by surface diffusion and subsequent integration of the building units into the kinks of spiral steps, and the crystallization by formation of two-dimensional nuclei between the arm of the steps.[5]) Those two-dimensional nuclei are not expected to be integrated very rapidly. From the expression for X_s derived by Burton et al. [19],

$$X_s = (D_s\, t_s)^{1/2} = A \exp \frac{W_s - U_s}{2\, kT},$$

where D_s is the surface diffusion coefficient, t_s is the mean life of an absorbed molecule before being evaporated again, W_s is the evaporation energy of an isolated molecule on the crystal surface, U_s is the activation energy for diffusion between two neighbouring equilibrium positions on the surface which are a distance A appart from each other, one can see that surface diffusion can play an important role when $X_s > A$ or $W_s > U_s$. This implies that X_s decreases with increasing T and increasing U_s. The fact that needles grow on plates can then be explained as a result of a change in the basal X_s; on the initially formed platelets growth in the length direction can start suddenly on account of temperature fluctuations in the growth region. This causes fluctuations in the supersaturation and hence in the formation of two-dimensional nuclei between the spiral steps. The change of lateral into needle growth is thus associated with a decrease in the value of X_s. This disturbance of the surface diffusion mechanism can be caused by the nuclei; if their number, condensated between the arms of the screw, is large, due to a large σ-value, it can disturb the surface diffusion [20]. At higher temperatures in the growth region (see Table 1), an increased amount of platelets is observed as well as an increase in the cross section of the needles.[6]) We suppose that at higher values of T_L the number of nuclei condensed on the spirals decrease owing to a decrease of σ. In that case the disturbing effect

---

[5]) The fact that many small S.G. platelets have been observed on which no screw dislocations could be detected, suggests that crystallisation by formation of two-dimensional nucleation can be the main mechanism in plate growth.

[6]) With higher values of T_L the plates obtained have larger areas, and on the needle-bearing plates the change from plate to needle can be seen to go via a series of hexagonal terraces which are slowly decreasing in area until eventually the cross section of the needles is reached. From then on the prism faces are smooth and in general a constant cross section is maintained. The change from lateral growth to needle growth is thus not so sudden as the change observed at lower T_L-values.

caused by those nuclei decreases also, so that X_s can increase with higher T_L — an observation seemingly in contradiction with the B.C.F. relation.

It must be noted that the growth of twinned prisms is explained by a mechanism which is quite different from that given for the common needles and rods [21]. However, in this case $T_L > 900$ °C (Table 1), with means a quite different values for the relative supersaturation of the vapour.

GaSe

Some additional remarks can be made about the analogous compound GaSe:

1. The two techniques for growth from the vapour have also been employed for GaSe. The crystals obtained show the same differences in habit as observed in the case of GaS; chemical transport gives large flat plates with hexagonal and triangular growth lines, while the sublimation method gives thin triangular plates and triangular needles [22]. The two temperatures T_H and T_L used in S.G. work do not differ much from those used for GaS.

2. Although parameters like tube geometry, iodine concentration, and undercooling ΔT have not yet been fully studied, it is reasonable to assume that analogous results concerning convection, ampoule diameter, and transport rate will be obtained. Nothing is known about the degree of dissociation of GaSe vapour either.

3. C.T.-grown GaSe flakes examined under large magnifications ($\geqq 400\times$) show contamination with free gallium in the form of small droplets, condensed along the growth lines as a kind of decoration. This phenomenon has also been observed in the case of C.T.-grown flakes of the sulpho-selenides $GaSe_{1-x}S_x$ in the range $0 \leqq x \leqq 0.3$.

4. The type of conductivity is also different for the two methods. C.T.-produced flakes proved to be n-type with 4 mg/cm³ of iodine; a change into p-type is observed when lower iodine concentrations are used. The exact limit is not yet investigated. Sublimation-grown plate-like material is p-type.

7. Conclusions

Employing C.T., large n-type single crystals can be grown as long as the halogen concentration exceeds ≈ 1 mg/cm³. Above this concentration the rate of transport in C.T. seems to be convection-determined, while at lower concentrations the influence of diffusion becomes important. With decreasing iodine content, the situation becomes comparable to sublimation and the crystals become p-type. The habit of C.T.-grown crystals is exclusively plate-like as long as the halogen concentration is above the 1mg limit; needle-like crystals can be obtained at lower concentrations and the dimensions of the plates decrease.

Nucleation and growth in C.T. seem to be favourable to enable a few crystallites to grow out; heat dissipation, the type of building units, and the degree of supersaturation favour nucleation control and the subsequent lateral outgrowth into large plates.

The S.G. technique is rather unfavourable for growing plate-like crystals, most of the yield being a microcrystalline condensate with only a small amount of suitable p-type crystals per batch. This technique is only of interest because needle-like crystals can be obtained which enable electrical and optical work to be carried out in a direction normal to the c-axis [23]. The rate of transport in S.G. is higher than in C.T. and completely diffusion-controlled in the temperature ranges used.

Habit differences as a function of temperature of the growth zone are observed in S.G. Two crystallization mechanisms are assumed to occur simultaneously; two-dimensional nuclei are formed on the spiral steps. The formation of a large number of such nuclei is probably responsible for a decrease in the mean surface migration distance X_s of the basal plane so that length grow becomes possible.

Acknowledgements

The author thanks Prof. Dr. E. Mooser from the Ecole Polytechnique of Lausanne and Dr. P. Hartman from the Department of Mineralogy and Crystallography of the Leyden University for their interest and for the stimulating discussions. Thanks are also due to Mr. C. A. Copray for this assistance with microphotography and to Mr. G. E. van Egmond for his assistance with the experiments.

Appendix

Using the relation

$$-R\,T \ln K = \Delta G\,, \tag{A1}$$

the equilibrium constants of the reactions (1) to (4) can be calculated from available data. Enthalpy and entropy values for the gallium iodides are taken from reference [24], the values for GaS, iodine, and sulphur are taken from references [5], [25], and [26].[7]) No heat capacity data are available to calculate the temperature-dependent terms under the integral signs. On the other hand, these two terms will cancel each other in a great many cases, since they have opposite signs and since ΔC_p is usually small (see [27] and [28]), so that it is possible to use the approximate formula

$$\Delta G_T = \Delta H_{298} - T\,\Delta S_{298}\,. \tag{A2}$$

For the reactions (1) to (4) the following ΔH_{298} and ΔS_{298} values are obtained:

	(1)	(3)	(4)	(2)
ΔH_{298} (kcal/mol)	91	64.5	46.5	11.5
ΔS_{298} (cal/deg mol)	63.5	47.5	35.5	15.5

This results in the following relations:

$$\left.\begin{aligned} K_{(1)} &= -\frac{19913}{T} + 13.9\,, & K_{(3)} &= -\frac{14114}{T} + 10.4\,,\\ K_{(4)} &= -\frac{10175}{T} + 7.8\,, & K_{(2)} &= -\frac{2516}{T} + 3.3\,, \end{aligned}\right\} \tag{A3}$$

which are presented in a K vs. $10^3/T$ plot in Fig. 1.

[7]) Above 850 °C, sulphur vapour consists almost entirely of S_2 [26], above 600 °C the dissociation of iodine becomes appreciable [29].

References

[1] H. J. M. HEIJLIGERS, T.H.E. Rep. (in Dutch).
[2] R. M. A. LIETH, C. W. M. VAN DER HEIJDEN, and J. W. M. VAN KESSEL, J. Crystal Growth **5**, 251 (1969).
[3] R. M. A. LIETH, Thesis, Eindhoven 1969.
[4] R. M. A. LIETH, H. J. M. HEIJLIGERS, and C. W. M. VAN DER HEIJDEN, J. Electrochem. Soc. **113**, 798 (1966).
[5] R. M. A. LIETH, H. J. M. HEIJLIGERS, and C. W. M. VAN DER HEIJDEN, Mater. Sci. Engng. **2**, 193 (1967).
[6] P. G. RUSTAMOV and B. A. GEIDAROV, paper given at Conf. Crystal Growth and Epitaxy from the Vapour Phase, Zürich 1970.
[7] V. J. SILVESTRI and V. J. LYONS, J. Electrochem. Soc. **109**, 963 (1962).
[8] F. JONA and G. MANDEL, J. Phys. Chem. Solids **25**, 187 (1964).
[9] F. JONA, J. Phys. Chem. Solids **23**, 1719 (1962).
[10] T. ARIZUMI and T. NISHINAGA, Japan. J. appl. Phys. **5**, 21 (1966); **5**, 588 (1966); **4**, 165 (1965).
[11] R. M. A. LIETH and F. VAN DER MAESEN, phys. stat. sol. (a) **9**, 483 (1972).
[12] P. HARTMAN and W. G. PERDOK, Acta cryst. **8**, 49 (1955); **8**, 521 (1955); Z. Krist. **119**, 65 (1963).
[13] I. SUNAGAWA, Acta cryst. (Suppl.) **25 A**, S_3, S238 (1969).
[14] P. HARTMAN, Acta cryst. (Suppl.) **25 A**, S_3, S237 (1969).
[15] H. U. BOELSTERLI and E. MOSER, Helv. phys. Acta **35**, 538 (1962).
[16] E. KALDIS, J. Crystal Growth **5**, 376 (1969).
[17] J. C. J. M. TERHELL and R. M. A. LIETH, phys. stat. sol. (a) **5**, 719 (1971).
[18] A. KREMHELLER, Sylvania Technologist **8**, 11 (1955).
[19] W. K. BURTON, N. CABRERA, and F. C. FRANK, Phil. Trans. Roy. Soc. (London) **243**, 209 (1951).
[20] P. BENNEMA, R. KERN, and B. SIMON, phys. stat. sol. **19**, 211 (1967).
[21] E. MOOSER, M. SCHLÜTTER, and R. M. A. LIETH, J. Crystal Growth, to be published.
[22] J. C. J. M. TERHELL and R. M. A. LIETH, J. Crystal Growth, to be published.
[23] A. H. M. KIPPERMAN and G. A. VAN DER LEEDEN, Solid State Commun. **6**, 657 (1968).
[24] E. SIRTL, Z. Naturf. **21a**, 2001 (1966).
[25] O. KUBASCHEWSKI, E. L. L. EVANS, and C. B. ALCOCK, Metallurgical Thermochemistry, 4th edition, Pergamon Press, Oxford 1967.
[26] H. BRAUNE, S. PETER, and V. NEVELING, Z. Naturf. **6a**, 32 (1951).
[27] F. E. C. SCHEFFER, Toepassingen van de thermodynamica op chemische processen, Waltman, Delft 1945 (p. 38).
[28] J. D. FAST, Interaction of Metals and Gases, Vol. I, Philips Technical Library, Eindhoven 1968.
[29] H. SCHÄFER, H. JACOB, and K. ETZEL, Z. anorg. allg. Chem. **286**, 42 (1956).

(Received February 23, 1972)

Note added in proof

Computer calculations on the transport rate in the system GaS–I_2 carried out by Dr. T. Nishinaga of the Nagoya University, Japan, are in good agreement with the results shown in Fig. 3. These theoretical results will be published in the future.

phys. stat. sol. (a) **12**, 413 (1972)

Subject classification: 18.2.1

Materials Section, Department of Electrical Engineering, Imperial College, London

A Second Domain Wall Parameter Measurable by Lorentz Microscopy

By

C. G. Harrison and K. D. Leaver

The measurement of domain wall parameters in thin magnetic films by conventional Lorentz electron microscopy in the Fresnel mode is hampered by the necessary similarity of all domain wall models, when viewed along the normal to the film plane. The flux density normal to the film is found to be more strongly characteristic of the wall structure. It is shown that by tilting the film with respect to the electron beam the theoretical Lorentz image is altered in such a way that the components of normal flux can be investigated quantitatively.

Die notwendige Ähnlichkeit aller Bereichswände in dünnen magnetischen Schichten bei Beobachtung senkrecht zur Schichtebene behindert die Bestimmung ihrer Parameter durch die defokussierte Lorentzelektronenmikroskopie. Die Betrachtung der senkrechten Flußdichte ergibt jedoch eine genauere Charakterisierung der inneren Struktur. Es wird weiter gezeigt, daß durch eine Kippung der Schicht das theoretische Lorentzbild so verändert wird, daß die senkrechte Flußdichte quantitativ untersucht werden kann.

1. Introduction

Notwithstanding the success of Lorentz electron microscopy as an observational technique, it has been found disappointingly ineffective in providing detailed information on the microstructure of magnetic objects. Chief among the objects of interest are magnetisation ripple, domain walls and stripe domains. The purpose of this paper is to discuss the use of Lorentz microscopy in investigating the two-dimensional nature of domain walls and to propose a new and simple quantitative measurement of one of their features.

Unfortunately, the conventional, defocussed Lorentz image calculable from the most recent, two-dimensional models [1, 2] is not significantly different, (as will be shown) except in respect of the effective wall width, from that of earlier one-dimensional models. Other differences are too small to be easily perceived owing to 'noise' introduced by a finite electron source size, inelastic scattering, the Fresnel interference fringes, and by the photographic plate. Indeed the only parameter to be measured reasonably accurately with Lorentz microscopy is the effective wall width. Wall widths measured experimentally correlate [3 to 6] broadly with those obtained from a number of theoretical studies, indicating the general validity of the latter, but providing no means of distinguishing between them.

To differentiate the structures of domain walls, still more those of stripe domains, Lorentz microscopy must develop new ways of coming to grips with them.

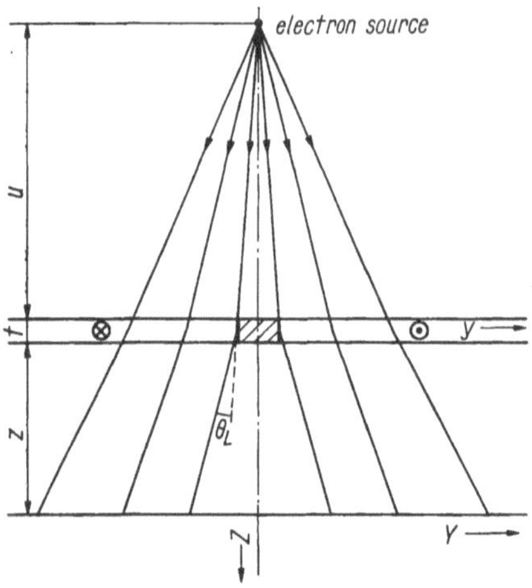

Fig. 1. Geometry of the Fresnel mode of Lorentz microscopy

2. Lorentz Images of Domain Walls

Domain wall studies by Lorentz microscopy are usually carried out in the defocussed or Fresnel mode, illustrated in Fig. 1. The intensity profile $I(Y)$ of the image observed in the XY plane may be described theoretically by wave optics [7] or geometrical optics [8]. The validity of geometrical optics in this respect has been the subject of much discussion [9, 10]. It will be convenient here to discuss the problem in terms of geometrical optics, since this gives a clearer understanding of the object–image relationship, but it is stressed that all the profiles given as examples have been calculated using wave-optical theory, as outlined in the Appendix.

The object for imaging by Lorentz microscopy is the flux density normal to the electron beam, integrated along its path. In the arrangement of Fig. 1, this is effectively the flux density B_x, since a component B_y could only cause an x-axis shift of the image which is undetectable. A function $A_x(y)$ may be defined, such that

$$A_x(y) \equiv \frac{1}{4\pi M_s t} \int_{\text{path}} B_x(y, z)\,\mathrm{d}z\,, \tag{1}$$

where M_s is the saturation magnetisation of the specimen. $A_x(y)$ is the normalised object function for this arrangement. If we assume that there is no x-component of stray field H_x, (1) may be simplified to

$$A_x(y) = \frac{1}{M_s t} \int_0^t M_x(y, z)\,\mathrm{d}z\,. \tag{2}$$

The normalised intensity profile $I(Y)$ as described by geometrical optics may thus be written in the form

$$\left.\begin{aligned} I(Y) &= \frac{1}{1 + z\,\theta_L(\mathrm{d}A_x/\mathrm{d}y)}\,, \\ Y &= y + z\,\theta_L\,A_x(y)\,. \end{aligned}\right\} \tag{3}$$

θ_L is the Lorentz deviation angle,

$$\theta_L = 4\,\pi\,M_s\,t\,\lambda\left(\frac{e}{h}\right), \tag{4}$$

where λ is the electron wavelength, e the charge of the electron, and h is Planck's constant.

If the transformation described by equations (3) could be carried out without loss of information, then the resulting image would contain a full description of $A_x(y)$. In practice this transformation cannot be performed as suggested, first because wave optical effects necessarily limit the resolution of $A_x(y)$ which can be obtained and secondly because of several practical limitations [11]. Loss of coherence in the beam, an extended source size, spherical aberration of the objective, and inelastic scattering within the specimen all combine to obscure information in the image. These effects can be minimised by careful selection of the imaging conditions and it has been shown [10, 12] that wave optical phenomena have least effect on the object transformation in the case of divergent (dark) wall image.

3. Distinguishing between Wall Models

To illustrate the problem of learning more about a wall structure from its Lorentz image, we may examine the theoretical images of two complex models due to Hubert, which are shown in Fig. 2 as magnetisation vector maps. These are typical of recent two-dimensional models as published by Hubert and LaBonte which are theoretically well justified.

In Fig. 3 are shown the $A_x(y)$ curves for these two models and it will be seen that despite the considerable differences in the wall structures, there is little difference of shape in the object functions. The chief difference of shape lies in the slight asymmetry of the Bloch-type curve. On this basis, even if the ideal

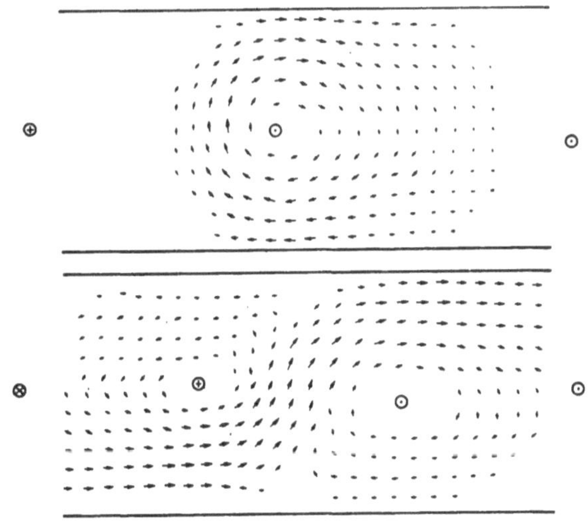
a

b

Fig. 2. Magnetisation vector maps of a) asymmetric Bloch- and b) asymmetric Néel-wall models due to Hubert

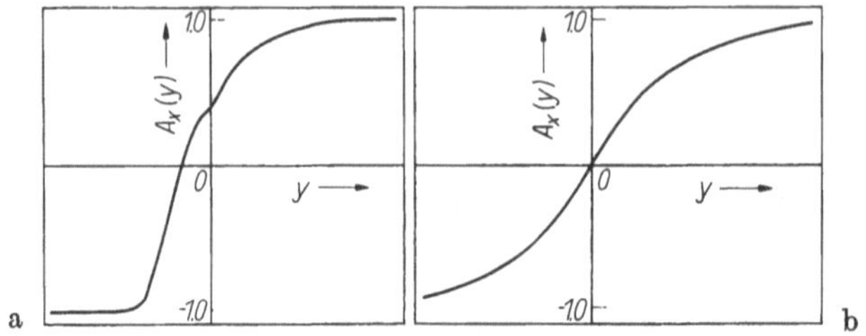

Fig. 3. Conventional magnetic object functions $A_x(y)$ for the models of Fig. 2. a) Bloch-type model; b) Néel-type model

transformation by equations (3) were to hold, it would not be easy to distinguish between the two models (except as to their effective wall widths), still less to gain insight on their complex vortex structures. This is illustrated in Fig. 4 by their divergent image intensity profiles calculated under identical, realistic, conditions (see Appendix). The large peak on the Néel-type image is anomalous and results from the model not representing a full 180° wall. Apart from differences due to the differing effective wall widths, only a slight asymmetry of the Bloch-type image distinguishes the two. Certainly it is impossible from these images to learn anything about the structures (vortex or otherwise) that lie behind them.

The failure of Lorentz microscopy as an analytical technique lies here, in the fact that the $A_x(y)$ curve does little to characterise the internal structure of a domain wall.

4. The M_z-components

Differences between the Bloch-type and Néel-type models of Fig. 2 become more distinct when we consider the z-components of the flux B. We define the curve $A_z(y)$ by the equation

$$A_z(y) \equiv \frac{1}{4\pi M_s t} \int_{\text{path}} B_z(y, z) \, dz . \tag{5}$$

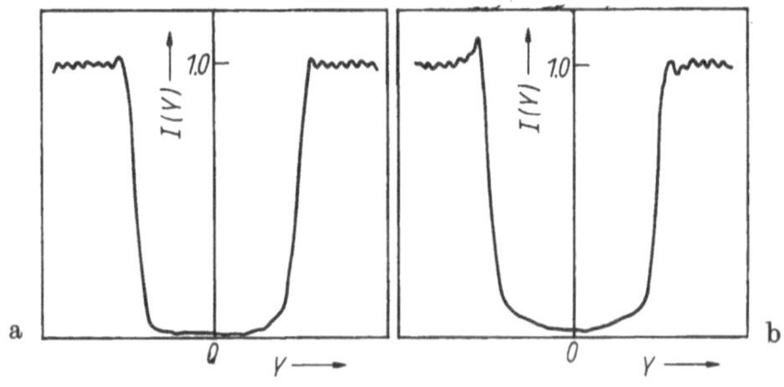

Fig. 4. Divergent image intensity profiles calculated from the $A_x(y)$ functions of Fig. 3. a) Bloch type; b) Néel type

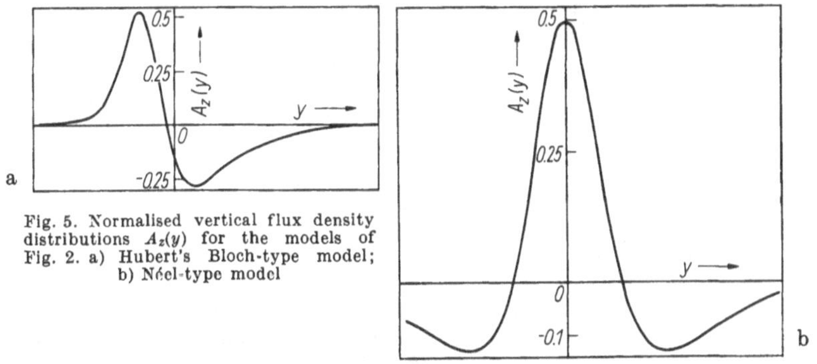

Fig. 5. Normalised vertical flux density distributions $A_z(y)$ for the models of Fig. 2. a) Hubert's Bloch-type model; b) Néel-type model

Since we have chosen stray-field-free models in which we may replace B_z by M_z this is simplified to

$$A_z(y) = \frac{1}{M_s t} \int_0^t M_z(y, z) \, dz \,. \tag{6}$$

The $A_z(y)$ curves for the two models are shown in Fig. 5, where it is seen that they correspond to the two possible symmetries for this flux curve. It follows from Maxwell's equations that

$$\int_{-\infty}^{\infty} A_z(y) \, dy = 0 \,. \tag{7}$$

That is, flux closure must take place. As a broadly symmetric function, the simplest forms available for $A_z(y)$ are those resulting from the two Hubert models illustrated. It remains only to cause the electron beam to interact with the M_z-components of a domain wall, for images more characteristic of the internal structure to appear.

5. Tilting the Specimen

Most electron microscopes are equipped with facilities for tilting the specimen under observation, as this has long been a standard technique in metallurgical studies. Modern side entry stages can permit tilt angles quite close to the vertical. It is possible to extend these facilities to the Lorentz stage and so to consider tilting a magnetic specimen [13]. Taking the coordinate axes x and y as the possible tilt axes, there are three sets of manipulations possible.

(i) Tilting about the x-axis; this changes the beam path through the specimen and effectively modifies the $A_x(y)$ function. The net result is to increase the effective wall width and the effective specimen thickness as seen by the beam.

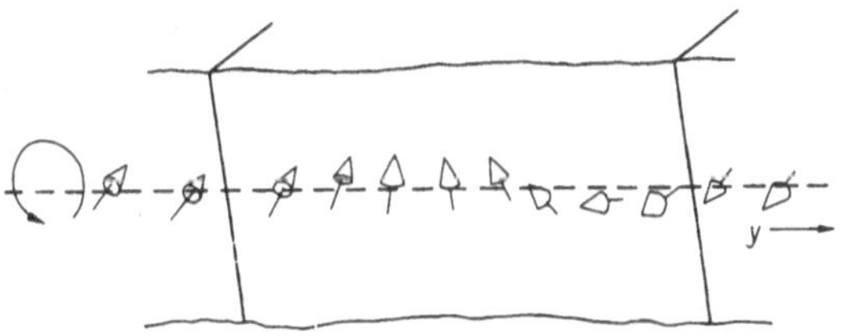

Fig. 6. Pictorial representation of a one-dimensional model illustrating tilt axis

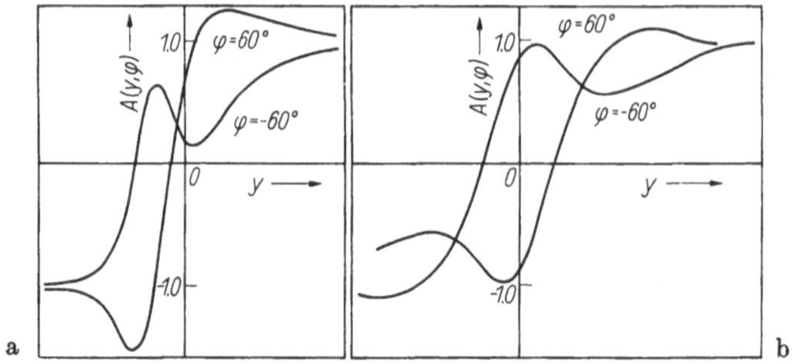

Fig. 7. Magnetic object functions $A(y, \varphi)$ for the models of Fig. 2 when tilted to $\varphi = \pm 60°$. a) Hubert's Bloch-type model; b) Néel-type model

(ii) Tilting about the y-axis; as is illustrated in Fig. 6 (which represents a one-dimensional wall model), tilting through an angle φ about the y-axis, brings a fraction $\sin \varphi$ of the B_z-components normal to the beam and reduces the projection of the B_x-component normal to the beam in the ratio $\cos \varphi$. At the same time the effective specimen thickness is increased in the ratio of $\sec \varphi$. Thus the net object function $A(y, \varphi)$ is given by

$$A(y, \varphi) = A_x(y) + A_z(y) \tan \varphi \tag{8}$$

for tilting about the y-axis.

(iii) Compound tilting about both axes combines the effects of (i) and (ii) and also brings the B_y-components into useful interaction with the beam.

Though the manipulations (i) and (iii) may eventually prove useful, method (ii) offers immediately a simple means of adding to the object function information which characterises the wall structure much better than $A_x(y)$ alone. In Fig. 7 are shown the $A(y, \varphi)$ curves for both the Bloch-type and Néel-type models when $\varphi = \pm 60°$. (N. B. There is an ambiguity of sign here, since reversing the signs of the M_y- and M_z-components of a domain wall yields an equally valid structure.) There are now marked differences between them. The Bloch-wall function for one direction of tilt changes rapidly from $|A| = 1$ and acquires a central hump, while in the opposite direction it acquires peaks on both sides. This behaviour is characteristic of the odd symmetry of the $A_z(y)$ curve. The function for the Néel-type wall has a symmetric behaviour, acquiring in both directions of tilt, a peak on one side and a central hump. This is characteristic of even symmetry in the $A_z(y)$ function.

6. Image Profiles of Tilted Specimens

To determine whether these object function characteristics carry over into the Lorentz image, we have performed calculations of image intensity profiles. Profiles were obtained for both Néel-type and Bloch-type models, imaged under identical, realistic conditions, when tilted about the y-axis at angles up to 60°, at which point the effective specimen thickness is doubled. Examples of these are shown in Fig. 8 (cf. Fig. 4). It can be seen that some of the information contributed by the $A_z(y)$ curve has carried over. For the Bloch-type image (Fig. 8a), the profile falls away sharply from $I(Y) = 1$ for $\varphi = -60°$ and acquires enhanced diffraction fringes for $\varphi = 60°$ showing good correspondence with the $A(y, \varphi)$

Fig. 8. Divergent image intensity profiles calculated from the $A(y, \varphi)$ curves of Fig. 7

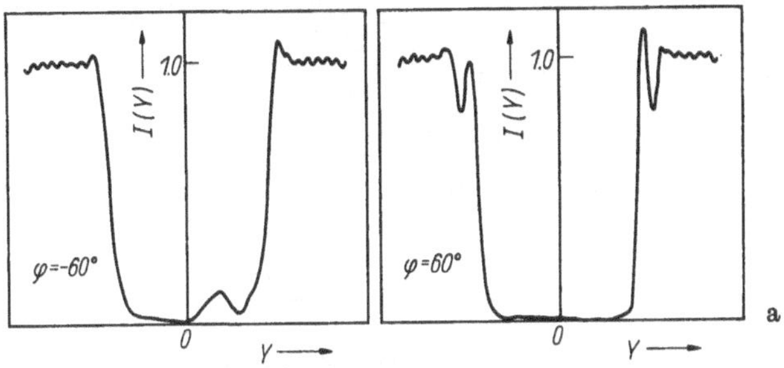

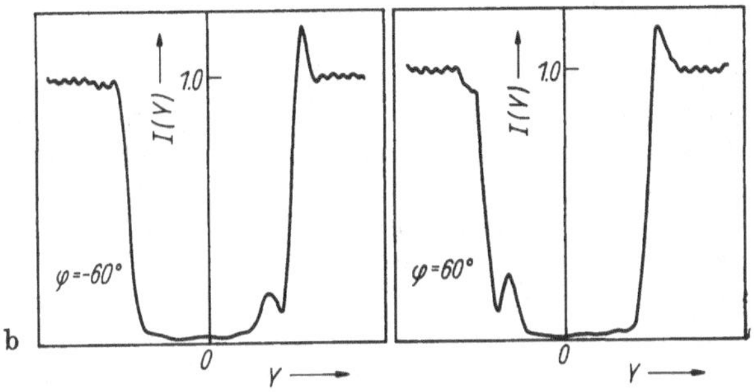

curves in Fig. 7. For the Néel-type image (Fig. 8b), the side peaks are (as noted previously) anomalous and do not relate to the $A(y, \varphi)$ curves.

The central humps developed by the $A(y, \varphi)$ curves do not carry over as accurately as the geometric–optical transformation suggests. The transformation in equations (3) predicts that, whenever the gradient of the object function ($\mathrm{d}A/\mathrm{d}y$) is negative, the normalised intensity will exceed unity. Partial agreement with this is found in the enhancement of the side diffraction peaks as the result of side peaks in the object functions. The central humps do not seem to constitute sufficiently strong inhomogeneities for the $I(Y)$ curve to exceed unity but give rise only to the distinct central peaks of the images. These various peaks in the image are strongly affected by such factors as the effective source size and under ideal conditions are somewhat larger. However, even in the ideal image, the central peaks do not exceed unity, indicating a lack of resolution imposed by wave optical effects. It is possible, that for small values of the defocussing distance, a more exact transfer of information may take place. This is borne out qualitatively by calculations at smaller z.

In view of this defect it does not appear that inversion of an image would be immediately rewarding in respect of extracting the $A_z(y)$ curve. Sufficient information should be transferred, however — the difficulties of photographic recording notwithstanding — for the $A_z(y)$ curve to be characterised as being of odd or even symmetry. Note that this applies whether the actual wall is stray-field-free or not.

7. The Second Parameter

If we consider walls for which $A_z(y)$ has odd symmetry, i.e. structures similar to the Bloch-type model of Fig. 2, then because the major Bloch-type regions (regions of large M_z) are to either side of the wall centre, we may expect a cor-

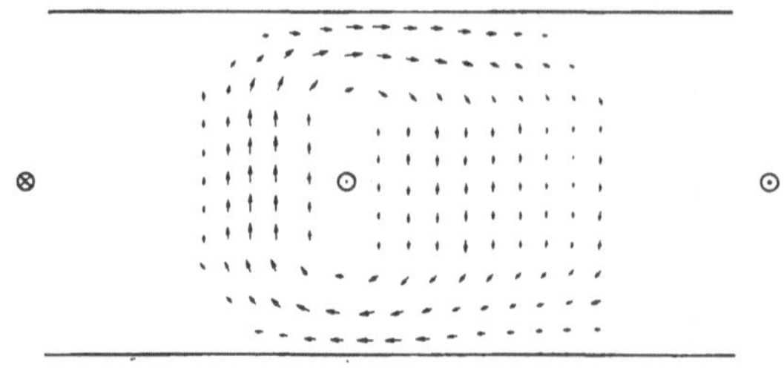

Fig. 9. Bloch-type wall model derived from Hubert's asymmetric Bloch-type wall of Fig. 2

relation between the relative volume of these regions and the degree of enhancement of the diffraction peak heights observed when the specimen is appropriately tilted. To test this hypothesis, profiles have been calculated for walls of differing Bloch content. These were derived from the original Bloch-type wall due to Hubert by transforming it in such a way that its stray-field-free condition is retained. This was achieved simply by repeated re-insertion of the central row of the vector map and then scaling to the same thickness. The most extreme wall shape actually generated in this way is shown in Fig. 9, where the central layer now occupies half the specimen thickness. These derived models will not be energy minima, but they are stray-field-free and hence useful for profile calculations in the tilted case.

To serve as a measure of the Bloch content of a domain wall, we define a parameter γ, such that

$$\gamma \equiv \left\langle \frac{M_z^2}{M_s^2} \right\rangle_{\text{wall}} . \tag{9}$$

For each of these derived models, profiles were calculated for the tilted and untilted cases and a measurement η taken from the profiles. η is arbitrarily defined as the increase of intensity in the first right-hand side diffraction peak, on tilting through 60° from the horizontal. The two parameters are plotted on Fig. 10, where it is seen that a strong relationship exists between enhancement and Bloch content defined in this way.

The exact relationship here depends of course on the imaging parameters selected, though the trend is similar for other combinations of parameters. Using these conditions and the assumption that $A_z(y)$ has odd symmetry, however, it

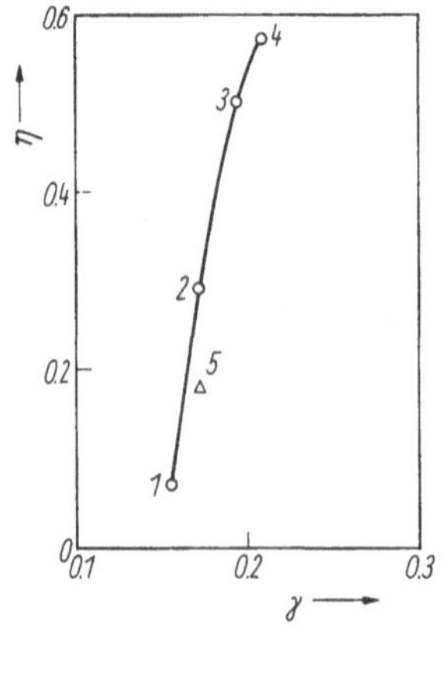

Fig. 10. Variation of the intensity profile peak height enhancement $\eta = \Delta_{\text{RHS}}$ peak height on tilting to $\varphi = 60°$ with the Bloch content $\gamma = \langle M_z^2/M_s^2 \rangle_{\text{wall}}$ of the wall model.
1 Hubert's Bloch-type model; 2 to 4 Hubert's Bloch-type model (variants); 5 Hubert's Néel-type model

should prove possible to use this technique to obtain experimental values for the parameter γ as a function of the specimen thickness. It has been assumed for the purposes of profile calculation that the total angular source size, i.e. including inelastic scattering, remains or can be held constant as the effective specimen thickness is increased. This can be checked on the low-angle diffraction pattern.

8. Conclusion

It has been demonstrated that the failure of Lorentz electron microscopy to provide detailed knowledge of the structure of domain walls can be attributed at source to the inadequacy of the conventional magnetic object $(A_x(y))$ as a means of characterising the wall structure. More characteristic of a wall are the Bloch or M_z-components, which when summed through the specimen thickness yield functions $(A_z(y))$ having broadly odd or even symmetry along an axis normal to the wall (y-axis). By tilting the specimen about this axis, these Bloch components can be incorporated in the object function. We have shown by wave theory calculations, that enough of the additional information about the wall can be transferred to the image of the tilted specimen to characterise the function $A_z(y)$ as having even or odd symmetry. Further, it has been found in the case of $A_z(y)$ having odd symmetry, and the wall being substantially stray-field-free, that a relation can be established between the Bloch content of a wall and a measurable characteristic of its image intensity profile, namely the enhancement of the Fresnel diffraction peaks obtained on tilting about the y-axis. It is proposed that in this way a measure of the Bloch content of the wall can be obtained experimentally.

Acknowledgements

We are grateful to Dr. A. Hubert for supplying the detailed data for his wall models. One of us (C. G. H.) acknowledges the assistance provided by the Science Research Council in the form of a maintainance grant, and we must thank the Imperial College and London University Computer Centres for their services, and the S.R.C. for the provision of these computing facilities.

Appendix

Wave-optical calculations of image intensity profiles

The calculation of the image intensity profiles described in this work was accomplished by a digital computer program. This accepts as data all the imaging parameters as absolute (non-reduced) values and a wall model described by 26×11 $\boldsymbol{M}$-vectors. The output is a profile of normalised image intensity against an absolute value of the Y-coordinate. The computation may be seen as three distinct parts: (i) generation of the object function $A(y, \varphi)$, (ii) transformation of the object into an ideal image, and (iii) degradation of the image by convolution with an assumed illumination function.

Initially the wall model is interpolated linearly to give 51×11 points which then form a 'window' to either side of which we put $M_y = M_z = 0$ and $M_x = \pm M_s$. The program is able to deal then with tilting about either or both axes (x and y). This is accomplished first by simple geometric transformations equivalent to rotation of the coordinate system. This is followed by the summation through the thickness. Secondly, in the case of x-axis tilt the summation takes place along the paths which do not generally correspond to columns of the wall array and interpolation is then performed between the points. x-axis and y-axis tilts may be combined. This process yields the function A which is then itself integrated to give the phase function $\Phi(y)$ used in the wave optical calculation:

$$\Phi(y) = 4\,\pi\, M_s\, t \left(\frac{e}{h}\right) \sum_0^y A(y)\,\Delta y\ ; \qquad -y_l \leqq y \leqq y_u\,. \tag{10}$$

When $\Delta y =$ mesh size and $y_l = y_u = 1000\,\Delta y$. $A(y)$ is represented by an array of 2001 points, the central 51 points corresponding to the wall array so that $\Phi(y)$ also has 2001 points.

The wave-optical transformation used to relate the amplitude at a point Y to the phase distribution of the electron beam is the well-known modified Fresnel approximation to the Kirchhoff integral.

$$\psi(Y) = c \int_{-y_l}^{y_u} \exp\left(-\,i\left\{\frac{k}{2}\left(\left[\frac{1}{u}+\frac{1}{z}\right] y^2 + y\,\frac{Y}{z}\right) + \Phi(y)\right\}\right) \mathrm{d}y\,, \tag{11}$$

where k is the magnitude of the electron wave vector, and u is the effective source–specimen distance.

It is customary to write the limits of this integral and of equation (10) as $\pm\,\infty$, i.e. for a specimen of infinite extent in the xy plane. Restriction of the limits to finite values is equivalent to considering a specimen of finite width. The effect of the finite limits of this integral is to create Fresnel diffraction fringes due to these artificial edges of the specimen. The limits are chosen to prevent overlaps of the fringes with the wall image.

The intensity distribution is normalised and then the effects of an extended source and inelastic scattering within the specimen are simulated by convolution with an assumed illumination profile [14].

$$I(Y) = c \int_{-Y_l}^{Y_u} I(Y') \exp\left\{-\left(\frac{Y - Y'}{z\,\beta}\right)^2\right\} \Delta Y'\ ; \qquad -\,Y_l \leqq Y \leqq Y_u\,, \tag{12}$$

where β is the effective angular source size. Here again the limits are restricted for convenience of computation and also to avoid including the 'edge' diffraction fringes. This will introduce negligible error if the condition

$$z\,\beta < \frac{1}{3}\,(Y_l + Y_u) \tag{13}$$

is maintained. In practice $z\,\beta$ is typically $(1/30)\,(Y_l + Y_u)$. Note that this degradation process simulates the effect of a total finite source size. It does not necessarily reproduce the effects of loss of beam coherence. This completes the profile computation.

For all of the profiles discussed in the text, the following values of the image parameters were used:

film thickness	t	$= 1100$ Å ,
mesh size	Δy	$= 100$ Å ,
saturation magnetisation	M_s	$= 800$ G ,
source distance	u	$= 20$ cm ,
defocussing distance	z	$= 5$ cm ,
beam voltage	U	$= 1$ MV ,
effective source size	β	$= 2 \times 10^{-6}$ rad giving $S = \beta/\theta_L = 0.1$.

The program was written in FORTRAN IV and requires 20 k words of core storage. The calculation of a profile requires about 90 s on a CDC 6600 computer. This rather lengthy method of computation has been adopted because the program is easily adapted for use with stripe domains and other extended, non-analytic magnetic structures.

References

[1] A. Hubert, phys. stat. sol. **32**, 519 (1969).
[2] A. E. LaBonte, J. appl. Phys. **40**, 2450 (1969).
[3] E. Fuchs, Z. angew. Phys. **15**, 206 (1963).
[4] D. C. Hothersall, Phil. Mag. **20**, 89 (1969).
[5] H. Kappert and P. Schmiesing, phys. stat. sol. (a) **4**, 737 (1971).
[6] T. Suzuki and C. H. Wilts, J. appl. Phys. **40**, 1216 (1969).
[7] H. Boersch, H. Hamisch, K. Grohmann, and D. Wohlleben, Z. Phys. **164**, 55 (1961).
[8] H. W. Fuller and M. E. Hale, J. appl. Phys. **31**, 238 (1960).
[9] D. Wohlleben, J. appl. Phys. **38**, 3341 (1967).
[10] J. P. Guigay and R. H. Wade, phys. stat. sol. **29**, 799 (1968).
[11] D. H. Warrington, Phil. Mag. **9**, 261 (1964).
[12] L. Reimer and A. Kappert, Z. angew. Phys. **26**, 58 (1969).
[13] C. G. Harrison, K. D. Leaver, and P. R. Swann, Proc. 5th Europ. Conf. Electron Microscopy, Manchester 1972.
[14] T. Suzuki and M. Wilkens, phys. stat. sol. (a) **3**, 43 (1970).

(Received April 18, 1972)

phys. stat. sol. (a) **12**, 425 (1972)

Subject classification: 1.5; 20.3; 22.8.2

Zentralinstitut für Festkörperphysik und Werkstofforschung der Deutschen Akademie der Wissenschaften zu Berlin, Institutsteil für magnetische Werkstoffe, Jena

Compositional Determination of Flash-Evaporated Ferrite Thin Films by X-Ray Fluorescence

By

M. WENDT

The composition of films prepared by flash-evaporation of $Mg_{0.75}Mn_{0.54}Fe_{1.71}O_4$-particles from tungsten filaments was determined by X-ray fluorescence with the result that the cation ratio Fe:Mn:Mg of the films was the same as that of the evaporant. But the films were heavily contaminated (about 20 at%) by the material of the filament.

Durch flash-Verdampfung von $Mg_{0,75}Mn_{0,54}Fe_{1,71}O_4$-Teilchen aus Wolframspiralen wurden Schichten hergestellt, deren Zusammensetzung durch Röntgenfluoreszenzanalyse bestimmt wurde. Dabei zeigte sich, daß das Kationenverhältnis Fe:Mn:Mg der Schichten mit dem des Ausgangsmaterials zwar übereinstimmt, die Schichten jedoch stark (etwa 20 At%) mit dem Tiegelmaterial Wolfram verunreinigt sind.

1. Introduction

Recently Nosé et al. [1] have reported on flash-evaporated ferrite thin films. There were found appreciable differences between the film properties and that of the bulk evaporant material. These differences are assumed to be caused by several defects but little is known about the nature of these defects. Especially no search for impurities was undertaken although it is known [2] that evaporated metal oxide films can be contaminated by the source material. In this paper we will report on the composition determination of films prepared under similar conditions.

2. Film Preparation

Evaporations were carried out in a conventional high-vacuum apparatus with oil diffusion pump and liquid nitrogen filled baffle. The base pressure was $< 10^{-5}$ Torr. Presintered $Mg_{0.75}Mn_{0.54}Fe_{1.71}O_4$ particles with an average diameter of 0.3 mm were brought into a conical tungsten filament at a rate of ≈ 5 mg/s. The filament temperature of 2000 °C was controlled with an optical pyrometer. Clean fused quartz plates were used as substrates and held at 300 °C by means of a radiation heater. The distance between the filament and the substrates was 15 cm, the deposition rate in the order of 10 Å/s.

Special examinations showed the as-evaporated films to be smooth, amorphous, and of oxidic nature but with oxygen deficit as compared with stoichiometric ferrite films. The films show ferrimagnetic properties only after annealing at elevated temperature and oxygen partial pressure.

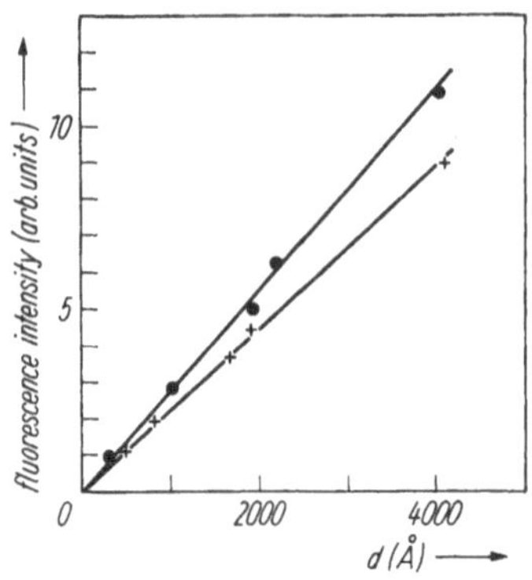

Fig. 1. X-ray fluorescence intensity of iron (●) and manganese (+) thin films as a function of the film thickness. The intensities are measured under the same experimental conditions

3. X-Ray Fluorescence Analysis

It was shown e.g. by Weyl [3] and Hoffmeister and Zuegel [4] that X-ray fluorescence spectroscopy is a very useful method for thin film analysis. The Mg-, Mn-, Fe-, and W-content of the as-evaporated films was determined by comparison with evaporated pure element films of known thickness. Mg was evaporated from tungsten baskets, Mn and Fe from indirect heated Al_2O_3-baskets, and W using an electron-gun. Fused quartz plates were used as substrates. In the case of Mg this substrates have to be prenucleated to obtain uniform films. This was achieved by pre-evaporation with Al. As shown in Fig. 1 for the K_α-radiation from Fe and Mn the fluorescence intensity is linearly proportional to the film thickness. Film thicknesses were measured by the Tolansky-method [5]. X-ray fluorescence measurements were carried out using a Zeiss Jena spectrometer VRA 2. Excitation and detection conditions are

Table 1

Experimental conditions for X-ray fluorescence analysis

element	primary radiation (kV)	fluorescent radiation	soller-collimator	analyzing crystal
Mg	Cr — 30	K_α	0.7°	KAP
Mn	W — 50	K_α	0.15°	LiF
Fe	W — 50	K_α	0.15°	LiF
W	Mo — 50	L_α	0.15°	LiF

given in Table 1. All measurements were carried out with gas flow proportional counter and pulse height analysis. A typical plot of the Fe and Mn K_α- and K_β-lines of a ferrite film is shown in Fig. 2.

In principle the analysis was carried out as follows. The intensity of the characteristic radiation of a sample element i was measured under the same experimental conditions as that of the standard films for the same element. This intensity shall be designated as I_i''. After a correction one obtains the intensity I_i and with the help of the calibration curve (e.g. Fig. 1) an equivalent thickness d_i. Then the composition can be calculated in the usual manner.

Fig. 2. Fluorescence spectrum of a typical ferrite thin film ($d \approx 4000$ Å) as measured with an LiF analyzing crystal between $2\theta = 50°$ and $2\theta = 65°$. The arrows indicate the expected positions of the characteristic lines

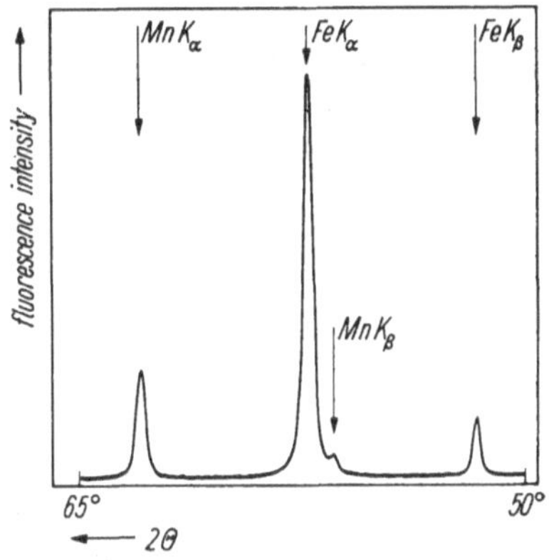

Because of absorption and secondary excitation the measured intensities I_i'' have to be corrected. In our case the secondary excitation by radiation of other sample elements can be neglected. But we have to consider the different influence of the Si K-radiation from the substrate on the Mg K_α-intensity of the standard and ferrite films. In the ferrite films this characteristic substrate radiation will be absorbed in particular by constituents other than Mg. Therefore the secondary excitation here is smaller than in pure Mg standards. According to measurements of Dahl and Schulz [6] this effect can be estimated to be as high as 10% resulting in $I'_{\mathrm{Mg}} = 1.1\, I''_{\mathrm{Mg}}$, where I'_i means that this intensity is corrected with respect to secondary excitation.

The influence of absorption was corrected in a first approximation similar to that described by Birks [7]. If we consider an element j then the intensity I_j is given by

$$\frac{I_j}{I'_j} = \frac{\mu_{\mathrm{eff}}\, D}{1 - \exp\left(-\mu_{\mathrm{eff}}\, D\right)}\,, \tag{1}$$

with

$$\mu_{\mathrm{eff}} D = \sum_i p_i\, d_i\, \mu_{ij}\,, \tag{2}$$

where d_i is the equivalent thickness of the element i, μ_{ij} is the absorption coefficient of the element i for characteristic radiation of the element j, p_i is a factor correlated firstly to the angles of the incident and fluorescent beam and secondly to the absorption of the primary beam. The factors p_i can be measured. Liebhafsky and Zemany [8] have shown that for pure element films the fluorescence intensity I_{d} is connected with the film thickness d by

$$I_{\mathrm{d}} = I_\infty \left[1 - \exp\left(-\mu_{\mathrm{avg}}\, d\right)\right], \tag{3}$$

where I_∞ is the intensity of an infinitely thick sample of the same element. For $\mu_{\mathrm{avg}}\, d \ll 1$ one obtains

$$\mu_{\mathrm{avg}} = \frac{I_{\mathrm{d}}}{d I_\infty}\,. \tag{4}$$

Then p_i is given per definitionem by

$$p_i = \frac{\mu_{\mathrm{avg},\, i}}{\mu_{ii}}\,. \tag{5}$$

Beside the measurements with pure element films one has to measure only the intensities of bulk pure element standards. The absorption coefficients μ_{ij} are calculated using the tables of Leroux [9]. Some p_i-values which are especially important for the analysis are given in Table 2. After all in Table 3

Table 2

p_i-values determined by the procedure described in the text

element	primary radiation (kV)	p_i
Fe	Cr — 30	2.90
W	Cr — 30	1.85
Fe	W — 50	5.02

Table 3

Results of the analysis of film ferrite 18 with a total thickness of (3960 ± 50) Å

element	I'/I''	I/I'	d_i(Å)
Mg	1.1	2.75	660
Mn	1	1.23	270
Fe	1	1.17	800
W	1	1.06	1175

one can find the results of the analysis of a typical ferrite film. As discernible from these values the corrections are small for W, Fe, and Mn. The relatively large absorption loss of the Mg K_α-radiation is mainly caused by tungsten and iron present in the films. The d_i-values result in the cation ratio Fe : Mn : Mg = = 1.71 : 0.55 : 0.72 in good agreement with the evaporant material. The error of the ratio Fe/Mn can be estimated to be smaller than 5%. Calibration incorrectness, instabilities of the apparatus, and correction uncertainties are taken into account. The error in the Mg determination can be as high as 20% caused mainly by the corrections and also by the uncertainty of the absorption coefficients [9]. The sum of the equivalent thicknesses d_i is smaller than the total thickness of the sample in agreement with other observations of the oxidic nature of the films.

4. Conclusions

The experiments confirm the concept of flash-evaporation in the view of stoichiometric deposition of the metallic constituents of such a complex compound as a ferrite. But the principal disadvantage will be the very high contamination by the source material. If one relates the number of the tungsten atoms to the total number of the film atoms then the tungsten content is as high as 20 at%. Therefore it seems to be doubtful to compare the properties of so heavily contaminated films with that of the tungsten-free evaporant material.

It may be interesting to note that flash-evaporated pure iron films prepared under the same evaporation conditions are contaminated by only about 2 at% tungsten. Therefore the oxygen present in the ferrite particles seems to be the main reason for the contamination.

Furthermore the usefulness of X-ray fluorescence spectrometry for thin film analysis was confirmed. For instance detection limits in the order of 0.1 Å in the equivalent thickness were achieved for manganese and iron. Further results of fluorescence-spectroscopic investigations of thin films we will report elsewhere.

Acknowledgements

The author wishes the express his thanks to Prof. F. Voigt and Dr. G. Vogler for stimulating discussions. He is indebted also to Mr. R. Furcht for his help in carrying out the fluorescence measurements and to Mrs. R. Euler for technical assistance. He especially wishes to thank Dr. H. Gengnagel for a critical reading of the manuscript.

References

[1] H. Nosé, M. Hashimoto, and R. Kimura, Trans. Nat. Res. Inst. for Metals **11**, 1 (1969).
[2] L. Holland, Vacuum Deposition of Thin Films, Chapman & Hall, London 1956.
[3] R. Weyl, Z. angew. Phys. **13**, 283 (1961).
[4] W. Hoffmeister and M. Zuegel, Thin Solid Films **3**, 35 (1969).
[5] S. Tolansky, Multiple Beam Interferometry of Surfaces and Films, Clarendon Press, Oxford 1948.
[6] H. Dahl and G. Schulz, Z. angew. Phys. **29**, 117 (1970).
[7] L. S. Birks, X-Ray Spectrochemical Analysis, Interscience Publ., New York 1959 (p. 58).
[8] H. A. Liebhafsky and P. D. Zemany, Ann. Chem. **28**, 455 (1956).
[9] J. Leroux, in: Encyclopedia of X-Rays and γ-Rays, Ed. G. L. Clark, Reinhold Publ. Corp., New York 1963 (p. 9).

(Received June 2, 1972)

phys. stat. sol. (a) **12**, 431 (1972)

Subject classification: 10.1; 1.4; 21.1; 21.6

United States Steel Corporation, Research Laboratory, Monroeville, Pennsylvania

Behavior of Misfit Dislocations during Diffusion of Gold into Copper–Gold Alloys

By

M. J. MARCINKOWSKI[1]), A. J. BAKER[2]), and R. M. FISHER

Upon electrochemical thinning of thin foils of the ordered Cu_3Au alloy prior to examination in the electron microscope, a thin layer of gold is redeposited epitaxially on both surfaces of the foil. In order to relieve the long range stresses due to the difference in cell size between the two crystals, an array of misfit dislocations is formed in the Cu_3Au–Au interface. These dislocations are generally of pure edge type, i.e. the extra half plane lies perpendicular to the foil surface while the $1/2\, a_0 \langle 110 \rangle$ Burgers vector lies perpendicular to the extra half plane. When two $\langle 110 \rangle$ directions lie in or close to the plane of the foil a crossed grid of misfit dislocations is generated. If on the other hand a single $\langle 110 \rangle$ direction lies in or close to the plane of the foil only one parallel set of dislocations perpendicular to this direction is observed. These results are satisfactorily accounted for on the basis of the anisotropy of strain energy in the foil. During annealing of the foils in the electron microscope, so as to allow diffusion to occur, the misfit dislocations climb into the foil from both sides and eventually annihilate one another as the foil composition becomes homogeneous. If the foil is annealed below the critical ordering temperature antiphase boundaries are generated as a result of dislocation climb. Because of the relatively high energies possessed by these boundaries, they can modify the misfit dislocation configuration rather markedly.

Nach elektrochemischer Abtragung von dünnen Schichten der geordneten Cu_3Au-Legierung wurde vor der Untersuchung im Elektronenmikroskop eine dünne Goldschicht epitaktisch an beiden Oberflächen der Schicht aufgebracht. Um die weitreichenden Spannungen, die durch den Unterschied der Zellengröße der beiden Kristalle hervorgerufen werden, zu entlasten, wird eine Anordnung von Misfit-Versetzungen in der Cu_3Au–Au-Grenzfläche gebildet. Diese Versetzungen sind im allgemeinen vom reinen Stufentyp, d. h. die zusätzliche Halbebene liegt senkrecht zur Schichtoberfläche, während der $1/2\, a_0 \langle 110 \rangle$ Burgersvektor senkrecht zur zusätzlichen Halbebene liegt. Wenn zwei $\langle 110 \rangle$-Richtungen in oder in der Nähe der Schichtebenen liegen, wird ein Kreuzgitter von Misfit-Versetzungen erzeugt. Wenn andererseits eine einzelne $\langle 110 \rangle$-Richtung in oder in der Nähe der Schichtebene liegt, wird nur ein paralleler Satz von Versetzungen senkrecht zu dieser Richtung beobachtet. Diese Ergebnisse können auf der Grundlage der Anisotropie der Verzerrungsenergie in der Schicht befriedigend erklärt werden. Während der Temperung der Schichten im Elektronenmikroskop, die das Auftreten von Diffusion ermöglicht, klettern die Misfit-Versetzungen in die Schicht von beiden Seiten her ein und annihilieren eventuell miteinander, wenn die Schichtzusammensetzung homogen wird. Wenn die Schicht unterhalb der kritischen Ordnungstemperatur getempert wird, werden als Ergebnis des Versetzungskletterns Antiphasengrenzen gebildet. Auf Grund der relativ hohen Energien dieser Grenzen, können diese die Misfit-Versetzungsanordnung ziemlich merklich modifizieren.

---

[1]) Presently with the Engineering Materials Group and the Dept. of Mech. Engr., University of Maryland, College Park, Maryland, USA.

[2]) Presently with the Dept. of Metallurgy, University of Leeds, United Kingdom.

1. Introduction

It is generally recognized that when one crystal is deposited epitaxially on another any long range stresses due to differences in dimensions of the two crystal lattices are greatly reduced by formation of a grid of misfit dislocations at the interface [1]. These misfit dislocations have also been termed epitaxial dislocations. They have been observed, using transmission electron microscopy to occur between single crystal films of PbS and PbSe by Matthews [2] as well as in Si into which a layer of phosphorus was diffused by Washburn et al. [3]. Embury and Duff [4] have also examined composite films of electro-deposited copper on copper–4% indium substrates and observed that misfit dislocations were generated in the process. Upon heating such composites to a temperature where diffusion can occur, the change in composition across the interface is replaced by a zone containing a continuous distribution of varying composition. The misfit dislocations correspondingly redistribute themselves throughout this zone to accomodate the gradient in strain produced by the gradient in composition. A detailed theory for this redistribution process has been formulated by Vermaak and van der Merve [5].

Contrast effects, subsequently recognized as due to incipient misfit dislocations, were observed in an earlier investigation of antiphase domain boundaries in Cu_3Au [6]. The basic origin of the effect became clear from a study of antiphase boundaries in Cu_3Au superlattices by Marcinkowski and Zwell [7] who showed by electron diffraction techniques that an epitaxial gold layer exists on both surfaces of foils electrochemically thinned for examination by transmission electron microscopy.

This gold layer apparently results from preferential dissolution of copper from the surface of the foil [8] or by redeposition of gold from the chromic-acid electrolyte and since its lattice parameter is 9% larger than the alloy substrate, formation of misfit dislocations is favored. In the course of an electron metallographic study of copper–gold alloys susceptible to stress corrosion cracking, Pickering and Swann [9] also noticed the formation of misfit dislocations and utilized this effect as a technique to detect the presence of gold layers on the surfaces of thin specimens.

Thus the copper–gold system affords the interesting possibility of being able to study the behavior of misfit dislocations in an ordered alloy where the effect of antiphase boundary energy should be important. The systematic observations described in this paper were presented in a preliminary account [10].

2. Experimental Procedure

The alloys used in this investigation consisted of fully ordered and disordered 24.4 at% Au (remainder Cu) foils, prepared as described in an earlier paper [6]. Particulars pertaining to the electrochemical thinning of the present alloys are also contained in the above mentioned paper and have been described in even greater detail by Glenn and Raley [11]. Some foils were examined during and after heating either on a hot stage consisting of coincident stainless steel grids between which the specimen was placed and through which a given current was passed [12] or else on the regular Siemens hot stage. In either case the power supply was calibrated so that a given temperature was attained by means of a selected power output setting. Other foils were heated after thinning in a small

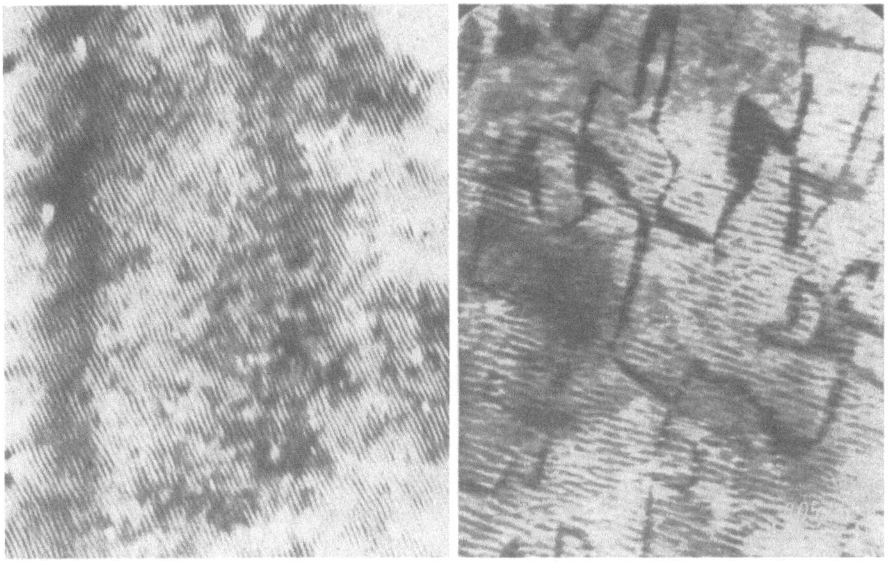

a b

Fig. 1. Misfit dislocations in as-thinned unheated foils of Cu_3Au.
a) Dislocation contrast; b) dislocation and domain boundary contrast

vacuum furnace and then examined in the microscope. A Siemens Elmiskop I operating at 100 kV was used for all the electron microscope and diffraction observations.

3. Experimental Results and Discussion

The transmission electron micrograph in Fig. 1a shows the fine striated structure observed in as-thinned unheated foils of Cu_3Au with orientations close to $\langle 110 \rangle$. The dark parallel lines which frequently terminate within the area shown in Fig. 1a are dislocations with an average spacing of 32.3 Å. When the condensor is focussed so that multiple beam contrast conditions are in effect, both the misfit dislocations and the thermally produced antiphase domain boundaries (APB's) are visible as shown in Fig. 1b.

It was noted that the contrast of the misfit dislocations appeared to increase during the course of observation with the electron microscope indicating that the structure was changing. This effect was greatly enhanced by removing the condensor aperture and heating the foil with the beam as with the example shown in Fig. 2a. Even more striking changes in the appearance of the dislocations are obtained by heating the foil in the microscope using a heating stage. An example is shown in Fig. 2b of a foil which was heated for 7 min at 350 °C, i.e. below the critical ordering temperature of 388 °C for the present alloy. It is clear the dislocation configuration becomes much coarser as a result of heating. The dislocation spacing in Fig. 2a is about 150 Å and in 2b it is about 1500 Å.

The sequence of the changes in the dislocation configuration during heating at 350 °C for periods of 9, 12, and 16 min is shown in Fig. 3a, b, and c, respectively, where it may be seen that the foil becomes almost completely devoid of dislocations. In all of these figures the dislocation lines are essentially along [001]. The white blotches in Fig. 3a, b, and c are regions where an unusual corrosion

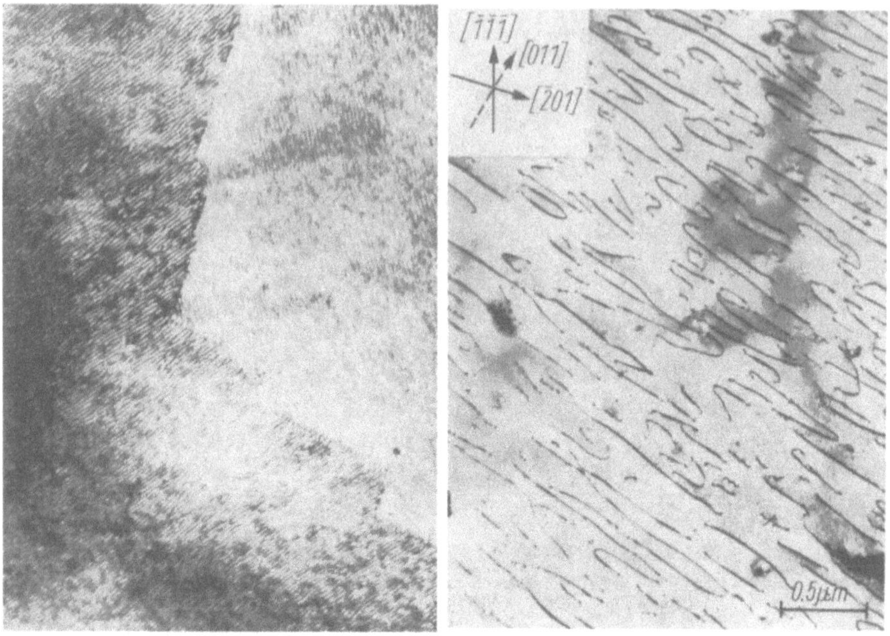

a b

Fig. 2. Misfit dislocations in heated foils of Cu_3Au. a) Heated with electron beam; b) heated 7 min at 350 °C with specimen hot stage

attack occurred during electrochemical thinning [9] and which serve as useful markers.

The previously mentioned electron diffraction observations of Marcinkowski and Zwell [7] showed that although the gold diffraction spectra were rather

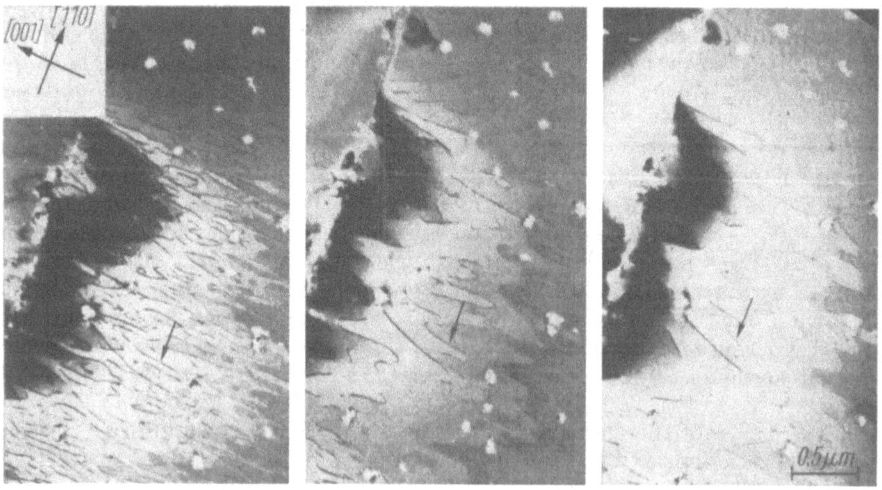

a b c

Fig. 3. Change in misfit dislocation configuration by heating at 350 °C. a) 9 min; b) 12 min; c) 16 min

Fig. 4. Schematic illustration of misfit dislocations. a) and b) Initial configuration at interface prior to heating (orthogonal views); c) antiphase boundaries generated by thermally induced climb of dislocations

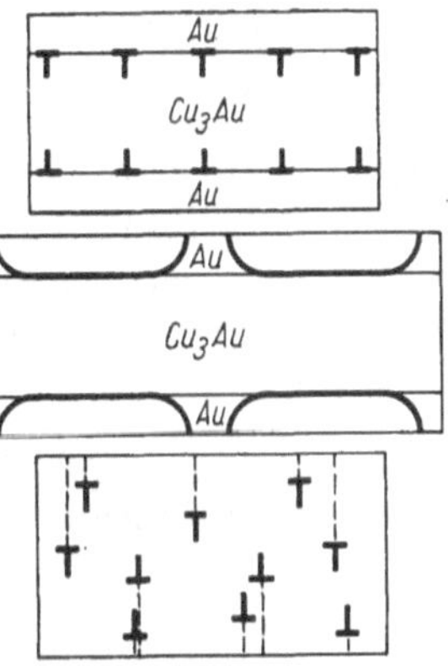

a b c

diffuse, they were nevertheless indicative of a single crystal layer of gold on the Cu_3Au substrate. This situation is shown schematically in Fig. 4a. Since the lattice constant of gold is nearly 9% greater than that of Cu_3Au, the misfit between the three layers will be accommodated by edge dislocations of opposite sign which lie in the plane of the interface. In the case of Cu_3Au, which is derived from the face centered cubic lattice, the Burgers vector of the ordinary dislocation is of the type $\frac{1}{2} a_0 \langle 110 \rangle$. Referring back to Fig. 3 it will be noted that one direction of the type $\langle 110 \rangle$ lies in the plane of the foil, in particular $[\bar{1}10]$. Furthermore since it is perpendicular to the dislocation lines, the dislocations in Fig. 3 must be of pure edge type. In addition since the dislocations do not lie in

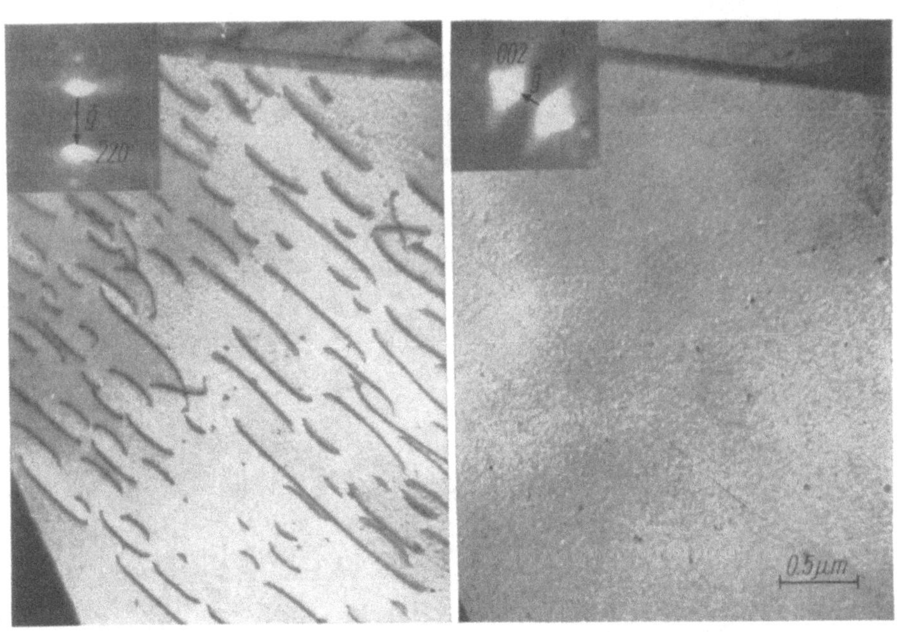

a b

Fig. 5. Demonstration that misfit dislocations are prismatic pure edge type the plane of the foil along their entire length, but emerge at the surface near their ends, they must be prismatic. This is shown by the schematic view in Fig. 4b which is at right angles to Fig. 4a.

Direct confirmation of the nature of the misfit dislocations was obtained by contrast experiments where in foils were tilted into orientations where dislocation contrast vanished, i.e. where $\boldsymbol{g} \cdot \boldsymbol{b} = 0$. An example of this is shown in Fig. 5. The extinction effect demonstrates that $\boldsymbol{b} = \frac{1}{2} a_0 [\bar{1}10]$, i.e. is perpendicular to the dislocation lines and parallel to the surface of the foil.

In order to completely relieve the long range internal stresses due to the misfit of the two superimposed crystal lattices the dislocations in Fig. 1 should be spaced at a distance p given by [1]

$$p = \frac{b^2 \cos \theta}{|b - b_1|}, \qquad (1)$$

where b is the Burgers vector of a dislocation in Cu_3Au, b_1 is the Burgers vector of a dislocation in Au, and θ is the angle between the Burgers vector and the direction which lies in the plane of the foil and which is also perpendicular to the dislocation line. For this epitaxial composite $a_0 = 3.746$ and 4.078 Å for

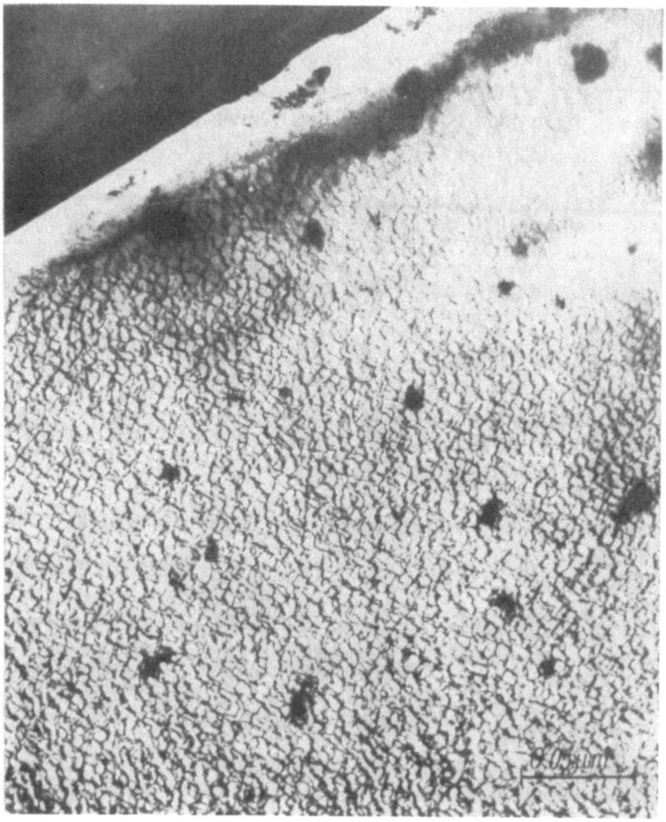

Fig. 6. Example of infrequently observed cross grid of misfit dislocations

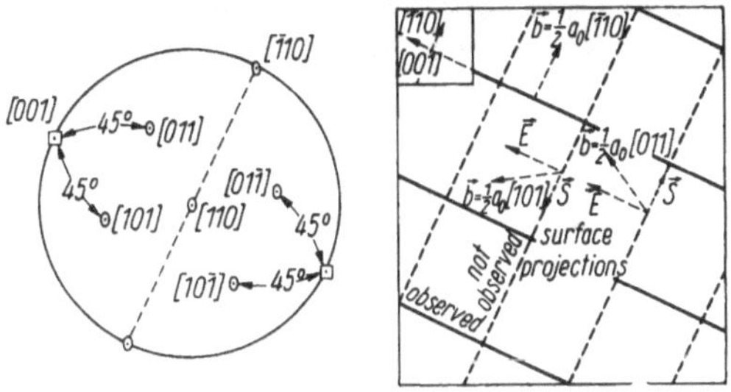

a b

Fig. 7. Schematic illustration of second set of misfit dislocations expected but unobserved in Fig. 3. a) Dotted in stereographic analysis ; b) possible configuration

Cu_3Au and Au, respectively, $b = 2.65$ Å and $b_1 = 2.88$ Å, while θ in the case of Fig. 1 is 0°. Substituting these values in (1) gives $p = 30.4$ Å which is in good agreement with the spacing of 32.3 Å shown in Fig. 1.

A rather surprising feature of the dislocation configuration observed during this study is the usual absence of a crossed grid of misfit dislocations, i.e. a second set running at a oblique angle to those in the micrographs in Fig. 1 to 3. Occasionally a very imperfect cross grid such as that shown in Fig. 6 was observed but these were quite rare.

The crystallographic analysis of the foil shown in Fig. 3 is presented in Fig. 4a and the expected second set running parallel to $[\bar{1}10]$ is shown dotted on the stereographic projection. As shown in the figure, these dislocations could not be pure edge with the extra half plane perpendicular to [001] since the closest $\frac{1}{2}a_0 \langle 110 \rangle$ Burgers vectors to [001] all make angles of $\theta = 45°$ with it. If such dislocations did occur they would have a strong screw component equal to $b \cos 45°$. However, there would not need to be long range stresses due to such a wall [13] since the Burgers vectors of neighboring dislocations could alternately be $\frac{1}{2}a_0 [101]$ and $\frac{1}{2}a_0 [011]$ (or $\frac{1}{2}a_0 [01\bar{1}]$ and $\frac{1}{2}a_0 [10\bar{1}]$) as illustrated in Fig. 7b. Since the screw components are of opposite sign, their stress fields cancel at large distance. The edge components however are of the same sign and their effects are thus additive. In order to have this second set of dislocations completely relieve all the long range stress due to epitaxy, their spacing p according to equation (1) would have to be 21.5 Å. Since this second set of dislocations along $[\bar{1}10]$ is not observed, it must be concluded that the energy associated with it is not very different than for the undislocated state. An attempt will next be made to resolve this question in a quantitative manner.

Consider first the strain energy associated with a single crystal foil of Cu_3Au both faces of which possess an epitaxially deposited layer of Au. All the crystals are assumed distorted to a common lattice parameter and in a state of plane strain. This assumption becomes more valid as the thickness of the gold layers decreases [14]. Under these conditions the following relationship is readily found:

$$a_0 (1 + \varepsilon) = a_{01} (1 + \varepsilon_1) , \quad (2)$$

where ε is the strain in the particular crystal measured along a given direction in the foil, and as usual the subscript 1 refers to the Au lattice. Another relation that must hold is that the total compressive load exerted on the outer Au layers by the inner Cu_3Au layer must equal the tensile load exerted on the Cu_3Au layer by the Au layers. This together with Hooke's law leads to the following relation:

$$E\,\varepsilon\,t = 2\,E_1\,\varepsilon_1\,t_1\,, \tag{3}$$

where E refers to Youngs modulus and t to the thickness of the respective layers. From equation (2) and (3) we find

$$\varepsilon_1 = \frac{a_{01} - a_0}{[(2\,E_1\,t_1\,a_0/E\,t) - a_{01}]} \tag{4a}$$

and

$$\varepsilon = \frac{a_{01} - a_0}{[a_0 - (a_{01}\,E\,t/2\,E_1\,t_1)]}\,. \tag{4b}$$

Combining equations (4) with the following expression for the strain energy per unit area of interface (both interfaces are counted),

$$E_s = \frac{1}{2}\,(E\,\varepsilon^2\,t + 2\,E_1\,\varepsilon_1^2\,t_1)\,, \tag{5}$$

we obtain

$$E_s = (a_{01} - a_0)^2 \left[\frac{E\,t}{2\,[a_0 - (a_{01}\,E\,t/2\,E_1\,t_1)]^2} + \frac{E_1\,t_1}{[(2\,E_1\,t_1\,a_0\,E\,t) - a_{01}]^2}\right]. \tag{6}$$

Because of the strong directional dependence of E it will be necessary to include this in our calculations based on equation (6). In particular it is given by Nye [15] as follows:

$$\frac{1}{E} = S_{11} - 2\,(S_{11} - S_{12} - 1/2\,S_{44})\,(l_1^2\,l_2^2 + l_2^2\,l_3^2 + l_3^2\,l_1^2)\,, \tag{7}$$

where the l's are the direction cosines of the particular direction in question and the compliance constants S are according to Zener [16] for Cu_3Au $S_{11} = 1.34$, $S_{12} = -0.565$, and $S_{44} = 1.51$ while for Au $S_{11} = 2.33$, $S_{12} = -1.07$, and $S_{44} = 2.38$ all multiplied by 10^{-12} cm²/dyn. Equation (7) leads to values of $E = 7.45 \times 10^{11}$ dyn/cm² and $E_1 = 4.29 \times 10^{11}$ dyn/cm² for the [001] direction.

Let us now examine the value of E_s given by (6) associated with the [001] direction. Also assume that the Au layers are on the order of p given by (1), i.e. 21.5 Å and the foil thicknesses are on the order of 2150 Å. Substituting these values in equation (6) gives $E_s^{[001]} = 542$ erg/cm². It is interesting to note that because of the large value of t/t_1 the first term of (6) is small and nearly all of the strain energy varies as the square of ε which is large in the Au layers.

In order to see whether an array of misfit dislocations lying along $[\bar{1}10]$ would have less energy than that just calculated, the following relation will be used:

$$E_m = \frac{2\,\mu\,b^2}{4\,\pi\,k\,p}\left(\ln\frac{p}{2\,b} + 0.1\,\ln\frac{r_1}{2\,b}\right), \tag{8}$$

where μ is taken to be the shear modulus associated with shear in a $\langle 110\rangle$ director, i.e. the direction of the Burgers vector on a $\{111\}$ plane, i.e. the slip plane. Using the relations of Hearman [17, 18], μ is readily found to be

$$\mu = \frac{1}{3}\,(c_{11} + c_{44} - c_{12})\,. \tag{9}$$

Values of $c_{11} = 1.86$, $c_{12} = 1.57$, and $c_{44} = 0.420$, all multiplied by 10^{12} dyn/cm^2 were used for Au while values of $c_{11} = 2.25$, $c_{12} = 1.73$, and $c_{44} = 0.663$ were taken for $AuCu_3$ [16]. The first term in equation (8) is the strain energy stored in a dislocation within radius p, while the second term takes account of the core energy of the dislocation [19]. The term r_1 is on the order of 1 cm while

$$\frac{1}{k} = \cos^2 \varphi + \frac{\sin^2 \varphi}{(1 - \nu)}\,.$$

The angle φ is that which the Burgers vector makes with the dislocation line. The factor of 2 takes account of the two interfaces. Thus $\mu = 3.94 \times 10^{11}$ while $\mu_1 = 2.37 \times 10^{11}$ dyn/cm^2. The average of these two values is $\mu_{av} = 3.16 \times 10^{11}$ dyn/cm^2. Upon substituting this value into equation (8) $E_m^{[001]}$ is found to be 435 erg/cm^2 as compared to $E_s^{[001]} = 542$ erg/cm^2 as discussed above. Considering the approximate nature of the calculations for E_s and E_m it suffices to say that both are about equal to 500 erg/cm^2 and thus for this orientation there is no strong driving force favoring the nucleation of interface dislocations. This however would not be the case as t_1 increases since while E_m from equation (8) will remain constant, E_s from (6) will increase continually with increasing t_1. However, the fact that interface dislocations were never found to lie along $\langle 110 \rangle$ in foils which possessed the [110] orientation suggests that the thickness of the Au layer rarely exceeds 25 Å.

Let us now calculate E_s for the $\langle 110 \rangle$ direction, i.e. $E_s^{[\bar{1}10]}$. For this orientation $E = 13.2 \times 10^{11}$ dyn/cm^2 and $E_1 = 8.2 \times 10^{11}$ dyn/cm^2. Upon substituting these values into equation (6), $E_s^{[\bar{1}10]}$ is found to be 1210 erg/cm^2. This large value for the strain energy is due to the large values of E along $\langle 110 \rangle$. On the other hand $E_m^{[\bar{1}10]}$ from (8) and (9) is found to be 318 erg/cm^2. This low value results from the relatively large value for p. In fact $E_m^{[\bar{1}10]}$ as calculated above is somewhat of an overestimate since $p > t_1$. Thus, the stress field of the dislocations is cut off by the foil thickness. Nevertheless it is seen that $E_s^{[\bar{1}10]}$ is nearly four times that of $E_m^{[\bar{1}10]}$. A substantial driving force is thus provided for the formation of an array of misfit dislocations along [001] in agreement with the observed results which showed them always to be present in foils with $\langle 110 \rangle$ orientations.

It is interesting to consider orientations where no $\langle 110 \rangle$ directions lie in the plane of the specimen such as the foil possessing a $[\bar{1}23]$ orientation shown in Fig. 2b. It is apparent from this micrograph that the dislocations lie for the most part along a single crystallographic direction. Stereographic analysis shows that a [011] direction which is perpendicular to these dislocations lies only 11° out of the plane of the foil, i.e. $\theta = 11°$. The arrow is shown dotted in the inset of Fig. 2b to indicate that it represents the projection of that direction onto the plane of the paper. Substituting $\theta = 11°$ into equation (1) indicates a value for p of 29.8 Å. Inasmuch as the values for E in this direction are nearly the same as those for [110], E_s and E_m are nearly identical to those calculated previously so that the arguments are essentially the same as that used to account for the configuration of misfit dislocations in foils with [110] orientations. Thus, it can be concluded that if a foil is oriented such that $\langle 110 \rangle$ lies very nearly in the plane of the foil, then a linear array of misfit dislocations will be generated perpendicular to it. Similarly, the absence of misfit dislocations at right angles to those shown in Fig. 2b can be explained by similar reasoning to that used for Fig. 3.

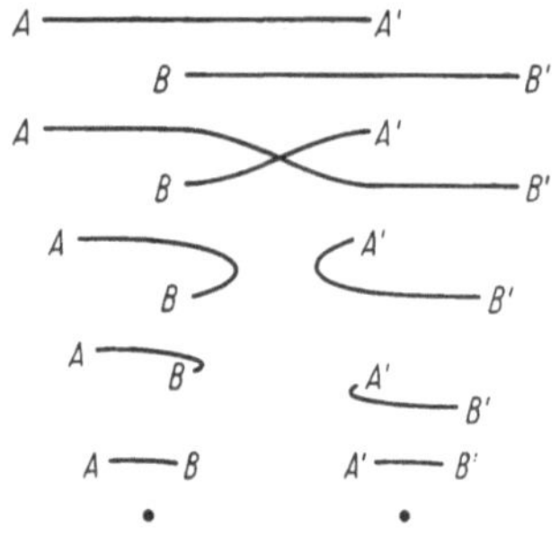

Fig. 8. Schematic illustration of sequence of events during interaction of edge dislocations by climb

The rearrangement of the misfit dislocation which occurs when a composite film is annealed at a sufficiently high temperature for diffusion to occur is shown schematically in Fig. 4c. A vertical composition gradient is developed which can be considered to consist of thin slabs of alloy, each of different composition, thus the misfit dislocations shown in Fig. 4a must climb into the foil and distribute themselves at the interface between each slab to accommodate the strain as illustrated in Fig. 4c. As the dislocations climb from both surfaces to the center of the foil, they will combine with one another since they are of opposite signs. One such reaction somewhat similar to that shown in Fig. 3a, b, c is illustrated schematically in Fig. 8. The line AA′ in Fig. 8a represents an edge dislocation in the upper half of the foil which emerges at the surface at points A and A′. The line BB′ on the other hand lies in the bottom half of the foil and emerges at the bottom surface at points B and B′. Obviously both are of opposite sign and therefore the stress fields between them are attractive. In the region where both dislocations overlap they will be attracted to one another as shown in Fig. 8b and finally pinch off as shown in Fig. 8c acquiring a typical hooked shape. As diffusion proceeds, the concentration gradient throughout the foil is reduced with the result that fewer and fewer edge dislocations are required to relieve the epitaxial stresses. The dislocations shown in Fig. 8c then begin to shorten as in Fig. 8d and e until they finally align themselves vertically with the surface of the foil. These may move further by climb and glide and be completely eliminated from the foil when they meet a dislocation of opposite sign. All of these processes can be seen occurring in Fig. 3a through c. In particular the arrows pointing to the dislocation configurations in Fig. 3a, b, c correspond to the processes depicted schematically in Fig. 8a, c, and e respectively.

The dislocation interaction and annihilation processes discussed above refers to dislocations which have climbed to appreciable depths within the foil. However, dislocations must climb from the original planar interface toward both the surface and the interior of the composite as it is homogenized by diffusion. The center of gravity of the dislocation array moves towards the center as those approaching the surface disappear.

If the epitaxial composite is heated at temperatures below the critical ordering temperature then antiphase boundaries are generated by the climb of dislocations. This is illustrated schematically in Fig. 4c by the dotted lines and is shown in actuality in the transmission electron micrograph of Fig. 9 of a sample which was annealed for 1.5 h at 275 °C. Antiphase boundaries are visible only within the nearly vertical [110] superlattice extinction contour whereas just outside this contour the misfit dislocations are in contrast. One set of boundaries is seen to lie almost horizontally with a mean spacing of about 1500 Å. They are

Fig. 9. Antiphase boundaries in Cu_3Au foil annealed 1.5 h at 275 °C after thinning

believed to lie on (100) planes whose projection, shown dotted in Fig. 9 is 24° out of the plane of the foil. As discussed by Fisher and Marcinkowski [6], antiphase boundaries in Cu_3Au can have essentially zero energy when they lie on these planes. The nearly vertical antiphase boundaries, on the other hand, are produced by climb of misfit dislocations, and as expected, are more closely spaced, i.e. on the order of 180 Å. It is interesting to note that although a [$1\bar{1}0$] direction lies in the plane of the foil, the misfit dislocations are oriented such that their Burgers vector is along [$0\bar{1}1$]. This direction is only 17° out of the plane of the foil so that the values of E_s and E_m will be nearly the same as those for [$0\bar{1}1$] resulting in very little preference for one orientation over the other. Residual elastic stresses developed during mounting of the foil in the specimen holder may be the controlling factor in this case.

The values of E_s and E_m' associated with the direction of the dislocations in Fig. 9 are 860 and 326 erg/cm² respectively if one again assumes an Au layer on the order of 21.5 Å thick. Under these conditions a second set of interface dislocations running perpendicular to those shown in Fig. 9 is to be expected. Indeed other foils with ⟨112⟩ orientation frequently showed crossed grids of dislocations. The most likely reason for this discrepancy is associated with the varying thickness of the Au deposits from specimen to specimen. In those cases where the Au layer is very thin, i.e. somewhat less than 20 Å it is quite possible that E_s is smaller than E_m for such orientations as discussed above. This can be seen qualitatively by noting from (5) and (8) that for large values of t/t_1, E_s varies roughly linearly with t_1 while E_m is expected to vary logarithmically with t_1. An accurate estimate of E_m for these thicknesses in which $t_1 < p$ is a rather formidible problem involving image forces.

Perhaps the most interesting foils are those possessing a ⟨100⟩ orientation. Since there are two mutually perpendicular ⟨110⟩ directions lying in the plane of the foil, crossed grids, of mutually perpendicular misfit dislocations lying along these same two directions may be expected. Washburn et al. [3] have indeed observed this to be the case in phosphorous doped silicon. However, reference to the transmission electron micrograph of Fig. 10 obtained from a foil that was annealed for 4 min at 300 °C and which possessed the [001] orientation shows this not to be the case with the present Au–Cu_3Au epitaxial composites. Instead the dislocations generally lie along [010] and contain numerous nodes. It is apparent that some type of reaction is taking place which can be most readily understood by reference to Fig. 11. In particular Fig. 11a shows a three dimensional view of the crossed grid of dislocations lying along ⟨110⟩

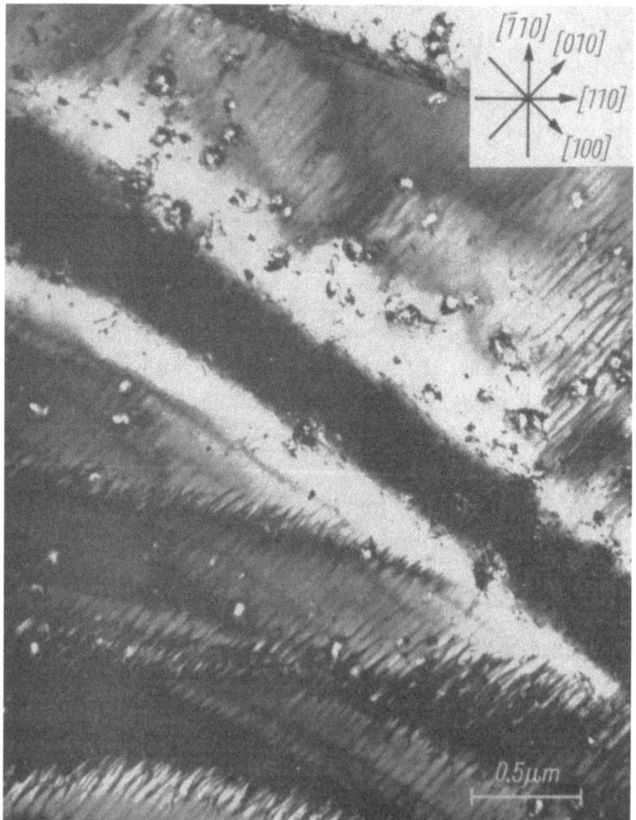

Fig. 10. Coalescence of misfit dislocations during annealing of Cu_3Au foil for 4 min at 300 °C

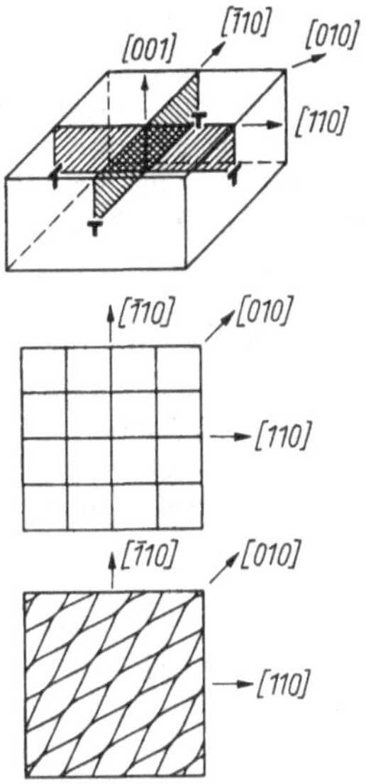

Fig. 11. Schematic illustration of coalescence of climb produced antiphase boundaries. a) and b) Boundaries as generated by dislocation climb; c) after coalescence

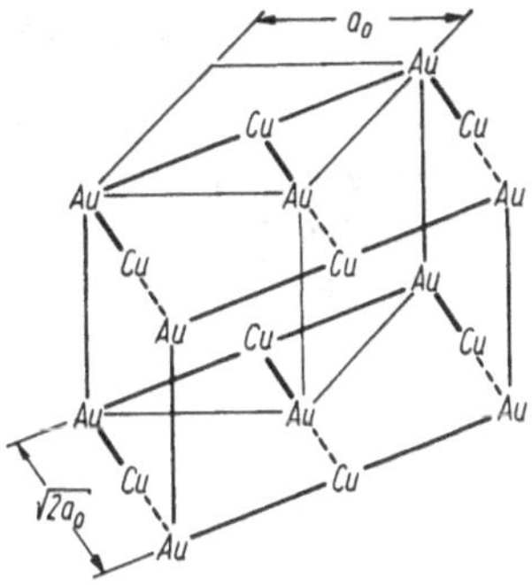

Fig. 12. Atom bonding across {110} planes in fully ordered Cu_3Au

directions while Fig. 11b is the same figure in two dimensions when viewed along [00$\bar{1}$]. This is the probable configuration taken up by the dislocations after the gold layer is deposited on the Cu_3Au foil and before annealing. After the foil is heated to a sufficiently high temperature however the dislocations climb into the foil leaving behind ribbons of antiphase boundary as shown by the shaded areas in Fig. 11a. It is apparent by looking at the following reaction that there is no reduction in strain energy due to dislocation combination:

$$\frac{1}{2\,a_0}\,[\bar{1}10] + \frac{1}{2\,a_0}\,[110] \rightarrow a_0\,[010]\,.$$

Strain energy considerations alone therefore will not cause the reaction to occur. However, it is apparent by reference to Fig. 11 that the climb produced antiphase boundaries with the displacement vectors given above can be annihilated by coalescence since the a_0 [010] displacement produces no antiphase boundary. The boundaries are thus attracted to one another and as a result drag the attached dislocation along with them, leading eventually to the configuration shown in Fig. 11c. The driving force for the above reaction must surmount that due to the long range epitaxial stresses. The antiphase boundary energy γ per unit area for the situation shown in Fig. 10 and 11 can be obtained readily by reference to Fig. 12. It is seen that the boundary formed by the insertion or removal of an extra {110} plane will create one wrong Au–Au bond and one wrong Cu–Cu bond over an area of $\sqrt{2}\,a_0^2$. The expression for γ is thus simply equal to $2\,V/\sqrt{2}\,a_0^2$ where V is the energy necessary to make a wrong atom bond and is found for Cu_3Au to be 11.1×10^{-14} erg [20]. γ is then readily found to be 112 erg/cm^2, i.e. a respectable energy to serve as a driving force for this type of dislocation coalescence.

4. Summary and Conclusions

During electrochemical thinning of a Cu_3Au alloy in a chromic-acetic acid electrolyte prior to examination by transmission via electron microscopy, gold redeposits epitaxially on the surface of the foil. In order to relieve the long range stresses due to the gold layer, misfit dislocations are formed in the Cu_3Au–Au interface. The misfit dislocation configuration is of lowest energy if it is composed of pure edge types, i.e. when the $\frac{1}{2} a_0 \langle 110 \rangle$ type Burgers vectors lie in the plane of the foil and are perpendicular to the dislocation line which may be thought to terminate on the vertical extra half plane. It frequently happens that the foil is oriented so that only one $\langle 110 \rangle$ direction lies in or close to the plane of the foil in which case a single set of parallel interface dislocations is formed. For some foil orientations such as $\langle 100 \rangle$, two $\langle 110 \rangle$ directions lie in the plane of the foil so that the formation of a crossed grid of interface dislocations is possible.

Annealing as-thinned foils causes the density of misfit dislocations to decrease. This is due to diffusion of Au and Cu atoms within the composite film because accompanying the approach to homogeneity of composition and lattice parameter there is a corresponding redistribution of misfit dislocations. Thus, dislocations move by climb from the original planar interface toward both the surface and the interior of a foil. The center of gravity of the dislocation array moves towards the center as those reaching the surface disappear. The dislocations approaching the center of the foil are attracted towards their counterparts from the other side and often annihilate them. If the foil is completely ordered, the climbing dislocations generate antiphase boundaries. In the case of certain foil orientations where crossed grids of dislocations are possible, such as those with $\langle 100 \rangle$ orientations, a corresponding crossed grid of antiphase boundaries is generated. These boundaries which possess relatively high energies can be eliminated by coalescence. In so doing they markedly alter the resulting misfit dislocation configurations.

Acknowledgements

The assistance of several members of this Laboratory is gratefully acknowledged. Many of the thin foils were prepared by Mr. A. Szirmae who also helped to carry out experiments using the heating stage. Dr. J. S. Lally made the Burgers vector identifications illustrated in Fig. 5 and helpful discussions were held with Dr. P. R. Swann of the Department of Metallurgy of Imperial College in London.

References

[1] J. Friedel, Dislocations, Addision-Wesley Publ. Co. Inc., Reading (Mass.) 1964.
[2] J. W. Matthews, Phil. Mag. 8, 711 (1963).
[3] J. Washburn, G. Thomas, and H. J. Queisser, J. appl. Phys. **35**, 1909 (1964).
[4] J. D. Embury and W. R. Duff, Electrochim. Acta (London) **11**, 1491 (1966).
[5] J. S. Vermaak and J. H. van der Merve, Phil. Mag. **10**, 785 (1964).
[6] R. M. Fisher and M. J. Marcinkowski, Phil. Mag. **6**, 1385 (1961).
[7] M. J. Marcinkowski and L. Zwell, Acta metall. **11**, 373 (1963).
[8] H. W. Pickering, private communication (1966).
[9] H. W. Pickering and P. R. Swann, Corrosion **19**, 374 (1963).

[10] M. J. MARCINKOWSKI and R. M. FISHER, VI. Internat. Congr. Electron Microscopy, Kyoto, Maruzen Press, Tokyo 1966 (p. 299).
[11] R. C. GLENN and J. C. RALEY, Amer. Soc. Testing Materials, Special Techn. Publ. **339**, 60 (1962).
[12] M. J. WHELAN, IV. Internat. Conf. Electron Microscopy, Springer-Verlag, 1958 (p. 96).
[13] J. C. M. LI, Electron Microscopy and Strength of Crystals, Ed. G. THOMAS and J. WASHBURN, Interscience Publishers, New York 1963 (p. 713).
[14] M. J. MARCINKOWSKI, Microscopie Electronique, Proc. VII. Internat. Congr. Electron Microscopy, Grenoble 1970, Ed. P. FAVARD, Vol. I, Société de Microscopie Electronique, Paris 1970 (p. 51).
[15] J. F. NYE, Physical Properties of Crystals, Clarendon Press, Oxford 1957.
[16] C. ZENER, Elasticity and Anelasticity of Metals, The University of Chicago Press, Chicago 1948.
[17] R. F. S. HEARMAN, Adv. Phys. **5**, 323 (1956).
[18] R. F. S. HEARMAN, Acta cryst. **10**, 121 (1957).
[19] A. H. COTTRELL, Dislocations and Plastic Flow in Crystals, Clarendon Press, Oxford 1953.
[20] M. J. MARCINKOWSKI, Electron Microscopy and Strength of Crystals, Ed. G. THOMAS and J. WASHBURN, Interscience Publishers, New York 1963 (p. 333).

(Received March 9, 1972)

phys. stat. sol. (a) **12**, 447 (1972)

Subject classification: 20.3; 22.8

Institute of Technical Physics, University of Oulu

X-Ray-Excited Optical Fluorescence of Ten Rare Earth Ions in Y_2O_2S, La_2O_2S, and Gd_2O_2S

By

H. Ratinen

The optical fluorescence of ten trivalent lanthanide ions, Pr, Nd, Sm, Eu, Gd, Tb, Dy, Ho, Er, and Tm in Y_2O_2S, La_2O_2S, and Gd_2O_2S have been measured by X-ray excitation at 300 K. Some lines of Eu^{3+}, Gd^{3+}, and Tb^{3+} in La_2O_2S at short wavelengths were found to be quenched at 300 K. La_2O_2S appeared to be a more efficient host than Y_2O_2S and Gd_2O_2S for all the lanthanides. The background emission from non-doped oxysulfides was measured at 100 K. The background emission was much weaker than the fluorescence of lanthanide ions.

Die optische Fluoreszenz von zehn dreiwertigen Lanthanid-Ionen, Pr, Nd, Sm, Eu, Gd, Tb, Dy, Ho, Er und Tm, in Y_2O_2S, La_2O_2S und Gd_2O_2S, wurde durch Röntgenstrahlenanregung bei 300 K gemessen. Es wurde gefunden, daß einige Linien von Eu^{3+}, Gd^{3+} und Tb^{3+} in La_2O_2S bei kurzen Wellenlängen verschwinden. La_2O_2S scheint für alle diese Lanthaniden ein wirksameres Grundgitter als Y_2O_2S und Gd_2O_2S zu sein. Die Grundgitteremission von reinen Oxysulfiden wurde bei 100 K gemessen. Die Grundgitteremission war viel schwächer als die Fluoreszenz von Lanthanid-Ionen.

1. Introduction

Y_2O_2S, La_2O_2S, and Gd_2O_2S crystals all have the trigonal space group D^3_{3d}, the point symmetry of the cation site being C_{3v}. Because a lanthanide ion is not at a centre of inversion, both magnetic and electric dipole transitions occur. The crystal structure parameters of these oxysulfides differ from each other in such a way that La_2O_2S has the largest interionic distances and Y_2O_2S the smallestones [1]. Because of the different interionic distances, the lanthanide activated rare-earth oxysulfides have been used for investigating crystal field parameters [1] and host cation effects [2]. These oxysulfides have also been used for studying excitation and quenching effects [3, 4].

Several lanthanide activated oxysulfides are known to be efficient phosphors under uv and cathode-ray excitation [3, 5]. The present paper gives a general survey of the luminescence properties of lanthanides in Y_2O_2S, La_2O_2S, and Gd_2O_2S when X-rays are used for the excitation.

2. Experimental

The characteristic emission of lanthanides was measured with the resolution of 5 Å at 300 K. The background emission from non-doped oxysulfides was measured with the resolution of 20 Å, the temperature of the samples being 100 K. The experimental set-up has been described earlier [6, 7].

0.5% of host cations were replaced by activator ions in the samples. Only one activator was used for each sample.

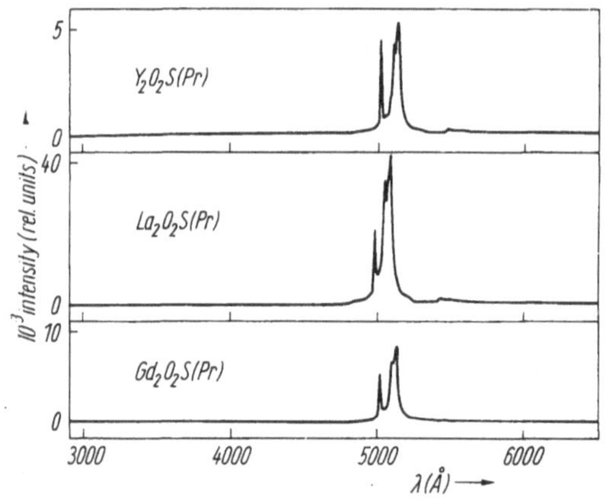

Fig. 1. The spectra of Pr^{3+} in Y_2O_2S, La_2O_2S, and Gd_2O_2S

The powders of rare-earth oxysulfates were prepared as Haynes and Brown [5] have reported. Oxysulfates were reduced to oxysulfides by heating at 850 to 900 °C, first for 10 min in an argon atmosphere and then for 30 min in a CO atmosphere. The X-ray diffraction patterns of non-doped La_2O_2S and Gd_2O_2S agreed well with the results of Haynes and Brown, but Y_2O_2S showed more peaks than reported by these authors. Also some lanthanides had more peaks in Y_2O_2S than in La_2O_2S and Gd_2O_2S. The extra peaks were due to Y_2O_3.

3. Results and Discussion

The spectra of a given lanthanide ion in Y_2O_2S, La_2O_2S, and Gd_2O_2S are very similar. Yet small differences occur with respect to positions, shapes, and intensities of lines. The spectra of Pr^{3+} in these hosts are shown in Fig. 1. The lines measured from Pr-activated La_2O_2S have shifted towards higher energies with respect to the corresponding lines of Pr-activated Y_2O_2S and Gd_2O_2S. Differences in the positions of the lines are easily understood qualitatively: as the inter-

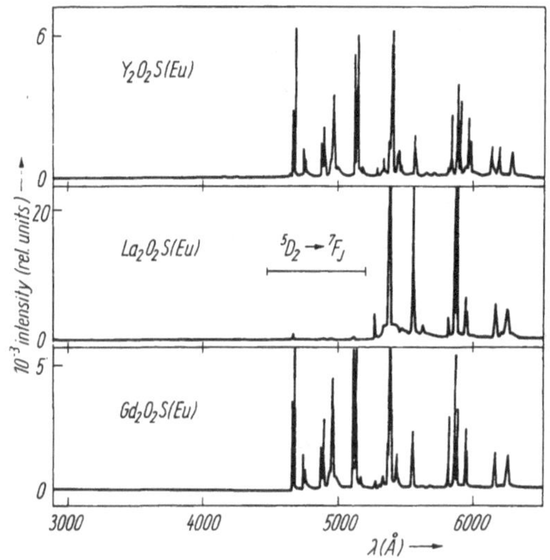

Fig. 2. The spectra of Eu^{3+} in Y_2O_2S, La_2O_2S, and Gd_2O_2S

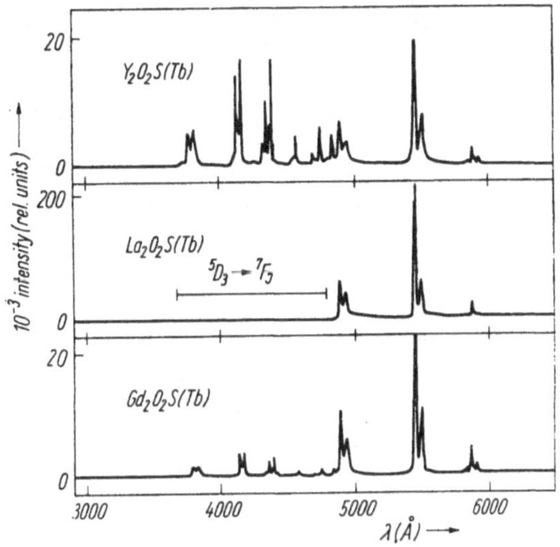

Fig. 3. The spectra of Tb^{3+} in Y_2O_2S, La_2O_2S, and Gd_2O_2S

ionic distances increase, the energy levels of a lanthanide ion tend to approach the energy levels of a free ion [2].

Fig. 2 shows the spectra of Eu^{3+}, and Fig. 3 the spectra of Tb^{3+} in the three host crystals. An interesting feature in the spectra of Eu^{3+} is the fact that the peaks in the range 4200 to 5300 Å in the case of La_2O_2S are very weak compared with those measured from Y_2O_2S and Gd_2O_2S. In the spectrum of Tb^{3+} in La_2O_2S the peaks at short wavelengths are missing completely. The same peaks of Tb^{3+} in Gd_2O_2S are weaker than those in Y_2O_2S. The weak peaks correspond to the transitions from the $^5D_2$ level to the ^7F levels in the case of Eu^{3+} and from the $^5D_3$ level to the ^7F levels in the case of Tb^{3+}. The peaks corresponding to the transition $^6P_{7/2} \rightarrow {}^8S_{7/2}$ of Gd^{3+} (3100 to 3200 Å) were missing almost completely in the case of La_2O_2S at 300 K, but at 100 K they were stronger in La_2O_2S than in Y_2O_2S.

Fig. 4 shows the spectra of Nd^{3+}. In this case the peaks are stronger in La_2O_2S

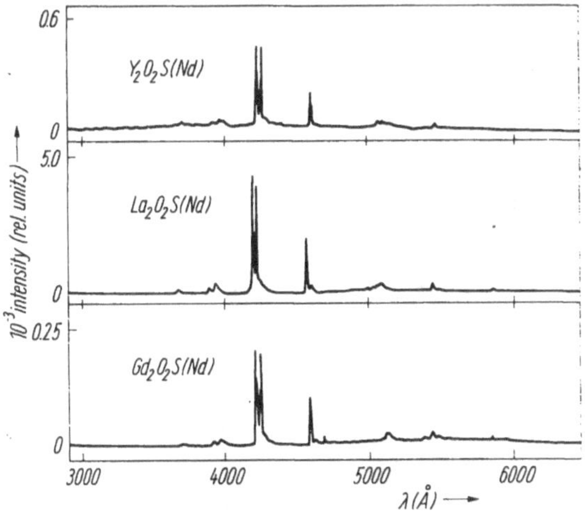

Fig. 4. The spectra of Nd^{3+} in Y_2O_2S, La_2O_2S, and Gd_2O_2S

even at short wavelengths. The lowest energy level from which the transition corresponding to the peaks at 4200 Å can originate is $^2D_{5/2}$. This energy level is above the $^5D_2$ level of Eu^{3+}. Er^{3+} has also stronger peaks at short wavelengths (about 4100 Å) in La_2O_2S than in Y_2O_2S and Gd_2O_2S.

Fonger and Struck (FS) [4] have explained the quenching of the transition $^5D_2 \rightarrow {}^7F$ of Eu^{3+} in La_2O_2S with the help of thermally excited transitions from the characteristic ^5D states of Eu^{3+} to charge-transfer states (CTS) of a host crystal. The feeding from upper ^5D states to CTS causes a decrease of the peaks originating from those states. The feeding from CTS to lower ^5D states, on the other hand, increases the intensity of the peaks corresponding to the transitions from lower ^5D states to ^7F states. According to FS the bottom of the CTS in La_2O_2S is at about the same level as the $^5D_3$ state of Eu^{3+}. The $^5D_3$ level is for this reason well quenched even at 100 K. The $^5D_2$ level is below the $^5D_3$ level, and hence the former level quenches at higher temperatures. The bottom of the CTS in Y_2O_2S lies higher than in La_2O_2S, and thus the quenching at the same temperature is also less complete. This model explains well the quenching of the $^5D_2$ state of Eu^{3+} in La_2O_2S. The $^5D_3$ level of Tb^{3+} lies about 5000 cm^{-1} higher than the $^5D_2$ level of Eu^{3+}. Therefore it is clear that the $^5D_3$ state of Tb^{3+} in La_2O_2S is completely quenched at 300 K. The fact that the peaks of Tb^{3+} in Gd_2O_2S at short wavelengths are more quenched than those in Y_2O_2S indicates that the bottom of the CTS in Gd_2O_2S lies lower than in Y_2O_2S.

The emitting level of Nd^{3+} is above the $^5D_2$ level of Eu^{3+}, and thus it should also be quenched. From Fig. 4 it can be seen, however, that the transitions originating from this level are not quenched even in La_2O_2S. The situation is the same in the case of Er^{3+}. It seems that the quenching effect is stronger for the lanthanide ions in the middle of the lanthanide series than for the others.

The intensities of the strongest peak for each lanthanide in Y_2O_2S, La_2O_2S, and Gd_2O_2S have been plotted in Fig. 5. These intensities are given in the same units as those from MgO and CaO [8]. The intensities have been corrected by taking into account the non-linearity of the photomultiplier and of the grating. The intensity of Gd^{3+} has been omitted, because the intensity of Gd^{3+} in Gd_2O_2S

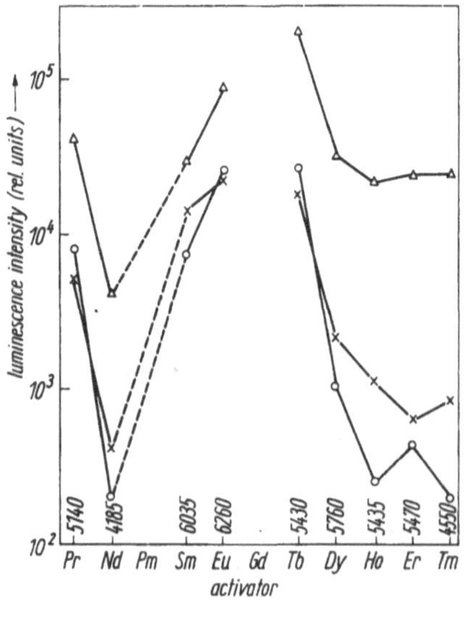

Fig. 5. The relative intensities of the strongest peaks of lanthanides in Y_2O_2S (×), La_2O_2S (△), and Gd_2O_2S (○) with 0.5% RE. Owing to the similarity of the spectra the strongest peaks in all the spectra are in almost the same positions. The wavelengths (Å) of these peaks have been denoted in the figure

Fig. 6. The background emission from Y_2O_2S, La_2O_2S, and Gd_2O_2S. T = 100 K

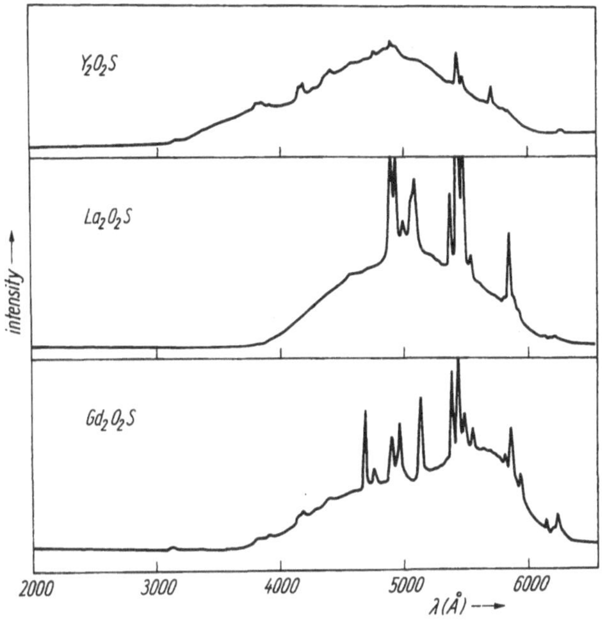

is very weak and the intensity of Gd^{3+} in La_2O_2S is very weak at 300 K. The figure shows that Eu^{3+} and Tb^{3+} at long wavelengths have greater intensities in La_2O_2S and Gd_2O_2S than in Y_2O_2S. La_2O_2S is the most efficient host of the three oxysulfides for all the lanthanides presented, but especially for the heavy ones.

Fig. 6 shows the broad emission bands from non-doped oxysulfides. The spectra have been measured with ten times greater slits than the characteristic emission of lanthanides. The original oxides for the oxysulfides were 99.999% pure. Owing to the weakness of the background emission, however, the characteristic emission of the impurity lanthanides appears clearly. The background emission is so weak that it did not disturb the measurement of the fluorescence of lanthanides when the activator concentration was 0.5%.

Acknowledgement

The author thanks Prof. E. Lähteenkorva for many valuable discussions during these experiments.

References

[1] D. J. Newman and G. E. Stedman, J. Phys. Chem. Solids **32**, 535 (1971).

[2] O. J. Sovers and T. Yoshioka, J. chem. Phys. **51**, 5330 (1969).

[3] K. A. Wickersheim and R. A. Buchanan, Appl. Phys. Letters **17**, 184 (1970).

[4] W. H. Fonger and C. W. Struck, J. chem. Phys. **52**, 6364 (1970).

[5] J. W. Haynes and J. J. Brown, Jr., J. Electrochem. Soc. **115**, 1060 (1968).

[6] H. Ratinen, Acta Polytechn. Scand., Ser. Ch. No. 107 (1971).

[7] H. Ratinen, Research Reports of the Institute of Technical Physics, University of Oulu, No 7 (1972).

[8] H. Ratinen, phys. stat. sol. (a), **12**, 175 (1972).

(Received May 8, 1972)

phys. stat. sol. (a) 12, 453 (1972)

Subject classification: 1.1; 22.1.2

Department of Applied Physics, Faculty of Engineering, University of Tokyo, Hongo, Bunkyo-ku, Tokyo

Improvement in the Visibility of Crystal Moiré Fringes with the Use of a Parallel X-Ray Beam

By

H. Hashizume, H. Ishida, and K. Kohra

Crystal moiré fringes in monolithic bi-crystals of silicon have been observed with a highly parallel X-ray beam obtained from an asymmetric diffraction. An appreciable improvement in fringe visibility has been achieved over the scanning method utilizing a slit-collimated X-ray beam. This is explained to be due to a high contrast of the standing wave reproduced outside the crystal and a weak intensity of the beams carrying no moiré information. A discussion is also given on the spatial coherence of the X-ray beam obtained from the asymmetric Bragg-case diffraction of a single crystal.

Kristallmoiréstreifen in monolithischen Silizium-Bikristallen wurden mit einem durch asymmetrische Beugung gewonnenen, sehr parallelen Röntgenstrahl beobachtet. Eine merkliche Verbesserung der Streifensichtbarkeit wurde gegenüber der Abtastmethode durch Benutzung des mit einem Spalt kollimatisierten Röntgenstrahls erhalten. Dies wird dadurch erklärt, daß ein stehendes Wellenfeld mit hohem Kontrast außerhalb des Kristalls gebildet wird und daß die keine Moiréinformationen tragenden Strahlen schwach sind. Es wird die räumliche Kohärenz des Röntgenstrahls diskutiert, der durch die asymmetrische Beugung des Einkristalls im Braggfall gewonnen wird.

1. Introduction

Recently the X-ray moiré technique has become of practical use to detect extremely small changes in orientation and interplanar spacing of crystal lattices. The sensitivity of the moiré method is so high that a lattice rotation $\varphi \approx 10^{-8}$ rad and dilatation (or compression) $\delta d/d \approx 10^{-8}$ can be detected. This high sensitivity, on the other hand, makes it quite unusual to observe moiré patterns in X-ray diffraction topographs obtained from two incidentally superposed crystals [1]. The moiré method was once utilized by Chikawa [2] to study small changes in the lattice spacing of epitaxially grown crystals. Artificial production of X-ray moiré fringes was made possible with the invention of the X-ray interferometers by Bonse and Hart [3]. Extensive studies have since been made on the measurement of minute lattice distortions, introduced by external or internal means, by making use of various types of X-ray interferometers [4, 5]. It has recently become possible to align two independent crystals highly parallel as needed for the production of X-ray moiré fringes of detectable spacings [6 to 8]. In particular, the method developed by Brádler and Lang [6] is adequate for X-ray moiré topography of single crystals.

Observations of crystal moiré patterns in the previous works [1 to 8] were exclusively made with the scanning method utilizing a slit-collimated X-ray beam [9]. Because the angular spread of a slit-collimated beam is much wider than the reflective range of the crystals under investigation, such an incident beam is not quite adequate for obtaining moiré fringes of high visibility, except when the crystals to be examined are thick enough for the anomalous transmission to occur. It is expected that, if we use a nearly parallel X-ray beam in place of a slit-collimated beam, moiré fringes would be observed with higher visibility.

In the present work, we study the improvement of fringe visibility by using a parallel X-ray beam obtained from an asymmetric diffraction of a single crystal. In order to produce moiré fringes in a simple way, we have employed the monolithic bi-crystal method, which is easy to be performed and requires no delicate adjustment of crystal orientations.

2. Experimental Method

Bi-crystals as schematically shown in Fig. 1 were fabricated from a crystal of silicon grown with the pull-method. The two crystal slices (1) and (2) are monolithic and preserve the original crystal coherence. The narrow gap between them was milled away with a diamond blade as thin as 0.15 mm. After a final chemical polishing the gap width s was about 0.2 mm. Two bi-crystals were studied: one with thicknesses of crystals (1) and (2), t_1 and t_2, 0.54 and 1.75 mm, respectively, and the other with $t_1 = t_2 = 0.5$ mm. The surfaces of the crystal slices were nearly normal to the diffracting plane (220). The bi-crystal and the collimator crystal, which utilizes an asymmetric Bragg-case diffraction, were set in the double-crystal arrangement of parallel setting for the 220 reflection with MoKα (Fig. 2).

The collimator was prepared from a high-purity and dislocation-free crystal of silicon. Its surface was cut so as to make an angle $\alpha = 10°26'$ to the diffracting plane (220). The X-ray beam obtained from the collimator was about 0.22″ in angular spread and 10 mm in spatial width.

For the X-ray source a conventional sealed-off X-ray tube with a Mo target was operated at 40 kV and 16 mA. The apparent focal size was 0.4 mm high and 0.8 mm wide. Moiré patterns in the O (transmitted) and H (diffracted)

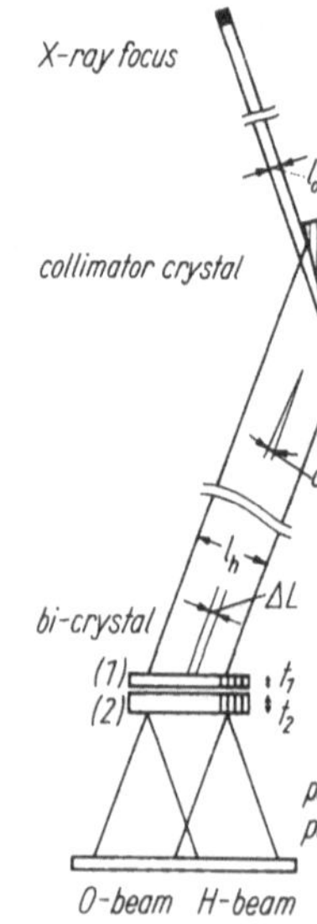

Fig. 2. Double-crystal arrangement of parallel setting employed for observing moiré fringes in the bi-crystal. The collimator crystal utilizes the asymmetric Bragg-case diffraction

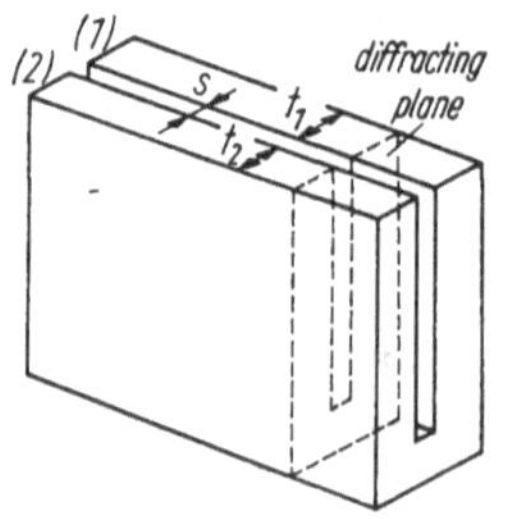

Fig. 1. Schematic diagram of the bi-crystal

beams were recorded on Sakura nuclear plates, emulsion El, 50 μm thick, placed about 2 cm behind the bi-crystal. The distances from the X-ray source to the collimator crystal and from the collimator to the specimen were 35 and 20 cm, respectively. No attempt was made to separate the two components of the Kα doublet.

For comparison, observations were also made with the scanning method utilizing a slit-collimated indicent beam. A fine-focus X-ray source with a Mo target was operated at 50 kV and 1.2 mA.

3. Experimental Observations

Fig. 3 reproduces moiré patterns obtained from a bi-crystal with $t_1 = 0.54$ mm, $t_2 = 1.75$ mm, and $s = 0.26$ mm. Fig. 3a and 3b, which correspond to the O and H beams respectively, were obtained with the parallel incident beam at the

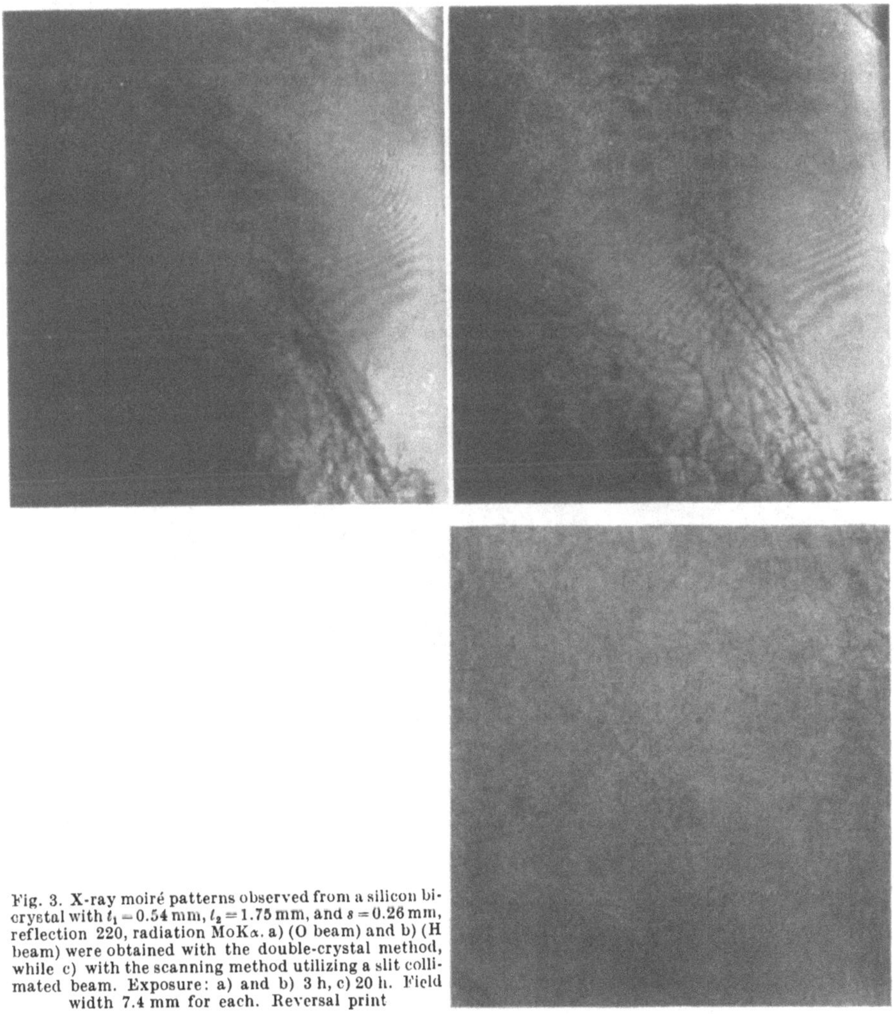

a b c

Fig. 3. X-ray moiré patterns observed from a silicon bi-crystal with $t_1 = 0.54$ mm, $t_2 = 1.75$ mm, and $s = 0.26$ mm, reflection 220, radiation MoKα. a) (O beam) and b) (H beam) were obtained with the double-crystal method, while c) with the scanning method utilizing a slit collimated beam. Exposure: a) and b) 3 h, c) 20 h. Field width 7.4 mm for each. Reversal print peak position of the rocking curve, while Fig. 3c was taken with the scanning method. It is evident that a higher fringe visibility is obtained in Fig. 3a and 3b. This moiré pattern originates presumably from macroscopic elastic strains introduced on fabricating the bi-crystal: the lattice of one crystal slice is slightly rotated and/or dilatated with respect to the other one. Some dislocation images in Fig. 3a and 3b are blurred or doubled because of the dispersion of the $K\alpha$ doublet, but the moiré fringes are observed with a sharpness comparable to that in Fig. 3c, which was obtained with the $K\alpha_1$ only.

It is to be noted in Fig. 3a and 3b that the positions of black and white fringes are the same in O and H beams. When the bi-crystal was turned and the X-ray beam was first incident on the slice with a thickness 1.75 mm ($= t_1$), the same fringe positions were observed for the H beam, but the reversed ones for the O beam. This observation is adequately explained by the diffraction theory for crystal moirés [10]. Incidentally, for the $MoK\alpha$ radiation used in the experiments, the crystal thicknesses 0.54 and 1.75 mm correspond to $\mu t = 0.77$ and 2.50, respectively, where μ is the linear absorption coefficient of silicon.

Moiré fringes in this specimen were further studied at various angular positions of the diffraction by using the parallel incident beam. It was observed that the fringe positions in both the O and H beams remained unchanged with the deviation from the exact Bragg condition, but the fringe visibility gradually decreased. This observation confirms that the observed fringes are moiré fringes.

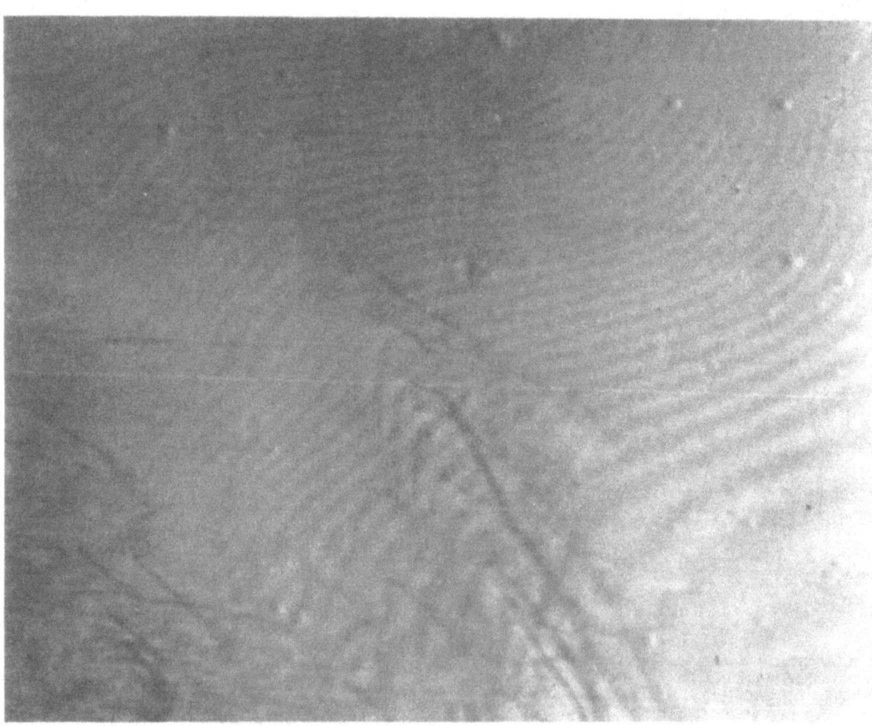

Fig. 4. Showing the central part of Fig. 3b in a large scale. Notice fringe-number jumps at dislocation outcrops. Field width 5.2 mm. Reversal print

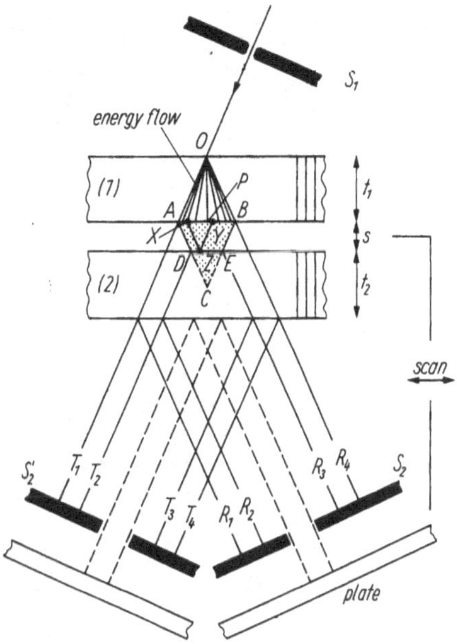

Fig. 5. Image formation process of moiré fringes in a bi-crystal for a slit-collimated incident beam

Fig. 4 shows the central part of Fig. 3b in a large scale. Some fringes in Fig. 4 run through dislocation lines in the central part of the field. At some but not all dislocation outcrops, one can recognize jumps in the number of fringes, which certainly correspond to the extra half plane(s) of the dislocations. Most dislocation outcrops which do not accompany fringe-number jumps are on the X-ray exit surface of crystal 2, which was confirmed in stereo-pair topographs taken by rotating the specimen about the normal to the diffracting plane. Similar observations were previously made by Lang on bi-crystals of natural quartz [7].

As will be explained in a later section of this paper, a poor fringe visibility in the topograph obtained with the scanning method is due in part to marginal beams involved, which do not carry moiré information. These beams are strong in intensity unless the specimen is thick enough, and smear out the moiré fringes in some cases. A way to eliminate such beams and improve the fringe visibility is to employ a narrow second slit like S_2 and S_2' in Fig. 5. Fig. 6 illustrates an improvement in fringe visibility thus achieved. Both topographs in Fig. 6 were obtained with the scanning method from another bi-crystal with $t_1 = t_2 = 0.5$ mm, but for Fig. 6a a second slit was used wide enough to pass the full width of the diffracted ribbon beam, while a narrower second slit 0.05 mm in width was used for Fig. 6b. Broad fringes running from top right to bottom left, which are more distinctly seen in Fig. 6b, are Pendellösung fringes due to the non-uniform thickness of the crystals. It is interesting to note that the moiré fringes are almost invisible in the centre of each Pendellösung fringe. Dislocations in the crystals are also observed with a higher contrast in Fig. 6b, although not all the dislocations seen in Fig. 6a are visible in Fig. 6b.

4. Visibility of Moiré Fringes

We now consider in some detail the image formation process of moiré fringes in a bi-crystal. The improvement of fringe visibility achieved by utilizing a crystal-collimated incident beam will be explained thereby. Suppose that two crystal slices (1) and (2), which have slightly different orientations and/or lattice spacings, are aligned to diffract the incident X-rays successively. The active diffracting planes in the two crystals are assumed to be normal to the crystal surfaces.

4.1 A slit-collimated beam

The angular divergence of a slit-collimated X-ray beam is not smaller than a few minutes of arc, which is much larger than the reflective range of the crystals

a

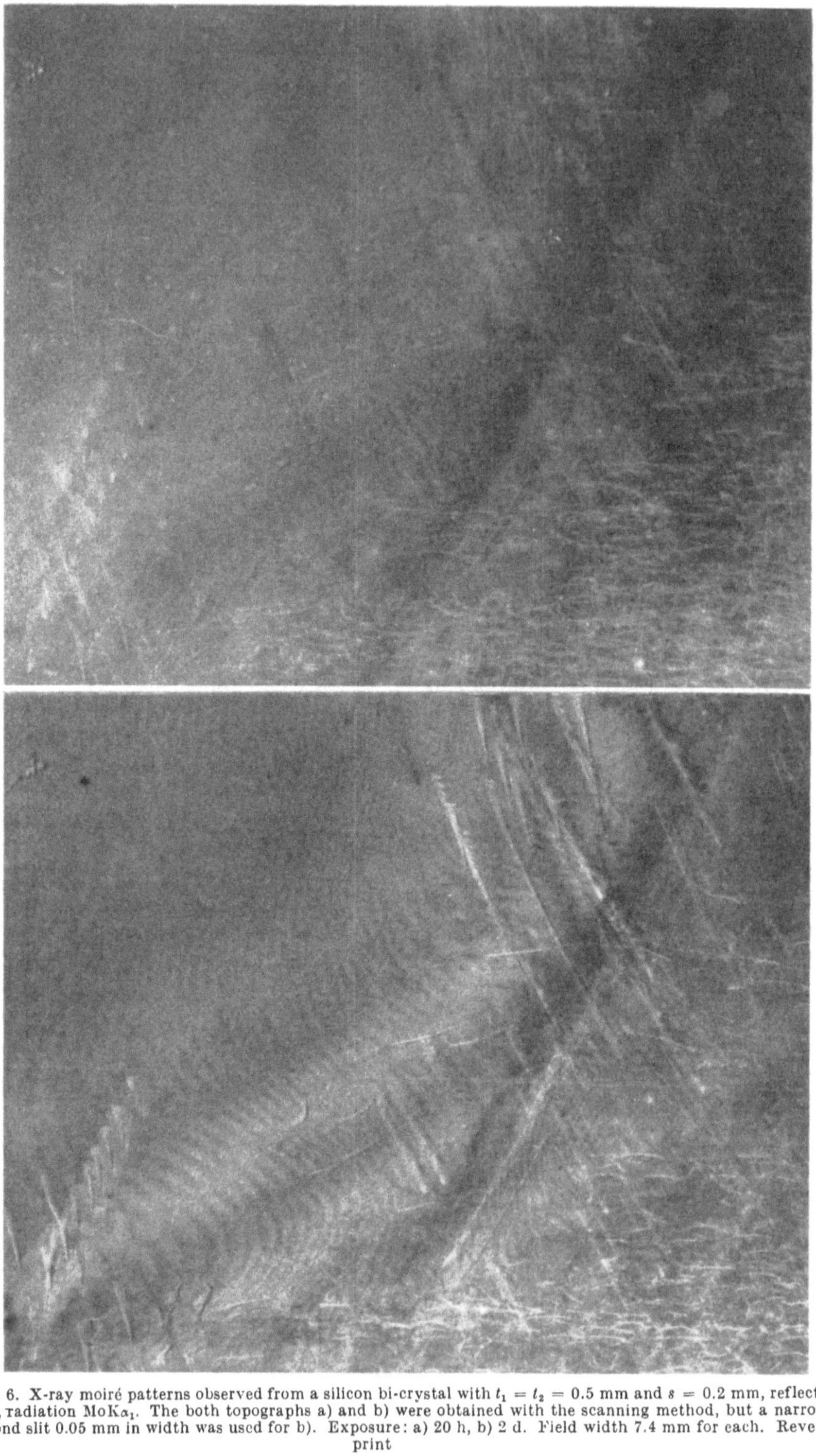

b

Fig. 6. X-ray moiré patterns observed from a silicon bi-crystal with $t_1 = t_2 = 0.5$ mm and $s = 0.2$ mm, reflection 220, radiation Mo$K\alpha_1$. The both topographs a) and b) were obtained with the scanning method, but a narrower second slit 0.05 mm in width was used for b). Exposure: a) 20 h, b) 2 d. Field width 7.4 mm for each. Reversal print under investigation. With such a beam incident on crystal 1, which is assumed to be nearly perfect, the X-ray energy flows in the crystal over the full Borrmann fan OAB as shown in Fig. 5. A small part of the incident beam satisfies approximately the exact Bragg condition and gives rise to energy flow around the diffracting plane OP, while most of the beam produces energy flows along OA and OB. Behind the crystal, the O and H waves from AB overlap in the triangle ABC and interfere with each other to form a standing wave, which reproduces the periodicity of atomic planes in the crystal 1. Because a lattice moiré is the result of the interference of this standing wave and the atomic planes in the crystal 2, the visibility of moiré fringes to be produced is closely related to the contrast of the standing wave. Here the contrast of the standing wave means the ratio of the amplitude of the antinodal planes to that of the nodal planes of the standing wave.

The amplitude of the standing wave at Z on the surface of crystal 2 is determined by the amplitudes of the O and H waves from Y and X, respectively. The latter amplitudes can be calculated from the spherical-wave theory of diffraction [11] as schematically shown in Fig. 7 unless crystal 1 is too thick. Oscillation of the amplitudes due to Pendellösung effects are neglected in Fig. 7 for simplicity. Obviously, the standing wave is high in contrast when the two interfering waves have comparable amplitudes. From the amplitude distributions in Fig. 7, it is seen that the contrast of the standing wave, and hence the visibility of the moiré fringes, is high in a middle region of DE, and low in the vicinity of D and E in Fig. 5. A topograph of a large specimen area is obtained in the scanning method by translating the specimen and the photographic plate with respect to the incident ribbon beam. With translation, the incident beam scans over the entrance surface of crystal 1, whereby the visibility of the moiré fringes becomes averaged. This results in a low fringe visibility in the topograph obtained with the scanning method.

When the gap width between crystals 1 and 2 is finite, another factor arises which reduces the fringe visibility. The O and H beams from AB are further enlarged in width on being diffracted through crystal 2, and ultimately extended over widths T_1T_4 and R_1R_4, respectively, as shown in Fig. 5. Within these widths, moiré information is contained between T_2T_3 and R_2R_3. The other parts of the beams do not carry moiré information and contribute only to the background. In particular, the marginal beams along T_1, R_1, and R_4 are strong in intensity except when one or both of the two crystals are sufficiently thick for the anomalous transmission to occur because they arise from energy flows along OA

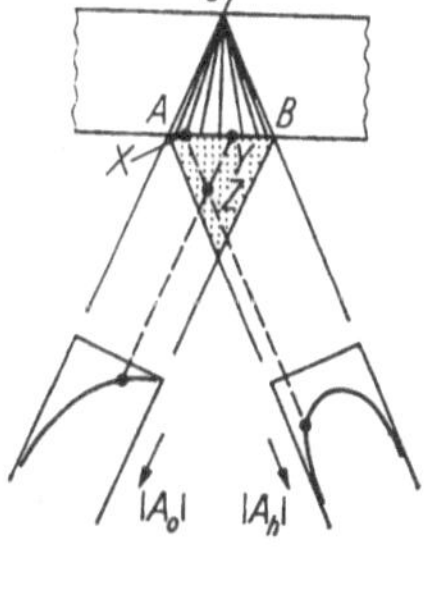

Fig. 7. Amplitude distributions of the O and H waves diffracted from a thin crystal for a slit-collimated incident beam

and OB in crystal 1, to which most of the incident X-ray energy is condensed. A reduction of the fringe visibility due to these beams can be suppressed by eliminating them with a narrow second slit. It must be noted that, if we use a still narrower second slit as used for taking Fig. 6b, a higher fringe visibility is obtained because only such moiré information as produced by a high-contrast part of the standing wave is selected. The exposure time, however, is prolonged considerably because only a small fraction of the incident X-ray energy reaches the photographic plate through the slit.

It was noticed in Fig. 6 that the moiré fringes are invisible in the centre of Pendellösung fringes. This phenomenon is explained in terms of the contrast of the standing wave, if we assume that the latter fringes are caused by the non-uniform thickness of crystal 1. When the crystal thickness corresponds to entire or half multiples of the extinction distance, the amplitude of either O or H wave becomes zero, so that no contrast of the standing wave is obtained at such crystal thicknesses of crystal 1, which results in a vanishing visibility of moiré fringes.

If crystal 1 is sufficiently thick for anomalous transmission, the X-ray energy flow in the crystal is limited in effect to a narrow angular region around the diffracting plane OP. Thus appreciable intensities of the O and H waves are obtained only from a small region around P in Fig. 5. This results in a considerable reduction of the size of the triangle where the standing wave is formed, although the standing wave thus produced is high in contrast because of a comparable amplitude of the O and H waves. Consequently, in order that the entrance surface of crystal 2 intercepts the triangle of the standing wave, which is a necessary condition for a moiré pattern to be produced, the gap width must be made much smaller than the thickness of crystal 1.

4.2 A nearly parallel beam obtained with the asymmetric diffraction

It is now evident that moiré fringes in a bi-crystal are observed with a higher visibility by making use of an incident beam of parallel X-rays. Discussions will then be made with special reference to the asymmetric diffraction utilized for the collimator crystal.

It is well established that an X-ray beam not only highly parallel but also wide in spatial cross section is obtained in the double-crystal arrangement of parallel setting which utilizes the asymmetric Bragg-case diffraction for the first or collimator crystal [12 to 15]. The angular spread ω_h and spatial width l_h are given by

$$\omega_h = \sqrt{b}\,\omega_s\,, \tag{1a}$$

$$l_h = \frac{l_0}{b}, \tag{1b}$$

respectively, where ω_s is the angular spread of the selective reflection for the symmetric Bragg-case diffraction of the crystal, l_0 the spatial width of the beam incident on the collimator, and b the asymmetry factor defined by

$$b = \frac{\sin(\theta_B - \alpha)}{\sin(\theta_B + \alpha)},$$

θ_B being the Bragg angle (Fig. 2). With such a parallel beam incident on a bi-crystal, the direction of energy flow excited in crystal 1 does not fan out

Fig. 8. The image formation process of a lattice moiré in a bi-crystal for a parallel incident beam of angular spread ω_h and spatial coherence width ΔL. Note that beams between T_1T_2, T_3T_4, R_1R_2, R_3R_4, which do not carry moiré information, are not so strong in intensity

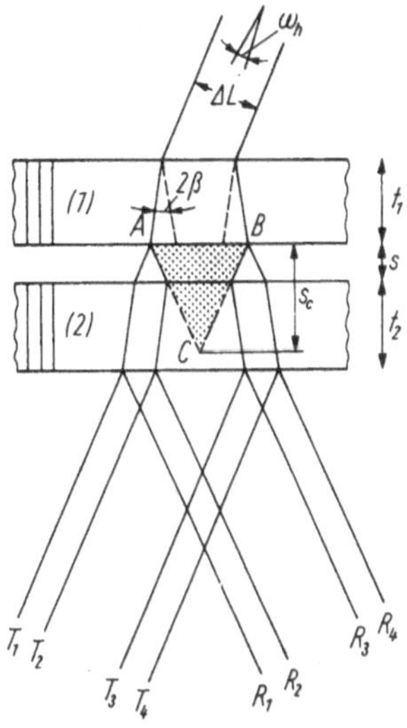

over the Borrmann fan, but is limited within a certain angular range. When crystal 1, which is assumed to be perfect, is set so as to satisfy the exact Bragg condition, the energy flow occurs only in an angular range $\pm\beta$ around the diffracting plane, where β is approximately given by

$$\frac{\tan\beta}{\tan\theta_B} = \sqrt{\frac{b}{1+b}}. \tag{2}$$

Consequently, the smaller the aysmmetry factor b of the collimator is, the smaller the angular range of energy flow becomes. This does not necessarily mean, however, that the triangular region where the standing wave is formed behind crystal 1 becomes correspondingly smaller in size, as the X-ray beam obtained from the collimator crystal possesses a finite spatial width of coherence, and the standing wave can also be created by interference of the O and H waves generated by two rays incident at different positions on the crystal surface as indicated in Fig. 8.

The coherence width ΔL of the X-ray beam used in the present experiment can be calculated from the dynamical theory of Bragg-case diffraction for a divergent incident beam [16]. It is given by

$$\Delta L = K\,\frac{\lambda}{\pi\,\omega_h}, \tag{3}$$

where λ is the X-ray wavelength and K a constant of the order of unity. This width corresponds to the width of the first Pendellösung fringe formed in a thick crystal by a divergent incident beam [17]. It must be noted that the coherence width given above does not depend on the distance from the X-ray source to the specimen, being in contrast with that deduced by Batterman and Hildebrandt [18]. Note also that (3) represents the uncertainty principle

$$\Delta p\,\Delta L \approx h$$

in wave mechanics, because we have $\omega_h/2\lambda = k\,\omega_h/2 = \Delta k$ and $h\,\Delta k = \Delta p$,

h being Planck's constant. Substituting (1a) into (3) we have

$$\Delta L = K \frac{\lambda}{\pi \sqrt{b}\, \omega_s}. \tag{4}$$

The height s_c of the triangle where the standing wave is formed is then given by

$$s_c = \frac{\Delta L}{2 \sin \theta_B} + t_1 \frac{\tan \beta}{\tan \theta_B} = K \frac{\lambda}{2 \pi \sqrt{b}\, \omega_s \sin \theta_B} + t_1 \sqrt{\frac{b}{1+b}}. \tag{5}$$

This determines the maximum permissible gap width between crystals 1 and 2, because a moiré pattern can be produced when crystal 2 intercepts the triangle of the standing wave, as noted before.

In the experimental conditions used to obtain Fig. 3a and b, we have $s_c = 58\,K + 54\ \mu m$ with the values $\lambda = 0.71$ Å, $b = 0.01$, $\omega_s = 1.1 \times 10^{-5}$, $\sin \theta_B = 0.19$, and $t_1 = 0.54$ mm. Consequently, in order that s_c becomes larger than 0.26 mm, which is the gap width of the bi-crystal employed, K should be greater than 3.5. On the other hand, $K = 3.5$ is almost the maximum value that can be expected from theory [16]. When we consider the high visibility with which the moiré fringes are observed in Fig. 3a and b, the K-value obtained above seems to be a little too large. Note that if $K = 3.5$ the triangle of the standing wave just reaches crystal 2 and, therefore, the ratio of the beam carrying moiré information to that contributing to the background is practically zero. Probably, the angular range of X-ray energy flow in crystal 1 would be greater than that expected from (2) because of elastic distortions of the crystal involved. As demonstrated by the authors [19, 20], the direction of X-ray energy flow in the diffracting crystal is highly sensitive even to elastic strains of the order of 10^{-6}. If we assume the second term in (5) to be three times larger owing to elastic strain, we have s_c appreciably greater than the gap width 0.26 mm even with $K = 2.0$.

5. Discussion and Conclusions

It is worth noticing in (4) that the spatial width of coherence of a crystal-collimated X-ray beam is inversely proportional to $\sqrt{b}$. This suggests the possibility that the gap width in a bi-crystal can be increased beyond the thickness of crystal 1 if we use a reduced b-value for the collimator crystal. Obviously, the gap width cannot exceed the thickness of crystal 1, as far as we employ a slit-collimated incident beam. A large gap width leads not only to an easy fabrication of the bi-crystal, but also to the interesting fact that a moiré pattern is observed from distant parts of a single crystal.

There is still another advantage that can be obtained from the use of an asymmetric diffraction for the collimator crystal. As indicated by (1b), the X-ray beam obtained possesses a large cross section, so that a wide area of the specimen is simultaneously observed without employing any scanning mechanism. This usually leads to a reduction of the exposure time for photography. It is also important when a dynamic observation of a moiré-fringe system is made with an X-ray video imaging device.

In conclusion, three advantages are obtained from the use of a crystal-collimated X-ray beam: (i) high visibility of moiré fringes, (ii) large permissible gap width in the bi-crystal which results from a large coherence width of the beam, and (iii) short exposure time for photography. These advantages are also available when moiré fringes are observed from an X-ray interferometer of Bonse-Hart types [3, 4].

References

[1] A. R. Lang and V. F. Miuscov, Appl. Phys Letters **7**, 214 (1965).
[2] J. Chikawa, Appl. Phys. Letters **7**, 193 (1965); J. Phys. Chem. Solids, Suppl. **1**, 817 (1967).
[3] U. Bonse and M. Hart, Appl. Phys. Letters **6**, 155 (1965); **7**, 99 (1965); Z. Phys. **188**, 154 (1965); **194**, 1 (1966); Acta cryst. **A24**, 240 (1968).
[4] U. Bonse and M. Hart, Z. Phys. **190**, 455 (1966).
[5] G. Christiansen, L. Gerwars, and A. L. Andersen, J. appl. Cryst. **4**, 370 (1971).
[6] J. Brádler and A. R. Lang, Acta cryst. **A24**, 246 (1968).
[7] A. R. Lang, Nature **220**, 652 (1968).
[8] U. Bonse and E. Te Kaat, Z. Phys. **214**, 16 (1968); **243**, 14 (1971).
[9] A. R. Lang, J. appl. Phys. **29**, 597 (1958); Acta cryst. **12**, 249 (1959).
[10] H. Ishida, to be published.
[11] N. Kato, Acta cryst. **14**, 526 (1961); **14**, 627 (1961).
[12] K. Kohra and S. Kikuta, Acta cryst. **A24**, 200 (1968).
[13] S. Kikuta and K. Kohra, J. Phys. Soc. Japan **29**, 1322 (1970).
[14] S. Kikuta, J. Phys. Soc. Japan **30**, 222 (1971).
[15] T. Matsushita, S. Kikuta, and K. Kohra, J. Phys. Soc. Japan **30**, 1136 (1971).
[16] H. Ishida and K. Kohra, to be published.
[17] T. Uragami, J. Phys. Soc. Japan **27**, 147 (1969).
[18] B. W. Batterman and G. Hildebrandt, Acta cryst. **A24**, 150 (1968).
[19] H. Hashizume and K. Kohra, J. Phys. Soc. Japan **31**, 204 (1971).
[20] H. Hashizume, J. Phys. Soc. Japan **31**, 1124 (1971).

(Received May 2, 1972)

phys. stat. sol. (a) **12**, 465 (1972)

Subject classification: 13.1; 19; 22.1.2

Physics Laboratory I, Technical University of Denmark, Lyngby

The Dependence of Hole Resonances in Si on the Wave Vector Component along the Direction of Magnetic Induction (II)

By

M. Owner-Petersen

An investigation is made of the anisotropy of the microwave absorption spectra from the valence band in Si, when the magnetic induction is in the (1$\bar{1}$0) plane. A comparison between calculated absorption spectra taking into account non-zero values of the wave vector component along the direction of magnetic induction, and experimentally observed cyclotron resonance spectra at 45 and 70 GHz is presented. The agreement seems to be satisfactory.

Cet article s'occupe de l'anisotropie des spectres de la résonance cyclotron dans la bande valence du Si, au cas où l'induction magnétique $\boldsymbol{B}$ est parallèle au plan symétrique (1$\bar{1}$0). L'article présent une comparaison entre des spectres calculés au condition que le composant parallele à $\boldsymbol{B}$ du vecteur d'onde est différent de zéro, et des spectres observés à 45 et à 70 GHz. La conformité paraît satisfaisante.

1. Introduction

In paper [1] a calculation of cyclotron resonance spectra in Si was presented for the magnetic induction $\boldsymbol{B}$ parallel to the [001], [111], and [110] directions. The calculations were performed allowing the wave vector k_B along the direction of $\boldsymbol{B}$ to be different from zero. The calculated spectra were compared with experimental cyclotron resonance spectra obtained at 45 GHz and approximately 1.2 K [1, 2]. For the choice of band parameters given in [1] good agreement was obtained. In this paper the calculations are extended to cyclotron resonance spectra for arbitrary angles between $\boldsymbol{B}$ and the [001] direction in the (1$\bar{1}$0) plane. The calculated spectra are compared with two sets of experiments, the one obtained at 45 GHz and 1.2 to 1.5 K [2], and the other at 70 GHz and 1.2 K.

2. Calculations

The procedure which has been followed when calculating the spectra is given in [1]. Using the well-known valence band Hamiltonian of diamond structure semiconductors [1 to 3], the lowest energy levels ε (in units of $\hbar e B/m_0$), the microwave-induced transition matrix elements, and the corresponding effective masses were calculated as functions of the normalized wave vector component ζ ($\zeta = k_B/\sqrt{e B/\hbar}$) along $\boldsymbol{B}$, and the angle θ between the [001] direction and $\boldsymbol{B}$ in the (1$\bar{1}$0) plane.

On Fig. 1 are shown the fifteen lowest central levels as functions of θ, taking $\zeta = 0$. The anisotropy is clearly seen to be very pronounced as should be expected since for the band parameters given in [1] the band term $(\gamma_3 - \gamma_2)$ mainly responsible for the anisotropic effects [3] is of the same order of magnitude as γ_1 and γ_2. Table 1 gives a labelling of the levels in the symmetry directions.

According to references [2] to [4] the four-dimensional Hamilton operator matrix is decoupled into two independent submatrices for **B** parallel to the [001] and [110] directions. Here the levels are written as $\varepsilon_p(n, r)$, where $p = 1$ corresponds to levels belonging to the upper left submatrix, and $p = 2$ corresponds to levels belonging to the lower right submatrix. The principal quantum number is n, which may take on positive integer values including zero, and r is a ladder classification number being either 1 or 2. For **B** parallel to the [111] direction the levels are written as $\varepsilon(n, r)$, where the principal quantum number n may take on integer values from -2 to $+\infty$, and where the ladder classification number r may be 1, 2, 3, and 4. The matrix elements M corresponding to transitions induced by a linearily polarized microwave field in the [1$\bar{1}$0] direction are governed by the following selection rules:

$$\zeta = 0$$

B || [001] and [110]

$$\varepsilon_p(n, r) \to \varepsilon_q(m, s):\ M \sim \delta_{p,q}\, \delta_{n-m,1+2l}$$

B || [111]

$$\varepsilon(n, r) \to \varepsilon(m, s):\ M \sim \delta_{n-m,\pm 1}$$

arbitrary θ

$$\varepsilon(n, r) \to \varepsilon(m, s):\ M \sim \delta_{n-m,1+2l}$$

$$\zeta \neq 0$$

B || [001] and [110]

$$\varepsilon_p(n, r) \to \varepsilon_q(m, s):\ M \sim \delta_{p,q}\, \delta_{n-m,1+2l} + \delta_{p-q,\pm 1}\, \delta_{n-m,2l}$$

B || [111]

$$\varepsilon(n, r) \to \varepsilon(m, s):\ M \sim \delta_{n-m,3l\pm 1}$$

arbitrary θ.

$$\varepsilon(n, r) \to \varepsilon(m, s):\ M \sim (\delta_{n,m}\, \delta_{r,s} - 1)$$

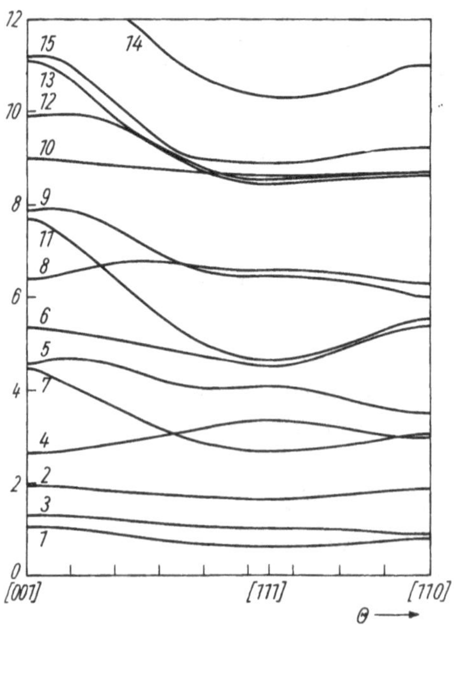

Fig. 1. The fifteen lowest central ($\zeta = 0$) levels ε in units of $\hbar eB/m_0$ as functions of the angle θ between **B** and the [001] direction in the (1$\bar{1}$0) plane

Table 1

n	[001]	[111]	[110]
1	ε_1 (0, 1)	ε (2,3)	ε_2 (0, 1)
2	ε_1 (0, 2)	ε (−2, 2)	ε_1 (0, 2)
3	ε_2 (0, 1)	ε (0, 1)	ε_1 (0, 1)
4	ε_2 (0, 2)	ε (0, 4)	ε_2 (0, 2)
5	ε_1 (1, 1)	ε (1, 1)	ε_1 (1, 1)
6	ε_1 (1, 2)	ε (−1, 2)	ε_1 (1, 2)
7	ε_2 (1, 1)	ε (3, 3)	ε_2 (1, 1)
8	ε_2 (1, 2)	ε (5, 3)	ε_2 (1, 2)
9	ε_1 (2, 1)	ε (2, 1)	ε_1 (2, 1)
10	ε_1 (2, 2)	ε (0, 2)	ε_2 (2, 2)
11	ε_2 (2, 1)	ε (4, 3)	ε_2 (2, 1)
12	ε_2 (2, 2)	ε (6, 3)	ε_1 (2, 2)
13	ε_1 (3, 1)	ε (3, 1)	ε_1 (3, 1)
14	ε_1 (3, 2)	ε (7, 3)	ε_2 (3, 2)
15	ε_2 (3, 1)	ε (1, 4)	ε_2 (3, 1)

Here l may take on all positive and negative integer values including zero. It should be mentioned that the anisotropy provides a strong mixing of the lowest levels, and for this reason the ladder classification r is rather arbitrary. Concerning the [001] and [110] directions $r = 1, 2$ is analogous to $+$, $-$ in [3].[1]) Given a value of r, s is chosen to make M numerically large when $s = r$.

From Fig. 1 to 3 in [1] it can be seen that in the three symmetry directions the variation of the effective masses as function of ζ is very dramatic. For the non-symmetry directions this turns out to be even more pronounced, since in principle all the degeneracies (cross-points) on the level versus ζ diagrams for the symmetry directions could be removed when departing from these directions. Fig. 2 shows the variation with θ of the large M extremal effective masses corresponding to transitions from the lowest levels, which according to [1] should correspond to large relative absorption amplitudes. Note that the dashed curves corresponding to non-central ($\zeta \neq 0$) extremal effective masses might change label several times. Knowing the levels, the effective masses, and the transition

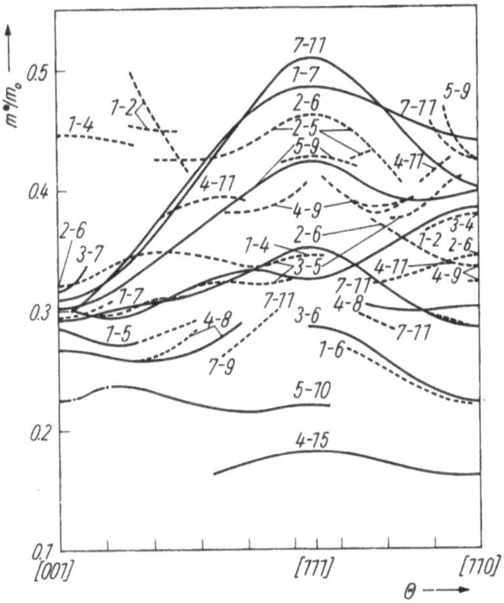

Fig. 2. The extremal ($dm^*/d\zeta = 0$) effective masses in units of m_0, corresponding to transitions from the six lowest levels as functions of the angle θ between $\boldsymbol{B}$ and the [001] direction in the (1$\bar{1}$0) plane. Full-line curves represent central ($\zeta = 0$) masses, and dashed curves represent non-central ($\zeta \neq 0$) masses

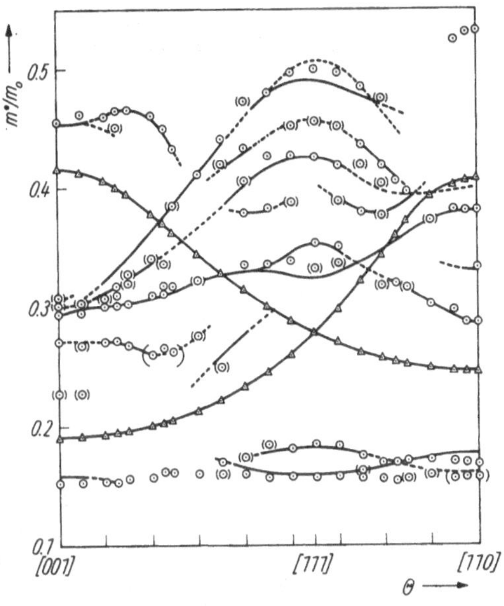

Fig. 3. A comparison between the peak masses in units of m_0 taken from a calculation of cyclotron resonance spectra using $F = 45$ GHz, $T = 2$ K, and $\omega\tau = 35$, and the experimentally obtained peak masses taken from cyclotron resonance spectra at 45 GHz and $T = 1.2$ to 1.5 K. The peak masses are shown as functions of the angle θ between $\boldsymbol{B}$ and the [001] direction in the (1$\bar{1}$0) plane. Solid curves correspond to reliable calculated peaks and dashed curves to more dubious calculated peaks. Points in circles correspond to reliable experimental peaks, and points in circles and paranthesis to more dubious peaks. Points in triangles correspond to electron resonance peaks

[1]) When comparing with [2], it should be mentioned that in the [001] and [110] directions the principal quantum number n here runs from 1 to infinity. Furthermore, it has turned out that concerning the [110] direction the calculations were incorrect, and hence a comparison should not be carried out here.

matrix elements as functions of θ and ζ, the cyclotron resonance spectra can be calculated for relevant values of θ, frequency F, temperature T, and relaxation time τ using formula (19) of [1].

3. Comparison between Calculations and Experiments

A set of cyclotron resonance spectra have been calculated as functions of θ for $F = 45$ GHz, $T = 2$ K, and $\omega\,\tau = 35$ ($\omega = 2\,\pi\,F$). The angle θ was running from 0° to 90° in steps of approximately 8°. From these spectra the effective masses of the peaks have been plotted as functions of θ shown by the solid and dashed lines on Fig. 3. The points represent experimental peaks. The experiments were performed at 45 GHz and $T = 1.2$ to 1.5 K [1, 2]. By comparing Fig. 2 and Fig. 3 the peaks can easily be identified and it is seen that $\zeta \neq 0$ transitions play an important role in the spectra. The "extra curve" appearing near $m^*/m_0 = 0.15$ on Fig. 3 is due to residual semi-classical light-hole transitions between high-lying levels, whereas the experimental points near $m^*/m_0 = 0.53$ in the [110] direction fit to the semi-classical heavy-hole transitions [2]. Fig. 4 shows a series of experimental and computer-calculated cyclotron resonance spectra. The experiments were performed at $F = 70$ GHz, $T = 1.2$ K, and at a microwave power level of approximately 10 μW. The corresponding calculated spectra were computed using $F = 70$ GHz, $T = 2.5$ K, and $\omega\,\tau = 40$.

4. Conclusion

In connection with the calculations of both the 45 and 70 GHz spectra the values of T and $\omega\tau$ were chosen to obtain the best agreement with the experiments. The relative heights of the peaks in the spectra are extremely sensitive to the value of $\eta = \hbar\omega/kT$, whereas the value of $\omega\tau$ determines the width of the peaks as can be seen from paper [1], formula (19). It is seen that in both cases the value of T chosen for the calculations had to be larger than the experimental value. However, concerning the 45 GHz measurements, the sample was placed in a cavity containing exchange gas. The cavity was placed in liquid He, and the temperature measurements were based upon a measurement of the vapour pressure of the He [1]. Hence there might have been a temperature gradient explaining the discrepancy in this case. The 70 GHz measurements were performed on the same sample, but this time the sample itself was in direct contact with the He bath, and so the lattice temperature must have been 1.2 K. The conclusion must be that the holes were heated above the lattice temperature by the microwave power, and the heating seems to have been of the order of magnitude of 0.1 deg/μW. At 45 GHz $\omega\tau = 35$ corresponds to $\tau = 1.2 \times 10^{-10}$ s (hole temperature 2 K), and at 70 GHz $\omega\tau = 40$ corresponds to $\tau = 0.9 \times 10^{-10}$ s (hole temperature 2.5 K), but there should not be placed too much emphasis on these values of τ. From Fig. 4 and Fig. 1 to 3 of [1] it is clearly seen that the overall appearance of the cyclotron resonance spectra is mainly due to effects of $\zeta \neq 0$ ($k_B \neq 0$), and with the above-mentioned remarks in mind it can be concluded that the Luttinger theory including $k_B \neq 0$ accounts very well for the experimental results.

Acknowledgements

The author wishes to thank lic. techn. M. R. Samuelsen and Prof. Dr. K. Særmark for their interest in the present work.

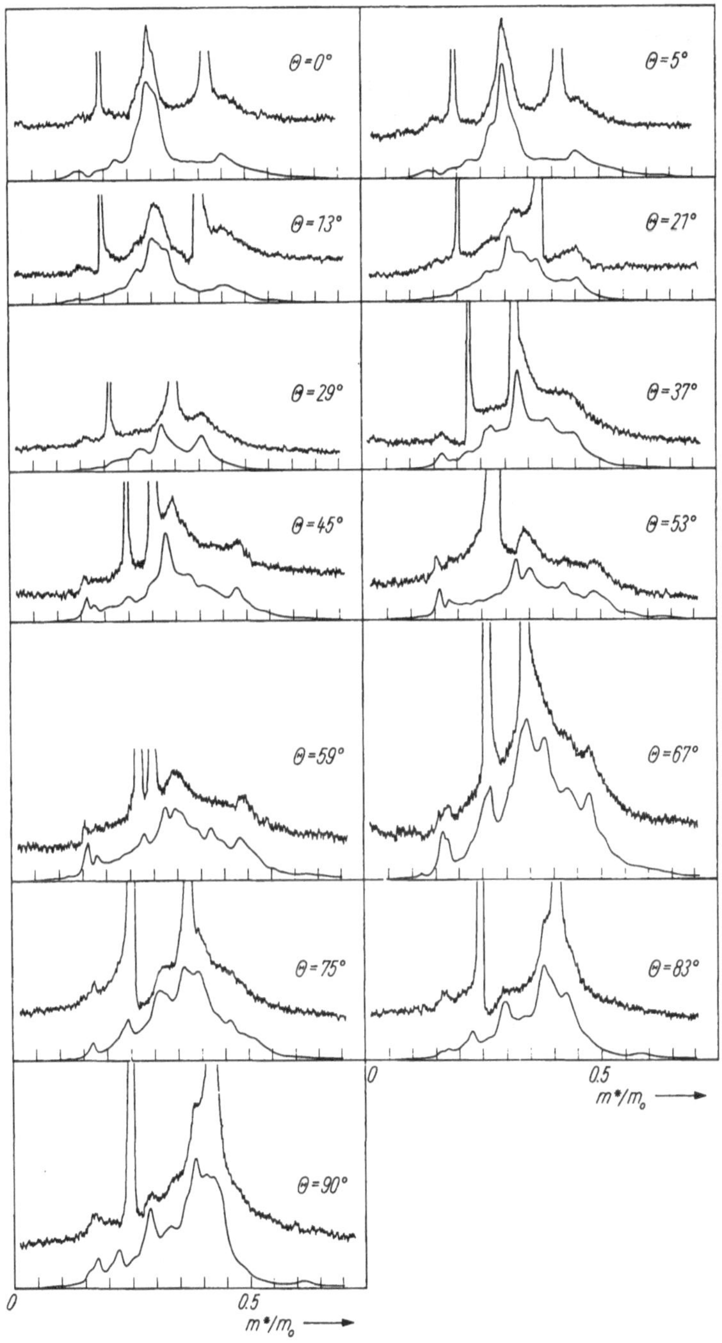

Fig. 4. A comparison between calculated spectra using $F = 70$ GHz, $T = 2.5$ K, and $\omega\tau = 40$ (lower curves) and experimental spectra obtained at $F = 70$ GHz and $T = 1.2$ K (upper curves). The very strong peaks are due to electron resonances. The abscissae are in units of m_0

References

[1] M. Owner-Petersen, phys. stat. sol. (a) **7**, 561 (1971).
[2] M. Owner-Petersen and M. R. Samuelsen, phys. stat. sol. **28**, 211 (1968).
[3] J. M. Luttinger, Phys. Rev. **102**, 1030 (1956).
[4] R. R. Goodman, Phys. Rev. **122**, 397 (1961).

(Received May 18, 1972)

phys. stat. sol. (a) **12**, 471 (1972)

Subject classification: 14.4; 10.2; 22

Zentralinstitut für Elektronenphysik der Deutschen Akademie der Wissenschaften zu Berlin, Abteilung Transportprozesse

ITC with Reciprocal Heating Scheme

By

P. Müller and J. Teltow

The advantages of the reciprocal heating scheme ($1/T \sim t$) in comparison to conventional linear heating ($T \sim t$), as applied to the ITC method, are clearly elucidated by direct confrontation. A simple relation is found between the current maximum and the activation enthalpy of dipole reorientation. The parameter evaluation in the case $1/T \sim t$ saves time and cost.

Die Vorteile des reziproken Aufheizprogramms ($1/T \sim t$) im Vergleich zum üblichen linearen Aufheizen ($T \sim t$) bei der ITC-Methode werden bei direkter Gegenüberstellung besonders deutlich. Es ergibt sich ein einfacher Zusammenhang zwischen dem Strommaximum und der Aktivierungsenthalpie der Dipol-Reorientierung. Die Parameterauswertung im Falle $1/T \sim t$ erspart Zeit und Kosten.

1. Introduction

The method of ionic thermo-currents (ITC), introduced by Bucci and Fieschi [1] in 1964, has found widespread use for the characterization of point defect dipoles in polar crystals, owing to its versatility and its unrivalled sensitivity.

Up to now, most workers have applied a time-linear increase of the sample temperature, T, following the original procedure of Bucci and Fieschi. This implies the occurrence of exponential integrals in the analysis which cannot be solved elementarily.

In the following we propose to vary $1/T$ linearly with time. This completely avoids the evaluation of such integrals during ITC-data processing. The subsequent direct confrontation of the two methods shows up additional possibilities of parameter evaluation and control in the case $1/T \sim t$.

2. Theory

We consider a plate-like sample polarized in an electric field $\boldsymbol{E}$. After switching off the field at $t = 0$, the polarization $P(t)$ will decay exponentially according to $\tau \dot{P} = -P$ or, for constant τ, $P = P_0 \exp(-t/\tau)$ with $P(0) = P_0$.

In the ITC-method applied to polar crystals with point defect dipoles (number N in cm^{-3}, dipole moment μ in C cm), P_0 is the frozen-in polarization imposed at a higher temperature T_p in the field E_p: $P_0 = N \mu^2 E_p/3\,kT_p$ due to partial dipole orientation in $\boldsymbol{E}_p$.

The relaxation time τ is supposed to have an Arrhenius-type dependence on temperature

$$\tau = \tau_0\, e^{H/kT}\,. \tag{1}$$

H is the activation enthalpy of reorientation.

The experiment consists in releasing the frozen-in polarization by slowly warming up the sample according to a temperature program

$$\frac{\mathrm{d}T}{\mathrm{d}t} = b \quad \text{or} \quad T = T_0 + b\,T \quad \Bigg| \quad -\frac{\mathrm{d}(1/T)}{\mathrm{d}t} = \frac{1}{T^2}\frac{\mathrm{d}T}{\mathrm{d}t} = a \quad \text{or} \quad \frac{1}{T} = \frac{1}{T_0} - a\,t\,. \tag{2}$$

This means that τ itself becomes dependent on time. Hence from $\tau\,\dot{P} = -P$ we get quite generally

$$P = P_0 \exp\left[-\int\limits_0^t \frac{\mathrm{d}t'}{\tau(t')}\right], \tag{3}$$

or, in the special cases (1) and (2), equation (4) (see Table 1). Introducing $\dot{T}$

Table 1

Equations for ITC-evaluation

quantity	with linear heating scheme	with reciprocal heating scheme	equation
$P(t)$	$P_0 \exp\left[-\frac{1}{\tau_0}\int\limits_0^t \mathrm{e}^{-H/(k(T_0+bt'))}\mathrm{d}t'\right]$	$P_0 \exp\left[-\frac{k}{\tau_0 a H}\mathrm{e}^{-H/kT_0}(\mathrm{e}^{aHt/k}-1)\right]$	(4)
$J(T)$	$\frac{P_0}{\tau_0}\exp-\left[\frac{H}{kT}+\frac{1}{b\,\tau_0}\int\limits_{T_0}^{T}\mathrm{e}^{-H/kT'}\,\mathrm{d}T'\right]$	$\frac{P_0}{\tau_0}\exp-\left[\frac{H}{kT}+ \frac{k}{\tau_0 a H}(\mathrm{e}^{-H/kT}-\mathrm{e}^{-H/kT_0})\right]$	(5)
J_{m}	not obtainable in closed form	$\frac{P_0 a H}{k}\exp\left[\frac{k}{\tau_0 a H}\mathrm{e}^{-H/kT_0}-1\right]$	(7)
J_{m}	not obtainable in closed form	$\frac{P_0 a H}{k}\exp\left[\mathrm{e}^{H(1/kT_{\mathrm{m}}-1/kT_0)}-1\right]$	(8)
$P(T)$	$P_0-\int\limits_0^t J\,\mathrm{d}t' = P_0-\frac{1}{b}\int\limits_{T_0}^{T}J\,\mathrm{d}T'$ not integrable in closed form	$P_0-\frac{1}{a}\int\limits_{T_0}^{T}\frac{J}{T'^2}\mathrm{d}T' =$ $P_0\exp-\left[\frac{k}{\tau_0 a H}(\mathrm{e}^{-H/kT}-\mathrm{e}^{-H/kT_0})\right]$	(12)

instead of t, we find the measured "ionic thermo-current" density $J = -\dot{P} = P/\tau$ (see (5)). Therefore its plot exhibits the typical ITC asymmetrical single-band structure shown in Fig. 1. In many cases, $H \gg kT_0$ so that (5) and the subsequent equations (but not (2) and (4)) can be simplified by taking $T_0 \to 0$. This means that T_0 experimentally can be chosen so low that any spontaneous depolarization near T_0 is avoided.

The problem is to calculate the atomistic parameters τ_0 and H from the ITC-curve (equation (5)). For this purpose we derive position and height of the band

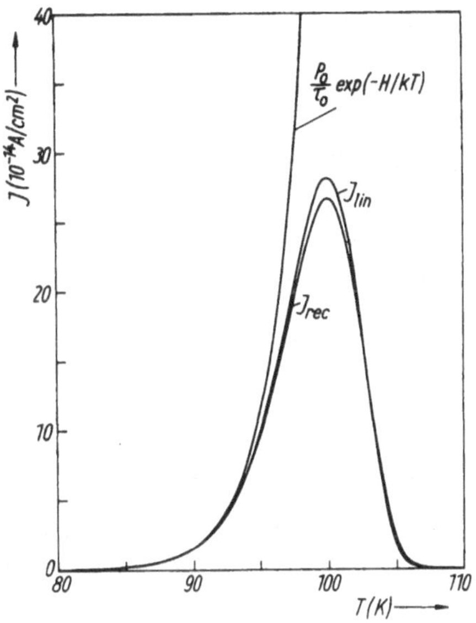

Fig. 1. Calculated ITC curves for linear (J_{lin}) and reciprocal (J_{rec}) heating schemes, dependent on temperature. Parameters: $P_0 = 6.224 \times 10^{-11}$ C cm^{-2}, $H = 0.30$ eV, $\tau_0 = 6.357\ 10^{-14}$ s (yielding $T_m = 100$ K), $b = 0.0333$ K s^{-1}, $a = 3.33 \times 10^{-6}$ K^{-1} s^{-1}. The simple approximative exponential curve of equation (14) is plotted for comparison

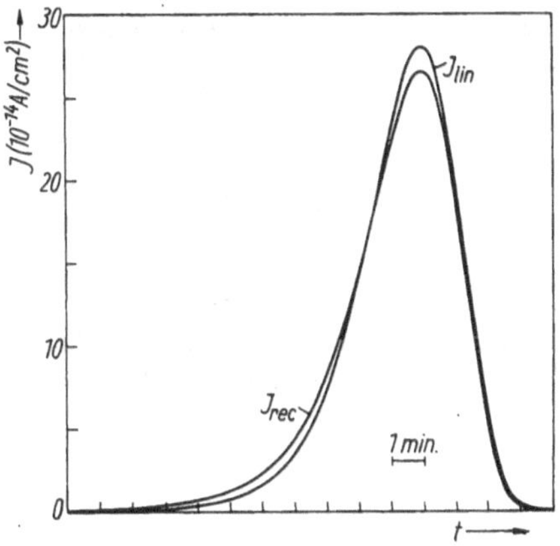

Fig. 2. The same as Fig. 1, in dependence on time (as usually measured)

maximum of (5). It is given by $\dot{\tau} = -1$, as evident from $\tau \dot{J} + \dot{\tau} J = -J$, the derivative of $\tau \dot{P} = -P$. Hence according to (1) and (2) with the subscript m pertaining to the maximum

$$\tau_0 \frac{b\,H\,e^{H/kT}}{kT_m^2} = 1 \qquad \Bigg| \qquad \tau_0 \frac{a\,H\,e^{H/kT_m}}{k} = 1 \tag{6}$$

or, with $\tau_m = \tau_0\,e^{H/kT_m}$

$$T_m^2 = \frac{H\,b\,\tau_m}{k} \qquad \Bigg| \qquad 1 = \frac{H\,a\,\tau_m}{k}. \tag{6'}$$

In (6) and (6′), H as well as τ_0 or τ_m enter as unknown quantities. By inserting (6) in (5), we obtain the ITC-maximum (equation (7)), another relation between H and τ_0, applicable only with the reciprocal heating scheme.

Using (6′) and (1), τ_0 can be expressed by the observed maximum temperature T_m, yielding equation (8). This equation allows to determine H directly from the ITC-maximum J_m. With $T_0 = 0$ the dependence on τ_0 or T_m vanishes in (7) and (8), and we get the remarkably simple form

$$J_m = \frac{P_0\,a\,H}{k\,e} \quad \text{or} \quad H = \frac{k\,e\,J_m}{a\,P_0}, \quad (e = 2.7183\ldots)\,. \tag{9}$$

Using (6):

$$\tau_0 = \frac{P_0}{e\,J_m} \exp\left[-\frac{e\,J_m}{a\,P_0\,T_m}\right]. \tag{10}$$

Again this simple direct determination of H and τ_0 ((8), or (9), (10)) is feasible only in the case of reciprocal heating.

In either case P_0 may be found graphically from the area below the ITC-curve:

$$P_0 = \int_0^\infty J\,\mathrm{d}t = \frac{1}{b}\int_{T_0}^\infty J\,\mathrm{d}T' \quad \Bigg| \quad P_0 = \int_0^\infty J\,\mathrm{d}t = \frac{1}{a}\int_{T_0}^\infty \frac{J}{T'^2}\,\mathrm{d}T' . \tag{11}$$

Hence in the $J(t)$ plot the areas below both curves are equal (Fig. 2).

Likewise, the time-dependent decay of the polarization is given by (12) (see Table 1). Therefore, a plot of

$$\ln \frac{J(T)}{P(T)} = -\ln \tau = -\ln \tau_0 - \frac{H}{kT} \tag{13}$$

offers another way to determine τ_0 and H from experiment. Note that for $T_0 = 0$ in the reciprocal heating regime, $P_m = P(T_m) = P_0/e$ from (12) and (6). This means that t_m divides the area below the ITC-curve in the ratio 0.632:0.368.

A third possibility arises from an initial development of (5) for small temperature increases:

$$\ln J \approx \ln \frac{P_0}{\tau_0} - \frac{H}{kT}, \tag{14}$$

which again yields H and τ_0 if P_0 is known. In the $J(T)$ plot, the initial parts of both ITC-curves must coincide (Fig. 1). It is evident from this outline, that the reciprocal heating scheme provides additional possibilities of parameter evaluation, in virtue of the avoidance of an unsolvable integral and a transcendental equation. We shall not dwell on further details which the interested reader will easily reveal himself.

3. Computer Adaptation

The desirability of computer fitting of complete measuring curves need not be stressed here. Only by this procedure the total information contents of the experiment can be fully accounted for, without over-accentuation of characteristic points or slopes of the curve, as in the proposals given above. Systematic deviations from the theoretical curve indicate insufficient suitability of the atom istic model which then has to be altered or refined.

Now in the ITC-case with linear heating, the integral form (equation (5)) requires to program a tedious numerical integration, whereas in the reciprocal heating scheme equation (5) can be programmed straightforward. The resulting shorter computing time may result in considerably economizing the cost in the course of an extensive ITC-investigation. According to our experience, P_0 can be found with high precision from the ITC-curves, whereas the calculated values of H and τ_0 are less exact, their effect on equation (5) being somewhat compensating. At any rate, (9) and (10) give good zero approach values for H and τ_0 to embark in the least squares minimization.

4. Concluding Remarks

After finishing this work the authors became aware of the fact that the idea of a reciprocal heating scheme had been developed earlier in the literature on thermoluminescence [2]. An experimental device for this purpose has been described [3]. Nevertheless, it seems useful at present to recall this method to the attention of ITC-workers hitherto applying the linear heating scheme, in virtue of the remarkable advantages outlined above. The proposal [3] to use the half-maximum points of the time-hyperbolic "glow curve" for additional parameter evaluation can easily be transferred to ITC-curves. Introducing the half-width t_h of the $J(t)$ curve

$$H = 2.447 \frac{k}{a\, t_h}$$

and, with $T_0 \to 0$:

$$P_0 = \frac{e\, J_m\, t_h}{2.447} = 1.111\, J_m\, t_h$$

2.447 being the difference of the two roots of $1 + x - e^{-x} = -\ln 2$. Thus, reciprocal heating allows P_0 to be found directly without graphical integration (11) as well.

For completion, it should be mentioned that a relation

$$H = \frac{kT_m^2\, J_m}{b\, P_m}$$

has been applied to the maximum of thermoluminescent glow curves with linear heating [4]. This kind of substitute for equation (9) follows from (6′) with $\tau_m = P_m/J_m$ and requires the (ill-defined) graphical integration of the ITC-curve from the maximum time to infinity (equation (12)).

Recently Bräunlich and Kelly [5] have given a general devaluating criticism of the informative value of thermostimulative experiments on carriers with a trap distribution. With full consideration of this scepticism, it is felt that in the present case of reorienting simple point defect dipoles in crystals, involving a single activation enthalpy or at least a few distinctly separated activation enthalpies, the improvement given above adds to the merits of the ITC-method.

References

[1] C. Bucci and R. Fieschi, Phys. Rev. Letters **12**, 16 (1964).
C. Bucci, R. Fieschi, and G. Guidi, Phys. Rev. **148**, 816 (1966).

[2] W. Arnold and H. Sherwood, J. phys. Chem. **63**, 2 (1959).

[3] A. Halperin, M. Leibovitz, and M. Schlesinger, Rev. sci. Instrum. **33**, 1168 (1962).

[4] F. Urbach, Sitzungsber. Akad. Wiss. Wien, Math.-Nat. Kl. Abt. IIa **139**, 363 (1930).
Ch. B. Lushchik, Dokl. Akad. Nauk SSSR **101**, 641 (1955).

[5] P. Bräunlich and P. Kelly, Phys. Rev. **B1**, 1596 (1970).

(Received June 6, 1972)

phys. stat. sol. (a) 12, 477 (1972)

Subject classification: 14.3.4; 22.2.3

II. Physikalisches Institut der Universität zu Köln (a) and Ludwig-Boltzmann-Institut für Festkörperphysik, Wien (b)

Recombination of Impact Ionized Excess Carriers in n-Type and p-Type Indiumantimonide

By

H.-D. BAUMGART[1]) (a), G. NIMTZ (a), and P. KOKOSCHINEGG (b)

The lifetime of impact ionized excess carriers were measured in n-type and p-type InSb samples of different carrier concentrations and carrier mobilities at 77 K. In contrast to the previous photoconductivity and photoelectromagnetic effect measurements these are direct lifetime observations via the decay of the microwave excess conductivity. The decay times of about 1 μs for one exponential decay in n-type material and of about 30 ns and 5 μs for two exponential decays in p-type material are in good agreement with the two level model of Laff and Fan. Studies of the recombination behaviour in electric fields of up to 200 V/cm and 200 ns length did not show any influence on the recombination rate.

In n- und p-dotiertem Indiumantimonid wurden die Lebensdauern stoßionisierter Überschußladungsträger bei 77 K gemessen. Im Gegensatz zu früheren Messungen der Photoleitfähigkeit und des photoelektromagnetischen Effektes liefert das Abklingen der Mikrowellenüberschußleitfähigkeit direkte Bestimmungen der Lebensdauern. Es ergeben sich Lebensdauern von ungefähr 1 μs in n-Material und zwei verschiedene Lebensdauern von 30 ns und 5 μs in p-Material. Diese Werte stimmen mit dem Zwei-Niveau-Modell von Laff und Fan überein. Elektrische Felder von 200 ns Länge und bis zu 200 V/cm Stärke, die während der Rekombination an die Proben angelegt wurden, hatten keinen Einfluß auf die Rekombinationsrate.

1. Introduction

As shown by Laff and Fan [1] the recombination behaviour of excess carriers in InSb is strongly dependent on the type of conduction of the material. In a stationary experiment they deduced carrier lifetimes from measurements of the photoconductivity and the photoelectromagnetic effect. Quite recently some doubt has arisen about the interpretation of these measurements. Rosenthal [2] suggested an explanation of the recombination behaviour by phonon assisted Auger-recombination rather than by the two level model by Laff and Fan. In this paper we report on an experiment, in which the excess carriers were produced by impact ionization and in which in contrast to the indirect measurements of Laff and Fan the lifetimes could be observed directly by microwave probes.

2. Experimental Procedure and Results

Impact ionized excess carriers were produced in n-type and p-type InSb samples with different carrier concentrations and different carrier mobilities by voltage pulses of about 300 V/cm at 77 K. The pulses had a duration of up to 100 ns and produced excess carriers of about 50% of the equilibrium concentration.

---

[1]) Present address: Siemens AG, 4 Düsseldorf (Germany), Lahnweg 10.

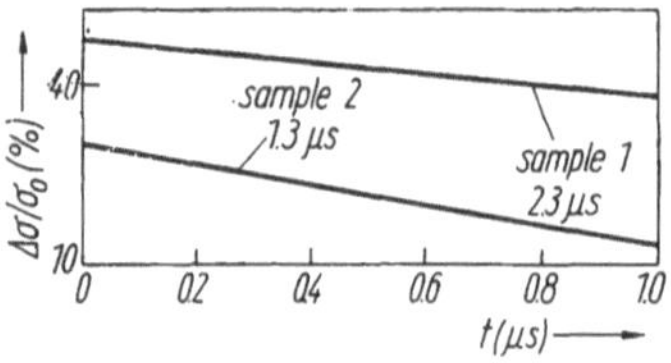

Fig. 1. Excess conductivity after end of impact ionization in n-type InSb at 77 K (logarithmic scale). The relative change in conductivity is proportional to the excess carrier concentration. Sample 1: Equilibrium carrier concentration $n_0 = 10^{13}$ cm^{-3}, electron mobility μ_e = 45 0000 cm^2/Vs. Sample 2: $n_0 = 2.6 \times 10^{12}$ cm^{-3}, μ_e = 24 0000 cm^2/Vs

The carrier lifetimes were deduced via the measurement of the microwave resistivity of the samples, the time constant of the apparatus being smaller than 5 ns. For a detailed description of the experiment see Bringer and Nimtz [3].

Fig. 1 shows the decay of the excess conductivity for two n-types samples, the carrier concentrations and the carrier mobilities being given in the figure. Because of the large ratio of electron to hole mobility the relative change of the conductivity corresponds directly to the change of the carrier concentration of the electrons. The decay is linear with decay times of 1.3 and 2.3 μs.

An oscilloscope pattern of the excess conductivity decay with a superimposed 200 ns voltage pulse of 200 V/cm is shown in Fig. 2. As we discuss later, this voltage pulse varies the carrier mobility but has obviously no influence on the recombination.

In Fig. 3 the results for the p-type samples are given. In contrast to the n-type material the p-type samples have decay curves with two different branches with decay times of approximately 30 ns and 5 μs.

3. Discussion

In n-type material our results are essentially the same as those of Laff and Fan. At 77 K our samples showed a linear decay up to our largest excess carrier concentration of 60% of the equilibrium concentration. From our measurements shown in Fig. 1 we could deduce decay times of 1.3 and 2.3 μs which are in good agreement with the theoretical value of 0.9 μs of the above authors.

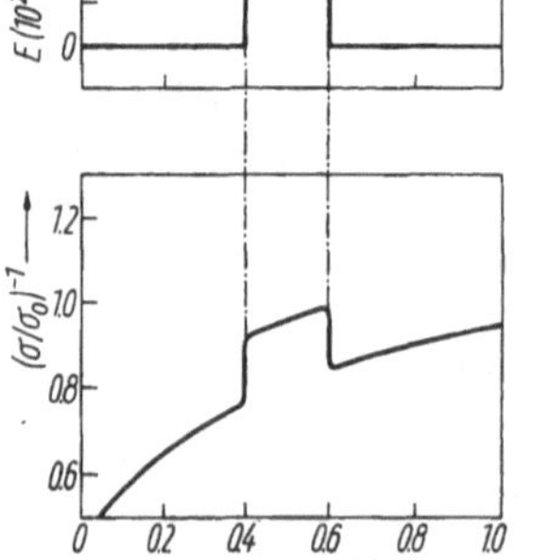

Fig. 2. Recombination of n-type InSb (oscilloscope display of sample resistivity $(\sigma/\sigma_0)^{-1}$). During the superimposed voltage pulse the resistivity of the sample increases because of hot carrier effects; afterwards unchanged recombination continues

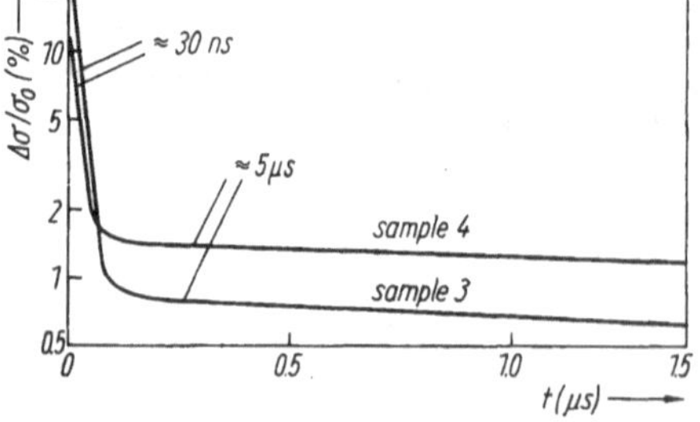

Fig. 3. Excess conductivity in p-type InSb at 77 K. The semilogarithmic plot shows two decay times for each of the curves. Data of both samples: Equilibrium hole concentration $p_0 = 1.6 \times 10^{13}$ cm^{-3}, hole mobility μ_h = 7400 cm^2/Vs

Because of the linear decay and the large electron to hole mobility ratio of $b = 25$ the excess conductivity can be written as

$$\Delta\sigma(t) = q\,\mu_e\,\Delta n(t) \tag{1}$$

with

$$\Delta n(t) = \Delta n_0\,\mathrm{e}^{-t/\tau_n}\,. \tag{1a}$$

The symbols have the following meaning:

$\Delta\sigma(t)$ time dependent excess conductivity,
q electronic charge,
$\Delta n(t)$ time dependent excess carrier concentration,
Δn_0 excess carrier concentration at the end of the ionization,
τ_n lifetime of electrons.

So $\Delta\sigma(t)$ is proportional to $\Delta n(t)$, since the mobility can be regarded as a constant because of the small energy relaxation time.

Our p-type material measurements show two branches of the decay, the fast ones exhibiting decay times of 25 and 30 ns and the slow ones exhibiting decay times of 4.6 and 5.8 μs. According to Laff and Fan's model the two branches are due to the decay of electrons and holes, respectively. This means that our experiments are the first direct observations of the lifetimes of minority carriers in InSb. Our lifetimes of holes are in complete accordance with Laff and Fan's model, whereas our electrons show lifetimes which are larger by a factor of five.

In p-type material the time-dependent excess conductivity can be written as

$$\Delta\sigma(t) = q\,\mu_e\left(\Delta n(t) + \frac{1}{b}\,\Delta p(t)\right) \tag{2}$$

with in addition to the above symbols b being the ratio of electron to hole mobility ($= 25$) and $\Delta p(t)$ being the time-dependent excess hole concentration.

Since the hole lifetime τ_p is largely different from τ_n, the decays for electrons and holes may be decoupled: The decay of the conductivity is controlled by electrons for $t < \tau_n$ and by holes for $t \gg \tau_n$ and can be written as

$$\Delta\sigma(t) = q\,\mu_e\,\Delta n_0\,\mathrm{e}^{-t/\tau_n} \qquad \text{for} \quad t < \tau_n \tag{2a}$$

and

$$\Delta\sigma(t) = q\,\mu_p\,\Delta p_0\,\mathrm{e}^{-t/\tau_p} \qquad \text{for} \quad t \gg \tau_n\,. \tag{2b}$$

The reason for the discrepancy between the theoretical lifetime of 6 ns and our experimental values of 30 ns may be that our samples contain a lower concentration of recombination centres than in Laff and Fan's theory is assumed. Comparing our measurements with those of Laff and Fan we want to recall the fact that our measured values are directly proportional to the concentration of the minority carriers, whereas in the stationary method of the photoconductivity and the photoelectromagnetic effect the lifetimes only can be deduced with the additional knowledge of the diffusion and the surface recombination rate.

An additional interesting fact is that the decay curves of Fig. 3 allow an approximate determination of the generation rate during the ionization. The zero point of the time scale is identical with the end of the ionization pulses. During the electron controlled decay the excess conductivity drops from 21 to 0.8% within 80 ns in sample 3 whereas the conductivity in sample 4 drops from 12 to 1.4% within 60 ns. Assuming the earlier mentioned ratio of electron to hole mobility of 25 this leads to the fact, that in sample 3 at $t = 0$ the excess carrier concentrations of electrons and holes are equal, while in sample 4 the number of excess electrons is only 34% of that of the holes. This suggests that in the second case a considerable amount of electron trapping has already taken place during the ionization.

Since Δn, $\Delta p \ll p_0$ is valid throughout our measurements and because of the large τ_p the net generation rate of the electrons can be written as

$$\frac{\mathrm{d}n}{\mathrm{d}t} = g_p(E)\, p_0 - \frac{1}{\tau_n} \Delta n \tag{3}$$

with the field-dependent generation rate $g_p(E)$ in p-InSb.

Evidently for sample 3 the recombination term in equation (3) can be neglected, i.e.

$$g_p(E) > \frac{1}{\tau_n} \frac{\Delta n}{p_0} \approx 3 \times 10^5 \text{ s}^{-1} ,$$

whereas for sample 4 the generation term and the recombination term are of the same order of magnitude, so that

$$g_p(E) \approx \frac{1}{\tau_n} \frac{\Delta n}{p_0} \approx 2 \times 10^5 \text{ s}^{-1} .$$

As a matter of fact sample 3 was ionized by a 50 ns pulse of 315 V/cm, while the field strength in sample 4 was 225 V/cm during a time of 100 ns, so that because of the exponential dependence of the generation rate on the electric field the generation term is dominant in the first case. These results are in agreement with field-dependent current–time measurements which we carried out with the same samples.

Recently an influence of electric fields on the recombination time was observed by Hongo et al. [4]. These authors found that the recombination rate of excess carriers in n-type InSb increases strongly with applied electric fields; for instance at fields of 200 V/cm the decay time of the excess conductivity decreased by the factor two. This effect was explained by hot carrier recombination as it is well known from germanium [5].

In our experiments we studied the field dependence of the recombination time in fields up to 200 V/cm, but we did not find any influence on the recombination behaviour as shown in Fig. 2. During the decay of the excess carriers a pulse of up to 200 ns length was applied to the sample. Due to the different carrier mobility in the electric field the excess conductivity is reduced during the time of the voltage pulse. After the voltage is switched off, the excess conductivity has the same value as in the case without any electric field. From these measurements follows that within 200 ns the voltage pulse has no influence on the recombintion behaviour in contradiction to the results of Hongo et al. [4]. An explanation for this discrepancy can be given by the different experimental methods. In contrast to our 200 ns the pulse length in Hongo's experiment was some microseconds so that because of the acoustoelectric amplification a large flux density may result in a strong emission of piezoelectric phonons. Nimtz [6] showed that in tellurium the recombination rates increased for some orders of magnitude under the condition of acoustoelectric amplification. This effect can be an explanation for the field dependence found by Hongo et al. [4].

4. Conclusion

Our direct observations of decay times of ionized excess carriers are in good agreement with the steady-state photoconductivity and photoelectromagnetic effect measurements of Laff and Fan and are described quite well by their two level model. There is no experimental evidence for the recently proposed phonon assisted Auger-recombination in this temperature range.

In p-type InSb our experiments were the first direct observations of the life times of minority carriers. We could estimate the generation rate of electrons and show that the generation rate is in the order of the recombination rate in the electric fields used for the ionization.

No influence of electric fields on the recombination rates can be found up to 200 V/cm.

Acknowledgements

We want to thank Prof. H. Happ and Prof. K.-H. Seeger for their support of this investigation.

References

[1] R. A. Laff and H. Y. Fan, Phys. Rev. **121**, 53 (1961).
[2] W. Rosenthal, Verh. DPG **7**, 588 (1972).
[3] A. Bringer and G. Nimtz, phys. stat. sol. (b) **46**, 235 (1971).
[4] S. Hongo, S. Panyakeow, J. Shirafuji, and Y. Inuishi, Japan. J. appl. Phys. **10**, 717 (1971).
[5] R. G. Pratt and B. K. Ridley, Proc. Phys. Soc. **85**, 293 (1965).
[6] G. Nimtz, Verh. DPG **7**, 589 (1972).

(Received June 6, 1972)

phys. stat. sol. (a) **12**, 483 (1972)

Subject classification: 18.3; 2; 22.8.2

Philips Forschungslaboratorium Hamburg

Local Magnetocrystalline and Induced Anisotropies in Liquid-Phase Epitaxial Garnet Films

By

J.-P. KRUMME, P. HANSEN, and J. HABERKAMP

The orientational dependence of the magnetization on an in-plane magnetic field in an LPE garnet film of nominal composition $Y_{2.5}Gd_{0.5}Fe_4Ga_1O_{12}$ has been measured in areas of $\geqq$ 10 μm diameter using the magnetooptic Faraday effect. From a fit of theoretical calculations to the experimental data the anisotropy parameters K_1 and K_u have been determined. In the temperature range from 278 to 318 °K K_1 varies from −4.2 to −2.7 and K_u from +3.6 to +1.0, expressed in 10^2 erg/cm^3.

Die Richtungsabhängigkeit der Magnetisierung von der Größe eines in der Schichtebene liegenden Magnetfeldes wurde an einer epitaxialen Granatschicht der vorgegebenen chemischen Zusammensetzung $Y_{2.5}Gd_{0.5}Fe_4Ga_1O_{12}$ in Bereichen von $\geqq$10 μm Durchmesser mit Hilfe des magnetooptischen Faraday-Effektes gemessen. Aus einer Anpassung theoretischer Berechnungen an die Meßwerte konnten die Werte der Anisotropiekonstanten K_1 und K_u ermittelt werden. Im Temperaturbereich von 278 bis 318 °K ändert sich K_1 von −4,2 bis −2,7 und K_u von +3,6 bis +1,0 in Einheiten von 10^2 erg/cm^3.

1. Introduction

Bubble-device applications require magnetic layers with carefully controlled saturation magnetization M_s, uniaxial anisotropy K_u, and magnetocrystalline anisotropy K_1 among other less stringent parameters.

Average values of K_1 and K_u of epitaxial garnet films have already been measured by ferromagnetic-resonance techniques [1 to 3], whereas local values in the 10 μm range of K_u are reported from simple in-plane saturation [4 to 5] and from bubble-collapse data [6] using the "material length" defined in [7]. These estimates, however, are inaccurate.

Present preparative techniques like liquid-phase epitaxy (LPE) of substituted rare-earth iron garnet films [8 to 10] ensure a considerable lateral uniformity of the relevant material parameters so that bubble stability and propagation can be made comparable over several millimeters of lateral extension of such films.

Nevertheless, the lateral uniformity of LPE films is still insufficient.

The purpose of this paper is to demonstrate a measuring technique which is capable of sensing the magnitude of K_u/M_s and K_1/M_s in magnetic film in areas of 10 μm apart from other parameters like stress-birefringence in the substrate in contrast to ferromagnetic resonance which gives average values and is limited by the linewidth.

2. Theory

The equilibrium position of the magnetic vector $\boldsymbol{M}$ is determined by the minimum of the total magnetostatic energy E. Excluding a domain wall energy term E can be expressed as

$$E = E_F + E_S + E_K + E_u\,. \tag{1}$$

E_F representing the energy associated with the external field $\boldsymbol{H}$

$$E_F = -\boldsymbol{M}\,\boldsymbol{H}\,. \tag{2}$$

E_S denoting the energy due to self-demagnetizing effects

$$E_S = 2\,\pi\,M_s^2\cos^2\gamma\,, \tag{3}$$

where γ is the angle between $\boldsymbol{M}$ and the normal to the sample plane.

The energy E_K from the cubic magnetocrystalline anisotropy can be expressed by

$$E_K = K_1\,(\alpha_1^2\,\alpha_2^2 + \alpha_2^2\,\alpha_3^2 + \alpha_1^2\,\alpha_3^2) + K_2\,\alpha_1^2\,\alpha_2^2\,\alpha_3^2\,, \tag{4}$$

where the α_i are the direction cosines of $\boldsymbol{M}$ referred to the cubic axes. Finally, E_u represents the energy term from the induced uniaxial anisotropy rendering the normal to the sample plane to be the easiest axis of magnetization [11, 12]

$$E_u = A\sum_{i=1}^{3}\alpha_i^2\,\beta_i^2 + \frac{1}{2}\,B\sum_{i\neq k=1}^{3}\alpha_i\,\alpha_k\,\beta_i\,\beta_k\,, \tag{5}$$

where the β_i denote the direction cosines of the direction of crystal growth in case of a growth-induced anisotropy [3, 4, 5, 13 to 17] or of the direction of the stress σ in case of a stress-induced anisotropy [2, 10, 18, 19]. Higher-order terms in (5) have been neglected as these effects are relatively small in the garnets. For pure stress-induced anisotropy planar and uniaxial stress along the plane normal lead to the same expression for E_u apart from a constant term [11] and the constant factors A and B in (5) are $\frac{3}{2}\,\sigma\,\lambda_{100}$ and $3\,\sigma\,\lambda_{111}$, respectively. The $\lambda_{hk}{}^{l}$ denote magnetostriction constants [2, 3, 10, 20, 21]. As this paper deals with the properties of LPE films with (111) orientation the present theory is restricted to a (110) plane rectangular to the (111) film. The theory treats the equilibrium position of $\boldsymbol{M}$ as a function of the external fields $\boldsymbol{H}_\perp$ normal to the film and $\boldsymbol{H}_\parallel$ in the film plane, the plane spanned by $(\boldsymbol{H}_\perp, \boldsymbol{H}_\parallel)$ being parallel to (110).

The total energy E, then, can be expressed in terms of the angles θ and φ of $\boldsymbol{M}$ referring to the cubic axis. However, as $K_1 < 0$ the total (110) plane is an easy plane of $\boldsymbol{M}$ concerning the cubic anisotropy. Since the fields $\boldsymbol{H}_\parallel$, $\boldsymbol{H}_\perp$, and $\boldsymbol{H}_{\text{eff}}$, where $\boldsymbol{H}_{\text{eff}}$ consists of the demagnetizing and uniaxial fields, also lie in this plane the condition $\partial E/\partial\varphi = 0$ remains valid for the total energy E of equation (1). Thus, the energy expression for this (110) plane can be expressed from equations (1) to (5) by

$$E = E_0 - M_s\,(H_\perp^2 + H_\parallel^2)^{1/2}\cos(\theta_0 - \theta) - \frac{1}{6}\,(B - 4\,\pi\,M_s^2)\times$$

$$\times[\sin^2\theta + \sqrt{2}\sin 2\,\theta] + \frac{K_1}{4}\,(\sin^4\theta + \sin^2 2\,\theta) + \ldots\,, \tag{6}$$

where $\theta_0 = 54.73° + \arctan(H_\parallel/H_\perp)$ and $E_0 = \frac{1}{3}\,(A + \pi\,M_s^2)$ with $A = \frac{3}{2}\,\sigma\,\lambda_{100}$ and $B = 3\,\sigma\,\lambda_{111}$ in the case of stress-induced uniaxial anisotropy.

The equilibrium condition $\partial E/\partial\theta = 0$ yields

$$-(H_\perp^2 + H_\parallel^2)^{1/2}\sin(\theta_0 - \theta) - H_{\text{eff}}\,(\sin 2\,\theta + 2\sqrt{2}\,\cos 2\,\theta) +$$

$$+\frac{1}{2}\,H_K\sin 2\,\theta\left(\frac{1}{2}\sin^2\theta + \cos 2\,\theta\right) = 0\,, \tag{7}$$

Fig. 1. Equilibrium position θ_E referred to the ⟨111⟩ film normal of the magnetic vector $\boldsymbol{M}$ as a function of the in-plane field $H_{||}$ oriented parallel to a (110) plane of the single-crystalline cubic magnetic layer

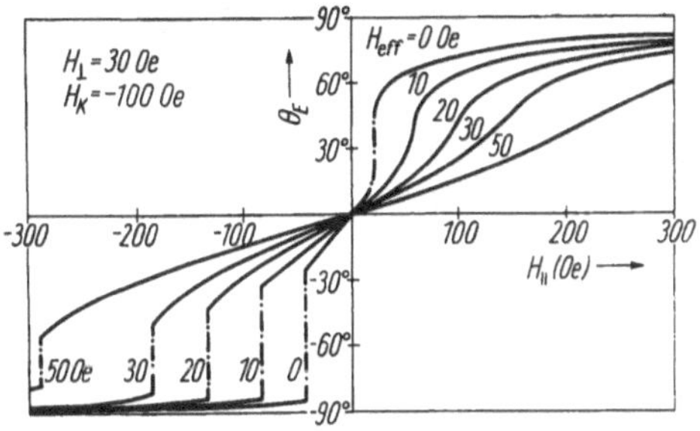

where $H_{\mathrm{eff}} = -1/6\,(B/M_s + 4\,\pi\,M_s)$ and $H_K = 2\ K_1/M$. For convenience the cubic reference frame is rotated by 54.73° such that $\theta_E = \theta + 54.73°$ denotes the angle between the film normal and the equilibrium position of $\boldsymbol{M}$. θ_E corresponds to angle γ in (3). K_2 is neglected as it is usually quite small. Then, the position θ of the magnetic vector $\boldsymbol{M}$ for given values of $H_\perp$ and $H_{||}$ depends only on the effective uniaxial field H_{eff} and the cubic anisotropy field H_K.

The effect of H_{eff} and H_K on the orientation θ_E of $\boldsymbol{M}$ in a (110) plane is shown in Fig. 1 and 2.

The problem is not symmetric with respect to the ⟨111⟩ film normal: For positive values of $H_{||}$ another ⟨111⟩ direction is present at +70.53°, whereas for negative $H_{||}$ the next ⟨111⟩ axis is under −109.47°. The structure of the $\theta_E = f(H_{||})$ curves (Fig. 1 and 2) reflects this property. For negative $H_{||}$ the dashed lines connect states of equal magnetostatic energy. The jump is expected to occur in a very narrow region of $H_{||}$ values due to the very steep crossover of the corresponding energies. However, in this range domains are expected to appear, rendering the given calculation invalid. With increasing fields $H_{||}$, $\boldsymbol{M}$ jumps into the new position. For positive $H_{||}$, $\theta_E = f(H_{||})$ either jumps (if H_{eff} is small) or displays a point of inflection. Both, the points of jump and inflection are prominently affected by H_{eff}, and not by H_K. From Fig. 2 the point of inflection turns out to be approximately independent of H_K except for very small H_K values and, therefore, represents the best feature of the curve to begin with the fitting of the theory to the experiments. An adjustment of H_K, subsequently, will improve the fit with respect to the approach to $\theta_E = \pm 90°$.

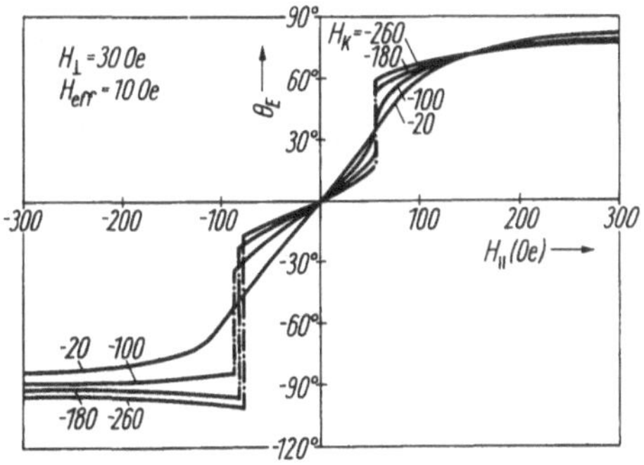

Fig. 2. θ_E as a function of $H_{||}$ calculated for a (110) plane of the (111) oriented cubic magnetic film

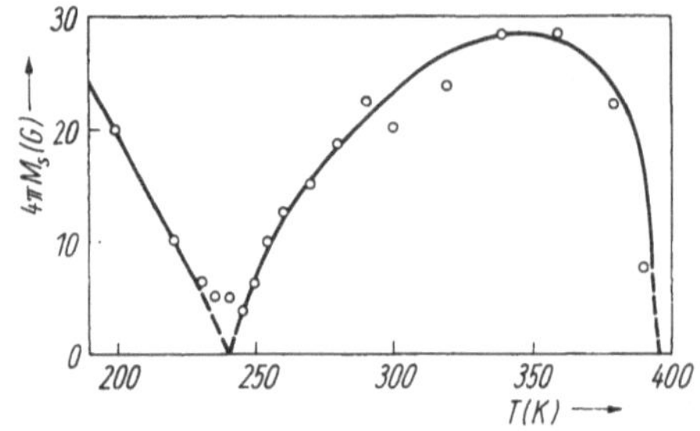

Fig. 3. Temperature dependence of the saturation magnetization M_s of a garnet film of nominal composition $Y_{2.5}Gd_{0.5}Fe_4Ga_1O_{12}$

3. Experimental

LPE films of the nominal composition $Y_{2.5}Gd_{0.5}Fe_4Ga_1O_{12}$ have been used in the experiments displaying a small temperature-varying M_s (Fig. 3) measured by means of a vibrating-sample magnetometer. This garnet composition was proposed as suitable material for the epitaxial growth on (111) $Gd_3Ga_5O_{12}$ (GGG) wafers with tensile stress induced uniaxial magnetic anisotropy normal to the film [9, 10]. Contrary to polycrystalline ceramic samples [9] LPE films of this nominal composition are found to have a higher compensation temperature (240 °K) and only 1/4 of the saturation magnetization (Fig. 3) at about the same Curie temperature ($\approx$400 °K). The average lattice mismatch $a_S - a_F = +0.009$ Å between the substrate and the film from X-ray reflection measurements compares well with the predicted value for this composition [9] as the lattice constant a_S of the Czochralski-grown substrate is 12.284.

The experimental set-up for the direct measurement of θ_E is outlined in Fig. 4 and the measuring procedure follows the line reported in [22]. Contrary to [22] the laser beam is expanded by a spatial filter whose pin-hole of 5 μm diameter is imaged to the sample by large-aperture lenses enabling the measurement of Faraday rotation in an area of the pin-hole dimension at a working distance of several hundred millimeters. Thus, sufficient space is available for the two Helmholtz coils around the sample which produce the mutually orthogonal magnetic fields $H_\perp$ and $H_{||}$, $H_\perp$ being oriented parallel and $H_{||}$ perpendicular to the incident test beam. The azimuth of the in-plane field $H_{||}$ can be varied with respect to the crystallographic axes of the magnetic film. An additional light source illuminates the whole sample through a semi-transparent mirror in the laser beam path so that the laser focus can be correlated with the position on the film.

The local Faraday effect gives rise to an azimuth change of the incident test beam and is measured by means of a phase sensitive detection scheme with 10^{-2} degrees

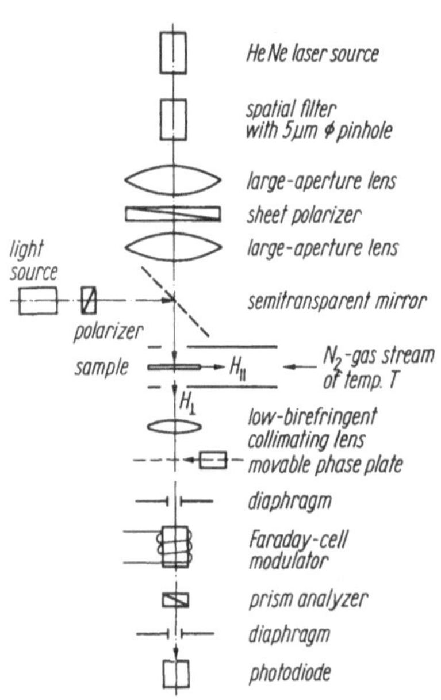

Fig. 4. Experimental set-up for the magnetooptical measurement of θ_E

Fig. 5. Configuration of $\boldsymbol{M}$, light wave vector $\boldsymbol{k}$, beam polarization $\boldsymbol{E}$ and the fields $\boldsymbol{H}_\perp$ and $\boldsymbol{H}_{||}$ during measurements

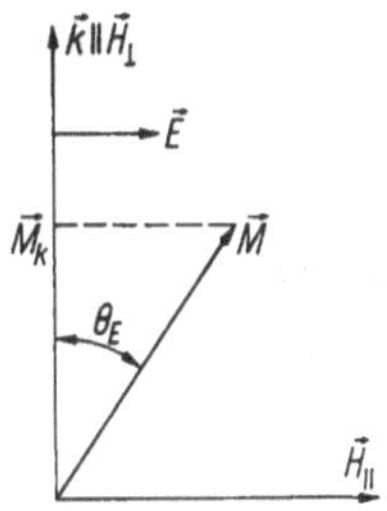

accuracy. Consider a light beam of wave vector $\boldsymbol{k}$ and linear polarization $\boldsymbol{E}$ vertically incident on the sample (Fig. 5). If $\boldsymbol{M}$ forms an angle θ_E referred to $\boldsymbol{k}$, then,

$$\cos \theta_E = \frac{M_k}{|\boldsymbol{M}|}, \tag{8}$$

where M_k is the component of $\boldsymbol{M}$ along the beam path and $|\boldsymbol{M}| = M_s$. In terms of Faraday-rotation angles (8) becomes

$$\cos \theta_E = \frac{\theta_F}{\theta_F^0}, \tag{9}$$

where θ_F^0 is equal to one half of the azimuth change measured for $\boldsymbol{M}$ parallel and antiparallel to $\boldsymbol{k}$, and θ_F is the effective azimuth with respect to $\boldsymbol{M}||\boldsymbol{k}$. This analysis neglects second-order effects from the transverse Faraday configuration [23] and from interaction with the surfaces of the absorbing medium since they have proved to be very small. This allows the incident beam azimuth to be chosen such that noise due to the ellipticity in the emerging beam from stress-birefringence in the imperfect substrate can be minimized.

By transverse motion of the sample after its alignment vertical to the beam path the area of interest is selected. Then, from arguments stated in Section 2 anyone of the three crystallographic (110) planes perpendicular to the film is adjusted parallel to the plane spanned by $(\boldsymbol{H}_\perp, \boldsymbol{H}_{||})$ through a rotation of the $\boldsymbol{H}_{||}$-field coils. This is accomplished by plotting the Faraday signal as a function of positive and negative $H_{||}$ at a given $H_\perp$ (that is sufficient to saturate the film for $H_{||} = 0$) with the angle φ between the plane $(\boldsymbol{H}_\perp, \boldsymbol{H}_{||})$ and the sample axes as parameter. From these curves the difference between the Faraday signals for equal, but oppositely signed and not too large $H_{||}$ is plotted as function of φ in Fig. 6 displaying the threefold symmetry of the (111) plane of the film. A (110) plane is, then, selected by aligning the $\boldsymbol{H}_{||}$-field coils along anyone of the peaks in Fig. 6. A variation by $\Delta\varphi = \pm 5°$ around $\varphi_{(110)}$ does not prominently affect the relation $\theta_E = f(H_{||})$ (Fig. 7a, b). Hence, $\varphi_{(110)}$ is sufficiently determined by the angle half between two zero points in Fig. 6. Associated with an exact $\varphi_{(110)}$, however, is a maximum of the field margin of negative $H_{||}$ at the jump of θ_E for which a demagnetized domain pattern exists.

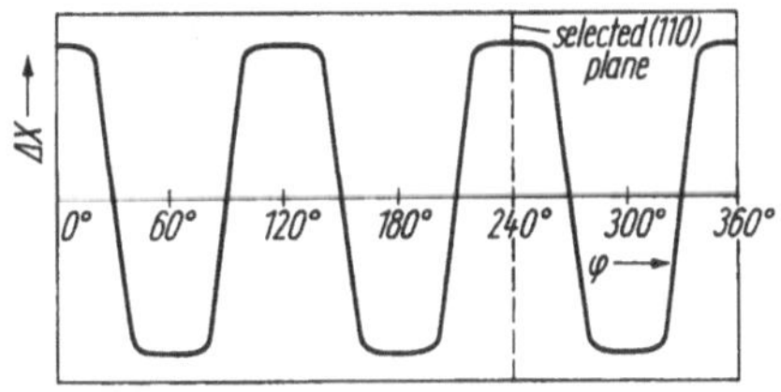

Fig. 6. Intensity difference ΔX between the detector output signals for a properly chosen positive and negative $H_{||}$ as a function of angle φ between a crystallographic (110) plane and the plane spanned by $\boldsymbol{H}_\perp$ and $\boldsymbol{H}_{||}$

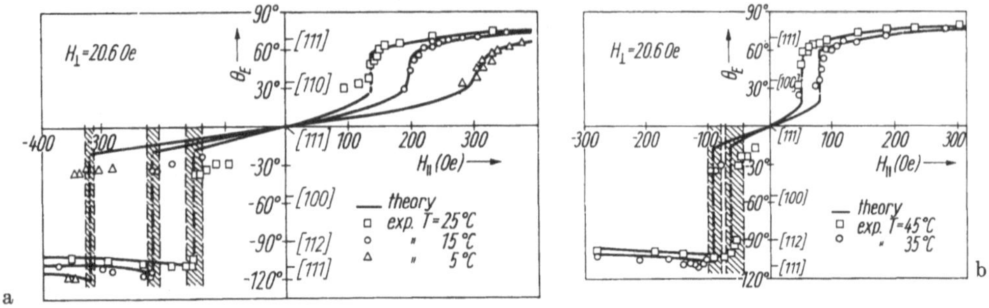

Fig. 7a, b. Theoretical curves fitted to the experimental data measured in a 10 μm diameter area of a 6 μm thick LPE garnet film of the nominal composition $Y_{2.5}Gd_{0.5}Fe_4Ga_1O_{12}$ on GGG with the temperature as parameter

Of course, transverse motion of the sample for the purpose of investigating other film areas preserves the alignment with respect to φ, though the incident beam azimuth has generally to be reset for an optimal signal-to-noise figure due to the varying orientation of the axes of stress-birefringence in different locations on the substrate.

The dependence $\theta_E = f(H_{||})$ is measured with temperature T and field $H_\perp$ as parameters and reproduced in Fig. 7a and b. Since $H_\perp$ adds constructively to K_u/M_s where $K_u = -B/2$, it is chosen just to saturate the film area of interest at $H_{||} = 0$ for keeping the in-plane field $H_{||}$ at the instability of θ_E (Fig. 7) as small as possible from practical considerations. As the Faraday rotation θ_F^0 is only 0.26° at room temperature in the 6 μm thick film, the accuracy of small values of θ_E from (9) is poor explaining one of the reasons of the scattering of data near the origin. However, two aids exist in the plotting of curves through the experimental data: One makes use of the theoretical plots (Fig. 1 and 2) displaying a nearly linear dependence of $\theta_E = f(H_{||})$ for small negative $H_{||}$, so that the curve may be extrapolated into the region of small positive $H_{||}$. The other aid involves optical observation of the local domain structure in a polarizing microscope, again establishing the fields $H_{||}$ where θ_E either "jumps" or has a point of inflection plus the additional information of the appearence of large-angle domain walls. In case the test region splits into domains at the jump or the point of inflection of θ_E two broken vertical lines demonstrate the field margin where domains are observed. This facilitates the plotting of $\theta_E = f(H_{||})$ through the experimental points in Fig. 7a and b. Thus, demagnetization assoziated with large-angle domain walls only occurs at the jumps of θ_E which is predicted from the instability of the energy relation at these fields $H_{||}$ (Fig. 1 and 2). Away from these characteristic field regions no domain walls have been observed.

4. Discussion

The theory developed in Section 2 has been assessed by fitting the theoretical plots to the experimentally derived dependence of the orientation θ_E of the magnetic vector $\boldsymbol{M}$, with respect to the film normal, on the two external magnetic fields $H_\perp$ along the film normal and $H_{||}$ in the film plane. Considering the features of the relation $\theta_E = f(H_{||})$ sensitive to each of the two anisotropy parameters, both K_u/M_s and K_1/M_s can reliably be determined by adjusting the theory to the experimental data. This has been done in Fig. 7a and b, the curves representing the theory and the marks the experimental data.

The first step of the fitting procedure is the search for the coordinates of the point of inflection or jump at positive $H_{||}$ from a graphical representation of the experimental results. From Fig. 2 this characteristic point is approximately independent of H_K but very sensitive to H_{eff}. It turns out that from this discrete fit the jump field at negative $H_{||}$ determined from measurements and optical observation is already well adjusted. The variation of H_K, then, leads to the correct adjustment concerning the approach to in-plane saturation (i.e. $\theta_E \rightarrow \rightarrow \pm 90°$) for positive and negative $H_{||}$ as found from the measurements.

The remaining discrepancy between experiment and theory for small negative $H_{||}$ is assumed to originate from magnetization ripple which transforms into a pattern with large-angle domain walls when $H_{||}$ approaches the region of the jump of θ_E.

The sensitivity of the fitting procedure is determined by the scattering of the experimental data in the ranges of prominent variations of θ_E.

The point of inflection (positive $H_{||}$) from which the fitting procedure starts can be evaluated from the measured data within $\pm 5\%$ accuracy. This, in turn leads to the same accuracy in H_{eff}. Subsequently, from variation of the parameter H_K an overall fit to the experimental plot can be accomplished where special care is taken of the approach to saturation. This yields an accuracy of $\pm 10\%$ in the determination of H_K.

Table 1

Local magnetic data of the LPE film of nominal composition $Y_{2.5}Gd_{0.5}Fe_4Ga_1O_{12}$

T (°C)	H_{eff} (Oe)	K_1/M_s (Oe)	$4\pi M_s$ (G)	K_1 (10^2 erg cm^{-3})	K_u (10^2 erg cm^{-3})
5	83	−300	176	−4.2	3.6
15	52	−230	201	−3.7	2.7
25	35	−190	226	−3.4	2.1
35	20	−170	251	−3.4	1.45
45	11	−130	264	−2.7	1.0

The parameters derived from the fitting procedure are summarized in Table 1. The temperature dependence of $K_1 = \frac{1}{2} H_K M_s$ and $K_u = M_s (3 H_{eff} + 2\pi M_s)$ is displayed in Fig. 8. K_1 and K_u are found to be linearly dependent on T in this range which is consistent with recent results [1]. The measured value at 60 °C was obtained from observation of the domain pattern ($H_{eff} = 0$) at zero external field.

From Fig. 8 K_u varies by a factor of 3 between +10 and +45 °C. The uniaxial anisotropy due to magnetostriction depends linearly on λ_{111}, which varies [21, 25] by a factor of 1.2 for similar garnet compositions. Thus, it may be concluded that the major contribution to K_u originates from growth-induced anisotropy and not from magnetostriction.

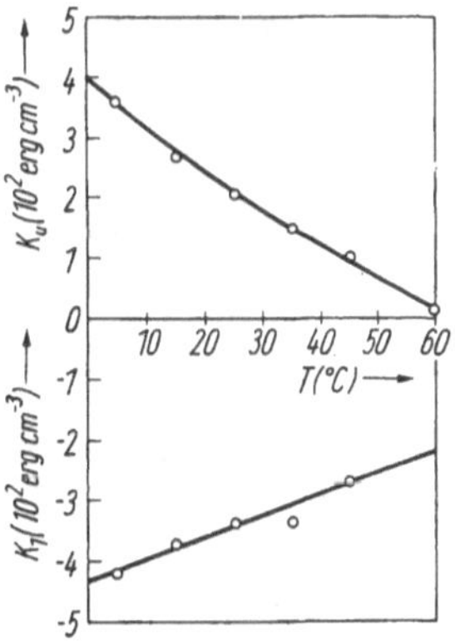

Fig. 8. Temperature dependence of the anisotropy constants K_1 and K_u as derived from the fit of the theory to the experimental data shown in Fig. 7a, b

This is consistent with the literature [1] where growth-induced anisotropy is reported to be predominant in such LPE films [4, 5, 9, 16, 24] contrary to CVD films [3, 5, 9] displaying stress-induced anisotropy, essentially.

5. Conclusion

A method has been described for the determination of the magnetocrystalline and uniaxial anisotropy fields H_K and H_u of magnetic thin films in areas as small as 10 μm diameter. The method has been assessed by application to an LPE layer of nominal chemical composition $Y_{2.5}Gd_{0.5}Fe_4Ga_1O_{12}$ and found to be of ±5% accuracy in $H_{eff} = \frac{1}{6}(H_u - 4\pi M_s)$ and ±10% accuracy in H_K.

Acknowledgements

The authors are grateful to J. Verweel and G. Winkler for helpful discussions and W. Tolksdorf for the preparation of garnet films.

References

[1] R. C. Le Craw and R. D. Pierce, AIP Conference Proceedings **5**, 200 (1971).
[2] P. J. Besser, J. E. Mee, P. E. Elkins, and D. M. Heinz, Mater. Res. Bull. **6**, 1111 (1971).
[3] D. M. Heinz, P. J. Besser, J. M. Owens, J. E. Mee, and G. R. Pulliam, J. appl. Phys. **42**, 1243 (1971).
[4] L. K. Shick, J. W. Nielsen, A. M. Bobeck, A. J. Kurtzig, P. C. Michaelis, and J. P. Reekstin, Appl. Phys. Letters **18**, 89 (1971).
[5] A. J. Kurtzig and F. B. Hagedorn, IEEE Trans. Magnetics **7**, 473 (1971).
[6] A. H. Bobeck, D. H. Smith, E. G. Spencer, L. G. van Uitert, and E. M. Walters, IEEE Trans. Magnetics **7**, 461 (1971).
[7] A. A. Thiele, J. appl. Phys. **41**, 1139 (1970).
[8] H. J. Levinstein, S. Licht, R. W. Landorf, and S. L. Blank, Appl. Phys. Letters **19**, 486 (1971).
[9] E. A. Giess, B. E. Argyle, B. A. Calhoun, D. C. Cronemayer, E. Klokholm, T. R. McGuire, and T. S. Plaskett, Mater. Res. Bull. **6**, 1141 (1971).
[10] E. A. Giess, B. A. Calhoun, E. Klokholm, T. R. McGuire, and L. L. Rosier, Mater. Res. Bull. **6**, 317 (1971).
[11] G. F. Dionne, Mater. Res. Bull. **6**, 805 (1971).
[12] R. D. Pierce, AIP Conference Proceedings **5**, 91 (1971).
[13] A. Akselrad and H. Callen, Appl. Phys. Letters **19**, 464 (1971).
[14] W. T. Stacy and C. J. M. Rooymans, Solid State Commun. **9**, 2005 (1971).
[15] A. Rosencwaig and W. J. Tabor, Solid State Commun. **9**, 1691 (1971).
[16] W. T. Stacy and W. Tolksdorf, AIP Conference Proceedings **5**, 185 (1971).
[17] W. T. Stacy and U. Enz, Internat. Conference on Magnetics, Kyoto 1972 (paper 1.1).
[18] E. Klokholm, J. W. Matthews, A. F. Mayadas, and J. Angilello, AIP Conference Proceedings **5**, 105 (1971).
[19] B. F. Stein and M. Kestigian, J. appl. Phys. **42**, 1806 (1971).
[20] A. H. Bobeck, E. G. Spencer, L. G. van Uitert, S. C. Abrahams, R. L. Barns, W. H. Grodkiewicz, R. C. Sherwood, P. H. Schmidt, D. H. Smith, and E. M. Walters, Appl. Phys. Letters **17**, 131 (1970).
[21] S. Iida, J. Phys. Soc. Japan **22**, 1201 (1967).
[22] J.-P. Krumme and J. Haberkamp, Internat. Conf. Thin Films, Venice 1972 (paper 13.3).
[23] M. J. Freiser, IEEE Trans. Magnetics **4**, 152 (1968).
[24] D. C. Cronemeyer, E. A. Giess, E. Klokholm, B. E. Argyle, and T. S. Plaskett, 3M-Conf., Chicago 1971 (paper 2C-6).
[25] G. A. Petrakovskii, E. M. Smokotin, and A. G. Titova, Soviet Phys.—Solid State **9**, 1820 (1968).

(Received May 23, 1972)

phys. stat. sol. (a) **12**, 491 (1972)

Subject classification: 14.3; 22.1.3

Fachbereich Physik der Universität Marburg

Field Effect on Trigonal Selenium Crystals

By

D. ENGEMANN and W. FUHS

The field effect of trigonal selenium crystals is investigated as a function of temperature. The field mobility μ_F has the same temperature dependence as the conductivity and is of similar magnitude as the Hall mobility. This shows that surface states are of minor importance. At high values of the transverse voltage, the surface conductance exhibits a minimum which seems to indicate the formation of an inversion layer. The results are discussed in terms of a barrier model. It is shown that the minimum of conductance can be explained by a reduction of potential barriers at the surface in high electric fields.

An trigonalen Selen-Einkristallen wurde der Feldeffekt in Abhängigkeit von der Temperatur untersucht. Die Feldbeweglichkeit μ_F hat die gleiche Temperaturabhängigkeit wie die Leitfähigkeit und stimmt in ihrer Größe annähernd mit der Hallbeweglichkeit überein. Daraus geht hervor, daß Oberflächenzustände bei Selen nur eine untergeordnete Rolle spielen. Bei hohen Werten der Querspannung ergab sich ein Minimum der Oberflächenleitfähigkeit, das auf die Entstehung einer Inversionsrandschicht hindeutet. Die Ergebnisse werden im Rahmen eines Barrierenmodells gedeutet. Dabei zeigt sich, daß das Minimum durch den Abbau von Potentialschwellen an der Oberfläche bei hohen elektrischen Feldern zu erklären ist.

1. Introduction

When, in a field effect measurement, an electric field is applied capacitively to a semiconductor, charge is influenced on the surface which can contribute to the surface conductivity. The change of surface conductivity depends upon the electronic structure of the surface, i.e. surface states, surface potential etc. For this reason, field effect has become a powerful tool for the investigation of semiconductor surfaces [1].

Most informative are measurements where, at high transverse field strength, a minimum of conductance is observed which indicates the formation of an inversion layer at the surface. This has been achieved in some cases for germanium and silicon [2] and, more recently, also for tellurium crystals [3]. Besides the calculation of the surface potential, such a minimum in principle allows the investigation of the carrier transport in the inversion layer. This would be of special interest for those materials which exhibit only one type of conduction and for which the knowledge about the transport properties of the other type of carriers is poor. Trigonal selenium is known to be p-type, possibly also in the intrinsic region [4]. Therefore, the formation of an inversion layer would in this case allow the investigation of the behaviour of electrons.

A further application of the field effect is the evaluation of the mobility of the carriers influenced in the surface region from the measured change of conductance and the externally induced charge. If trapping in surface states occurs, this field mobility is equal to the drift mobility of the majority carriers. Only in the absence of trapping, where the total induced charge contributes to the current, the field mobility is expected to have the same value as the Hall mobility. Thus, in some special cases where other methods for the measurement of the mobility fail, field effect can be used.

For trigonal selenium, field effect has been investigated to date only for polycrystalline material [5 to 7]. The main result has been that the field mobility and the Hall mobility have proved to be approximately equal. Therefore, in polycrystalline selenium, surface states do not play an important role. From this behaviour, it is to be expected that in single crystals, too, surface states are of minor importance, so that the field effect allows the evaluation of the mobility of majority carriers.

2. Experimental Procedure

Both vapour-grown platelets and crystals grown from the melt were investigated. The platelets had a thickness of about 100 μm, a length of 8 to 10 mm, and a width of 2 to 3 mm. They had good natural surfaces and therefore could be used without further surface treatment. The melt-grown crystals of about 0.5 mm thickness were cleaved from larger crystals. Here the surfaces were polished and etched with Na_2S. The electrode arrangement is shown in detail in Fig. 1. On one of the crystal faces, five Au strips were evaporated in a vacuum of 5×10^{-5} Torr. The electrodes 1 and 5 served as current carrying electrodes, whereas 2, 3, and 4 were used as potential probes. With probe 3 grounded, the symmetry as well as the polarity of the arrangement could be checked simply by measuring alternately the voltage drop across 2–3 and 3–4. On the opposite crystal face as field electrode, a sheet of mica, covered by an evaporated Au film, was fixed with an adhesive. Both the mica and the adhesive layer had a thickness of about 60 μm. For the investigation of the influence of surface quality in some cases the surface was damaged by grinding. In all samples, the c-axis was parallel to the field electrode. However, the current direction was parallel to the c-axis for melt-grown platelets and perpendicular to the c-axis for vapour-grown platelets.

The measurements were taken at constant current condition. The voltage drop across 2–3 or 3–4 was compensated so that the signal which appeared when the transverse dc field was switched off and on is a direct measure of the change of conductance. Measurements of the conduction of selenium crystals are often complicated by relaxation processes. Also when the transverse field was changed, the new value of resistance was reached only slowly within some minutes. This relaxation, however, is not due to surface states. The size of this effect, as well as the value of the time constant indicate that it originates from the small change of the drift voltage which occurs when the conductance of the crystal is varied by the transverse field.

3. Results

The conductance of trigonal selenium changes by application of the transverse electric field in a manner as is expected for a p-type semiconductor (Fig. 2). When the field electrode is negatively biased, the change of conductance is

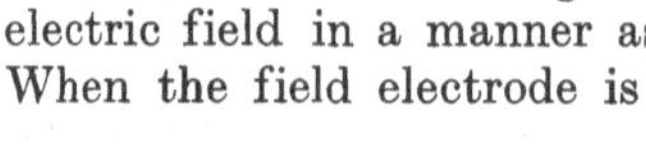

Fig. 1. Electrode arrangement

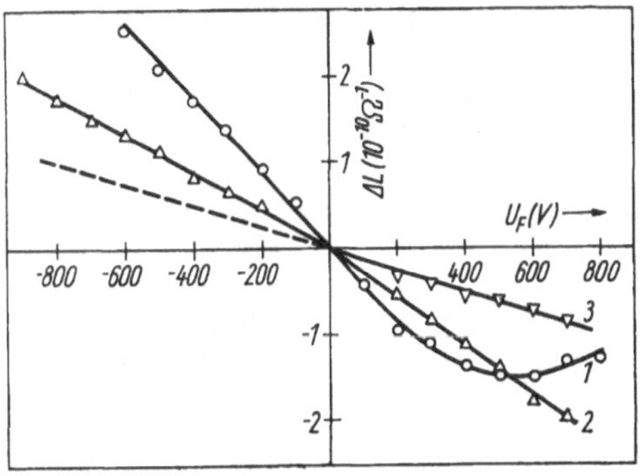

Fig. 2. Change of conductance ΔL as a function of the transverse voltage U_F. (1) vapour-grown platelet with natural surface; (2) melt-grown crystal with etched surface; (3) melt-grown crystal with ground surface

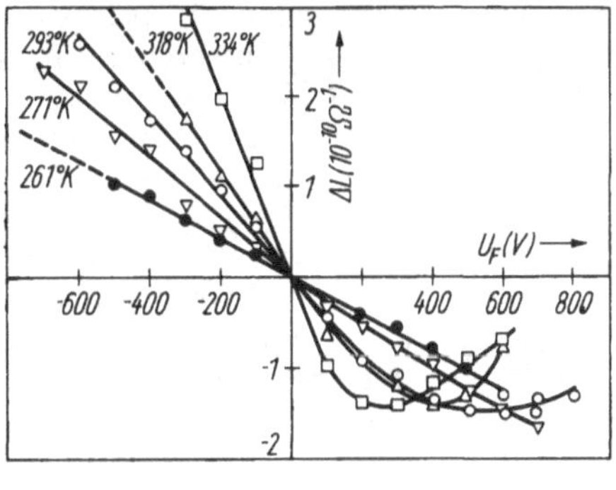

Fig. 3. Field effect at different temperatures for a vapour-grown platelet

positive ($\Delta L > 0$) because the surface of the crystal is positively charged so that the density of majority carriers, i.e. of holes, increases. For reversed polarity, negative charge is induced, resulting in depletion of holes which leads to $\Delta L < 0$. In Fig. 2 this behaviour is demonstrated for three selenium crystals: 1. a vapour-grown platelet with natural surface, 2. a melt-grown crystal with etched surface, and 3. a melt-grown crystal with a ground surface. Damaging of the surface somewhat reduces ΔL at low voltages U_F, a trend which is usually likewise observed for other semiconductors. However, the influence of surface quality here is relatively small in comparsion to germanium and silicon. The field effect of these semiconductors is reduced by orders of magnitude when the surface is ground because the produced lattice defects lead to surface states which trap the major part of the influenced charge [2]. For selenium crystals, however, the weak influence of surface treatment indicates that surface states produced by grinding are of minor importance.

A further difference in the behaviour of the three types of crystals is that for vapour-grown platelets ΔL exhibits a minimum (curve 1). Remarkably, for all melt-grown samples, ΔL varies proportionally to U_F without any indication of the occurrence of a minimum at high voltages. Possibly this minimum of conductance can be explained by the formation of an inversion layer at the surface of the samples as in the case of Ge, Si, and Te [2, 3]. For this, however, one must bend the energy bands at the surface considerably. This bending must be larger, the nearer the Fermi energy in the bulk is to the valence band. Since the Fermi energy shifts towards the middle of the gap with rising temperature, the minimum of ΔL lies at smaller values of U_F when the temperature is higher (Fig. 3).

Field effect measurements allow the evaluation of the mobility of the majority carriers at the surface. This field mobility is given by

$$\mu_F = \pm \frac{a^2}{C} \cdot \frac{\Delta L}{U_F}, \tag{1}$$

where the positive and negative sign is valid for n-type and p-type surfaces, respectively. Apart from the spacing a of the probes and the effective capacity C, only the slope $\Delta L/U_F$ at small values of U_F enters this formula. It is important that in many semiconductors a considerable part of the induced charge is trapped in surface states. For the evaluation of μ_F, however, the total induced charge has been inserted. That is why μ_F generally equals the drift mobility but not the Hall mobility of the majority carriers.

For trigonal selenium one finds from Fig. 2 and 3 that $\Delta L/U_F < 0$, as is expected for p-type conduction. When the measured value of C is inserted, at room temperature the field mobility amounts to $\mu_F = 6 \times 10^{-3}$ cm²/Vs. Values of similar size were reported for the Hall mobility of selenium crystals [8, 9]. This agreement demonstrates that surface states do not play an essential role in trigonal selenium. Like the Hall mobility, μ_F increases with temperature, as is evident from Fig. 3 where the slope $\Delta L/U_F$ clearly rises with temperature. It is interesting that this rise occurs exponentially with $1/T$ with the same activation energy as is found for the conductivity. This can be seen in Fig. 4, where μ_F and σ are plotted logarithmically vs. $1/T$ for two platelets. Both curves are parallel and yield an activation energy of 0.2 eV. According to this result, the temperature dependence of σ is mainly due to that of the mobility and the carrier concentration is constant. This behaviour is typical for selenium. It is also found when the hole concentration is determined from the measurement of Hall effect or thermoelectric power [8 to 10].

According to the equal temperature dependence of σ and μ in Fig. 4, a simple relation should exist between ΔL and the conductance $L = e\, p\, \mu\, (b\, d/a)$. ($d$ is the thickness and b the width of the crystal.) Making use of formula (1) one obtains

$$\Delta L = \frac{C\, U_F}{e\, a\, b\, d} \cdot \frac{\mu_F}{\mu} \cdot \frac{L}{p} \,. \qquad (2)$$

Since the first factor is constant and the ratio μ_F/μ is about 1, independent of temperature, ΔL is proportional to L, provided that the carrier concentration p remains unchanged. In Fig. 5 ΔL is plotted vs. L for five different crystals. The conductance L was changed here by varying temperature and by illuminating the crystals. For all samples in this plot, a straight line of slope 1 is obtained, i.e. $\Delta L \sim L$. The hole concentration p of trigonal selenium therefore practically does not depend on temperature. The result that different lines are obtained in Fig. 5 is possibly due to the somewhat different sample dimensions or to the hole concentration which may differ slightly from one crystal to another.

The conductivity of selenium can be considerably increased by illumination. Particularly at low temperature, a large part of this photoconductivity can be

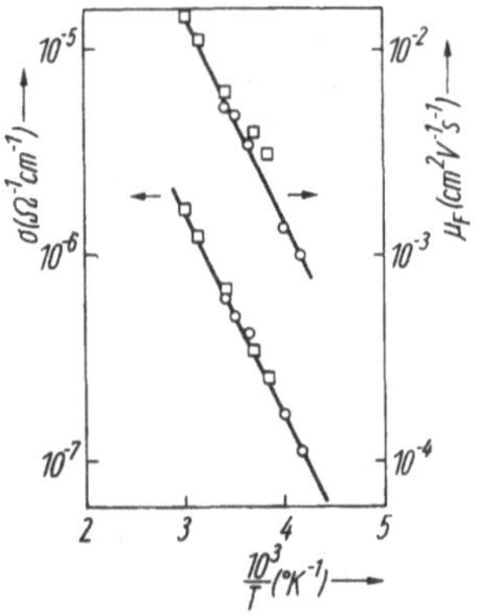

Fig. 4. Temperature dependence of conductivity σ and field mobility μ_F for two different platelets

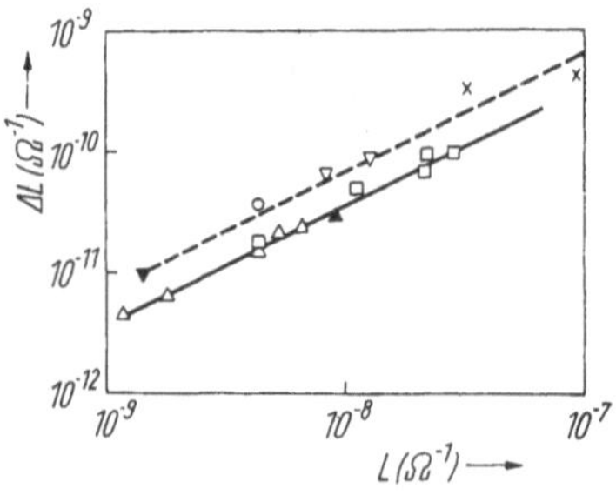

Fig. 5. ΔL as a function of conductance L for five different crystals (full points taken at 90 °K after illumination)

frozen in. The field effect measurements taken at 90 °K after illumination also lie on the straight lines in Fig. 5. Thus, relation (2) is also valid for illuminated samples with the same proportionality factor. The hole concentration therefore is also independent of illumination and the photoconductivity is mainly due to an increase in mobility.

4. Discussion

4.1 Field mobility

The results of these field effect measurements can be described in terms of the generally accepted model for the conduction mechanism in trigonal selenium single crystals. According to this model [11], the conductivity is determined by a strongly inhomogeneous distribution of holes which originates from a system of positively charged dislocation lines which are surrounded, for neutrality, by cylinders of negative space charge. These cylinders overlap, thus giving rise to a potential saddle midway between them, which, as the lowest potential barrier, controls current flow. If there is a closed network of overlapping space charge regions, the conductivity of the crystal becomes thermally activated, namely

$$\sigma = G\, e\, p_0\, \mu_0 \exp\left(-\Delta E/kT\right), \tag{3}$$

where μ_0 is the microscopic mobility, p_0 the hole concentration in the regions of high conductivity, ΔE the height of the potential barrier and G a geometry factor. The Hall effect and thermoelectric power in this barrier model are essentially determined by the regions of high conductivity. Their measurement, therefore, yields for the concentration p_0 a value of about 10^{14} cm^{-3} independent of temperature [11]. The combination of this concentration with the conductivity finally gives a thermally activated mobility.

For the above evaluation of the field mobility μ_F, a similar method was, in fact, used, since the concentration of influenced carriers, given by the external parameters capacity C and the applied voltage U_F, was combined with the measured change of conductivity. By this procedure, a mobility is obtained which depends exponentially on $1/T$, i.e.

$$\mu_F = \mu_0 \exp\left(-\Delta E/kT\right). \tag{4}$$

Inserting $\Delta E = 0.2$ eV, the lattice mobility μ_0 at 300 °K is found to be 10 cm^2/Vs for current flow perpendicular to the c-axis. Similar values were obtained by measurement of the magnetoresistance [12] and of the acousto-electric current saturation [13]. This accordance indicates that the conduction mechanism is the same both in the bulk and in the surface region. The reason for this behaviour is that the dislocation lines and hence the barrier regions end at the surface of the crystal, so that the surface conductance is also barrier controlled. When the transversal field is applied, these barriers are not destroyed but the potential energy of holes in the surface is changed on the whole by a constant amount. Then the influenced charge distributes according to this potential energy, the major part being in the regions of high conductivity. However, the observed change of conductance is determined by the excess charge density in the barrier regions which is given by

$$Z = Z_0 \exp\left(-\Delta E/kT\right), \tag{5}$$

Z_0 being the excess density in the regions of high conductivity. The latter quantity approximately equals the total induced charge density and thus can be calculated from C and U_F. The thermally activated form of the mobility (4) here is therefore due to the combination of the barrier controlled change of conductance ΔL with the charge density Z_0 in the regions of high conductivity.

The influence of illumination on the field effect can also be interpreted in this model. The main effect of illumination is to reduce the barrier height while the concentration p_0 between the barriers remains constant. Thus, at fixed temperature, the field effect should increase appreciably according to relation (5). Since the dark conductivity at low temperature was too small for field effect measurements, a direct comparison of results from illuminated crystals and from samples kept in the dark at constant temperature is not possible. However, one can compare the field effect at constant conductivity, namely at 90 °K after illumination and in the dark at a higher temperature. Then the exponent $\Delta E/kT$ in formula (5) has the same value and the field effect is expected to be equal in both cases. Indeed this is found experimentally as can be seen in Fig. 5, where $\Delta L \sim L$ with the same proportionality factor as in the dark.

In spite of the large electric fields applied to the surface of the crystals, no field quenching of the photocurrent frozen in at 90 °K was observed. In view of the pronou nced field quenching effects usually observed for selenium already at relatively small fields [14, 15], this result is astonishing. This discrepancy can be explained by the different arrangements. For the field effect, current and the high electric field are oriented perpendicularly. The barrier regions which limit the current flow are then mainly parallel to the direction of the transverse field. The effective field strength, therefore, is not expected to be significantly different from that applied externally. On the contrary, the effective field in the barriers is much enhanced when current and quenching field are parallel, i.e. in the longitudinal arrangement, because in this case the voltage drops mainly across the high resistive barrier regions. Thus, field quenching in this barrier model is more likely to occur in logitudinal rather than in transverse configuration.

4.2 Conductance minimum

The minimum of conductance is usually explained by the formation of an inversion layer at the surface [1]. Because surface states in most cases screen the external field, it is only occasionally possible to shift the surface potential over a sufficiently large energy range. That is why this minimum of conductance has been observed so rarely. It is important that the position of the minimum provides information about the surface potential. According to Many et al. [1], the minimum of conductance occurs when the band bending at the surface V_s becomes

$$e\,V_{sm} = -\,2\,e\,\Phi_b - kT \cdot \ln \mu_n/\mu_p\,. \tag{6}$$

In this formula $e\,\Phi_b$ is the energetic distance between the Fermi energy E_F and the middle of the gap E_i^b in the bulk. μ_n and μ_p represent the electron and hole mobilities, respectively. Due to the factor 2, the minimum appears at a larger band bending than is necessary to reach intrinsic conduction for which approximately $E_F = E_i^b$ is valid. When the minimum is reached, the surface is already inverted, i.e. it is n-type for trigonal selenium.

Now taking the barrier model into account, one has to insert the Fermi energy within the barriers $E_F = E_F^0 + \Delta E$ where ΔE is the height of the barriers and

$E_F^0 = kT \ln N_v/p_0$ is the Fermi energy in the regions of high hole density p_0. With the assumption $\mu_n/\mu_p = 1$ formula (6) yields

$$e\, V_{sm} = -\,2\,(E_F^0 + \Delta E - E_i^b)\,. \tag{7}$$

Obviously $e\, V_{sm}$ depends on the hole concentration p_0 because E_F^0 does. Taking $\Delta E = 0.2$ eV and a hole concentration $p_0 = 10^{14}$ cm^{-3} as is typically found from thermoelectric power measurements [10], one obtains $e\, V_{sm} = 0.75$ eV. A band bending at the surface of this magnitude is required in addition to the bending by the potential barriers in order to obtain the minimum of conductance for trigonal selenium by inversion.

This band bending is not necessarily produced entirely by the external field but can consist of several contributions: 1. the bending by the external field which is proportional to the applied voltage U_F and 2. a constant bending δ that possibly exists already at zero field:

$$e\, V_{sm} = \beta\, e\, U_{Fm} + \delta\,. \tag{8}$$

With the relation (7), one then arrives at the following expression for the voltage U_{Fm} where the conductance minimum occurs:

$$e\, U_{Fm} = -\,T \cdot 2\,\frac{k}{\beta} \ln \frac{N_v}{p_0} + \frac{2\,E_i^b - 2\,\Delta E - \delta}{\beta}\,. \tag{9}$$

Since the last term is constant, U_{Fm} should vary about linearly with temperature. This is found experimentally, as can be seen from Fig. 6 where the position of the minimum U_{Fm} is plotted versus temperature for two selenium platelets. In both cases straight lines are obtained. Their slopes and ordinate intersections allow the determination of the constants β and δ for a given hole concentration p_0. With $p_0 = 10^{14}$ cm^{-3}, the result for curve (a) is $e\, V_{sm} = 0.74$ eV in accordance with the value of 0.75 eV calculated directly from formula (7). The important result, however, is that already without external field a band bending of about $\delta = 0.5$ eV exists at the surface which corresponds to a strong depletion layer for holes.

This enormous constant band bending which is necessary to reach the minimum of conductance at relatively small external fields makes the above interpretation problematic. A bending of this magnitude would require a large concentration of surface states and probably would already lead to an intrinsic or even n-type surface, at least in the barrier regions relevant for conduction. For both properties there is, to our knowledge, no experimental evidence.

An alternative explanation for the observed minimum arises from the barrier model outlined in Section 4.1. These barriers are caused by positively charged donors which lie approximately 0.9 eV below the bottom of the conduction band, i.e. about in the middle of the gap [11]. When the band bending by the external field proceeds, the Fermi energy at the surface approaches the do-

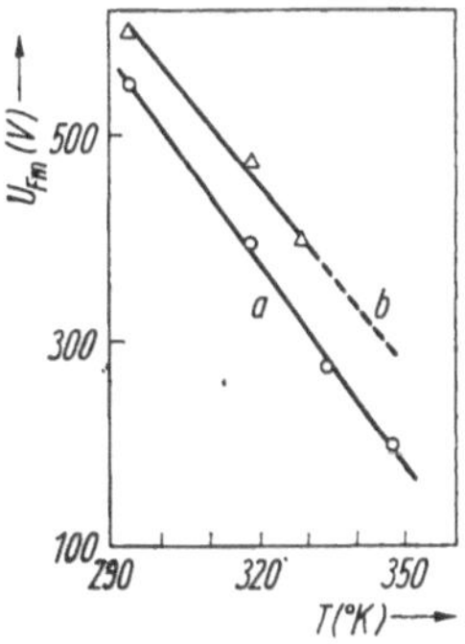

Fig. 6. Temperature dependence of the minimum of ΔL for two platelets

nor level. Since the donors are then partly neutralized, the height of the potential barriers decreases. This leads to a pronounced rise of conductance. On the other hand, by the transverse electric field majority carriers are pulled out of the surface region, so that the conduction decreases. Thus, there are two competing processes, of which, finally, at higher applied voltages, the reduction of the barrier height by partial neutralization of the donors will take over. Hence, in this barrier model, the surface conductance can be expected to pass through a minimum as a function of the transverse field.

From the present measurements, it is not possible to decide definitely which of these two interpretations for the minimum of conductance in the field effect applies. Also the linear variation of U_{Fm} with temperature does not allow further conclusions since in both cases the shift of the Fermi energy with temperature is responsible for this behaviour. However, it seems that the explanation in terms of the barrier model has to be preferred, because it does not need an unreasonably large band bending.

Acknowledgement

The authors gratefully acknowledge valuable discussions with Prof. Dr. J. Stuke.

References

[1] A. MANY, Y. GOLDSTEIN, and N. B. GROVER, Semiconductor Surfaces, North-Holland Publ. Co., Amsterdam 1965.
[2] G. G. E. LOW, Proc. Phys. Soc. **B68**, 10 (1954).
[3] R. SILBERMANN, G. LANDWEHR, and H. KÖHLER, Solid State Commun. **9**, 949 (1971).
[4] W. BEYER, H. MELL, and J. STUKE, phys. stat. sol. (b) **45**, 153 (1971).
[5] K. ZÜCKLER, Z. Phys. **136**, 40 (1953).
[6] H. GOBRECHT, A. TAUSEND, K. PAWLEK, and F. ZOERNER, Verh. DPG **4**(2), 55 (1964).
[7] K. SIEMSEN, Thesis, Berlin 1965.
[8] K. W. PLESSNER, Proc. Phys. Soc. **B64**, 671 (1951).
[9] J. HELESKIVI, T. STUBB, and T. SUNTOLA, J. appl. Phys. **40**, 2923 (1969).
[10] J. STUKE and K. WENDT, phys. stat. sol. **8**, 533 (1965).
[11] J. STUKE, phys. stat. sol. **6**, 441 (1964).
[12] H. MELL and J. STUKE, phys. stat. sol. **24**, 183 (1967).
[13] J. MORT, Phys. Rev. Letters **18**, 540 (1967).
[14] H. GOBRECHT and H. HAMISCH, Z. Phys. **148**, 218 (1957).
[15] W. FUHS and J. STUKE, phys. stat. sol. **27**, 171 (1968).

(Received June 12, 1972)

phys. stat. sol. (a) **12**, 499 (1972)

Subject classification: 18.2.1 and 18.4; 21.1; 22.6

Physics Department, University of York, Heslington

Transverse Susceptibility of an Exchange-Coupled Pair

By

K. O. Legg, W. D. Doyle[1]), and M. Prutton

A simple theory is described for the transverse susceptibility of an exchange-coupled ferromagnetic–antiferromagnetic pair. The theory is evaluated for parameters corresponding to the system Ni–NiO. The susceptibility as a function of the applied field shows large peaks at fields characteristic of the volumes of the two materials, their magnetic properties, and the interaction between them. The problems associated with the experimental derivation of an unambiguous value of the exchange interaction are discussed and it is concluded that precise definition of the interface is necessary and requires the use of in-situ ultra-high vacuum techniques of analysis.

Nous avons décrivé une simple théorie pour la susceptibilité transversale d'une paire ferromagnétique–antiferromagnétique accouplée par l'interaction d'échange. La théorie est évaluée pour les paramètres correspondants au système Ni–NiO. La susceptibilité en fonction du champ appliqué montre des pics élevés pour certains valeurs du champ correspondant aux volumes des deux matérieaux, à leurs propriétés magnétiques et à l'interaction entre eux. Nous avons envisagé les problèmes associés à la dérivation expérimentale d'une valeur definie pour l'interaction d'échange. En conclusion, une definition précise de l'interface est nécessaire et exique l'untilisation de techniques d'analyse „in situ“ dans le ultra vide.

1. Introduction

Since its discovery by Meiklejohn and Bean [1], the phenomenon of exchange anisotropy has been extensively studied. Although it occurs in various magnetically ordered systems [2], it is perhaps most easily visualised in the interaction across the interface between a ferromagnet and an antiferromagnet.

In their study of superficially oxidised cobalt particles, Meiklejohn and Bean observed a $\sin \theta$ torque component and a displaced easy axis hysteresis loop, both indicative of an unidirectional anisotropy. The samples were also characterised by a high field rotational hysteresis, in contrast to the low-field hysteresis expected for a simple ferromagnet in an applied field less than the anisotropy field. Further work [2] showed that both the unidirectional torque and the rotational hysteresis were fully established on a 300 Å cobalt film after 25 Å of oxide had been formed at its surface. These properties were explained in terms of exchange coupling across the interface between the cobalt and its oxide, in which the interaction was assumed to be ferromagnetic.

Bransky et al. [3] attempted to define the interface more precisely by depositing cobalt films on untwinned single crystals of CoO. They observed a hysteresis loop shift when cobalt was deposited on both the (111) uncompensated spin plane of CoO and the (100) fully compensated spin plane, the shift being smaller in the latter case. Since there should be no nett coupling on a perfectly compensated

---

[1]) Permanent address: Univac Division of Sperry Rand, P.O. Box 500, Blue Bell, Penn., U.S.A.

plane, they suggested that the loop shift in this case resulted from a slight departure from the ideal (100) plane. Schlenker and Buder [4] also found little difference between the two planes when they deposited low magnetostriction NiFe on CoO. These workers applied the technique of "oscillating hysteresis" to the problem, augmenting the well-established methods of Meiklejohn and Bean.

The Ni–NiO combination has been studied using polycrystalline nickel films by Lommel and Graham [5] and using nickel deposited on (100) and (111) NiO single crystals by Berkowitz and Greiner [6], who also investigated strain effects by depositing NiFe on NiO. The rotational hysteresis data were explicable in terms of strain for nickel on (100) NiO, and a combination of strain and exchange anisotropy on (111) NiO. The results of Schlenker and Buder [7], on the other hand, were independent of the NiO orientation for the NiFe–NiO system.

Another discrete ferromagnetic–antiferromagnetic pair which has been studied is NiFe–Cr [8], this experiment being conducted on films grown in ultra-high vacuum. The relative structural orientations of the components of various exchange coupled pairs have been studied by Schmid [9]. Practically all of these studies have used as probes the properties of exchange-coupled pairs discovered by Meiklejohn and Bean.

Despite this wealth of work, a number of basic problems remain unresolved, some of which have been noted by Jacobs and Bean [10], and more recently by Schlenker and Buder [7] and by Prutton et al. [11]. These problems lie in the definition of the interface and (in many ways inseparable from this) the arrangement of, and coupling between interfacial spins in each component.

A well-defined plane interface is unlikely to exist in practice. If the surface of antiferromagnet contains steps or faceting, as expected if it had been polished or cleaved to receive a ferromagnetic film [3, 4, 6, 7], spins would be exposed lying in opposite directions, even on a nominally uncompensated spin plane. Faceting or roughness would also expose planes other than the intended surface plane. Such considerations have been invoked by Schlenker and Buder [4, 7] to explain their observations. Some of the problems of interface definition are illustrated in Fig. 1.

A further problem, which is particularly relevant where films are deposited on antiferromagnetic substrates, is that of surface contamination [11]. Foreign matter between the two media could seriously affect the exchange interaction across the interface. Local interactions might also be modified by surface damage inflicted during polishing. These difficulties are largely overcome in experiments where films are oxidised [5], but in these experiments it is possible that the nature and position of the interface is not clearly defined. Yet, despite their importance to the problem, very few thorough investigations of surface structure and cleanliness have been made.

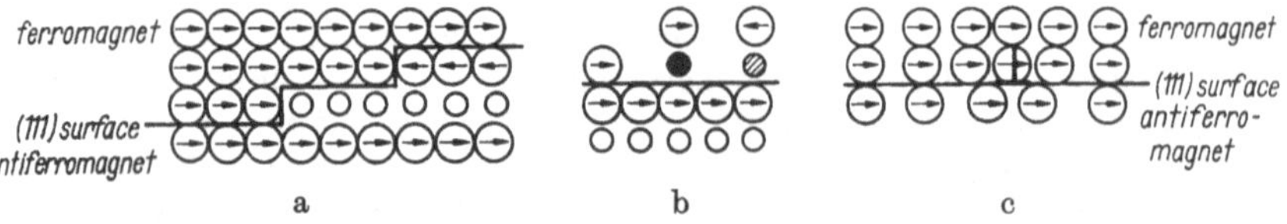

Fig. 1. Examples of non-ideal coupling between a ferromagnet and its antiferromagnetic substrate. encircled arrows — Ni, empty circles — O, full and hatched circles — contaminants. a) Direct exchange via a monatomic step in a Ni surface region and super-exchange via a monatomic step in an O-surface region of a NiO surface. b) Direct and super-exchange via different species of surface impurities. c) Inhomogeneous exchange due to a misfit dislocation at the interface of an epitaxial antiferromagnet–ferromagnet pair

The above points suggest that in order to carry out well-defined exchange anisotropy experiments, it is necessary to specify and control both the interfacial conditions and the spin state in each component. There is no known way of accomplishing this, and thus the exchange coupling cannot be unambiguously determined. But the experimental conditions can be more carefully characterized by the use of modern techniques of surface analysis. In an ultra-high vacuum environment, surface cleanliness and order may be investigated by low [12] and high energy electron diffraction and Auger electron spectroscopy [13], and to some extent controlled. Under these conditions, however, it would be difficult to make torque magnetometer measurements. A probe which could be used in situ, and which is suited to a study of surface spin orientation, is the transverse Kerr effect. An apparatus using this effect for the measurement of transverse susceptibility [14] has been described by Chambers et al. [15] and by Prutton [16]. With this type of instrument, measurements may easily be made from outside the chamber. It could also be used to examine small areas ($\approx$5 μm across) and would thus not average out large scale inhomogeneities. With the use of such equipment in mind, a theoretical study of the transverse susceptibility of an exchange-coupled pair has been carried out using the simple model outined below.

2. Theory

A model which is an extension of that put forward by Meiklejohn and Bean [1] and which is identical in principle to that of Berkowitz and Greiner [6] has been adopted for exchange coupling between a uniaxial antiferromagnet and an isotropic ferromagnet. The switching curves of a pair in which both components are anisotropic have recently been calculated by Brodkorb and Haubenreisser [17] using a similar approach. The assumptions used here are listed explicitly below:

(i) The pair comprises an isotropic, homogeneous ferromagnet coupled ferromagnetically (by convention) across a plane interface to a uniaxial, homogeneous antiferromagnet, whose uncompensated spin planes lie in the plane of the interface.

(ii) The spins are rigidly coupled within each component, so that the spins in each material move as rigid blocks.

(iii) Small deviations from collinearity of the sublattice moments of the antiferromagnet are neglected in formulating the magnetostatic energy terms, but are accounted for through its susceptibility.

(iv) Both materials are single domains.

(v) All magnetic changes are quasi-static.

(vi) All terms arising from magnetoelastic and magnetostatic effects are neglected.

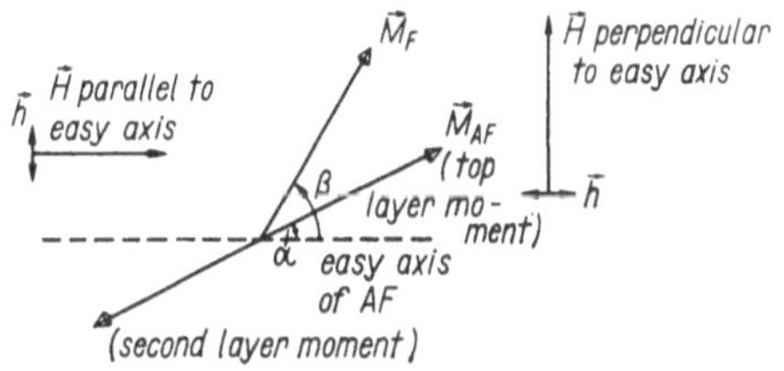

Fig. 2. Moment arrangement of ferromagnetic–antiferromagnetic pair

The situation thus defined is shown in Fig. 2. In experiments utilising the transverse Kerr effect for the measurement of the transverse susceptibility, a bias field H is applied normal to a small oscillating field h, both in the film plane. Accordingly, the energy equations are set up for two such perpendicular applied fields. The problem is to determine the magnetization orientation by energy minimisation, and hence to find the transverse susceptibility [14], defined as

$$\chi_t = \lim_{h \to 0} \frac{\partial M_t}{\partial h}, \tag{1}$$

where M_t is the component of the magnetisation M_F of the ferromagnetic film which is parallel to h. The treatment below is concerned solely with the cases where H is parallel to, or perpendicular to, the easy axis of the antiferromagnet.

When H is parallel to the easy axis, we obtain the energy expression

$$E = \tfrac{1}{2} [H_{SF}^2 - H^2 + h^2] (\chi'_\perp - \chi'_\parallel) V_{AF} \sin^2 \alpha - \tfrac{1}{2} V_{AF} (\chi'_\perp h^2 + \chi'_\parallel H^2) - \\ - M_F V_F (H \cos \beta + h \sin \beta) - J A \cos (\beta - \alpha) . \tag{2}$$

In this equation $\chi'_\perp$ and $\chi'_\parallel$ are the susceptibilities of the antiferromagnet perpendicular to and parallel to an applied field H, J is the exchange energy per unit area of interface, A the total interfacial area, and V_F and V_{AF} are the volumes of the ferromagnet and antiferromagnet respectively. H_{SF} is the spin-flop field [18] of the antiferromagnet, defined by

$$H_{SF}^2 = \frac{2 K}{\chi'_\perp - \chi'_\parallel}, \tag{3}$$

where K is the anisotropy constant per unit volume of the antiferromagnet.

The minimum energy conditions are,

$$\frac{\partial E}{\partial \alpha} = 0 \quad \text{and} \quad \frac{\partial E}{\partial \beta} = 0 , \tag{4}$$

subject to the restrictions [15]

$$\left(\frac{\partial^2 E}{\partial \alpha^2}\right) \left(\frac{\partial^2 E}{\partial \beta^2}\right) - \left(\frac{\partial^2 E}{\partial \alpha \, \partial \beta}\right)^2 > 0 , \qquad \frac{\partial^2 E}{\partial \alpha^2} > 0 . \tag{5}$$

Hence, in the limit of zero "tickle field", h,

$$\tan \beta = \frac{\sin \alpha}{N H + \cos \alpha}, \tag{6}$$

where α is given by the specific solution $\sin \alpha = 0$, or by the general solution

$$2 N H R^2 [H_{SF}^2 - H^2]^2 \cos^3 \alpha + [N^2 H^2 + 1] R^2 [H_{SF}^2 - H^2]^2 \cos^2 \alpha - H^2 = 0 , \tag{7}$$

whichever is stable under given conditions, and in which

$$N = \frac{M_F V_F}{JA},$$

$$R = \left(\frac{\chi'_\perp - \chi'_\parallel}{M_F}\right) \frac{V_{AF}}{V_F} = \varrho \frac{V_{AF}}{V_F},$$

and

$$\Delta = N^2 H^2 + 2\,NH \cos\alpha + 1\,. \tag{8}$$

Using the above equations the susceptibility is then given by

$$\chi_t = M_F \left\{ \frac{N + \dfrac{\partial \sin\alpha}{\partial h}}{\Delta^{1/2}} - \frac{N \sin^2\alpha}{\Delta^{3/2}} \left[1 - \frac{H}{\cos\alpha} \frac{\partial \sin\alpha}{\partial h} \right] \right\}, \tag{9}$$

where

$$\frac{\partial \sin\alpha}{\partial h} = \frac{\Delta \cos^2\alpha + N\,H \sin^2\alpha \cos\alpha}{N\,H^2 \sin^2\alpha + R\,\Delta^{3/2}\,(H_{SF}^2 - H^2)\cos 2\alpha + H\,\Delta \cos\alpha}. \tag{10}$$

If the static field is applied along the hard axis, H and h are merely interchanged in equation (2). This has the effect that equation (6) becomes

$$\tan\beta = \frac{N\,H + \sin\alpha}{\cos\alpha}, \tag{11a}$$

and Δ is given by

$$\Delta = N^2 H^2 + 2\,NH \sin\alpha + 1\,, \tag{11b}$$

while in all other equations, terms in $\sin\alpha$ are replaced by equivalent terms in $\cos\alpha$, and vice versa, while terms in $H_{SF}^2 - H^2$ are replaced by terms in $H_{SF}^2 + H^2$.

Given all the necessary parameters of the system, χ_t may be determined by finding values of α and β from equations (6) and (7), satisfying the inequalities (5), and complying with the equalities (4), which are substituted into equations (9) and (10).

Note that, as well as equation (7), the specific solution $\sin\alpha = 0$ is possible (or, equivalently, $\cos\alpha = 0$ for H parallel to the hard axis). Thus, provided the stability condition $\partial^2 E/\partial\alpha^2 > 0$ holds, the solution $\alpha = \beta = 0$ is possible if

$$-H^3 - \frac{1}{N} H^2 + \left(H_{SF}^2 + \frac{1}{NR}\right) H + \frac{H_{SF}^2}{N} \geqq 0\,, \tag{12}$$

while the solution $\alpha = \beta = \pi$ is possible if

$$H^3 - \frac{1}{N} H^2 - \left(H_{SF}^2 + \frac{1}{NR}\right) H + \frac{H_{SF}^2}{N} \geqq 0\,. \tag{13}$$

Similarly, $\alpha = \pi$, $\beta = 0$ is allowed if

$$-H^3 + \frac{1}{N} H^2 + \left(H_{SF}^2 - \frac{1}{NR}\right) H - \frac{H_{SF}^2}{N} \geqq 0\,. \tag{14}$$

For H perpendicular to the easy axis, the solution $\alpha = \beta = \pi/2$ is possible if

$$-H^3 - \frac{1}{N} H^2 - \left(H_{SF}^2 - \frac{1}{NR}\right) H - \frac{H_{SF}^2}{N} \geqq 0\,. \tag{15}$$

We shall for convenience designate these the $(0, 0)$, (π, π), $(\pi, 0)$, and $(\pi/2, \pi/2)$ solutions, respectively. No other special solutions of the type $\sin\alpha = 0$ or $\cos\alpha = 0$ are possible for $H > 0$. The fields at which these particular solutions become possible or cease to exist are found to be associated with singularities in the transverse susceptibility. It may readily be shown that when such a special

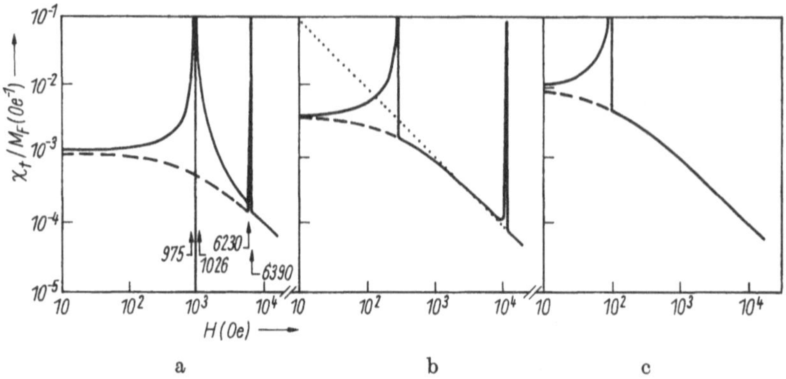

Fig. 3. Susceptibility curves for Ni–NiO ($\varrho = 10^{-7}$ Oe^{-1}, $N = 10^{-3}$ Oe^{-1}, $H_{SF} = 6325$ Oe) compared with uniaxial Ni ($H_K = 100$ Oe), H parallel to easy axis, and presaturated in negative H direction. a) $V_{AF}/V_F = 10^4$, ——— H increasing, – – – – H decreasing. b) $V_{AF}/V_F = 10^2$, ——— H increasing, $V_{AF}/V_F = 10^2$, – – – – H decreasing, $V_{AF}/V_F = 10^{-1}$ ········ c) Nickel

state is entered or left (i.e. when one of the inequalities (12) to (15) becomes an equality), the denominator in equation (10) becomes zero and an infinite transverse susceptibility peak results.

Examples of curves of χ_t/M_F computed on an Elliott 4100 computer by the method outlined above, are shown in Fig. 3 and 4. These may be compared to the accompanying curves for a uniaxial ferromagnet. The parameters used in computing these curves are roughly in accord with the work of Berkowitz and Greiner [6] for Ni–NiO. These parameters have been assigned the values $N = 10^{-3}$ Oe^{-1}, $\varrho = 10^{-7}$ Oe^{-1}, $K = 10^3$ erg/cm^3, and $\chi'_\perp - \chi'_\parallel = 5\times10^{-5}$ erg/cm^3 – hence $H_{SF} = 6325$ Oe, in the range 5 to 10 kOe, as suggested in [6]. The values of V_{AF}/V_F are given in the figure captions.

The curve of Fig. 3 for $V_{AF}/V_F = 10^4$ illustrates the most complicated case, where H is parallel to the easy axis of a moderately strongly coupled pair. With the sample initially in the (π, π) state, the field is increased from zero. When the initial state becomes unstable at a little less than $(1/N)$ Oe, β changes continuously from π to 0, while the spins in the antiferromagnet rise from the easy axis, before their high anisotropy forces them back onto it again. Infinite susceptibility peaks occur at the beginning and end of this change, the first as the (π, π) state is left and the second as the $(\pi, 0)$ state is entered. As H is further increased, a discontinuous switch to the (0, 0) configuration gives a further peak in χ_t/M_F.

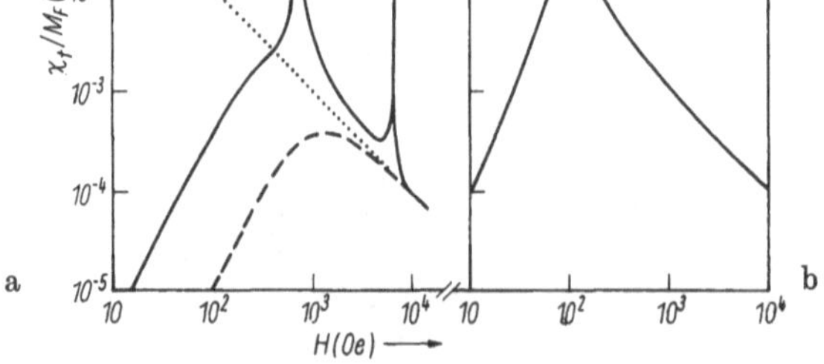

Fig. 4. Susceptibility curves for Ni–NiO compared with uniaxial Ni; H perpendicular to the easy axis. a) Ni–NiO ——— $V_{AF}/V_F = 10^4$, – – – – $V_{AF}/V_F = 10^2$, ········ $V_{AF}/V_F = 10^{-1}$. b) Nickel

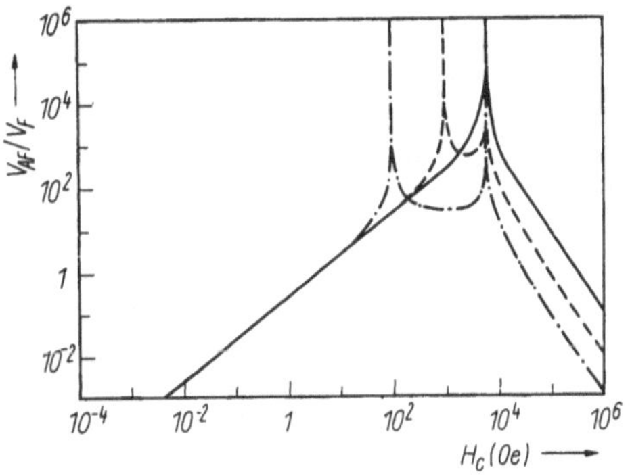

Fig. 5. Critical fields in Ni–NiO as functions of V_{AF}/V_F for H parallel to the easy axis. ——— $N = 10^{-4}$, – – – – $N = 10^{-3}$, –·–·– $N = 10^{-2}$

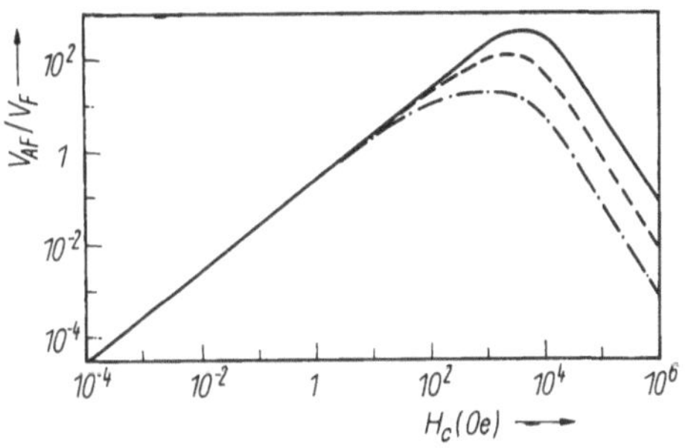

Fig. 6. Critical fields in Ni–NiO as functions of V_{AF}/V_F for H perpendicular to the easy axis

Finally, the antiferromagnet spin-flops, its spin axis rising towards $\pi/2$ while the magnetisation turns away from H before the increasing field realigns it. If the field is now reduced, the spin-flop transition occurs in reverse, leaving the sample in the stable (0, 0) configuration. If $V_{AF}/V_F = 10^2$, the ferromagnet is more dominant, so that the configuration changes discontinuously from (π, π) to (0, 0).

When H lies along the hard axis, the two possible types of behaviour are shown in Fig. 4. If V_{AF}/V_F is large, the $(\pi/2, \pi/2)$ configuration is never attained; the spins in the ferromagnet and antiferromagnet merely separate when the latter attempts to spin-flop. However, for $V_{AF}/V_F = 100$, the $(\pi/2, \pi/2)$ state is entered at about 700 Oe and left at about 7000 Oe, when the antiferromagnet spin-flops, giving rise to peaks in χ_t/M_F.

Fig. 5 and 6 can be thought of as phase diagrams for the spin configuration. They have been computed from the equalities of (30) to (33) and thus represent the fields at which transitions occur into or out of the specific spin states — i.e. the positions of susceptibility peaks. For a given set of values of J, N, etc., these curves define the values of H and V_{AF}/V_F at which transitions occur to or from the $\sin\alpha = 0$ or $\cos\alpha = 0$ states. In Fig. 5 the (π, π) state is stable at the left of the diagram, while the $(\pi, 0)$ configuration is stable above the centre branches. Beneath the curves, the spin orientation, determined by the stability conditions, is either (0, 0) or defined by equations (6) and (7). At the right, spin-flopping has occurred. In Fig. 6 the $(\pi/2, \pi/2)$ state is stable only beneath the curves. The curves are drawn for $\varrho = 10^{-7}$ and various N. Thus, for instance, by tracing the line $V_{AF}/V_F = 10^2$ for $N = 10^{-3}$ in Fig. 5 and 6, the suscpetibility peak fields shown in Fig. 3 and 4 can be obtained.

The longitudinal susceptibility is defined as $\chi_l = \partial M_l/\partial h$, where M_l is the component of M_F parallel to H. This exhibits some interesting effects, notably finite negative peaks at spin-flop and whenever spin reorientations pull M_F away from H as H is increased. These peaks generally have small amplitudes ($\chi_l/M_F \approx$ $\approx 10^{-4}$ to 10^{-6} Oe^{-1}) and are therefore unlikely to be seen in practice. Large longitudinal susceptibility peaks ($\chi_l/M_F \approx 10^{-2}$ to 1 Oe^{-1}) also exist in regions where the spins reorient continuously from the (π, π) to the $(\pi, 0)$ or (0, 0) configurations.

3. Discussion

It has been shown that the $\chi_t(H)$ curves display a wealth of features associated with transition fields related to the parameters of the pair. But the question of whether these could be observed experimentally is of prime importance, since the peaks will obviously not be infinite in practice. Peak fields will not be well defined owing to the inhomogeneity in magnetic parameters. In particular, J will not be uniform, for the reasons we have already noted. This will lead to broad peaks made finite by spatial averaging over regions comparable with the wavelength of the light used and by interactions between these regions. Work with Kerr effect susceptibility equipment [16] suggests that the minimum observable signal is $\chi_t/M_F \approx 10^{-3}$ Oe^{-1}. Only those peaks will be observed which are not reduced below this level by magnetic inhomogeneities.

A further point is immediately apparent from Fig. 5 and 6, namely that for a large relative volume of ferromagnet, the predicted peaks in χ_t/M_F are shifted to extremely low and extremely high fields. Therefore this method is not suited to the examination of such cases. It is unlikely to be usable in studies of oxidised ferromagnetic films.

It is quite probable that the high field behaviour of a pair will be dominated by domain wall motion and will not therefore fall within the ambit of this model. (S-wall motion takes place in NiO at fields above 2.4 kOe [19]). Nevertheless, this model has been found to give surprisingly good predictions of the rotational hysteresis for Ni–NiO [6], and so it should be capable of yielding at least a qualitative picture of what might be seen experimentally.

It has been noted earlier that the interface is far from well defined, and because of steps, faceting, etc., many of the antiferromagnetic surface spins are likely to be antiparallel to those in the ferromagnet (or parallel if the nett interaction is antiferromagnetic). It is readily shown that, if domain formation does not occur, the constant J in the equations above should be replaced by $(2r - 1) J$, where r is the ratio of the area containing correctly aligned spins to the total interfacial area. It follows that only an effective coupling can be measured, so that exchange anisotropy between discrete components is unsuitable for studying the exchange interaction itself.

4. Conclusion

Some of the basic problems inherent in exchange anisotropy studies have been considered, and it has been pointed out that some of them could be overcome by the use of the techniques of modern surface analysis to specify the experimental conditions as accurately as possible. In situ magnetic measurements could be made using a transverse Kerr effect apparatus to determine the transverse susceptibility of the ferromagnet in an exchange-coupled pair.

This susceptibility, as a function of the applied bias field, has been examined using a simple model. The results are encouraging, suggesting that this probe could well be useful in detecting peaks in the susceptibility arising from spin reorientations.

Acknowledgements

We are grateful to C. Schlenker of the Laboratoire d'Electrostatique et Physique du Metal, Grenoble, for useful and stimulating discussions, and to the Science Research Council, who supplied a Senior Visiting Fellowship to W. D. Doyle and a studentship to K. O. Legg.

References

[1] W. H. MEIKLEJOHN and C. P. BEAN, Phys. Rev. **102**, 1413 (1956); **105**, 904 (1957).
[2] W. H. MEIKLEJOHN, J. appl. Phys. **33**, 1328 (1962).
[3] J. BRANSKY, I. BRANSKY, and A. A. HIRSCH, J. appl. Phys. **41**, 183 (1970).
[4] C. SCHLENKER and R. BUDER, Czech. J. Phys. **B21**, 506 (1971).
[5] J. M. LOMMEL and C. D. GRAHAM, JR., J. appl. Phys. **33**, 1160 (1962).
[6] A. E. BERKOWITZ and J. H. GREINER J. appl. Phys. **36**, 3360 (1965).
[7] C. SCHLENKER and R. BUDER, phys. stat. sol. (a) **4**, K79 (1971).
[8] R. W. SHARP and P. C. ARCHIBALD, J. appl. Phys. **37**, 1462 (1966).
[9] H. SCHMID, Cobalt **6**, 8 (1960).
[10] I. S. JACOBS and C. P. BEAN, in: Magnetism, Ed. G. T. RADO and H. SUHL, Academic Press, 1963.
[11] M. PRUTTON, W. D. DOYLE, and K. O. LEGG, 17th Annual Conference on Magnetism and Magnetic Materials, Chicago 1971.
[12] M. PRUTTON, Metallurg. Rev., No. 152 (1971).
[13] T. E. GALLON and J. A. D. MATTHEW, Rev. Phys. and Technology, to be published.
[14] E. FELDTKELLER, Z. Phys. **176**, 510 (1965).
[15] A. CHAMBERS, D. POMFRET, and M. PRUTTON, J. sci. Instrum. **44**, 181 (1967).
[16] M. PRUTTON, J. appl. Phys. **39**, 1153 (1968).
[17] W. BRODKORB and W. HAUBENREISSER, phys. stat. sol. (a) **3**, 333 (1970).
[18] D. H. MARTIN, Magnetism in Solids, Iliffe, London 1967.
[19] W. L. ROTH and G. A. SLACK, J. appl. Phys. **31**, 352S (1960).

(Received May 31, 1972)

phys. stat. sol. (a) **12**, 509 (1972)

Subject classification: 17; 1.6; 21.1; 21.1.1

Department of Pure and Applied Physics, The Queen's University of Belfast

The Adsorption of Carbon Monoxide on Iron, Cobalt, and Manganese

By

G. K. HALL[1]) and C. H. B. MEE

Polycrystalline films of iron, cobalt, and manganese were evaporated at pressures of less than 10^{-9} Torr on to fused silica substrates. Changes in work function as carbon monoxide was adsorbed at room temperature were followed by a photoelectric technique. The curves of the change in work function versus exposure for both iron and cobalt showed an increase to saturation values of $\Delta\varphi$ of +0.13 eV for iron and +0.15 eV for cobalt. Any change in the work function of manganese was less than 0.01 eV. The results can be explained using a simple dipole theory, allowing for depolarization at high coverages.

Polykristalline Eisen-, Kobalt- und Manganschichten wurden bei Drücken von weniger als 10^{-9} Torr auf Quarzglassubstrate aufgedampft. Änderungen der Austrittsarbeit durch Absorption von Kohlenmonoxid bei Zimmertemperatur wurden mit einer photoelektrischen Methode verfolgt. Die Änderung der Austrittsarbeit mit der Belichtung zeigte sowohl für Eisen als auch für Kobalt einen Anstieg bis zu Sättigungswerten $\Delta\varphi$ von +0,13 eV für Eisen und +0,15 eV für Kobalt. Eine etwaige Änderung der Austrittsarbeit von Mangan ist kleiner als 0,01 eV. Die Ergebnisse lassen sich mit einer einfachen Dipoltheorie, die Depolarisation bei hohen Bedeckungen erlaubt, erklären.

1. Introduction

The adsorption of carbon monoxide on metals has been studied extensively. Surface potential measurements have shown that the type of bonding between the carbon monoxide molecule and the metal depends upon the electronic structure of the adsorbent. For example, the adsorption of carbon monoxide causes a reduction in the work function of the noble metals copper [1 to 3], silver [2], and gold [2], but an increase in the work function of transition metals such as tungsten [4], tantalum [5], iron [1, 2, 5], cobalt [2], and nickel [1, 2]. It has been suggested that the bonding of carbon monoxide to the noble metals is essentially due to electrostatic forces, with the oxygen atom towards the metal surface [2]. For the other metals, however, the bonding is covalent, the orientation being one in which the carbon atom is towards the metal surface [2]. Infrared absorption studies have been quoted as providing further evidence for these characteristic types of bonding. For example, in the case of carbon monoxide adsorption on copper, the frequency at which the optical absorption peak is observed differs little from that for the free carbon monoxide molecule [6], but for adsorption on cobalt [7, 9], iron [8, 9], and nickel [7, 9, 10] the frequencies are markedly lower. The metal samples employed in many of these infrared determinations were prepared under vacuum conditions which would be considered quite unsuitable for surface potential measurements; there is, however, evidence that the spectra obtained for carbon monoxide adsorbed on films of nickel prepared under ultra-

[1]) Present address: Vacuum Physics Division, Mullard Research Laboratories, Redhill, Surrey, England.

high vacuum conditions are similar to those reported for adsorption in samples prepared under poor vacuum [10]. Flash-filament spectra for carbon monoxide desorbed from tungsten [11] show the existence of two main phases of the adsorbed state. These phases are identified with the linear binding $M{-}C \equiv O$ and the bridged adsorption $\genfrac{}{}{0pt}{}{M}{M}\!\!>\!C = O$, forms which are thought to be responsible for characteristic features of the infrared absorption spectra. Measurements of the heat of adsorption of carbon monoxide on evaporated films of a number of metals [12] have been discussed in terms of the possibility of adsorption in the linear and bridged configurations on various crystal faces.

Table 1 summarizes the literature values of the surface potential of carbon monoxide on iron and on cobalt. Although the figures confirm the general observation that the adsorption of carbon monoxide on transition metals causes an increase in work function, there is no agreement between different workers on the magnitude of this increase. Information on the vacuum conditions in these experiments is insufficient to allow an assessment of the cleanliness of the metal films. There are considerable difficulties in the evaporation of uncontaminated films of these metals, as we have emphasized in a recent work function study [13]; it is unlikely that clean films will be obtained unless the pressure during evaporation is less than 10^{-9} Torr.

Table 1

The surface potential of carbon monoxide on iron and cobalt

author	surface potential (eV)		method	notes
	Fe + CO	Co + CO		
Baker and Rideal [5]	−1.15	−0.27	photoelectric threshold	no information on pressure during deposition
Bloyaert et al. [1]	−1.33		contact potential difference (vibrating capacitor)	no information on pressure during deposition; temperature during adsorption, 293 K
Culver et al. [2]	−1.64	−1.50	contact potential difference (shift of space-charge-limited characteristic)	background pressure 10^{-8} Torr; substrate temperature during adsorption, 78 K

In the present work we have investigated the adsorption of carbon monoxide on films of iron, cobalt, and manganese prepared under ultra-high vacuum conditions. The use of a photoelectric technique brought with it the advantage that the work functions could be measured absolutely, and the cleanliness of the initial state of the film could be confirmed. Our experiments give much smaller surface potentials of carbon monoxide on iron and cobalt than have been reported by previous workers. Within experimental error, it was not possible to detect a change in the work function of manganese films on exposure to carbon monoxide.

2. Experimental

The experimental tube, vacuum system, method of preparation of the metal films, and photoelectric technique have been described in detail elsewhere [13, 14]. Films of iron, cobalt, and nickel were deposited by double evaporation on fused silica substrates at pressures of less than 1×10^{-9} Torr; the background pressure in the mercury-pumped system was less than 2×10^{-10} Torr. Spectroscopically pure carbon monoxide was admitted into the experimental tube at a pressure of about 1×10^{-8} Torr through an ultra-high vacuum valve. All the adsorption measurements took place at about 290 K. The work function of the film was determined by measuring the photoelectric yield at a number of incident wavelengths and fitting the experimental results to Fowler's theoretical curve [15].

3. Results

Fig. 1 and 2 show the change in work function $\Delta\varphi$ caused by the adsorption of carbon monoxide on iron and cobalt films. In each case there is an increase in the work function to a saturation value. Three separate runs on iron gave a maximum increase in work function of 0.13 ± 0.01 eV above the clean value of 4.71 ± 0.02 eV. For cobalt, the maximum increase (again for three independent runs) was 0.15 ± 0.01 eV above the clean value of 4.89 ± 0.04 eV. Continued exposure of up to more than 10^{22} molecules m^{-2} produced no further change in work function. In the case of manganese, any change in work function on adsorption of carbon monoxide was less than 0.01 eV, the uncertainty in an individual work function determination, for exposures of up to 10^{22} molecules m^{-2}.

4. Discussion

An explanation of the observed changes in the work functions on adsorption of carbon monoxide may be attempted from two directions. In one method, a simple model of covalent adsorption (see Culver and Tompkins [16] for a summary) is applied. It is assumed that the system consisting of a certain coverage of adsorbate molecules on a metal surface is equivalent to the formation of an electrical double layer of uniform density. The surface potential $\Delta\varphi$ corresponding to a coverage θ is given by

$$\Delta\varphi = \pm \frac{\sigma_m \theta \mu}{2 \varepsilon_0}, \tag{1}$$

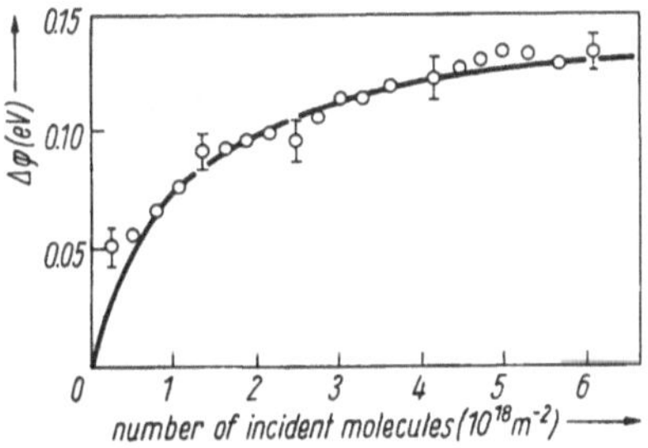

Fig. 1. The change in work function due to the adsorption of carbon monoxide on iron at 290 K

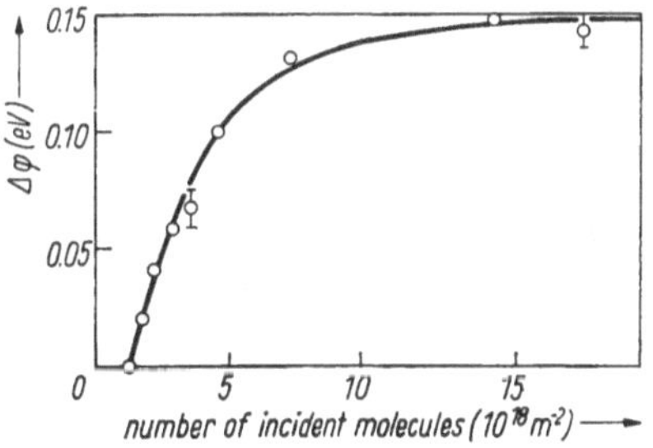

Fig. 2. The change in work function due to the adsorption of carbon monoxide on cobalt at 290 K

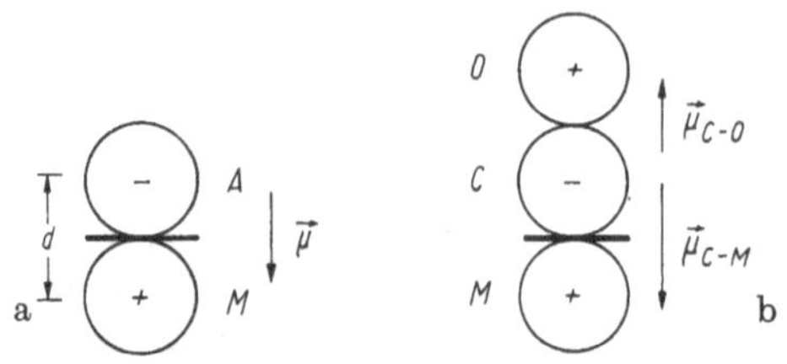

Fig. 3. a) Covalent adsorption (with partial ionic character) of an atom A on a metal M. The conventional dipole moment μ is ed, but because metallic conduction is assumed to start at the metal surface, the effective dipole length is $d/2$. With the dipole in the direction shown, there will be an increase in the work function of $\mu/2\,\varepsilon_0$ for each dipole adsorbed. b) Covalent adsorption of carbon monoxide on a metal. The resultant dipole moment μ, directed towards the surface, is given by $\mu = \mu_{C-M} - \mu_{C-O}$

where σ_m is the number of adsorption sites per unit area, θ is the fraction of these sites which are occupied, and μ is the dipole moment, as shown in Fig. 3a. This equation predicts that the surface potential should vary linearly with coverage, provided that the dipole moment μ is constant. This assumption is unlikely to be true, except at very low coverages; at higher coverages there will be a depolarizing effect due to the interaction between neighbouring dipoles. In a treatment due to Miller [17] the effective dipole moment μ_θ at coverage θ is given by

$$\mu_\theta = \frac{\mu}{1 + B\alpha\,\theta/r^3}\,, \tag{2}$$

where α is the polarizability of the adsorbent–adsorbate system and r is the distance between neighbouring dipoles at monolayer coverage. The quantity B is a geometrical factor which is to a good approximation equal to 9 for both square and hexagonal arrays of dipoles [18]. The consequence of the depolarizing interaction is that the dipole moment μ_θ decreases as the coverage increases, so that the variation of $\Delta\varphi$ with θ is a smooth curve of gradually decreasing gradient represented by

$$\Delta\varphi = \pm\frac{\sigma_m\,\theta\,\mu}{2\,\varepsilon_0\,(1 + B\alpha\,\theta/r^3)}\,. \tag{3}$$

In attempting to apply this treatment to the present results we are faced with a number of difficulties. The adsorbents in our experiments were polycrystalline films; the density of adsorption sites on a given crystallite depends upon the crystal faces exposed. Secondly, carbon monoxide is itself a polar molecule, and in such a case value the of μ deduced from equation (3) is the resultant of the dipole moment of the carbon monoxide molecule (about 0.4×10^{-30} C m) and the dipole formed by the bond between the carbon and the metal atoms. Further, as shown by flash-filament and infrared absorption studies, carbon monoxide is adsorbed in both linear and bridged forms, with corresponding differences in the densities of the adsorption sites and the resultant dipole moments. Finally, the coverage is required in equation (3), whereas the experimentally determined quantity is the exposure; conversion between the two requires a knowledge of the sticking coefficient of carbon monoxide on iron and cobalt, information which does not appear to be available in the literature.

Brennan and Hayes [12] have discussed the energetics of the possible configurations of carbon monoxide adsorbed on various metals. The arrangements of the carbon monoxide molecules on a number of high atom density, low-index faces are shown in Fig. 4. The maximum practicable density of the adsorbate is about 1.22×10^{19} molecules m^{-2} in a bridged form on the (110) face; denser arrangements which are possible geometrically are unacceptable on energetic grounds due to the very strong repulsive interaction between adjacent molecules.

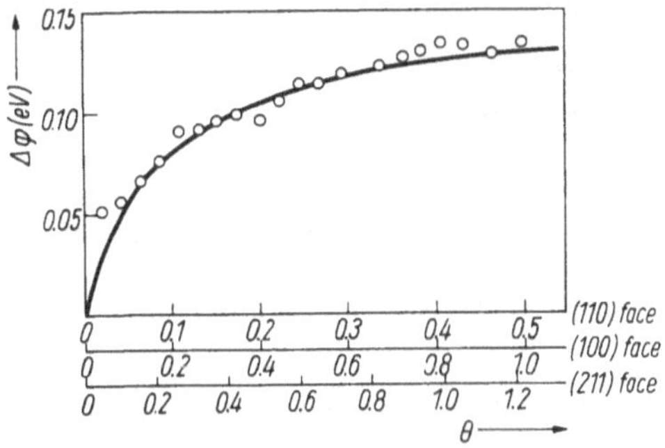

Fig. 5. Graph of $\Delta\varphi = k_1\,\theta/(1 + k_2\,\theta)$ for carbon monoxide adsorption on the (110), (100), and (211) faces of iron. The values of k_1 and k_2 required to fit the curve to the experimental points are given in Table 2

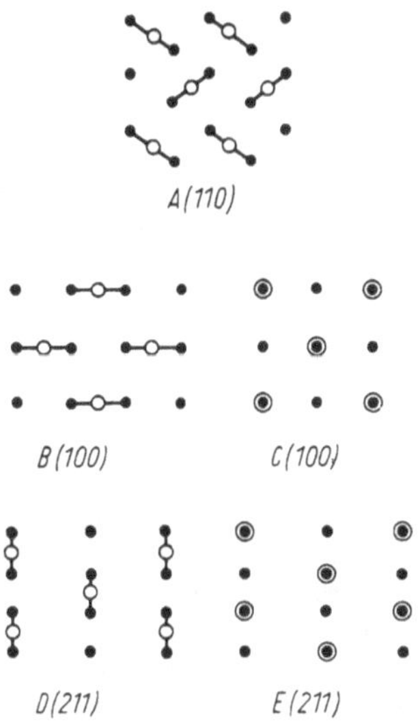

Fig. 4. Adsorption sites for carbon monoxide on various faces of iron crystals; ●, metal atoms; ●—○—●, bridge-bonded carbon monoxide; ⊙, linearly-bonded carbon monoxide. (After Brennan and Hayes [12])

The continuous curve in Fig. 5 is a plot of the equation

$$\Delta\varphi = \frac{k_1\,\theta}{1 + k_2\,\theta} \tag{4}$$

for the adsorption of carbon monoxide on various crystal faces of iron in the configurations shown in Fig. 4. We make the assumption that the sticking coefficient of carbon monoxide on each face is unity, so that, for example, for configuration A on the (110) face an experimental exposure of 6.1×10^{18} molecules m^{-2} corresponds to a coverage of half a monolayer. Comparison of equation (4) with equation (3) gives $k_1 = \sigma_m\,\mu/2\,\varepsilon_0$ and $k_2 = B\,\alpha/r^3$. The values of k_1 and k_2 required to fit the theoretical curve to the experimental points are given in Table 2.

Table 2

configuration	face	monolayer density of adsorbed CO, σ_m (m^{-2})	k_1 (eV)	k_2	r (nm)	μ (C m)	α (F m²)
A	(110)	1.22×10^{19}	1.64	10.6	0.35	2.4×10^{-30}	5.0×10^{-36}
B, C	(100)	0.61×10^{19}	0.82	5.3	0.45	2.4×10^{-30}	3.9×10^{-36}
D, E	(211)	0.50×10^{19}	0.67	4.3	0.50	2.4×10^{-30}	6.0×10^{-36}

If the sticking coefficient is not unity, the values of k_1 and k_2, and hence of μ and α, will be increased in proportion to the reciprocal of the sticking coefficient; the values tabulated thus represent minimum values of dipole moment and polarizability. The corresponding minimum value of the carbon metal dipole moment is given by $\mu = \mu_{C-Fe} - \mu_{C-O}$ (see Fig. 3b), i.e. $\mu_{C-Fe} = 2.8 \times 10^{-30}$ C m. This value is consistent with the partial ionic nature of the carbon–metal bond to the extent of approximately 10%. The average value of the polarizability of the carbon monoxide–iron system, about 5×10^{-36} F m^2, may be compared with the polarizability of the free carbon monoxide molecule, about 2×10^{-37} F m^2. A similar analysis for the adsorption of carbon monoxide on cobalt gives a value for the resultant dipole moment μ of about 0.8×10^{-30} C m, so that the dipole moment of the carbon–cobalt bond is about 1.2×10^{-30} C m. The polarizability is about 5×10^{-37} F m^2.

In a second approach to the interpretation of the experimental curves, the undoubted heterogeneity of the adsorbed system is recognized. It might be anticipated that the adsorption of carbon monoxide in linear and in bridged forms might lead to different dipole moments, and hence to different changes in work function for the same coverage of a given crystal face. Crowell and Norberg [19] have developed a model to predict the work function change due to the adsorption of a gas on a metal surface at two different types of site A and B, producing the dipole moments μ_A and μ_B. In the simple model considered by Crowell and Norberg, in which depolarizing effects at high coverages are neglected, the shape of the $\Delta\varphi$ versus coverage curve obtained in our experiments corresponds to the case where μ_A is finite and $\mu_B = 0$. This latter condition implies that on one set of adsorption sites the configuration of the surface metal atoms and the carbon monoxide molecule is such that the resultant dipole moment is zero. This situation could be visualized in an arrangement in which the carbon monoxide molecules are located in hollows in the surface plane. However, the most closely-packed planes of the iron lattice have such shallow hollows (of depth about 0.02 nm [14]) that this type of arrangement is not possible. An interpretation based solely on a simple two-site adsorption process seems less acceptable than one in which the shape of the experimental curve is explained by a contribution from the depolarizing interaction of neighbouring adsorbed molecules. There is little doubt that the true situation is represented by a combination of the effects of the heterogeneity of the surface, the different types of carbon monoxide–metal bonding, and the depolarization due to neighbouring dipoles. It is worth pointing out that individual work function measurements, both for the clean metals and for the metal–gas systems, were a good fit to the Fowler plot. It thus seems unprofitable to attempt to analyse the results in terms of a model of the surface consisting of patches of markedly different work functions.

The fact that any change in work function when carbon monoxide is adsorbed on manganese is less than a tenth of the magnitude of the changes observed for iron and cobalt is of interest. In a recent study of the manganese–carbon monoxide system, Bickley et al. [20] noted that at temperatures up to 350 K carbon monoxide adsorption is restricted to one monolayer, even for very large exposures. Above 350 K incorporation of carbon monoxide into the manganese lattice commences, the limiting coverage increasing very rapidly with increasing temperature. The suggestion is that the initial process of chemisorption to one monolayer is followed by diffusion within the lattice. It is significant that of the three metals studied in the present work manganese has the most open structure [14], and adsorption at sites which would give only a very small change in work function is much more likely than in the case of iron and cobalt.

5. Conclusions

The adsorption of carbon monoxide on polycrystalline films of iron and cobalt prepared under ultra-high vacuum conditions causes increases in work function of less than 0.2 eV in each case, considerably less than changes which have been reported in previous determinations in which the metal surfaces were produced under higher background pressures. Changes of this magnitude can be explained satisfactorily using a simple dipole theory, although a complete interpretation would require a knowledge of the distribution of the orientations of the crystallites making up the film and the way in which carbon monoxide is adsorbed on each metal face.

Acknowledgements

Thanks are due to Prof. D. J. Bradley for the provision of laboratory facilities, and to R. Greer for his skill in the construction of the experimental tubes. One of us (G. K. H.) wishes to thank the Science Research Council for the award of a research studentship.

References

[1] F. Bloyaert, L. d'Or, and J. Mignolet, J. Chim. phys. et Physico-Chim. biol. **54**, 53 (1957).
[2] R. Culver, J. Pritchard, and F. C. Tompkins, Z. Elektrochem. **63**, 741 (1959).
[3] M. A. Chesters and J. Pritchard, Surface Sci. **28**, 460 (1971).
[4] J. Eisinger, J. chem. Phys. **27**, 1206 (1957).
[5] M. McD. Baker and E. K. Rideal, Nature (London) **174**, 1185 (1954).
[6] R. P. Eischens and W. A. Pliskin, Adv. Catalysis **10**, 1 (1958).
[7] R. A. Gardner and R. H. Petrucci, J. Amer. Chem. Soc. **82**, 5051 (1960).
[8] G. Blyholder, J. chem. Phys. **36**, 2036 (1962).
[9] F. S. Baker, A. M. Bradshaw, J. Pritchard, and K. W. Sykes, Surface Sci. **12**, 426 (1968).
[10] A. M. Bradshaw and J. Pritchard, Surface Sci. **17**, 372 (1969).
[11] P. A. Redhead, Trans. Faraday Soc. **57**, 641 (1961).
[12] D. Brennan and F. H. Hayes, Phil. Trans. Roy. Soc. (London) **A258**, 347 (1965).
[13] G. K. Hall and C. H. B. Mee, phys. stat. sol. (a) **5**, 389 (1971).
[14] G. K. Hall and C. H. B. Mee, Surface Sci. **28**, 598 (1971).
[15] C. Lea, B. H. Blott, and C. H. B. Mee, Appl. Optics **8**, 203 (1969).
[16] R. V. Culver and F. C. Tompkins, Adv. Catalysis **11**, 67 (1959).
[17] A. R. Miller, Proc. Cambridge philos. Soc. **42**, 292 (1946).
[18] J. Topping, Proc. Roy. Soc. **A114**, 67 (1927).
[19] A. D. Crowell and A. L. Norberg, J. chem. Phys. **37**, 714 (1962).
[20] R. I. Bickley, M. W. Roberts, and W. C. Storey, J. Chem. Soc. (A), 2774 (1971).

(Received May 29, 1972)

phys. stat. sol. (a) **12**, 517 (1972)

Subject classification: 14.3; 22.6

Laboratoire de Spectroscopie et d'Optique du Corps Solide, Groupe de Recherche du C.N.R.S., Université Louis Pasteur, Strasbourg

Electrical Conductivity and Thermal Activation Energies in Cu_2O Single Crystals

By

M. TAPIERO, J. P. ZIELINGER, and C. NOGUET

A systematic investigation, performed on some fifty single crystals, between −40 and 250 °C shows that the values of the bulk resistivity ϱ_V (20 °C) range from 10^3 to 10^{13} Ωcm and that the activation energies lie between 0.1 and 1 eV. These values depend on the conditions of the cooling after oxydation and of high vacuum annealing. A quasi-continuous relation between E and $\lg \varrho_V$ (20 °C) is found.

Une étude systématique de la conductibilité électrique, entre −40 et 250 °C, effectuée sur une cinquantaine de monocristaux de Cu_2O, d'histoires différentes, montre que les valeurs de la résistivité interne ϱ_V (20 °C) s'étagent de 10^3 à 10^{13} Ωcm, et les énergies d'activation correspondantes de 0,1 à 1 eV. Une relation quasi-continue et biunivoque entre E et $\lg \varrho_V$ (20 °C) est mise en évidence.

The electrical properties of Cu_2O crystals, reported in numerous papers, show a great dispersion, in particular in regard to the conductivity and the thermal activation energies in the temperature range below 350 °C [1].

Several reasons have been suggested to explain the considerable variation in the data, but no real attempt has been done so far to analyse the various factors involved in the electrical conductivity of Cu_2O.

In order to get a general picture of the electrical properties of Cu_2O independently of the thermal history of the samples, we performed a systematic investigation on a great number of crystals submitted to various thermal treatments.

The experimental techniques used and the results obtained are described in two detailed publications which will soon appear [1, 2]. In this paper only the essential physical results are exposed.

The measurements were carried out on about fifty single crystals and ten polycrystalline samples in the temperature range −40 to +250 °C. They show that the electrical conductivity of Cu_2O is strongly influenced both by extrinsic and intrinsic factors. By intrinsic factors we mean inhomogeneities due to low or high resistivity surface layers, grain boundary leakages in polycrystalline samples, and bulk photomemory effects [3]. All these factors lead, even for the same specimen, to a great lack of reproducibility of the results. The values of the resistivity ϱ and the shapes of the curves $\lg \varrho = f(1/T)$ depend on the experimental conditions: whether the temperature is increased or decreased, whether the sample is in air or in vacuum, whether it is in darkness or not.

In order to eliminate to a large extent the influence of these extrinsic factors, we have performed our measurements on single crystals, in high vacuum ($p < 10^{-6}$ Torr), and in complete darkness and we used a method derived from the guard ring technique [1]. This method allows to investigate separately bulk

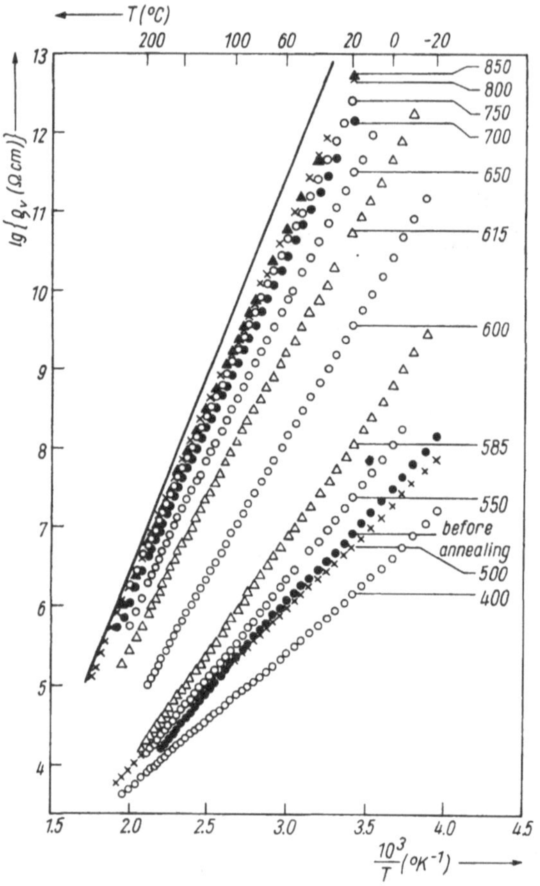

Fig. 1. Plots of $\lg \varrho_V = f_V(1/T)$, for two Cu_2O single crystals of the same platelet after high-vacuum annealings at various temperatures. Sample M'23: ○ ● × ▲; sample M'''23: △

and surface properties. It is then possible to analyse systematically the influence of intrinsic factors, i.e. modifications of the internal state, on the bulk conductivity. It appears that a good reproducibility of the bulk conductivity can be obtained for a given crystal, provided that the specimen has been preheated to about 200 °C in vacuum. The bulk resistivity depends then in general exponentially on $1/T$ in the whole temperature range explored, with a unique activation energy for both increasing or decreasing temperatures. However, the value of the activation energy varies from sample to sample.

The most striking parameter which seems to be responsible for this variation, as well as for the different values obtained for the bulk resistivity at 20 °C is the cooling velocity of the sample from the temperature of synthesis (1050 °C) to room temperature. This result can be established at least when the conditions of preparation are kept invariant [2] and when the pressure during the cooling lies between 10^{-4} and 5×10^{-2} Torr.

Quenched samples have lower bulk resistivities ($10^6 < \varrho_V < 10^8$ Ωcm) and lower activation energies ($0.4 < E < 0.6$ eV) than slowly cooled specimens ($10^8 < \varrho_V < 10^{11}$ Ωcm and $0.6 < E < 0.8$ eV). For quenched crystals the pressure has no effect. For the slowly cooled samples (cooling rate lower than about 100 °C per hour), the bulk resistivity and the activation energy seem to increase with decreasing pressure. However, this correlation is rather qualitative.

In order to have a better control of the parameters, we used a special program of annealing: we started with a quenched single crystal and annealed it at different increasing temperatures T_r between 350 and 850 °C in vacuum ($p < 10^{-5}$ Torr). After 16 h of annealing at each chosen temperature T_r, the sample was cooled down slowly (80 deg/h). The electrodes, which had been removed by etching before each thermal treatment, were again put on the platelet and measurements were carried out under the conditions specified above.

It appears that the bulk resistivity ϱ_V increases with the annealing temperature [2]. Fig. 1 shows that the curves $\lg \varrho_v = f_v(1/T)$ are fairly straight lines and that the corresponding activation energy E increases with ϱ_v (20 °C). If one plots E versus $\lg \varrho_v$ (20 °C), one gets a quasi-linear dependence: to each value of ϱ_v a well-defined activation energy E seems to correspond.

Furthermore, we have shown that this dependence obtained for one specimen (here M23) is, in fact, a characteristic property of the material. We found indeed that the bulk resistivities of all samples depend exponentially on $1/T$ (with a restriction for crystals having a bulk resistivity lower than 10^4 Ωcm). The values of ϱ_v at room temperature range, for all samples studied, from 10^3 to about 10^{13} Ωcm, whereas the corresponding activation energies lie between 0.1 and 1 eV. Fig. 2 shows that the plot of E versus $\lg \varrho_v$ (20 °C) (for the single crystal M23 and various other single crystals, annealed or not) is quasi-continuous.

These results contradict the interpretations of the conduction processes in Cu_2O. This leads to theoretical problems, which, in view of the fragmentary results known for some other oxides and large band gap compounds, are probably more general than the particular case of cuprous oxide. The fact that a nearly continuous resistivity variation, in so large a range, can be obtained for a given specimen by the annealing method, makes cuprous oxide particularly suitable for the detailed investigation of these questions.

The electrical properties of the surface are in general very different [3] from those of the bulk. They may differ considerably if the sample has been in contact with air or heated in high vacuum to temperatures above 60 °C. At a fixed temperature the ratio of the surface resistivities, with regard to the two states of the surface, increases with the bulk resistivity ϱ_v to values of about 10^6. In most cases, when ϱ_v (20 °C) is greater than about 5×10^7 Ωcm, the existence of high or low conductivity layers at the surface does not allow reliable Hall effect measurements, so that the mobility contribution to the electrical conduction cannot be determined. However, Hall effect measurements [4, 5] carried out on samples with $\varrho_v < 10^7$ Ωcm show that the mobility contribution to the thermal behaviour of the electrical conductivity is relatively weak, so that the important variation of the apparent activation energy with respect to the internal resistivity cannot be attributed to the mobility.

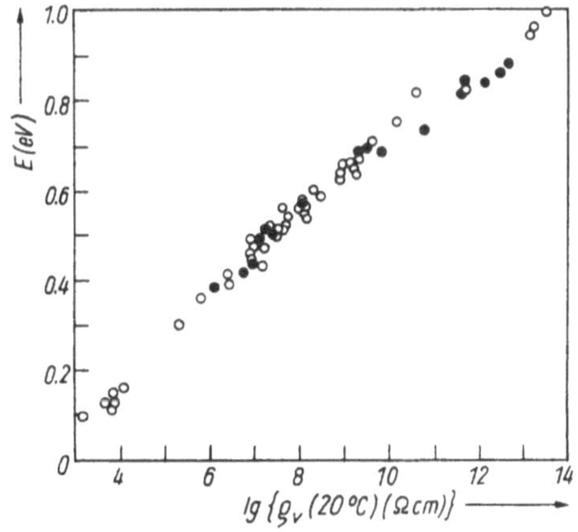

Fig. 2. Dependence of the activation energy E on $\lg \varrho_v$ (20 °C) for various Cu_2O single crystals. ●: M '23 and M'''23; ○ other samples

Conclusion. The results given in this paper show that, according to the thermal treatment, a crystal of Cu_2O can be prepared having any volume resistivity between 10^4 and 10^{13} Ωcm. The curves lg $\varrho_V = f(1/T)$ are approximately straight lines in a wide range of temperatures. The activation energy E is a nearly linear function of lg ϱ_V (20 °C). These simple relations are obtained provided that different experimental conditions are fulfilled in order to eliminate extrinsic effects.

Several authors found results which often seem to be relatively coherent: they obtain nearly the same values for the resistivity and approximately the same activation energies for their samples. This is probably due to the choice of a constant mode of preparation and cooling, and to the fact that the investigations were restricted to a limited number of samples. Their results can be, in many cases, interpreted in terms of a simple model.

Our results show that the electrical conduction of Cu_2O cannot be explained by means of a simple energy level diagram. Recent IR absorption and IR photoconductivity measurements are interpreted by means of a more elaborated energy level diagram [6] than that proposed by Bloem [7]. Further experiments on the photomemory effect and the infra-red absorption are still in progress. They will provide additional data and will probably allow an interpretation of the results obtained.

Acknowledgement

The authors wish to thank Prof. S. Nikitine for fruitful discussions during the course of this work.

References

[1] M. TAPIERO, J. P. ZIELINGER, and C. NOGUET, Ann. Phys. (France), in the press.
[2] J. P. ZIELINGER, M. TAPIERO, and C. NOGUET, Ann. Phys. (France), in the press.
[3] J. P. ZIELINGER, M. TAPIERO, C. ROUBAUD, and M. ZOUAGHI, Solid State Commun. 8, 1299 (1970).
[4] M. ZOUAGHI, M. TAPIERO, J. P. ZIELINGER, and R. BURGRAF, Solid State Commun. 8, 1823 (1970).
M. ZOUAGHI, Thesis, Strasbourg 1972.
[5] R. KUZEL and F. WEICHMANN, Canad. J. Phys. **48**, 2643 (1970).
[6] B. PREVOT, C. CARABATOS, and M. SIESKIND, phys. stat. sol. (a) **10**, 455 (1972).
M. ZOUAGHI, phys. stat. sol. (a) **11**, 219 (1972).
M. ZOUAGHI, B. PREVOT, C. CARABATOS, and M. SIESKIND, phys. stat. sol. (a) **11**, 449 (1972).
[7] J. BLOEM, Philips Res. Rep. **13**, 167 (1958).

(Received June 19, 1972)

phys. stat. sol. (a) **12**, 521 (1972)

Subject classification: 20.1; 13.4; 22.8.2

Physics Department, Sri Venkateswara University, Tirupati, India

Optical Absorption Spectrum of Iron in Garnet

By

S. V. J. LAKSHMAN, B. J. REDDY, and J. BHIMASENACHAR

The optical absorption spectrum of garnet has been studied at room and liquid air temperatures. For the first time the ion responsible for the bands has been identified from spectrochemical and Mössbauer spectral studies as Fe^{3+}. The cubic field approximation with $Dq = 1000\ cm^{-1}$, $B = 650\ cm^{-1}$, and $C = 4\ B$ is found to give a good fit to the observed band positions.

Das optische Absorptionsspektrum von Granat wurde bei Zimmertemperatur und der Temperatur der flüssigen Luft untersucht. Zum erstenmal wurde das Ion, das für die Banden verantwortlich ist, aus spektrochemischen Untersuchungen und Mößbauerspektren als Fe^{3+} identifiziert. Es wird gefunden, daß die kubische Feldnäherung mit $Dq = 1000\ cm^{-1}$, $B = 650\ cm^{-1}$ und $C = 4\ B$ gute Anpassung an die beobachteten Banden liefert.

1. Introduction

Clark [1], White and Keester [2], and Manning [3] reported some optical absorption bands obtained in certain garnets. Since no detailed and satisfactory analysis of these bands has been made by any of these authors, a detailed analysis of the bands of almandine garnet has therefore been taken up by the present authors.

2. Experimental

2.1 The absorption spectrum

The absorption spectrum of an orange coloured almandine garnet collected from Gudur Mica Mines, Andhra Pradesh, India, was photographed on a Hilger medium quartz spectrograph in the wavelength region $\lambda = 10000$ to 2000 Å. Using appropriate continuous sources such as incandescent and stabilized hydrogen lamps, the spectra were recorded in about two to three hours on Kodak IM, IN, Ilford H.P3, Astra, and Zenith plates at room and liquid air temperatures. In these studies, two crystals of thicknesses 0.5 and 0.1 cm were employed. Spectra were also recorded using polarized incident light.

2.2 Mössbauer study

Mössbauer spectral studies were kindly carried out for us at different temperatures at the National Physical Laboratory, New Delhi.

2.3 Wavelength and intensity measurements

Using microphotometric profiles recorded on a Moll microphotometer, the wavelengths of the band maxima were measured against iron arc standard lines in a similar way as has been done by Pappalardo [4].

3. Theory

The free ion Fe^{3+} has an electron configuration (A) $3d^5$, where (A) stands for the closed eighteen-electron argon shell. This configuration gives rise to ^{2}(S, P, D, F, G, H, I), ^{4}(P, D, F, G), and ^{6}S states of which ^{6}S is the ground state.

In a strong cubic crystalline field, the five valence d-electrons are distributed into the triply degenerate t_{2g} and doubly degenerate e_g orbitals. The ground state configuration for Fe^{3+} ion is then written as $t_{2g}^3 e_g^2$ which gives rise to the terms $^6A_{1g}$, $^4A_{1g}$, $^4A_{2g}$, ^4E_g, $^4T_{1g}$, $^4T_{2g}$, and some other doublet states of which $^6A_{1g}$ lies lowest.

4. Results and Analysis

The bands obtained at room and liquid air temperatures are shown in Fig. 1a and b. The bands are to be attributed to either manganese or iron ions as only these two transition elements are detected from spectrochemical analysis. Of the two, the element iron is chosen as responsible for the occurrence of the bands as its content is found to be about twenty to thirty times higher than that of manganese. Detailed Mössbauer studies carried out at different temperatures did not reveal even a trace of Fe^{2+} but confirmed the presence of Fe^{3+} ions in the sample. The analysis is therefore carried out with the assumption that the bands are due to Fe^{3+} ions.

Bands at 11806, 16293, 17510, 19040, 19829, 20694, and 21719 cm^{-1} are obtained with a crystal of thickness 0.5 cm while the bands at 23260, 23671, 24339, and 24457 cm^{-1} are recorded only with a crystal of thickness 0.1 cm. The first group of bands exhibited a sharp fall in intensity when the second group was developed. The intensity of the spectrum considerably increased when the crystal was cooled to liquid air temperature as is shown in Fig. 1b.

The spectrum consists of seven main bands located at 19040, 19829, 20694, 21719, 23260, 23671, and 24339 cm^{-1}. Of these, the bands at 19829 and 23260 are extremely sharp and those at 21719, 23671, and 24339 cm^{-1} a little less sharp. The ligand-field bands are sharp either when the energy expressions for the transitions are independent of Dq or when the number of t_{2g} electrons is the same [5] in both the excited and ground states. The extremely sharp bands at 19829 and 23260 cm^{-1} are attributed to ^4E_g (G) and ^4E_g (D) states as their energy expressions are independent of Dq. Further, no shift for such bands should be observed by cooling the crystal which indeed has been observed in the case of these two bands.

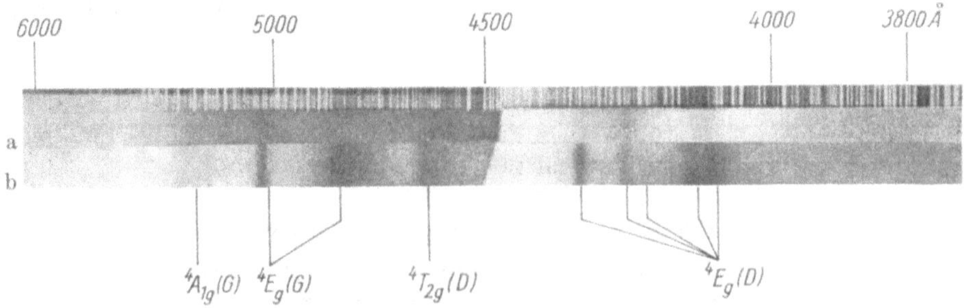

Fig. 1. Medium quartz spectrogram of Fe^{3+} in garnet at a) 300 °K and b) 90 °K

The electron configuration for Fe^{3+} in its ground state is written as $t_{2g}^3 e_g^2$ which gives rise to $^6A_{1g}$, $^4A_{1g}$, $^4A_{2g}$, ^4E_g, $^4T_{1g}$, and $^4T_{2g}$ states. Of these, the $^6A_{1g}$ state lies lowest. Next are $^4A_{1g}$ and ^4E_g both with the same energy. Above these lie in the order of increasing energy $^4T_{2g}$, $^4T_{1g}$, and $^4A_{2g}$ states. Since the $^4T_{2g}$ state of this configuration has the same number of t_{2g} electrons as that of the ground state, this band should be sharp. The sharp band at 21719 cm^{-1} is therefore attributed to the $^4T_{2g}$ (D) state.

Using the diagrams of Tanabe and Sugano [6] for the d^5 configuration, the weak band at 11806 cm^{-1} is attributed to the $^4T_{1g}$ (G) state and the bands at 16293 and 17510 cm^{-1} to the $^4T_{2g}$ state (both arising from the electron configuration $t_{2g}^4 e_g$). On cooling the crystal, these bands showed a small red shift indicating a negative slope for the *E–Dq* dependence. Since only $^4T_{1g}$ (G) and $^4T_{2g}$ (G) (as is seen from Tanabe-Sugano diagram [6]) levels have negative slopes, the assignments given above are confirmed.

The absorption spectrum taken with polarized incident light of mutually perpendicular orientations reveals the change in intensity for the band at 19040 cm^{-1}. This band therefore belongs to a transition of the type A–A [5]. The band is accordingly assigned to the transition $^4A_{1g} \rightarrow {}^6A_{1g}$.

The energy values are evaluated for $^4T_{1g}$, $^4T_{2g}$, ^4E_g, and $^4A_{1g}$ states for different values of B, C, and Dq both by diagonalizing the energy matrices and solving the energy expressions given by Lakshman and Reddy [7]. Several sets of plots have been drawn for different values of C between E and Dq. The interelectronic repulsion factor B and the crystal splitting factor Dq are then evaluated from the graph $C/B = 4$ to which the experimentally observed bands could be fitted very well. The energy positions thus calculated are presented in Table 1 along with the observed band maxima at liquid air temperature.

The oscillator strengths of the bands calculated using microphotometric profiles at liquid air temperature are also presented in Table 1.

Table 1

The observed energies, oscillator strengths, and the assignment of the bands for Fe^{3+} in garnet ($Dq = 1000$ cm^{-1}, $B = 650$ cm^{-1}, and $C = 4\,B$)

transitions from $^6A_{1g}$	band positions (cm^{-1})		oscillator strengths at 90 °K ($f \times 10^6$)
	observed at 90 °K	calculated	
$^4T_{1g}$ (G)	11806	11010	—
$^4T_{2g}$ (G)	16293	15400	—
	17510		
$^4A_{1g}$ (G)	19040	19500	1.3
^4E_g (G)	19829	19500	2.4
	20694		2.1
$^4T_{2g}$ (D)	21719	21980	6.6
^4E_g (D)	23260	24050	8.8
	23671		3.9
	24050		—
	24339		21.9
	24575		13.3

5. Discussion

The absorption bands of almandine garnet are proved for the first time to belong to an Fe^{3+} and not Fe^{2+} ion. The assignments of the bands are made assuming a cubic crystal field. However, the splitting of the band at 17510 cm^{-1} (component 16293 cm^{-1}) suggests that the Fe^{3+} ion is not entirely subjected to a cubic field but may also be under the influence of a trigonal or tetragonal field. Whenever a mulitplet E_g state is involved in either the upper or lower states, splitting of the bands may be due to the Jahn-Teller instability. The additional bands found at 20694 for 19829 cm^{-1} and 24457 for 23260 cm^{-1} assigned to ^4E_g (G) and ^4E_g (D) states, respectively, are thus attributed to the Jahn-Teller effect. The band at 20694, however, is found to be broader than the band at 24457 cm^{-1} which may probably be due to the coincidence of the energy levels of the ^4E_g (G) and $^4T_{2g}$ (D) states (equal to $(13\,B + 5\,C) = 21450$ cm^{-1} with respect to the ground state) in the absence of configurational interaction.

Further, at liquid air temperature, an additional band at 24050 cm^{-1} is developed and the band at 24457 cm^{-1} is split into two with peaks at 24575 and 24339 cm^{-1}. These and other results suggest that the field is slightly tetragonal rather than trigonal as in the former but not in the latter a splitting in the E_g state can occur.

Acknowledgements

The authors wish to express their thanks to Dr. V. G. Bhide of N.P.L. Delhi for Mössbauer spectral results and Prof. G. Ramachandran of Madras University (now at the Indian Institute of Science, Bangalore) for the kind help in solving the matrices on the Madras University Computer.

References

[1] S. P. CLARK, JR., Amer. Mineralogist **42**, 732 (1957).
[2] W. B. WHITE and K. L. KEESTER, Amer. Mineralogist **51**, 774 (1966).
[3] P. G. MANNING, Canad. Mineralogist **9**, 723 (1969).
[4] R. PAPPALARDO, Phil. Mag. **2**, 1397 (1957); **4**, 219 (1959).
[5] C. J. BALLHAUSEN, Introduction to Ligand Field Theory, McGraw-Hill Publ. Co., 1962.
[6] Y. TANABE and SUGANO, J. Phys. Soc. Japan **9**, 753 (1954).
[7] S. V. J. LAKSHMAN and B. J. REDDY, Spectrochim. Acta **26A**, 2230 (1970).

(Received May 10, 1972)

phys. stat. sol. (a) **12**, 525 (1972)

Subject classification: 18.1; 22

Institute for Low Temperature and Structure Research, Polish Academy of Sciences, and International Laboratory for High Magnetic Fields and Low Temperatures, Wroclaw

Magnetic Properties of the Substoichiometric Uranium α-Dichalcogenides UY_{2-x}

By

W. SUSKI, A. CZOPNIK, and T. MYDLARZ

Magnetic measurements on uranium α-disulfide and α-diselenide have been performed in the temperature range 4.2 to 1000 K and in magnetic fields up to 80 kOe. A sharp maximum of the magnetization of α-USe_2 is found at the temperatures 8.5 to 12 K in magnetic fields below 15 kOe. A diffuse maximum is obtained for α-US_2 in the whole interval of the magnetic field at temperatures 10 to 35 K. Both the value of the effective magnetic moment (2.83 BM) in the temperature range 100 to 300 K and the value of the previously measured magnetic entropy are explained on the assumption that the ground state and the next excited state are singlets.

Измерены магнитные свойства α-дисульфида и α-диселенида урана в температурном интервале 4,2 до 1000 К в магнитных полях до 80 кэ. Для α-USe_2 обнаружен четкий максимум намагниченности при температурах 8,5 до 12 К в магнитных полях до 15 кэ; в более высоких полях этот максимум исчезает. Для α-US_2 при температурах 10 до 35 К во всём диапазоне магнитных полей максимум является более широким. Значение эффективного магнитного момента равного 2,83 μ_B в интервале температур 100 до 300 К, а также как и низкотемпературная магнитная теплоемкость могут быть объяснены тем, что основный а также первый возбужденный уровни являются синглетами.

1. Introduction

Uranium dichalcogenides form an interesting class of compounds due to the different types of their crystal structures. Three crystal modifications of disulfides and diselenides have been previously reported: the tetragonal α-modification, the orthorhombic β-modification and, metastable at room temperature, the hexagonal γ-modification [1 to 5]. Uranium ditelluride forms the orthorhombic modification only but different from that of β-US_2 and β-USe_2 [6, 7]. Only recently it has been established that in fact the tetragonal α-modification contains more uranium than one would expect from stoichiometry. It constitutes another compound of the composition $UY_{1.9}$ [8, 9]. But, for the sake of simplicity, we will call this class of substoichiometric compounds dichalcogenides, too.

So far, the magnetic properties of the α-dichalcogenides have been investigated several times in the temperature region 88 to 720 K in a relatively weak magnetic field [8, 10]. These investigations have revealed the α-dichalcogenides to be paramagnetic obeying the Curie-Weiss law with small negative Weiss constants. The low-temperature heat capacity measurements of α-USe_2 have shown a λ-type anomaly at 13.1 K [11], and the authors have suggested that the Schottky transition near 25 K in $US_{1.9}$ results from a singlet ground state.

In this paper we report some new results obtained by extending the magnetic measurements to the temperature range 4.2 to 1000 K and increasing the magnetic field up to 80 kOe. The temperature dependence of the magnetic susceptibility and the value of the magnetic moment are discussed in terms of the crystal field effect on the four-fold ionized state of uranium. Attempts are made of explaining the magnetic properties and the magnetic entropy at low temperatures by assuming that the ground level and the first excited level are singlets.

2. Experimental and Results

Preparation of the samples was carried out by reacting the corresponding elements of previously described purity [12] in evacuated and sealed quartz ampules in stoichiometric proportions as described elsewhere [8, 10, 11]. In the case of the selenide the sample was prepared also by thermal decomposition of USe_3 [8, 12]. The chemical analysis of the uranium selenide was performed in a special apparatus described previously [13]. Uranium sulfide was analysed in the same equipment and sulphur was precipitated as $BaSO_4$. The results show less than 1% difference compared to the initial composition. The lattice constants were determined from X-ray powder diffraction patterns obtained by a 114 mm diameter camera at room temperature. The results are listed below:

	a (Å)	c (Å)
$US_{1.88}$	10.312	6.352
$USe_{1.88}$	10.721	6.612
$USe_{1.98}$ (therm. decomp. USe_3)	10.697	6.606

The above values are close to those given in [4, 5, 8]. The magnetic measurements were carried out on powder samples using the equipment and technique now standard in this laboratory [14, 15]. It should be noted that in all cases the samples were cooled in zero applied field and measurements were made while the sample was warmed up.

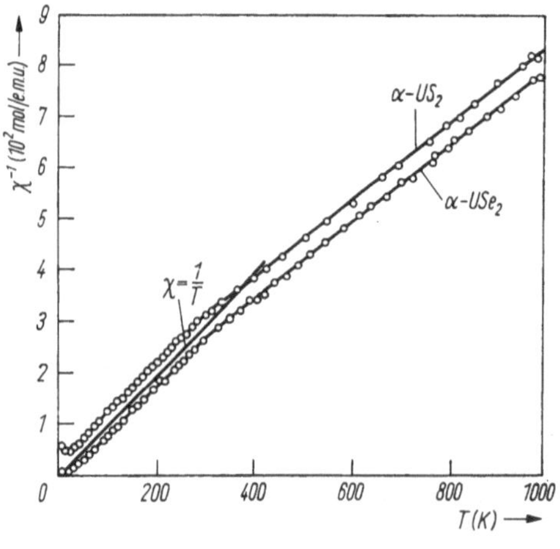

Fig. 1. The reciprocal susceptibility of α-US_2 and α-USe_2 as function of temperature

2.1 The paramagnetic region

The results obtained for both selenide samples in the paramagnetic region are exactly the same. As is seen from Fig. 1, the paramagnetic Curie temperatures are rather small for sulfide and selenide. Hence, one may consider the magnetic susceptibility as due exclusively to crystal field interactions.

2.1.1 The cation coordination

The proposed structure of α-US_2 [4] has the symmetry I4/mcm or D_{4h}^{18}. The two geometrically different uranium atoms in this structure have different environments. Of the ten uranium atoms in the unit cell approximately one-fifth are surrounded by the eight sulphur atoms at the corners of a right antiprism, with the uranium to sulphur distance equal to 2.81 Å, and four-fifths are surrounded by the six sulphur atoms: four in the cation plane and two in neighbouring planes, all at the distance of 2.77 Å. However, the configuration of the sulphur atoms is not octahedral. A detailed description of the structure and the atomic positions have been given earlier [4]. According to Khododad [5], α-USe_2 has the same structure, but the exact atom positions and the interatomic distances are unfortunately unknown as yet. For this reason the crystal-field calculations were based on the crystal parameters reported for α-US_2. We believe that there is no serious difference between the parameters, and that the results obtained for sulfide can be applied to selenide.

2.1.2 The splitting of the $^3H_4$ term and the magnetic susceptibility of the U^{4+} ion

The assumption that the U^{4+} ion has the pure $^3H_4$ ground term seems to be quite justified in the light of further considerations as well as on the basis of other papers (see, e.g., [16, 17]). We also assume that there are S^{2-} and Se^{2-} ions.

The energy level diagram for the $^3H_4$ term in a crystal field of D_{4d} symmetry (antiprismatic coordination) was reported by Mulak and Czopnik [18]. In this case the doublet $|\pm 3\rangle$ was found to be the ground state for c/a close to 1.

The formula for the magnetic susceptibility of the U^{4+} ion in an ideal antiprismatic coordination, within the low temperature range (0.1 of the total splitting Δ) in which one can assume population of the lowest doublet only, is given by [18]

$$\chi_{M,8} = \frac{0.72}{T} + \frac{3.89}{\Delta}. \tag{1}$$

The subscript 8 means the eight-fold coordination of the uranium ion. In Fig. 2 we present the energy-level splitting scheme of the $^3H_4$ term of the uranium ion surrounded by the six sulphur atoms. The expression for the crystal field potential of the first coordination sphere and some further details will be published in a separate paper [19]. As it is shown in Fig. 2, the nine-fold degenerate $^3H_4$ term splits in this case into nine singlets with a total splitting about 8200 cm^{-1}. For the paramagnetic region at temperatures which do not exceed 300 K, we assume that the ground state is composed of the two singlets

$$\varphi_1 = -0.072\,|\pm 4\rangle - 0.335\,|\pm 3\rangle - 0.471\,|\pm 2\rangle + 0.019\,|\pm 1\rangle + 0.567\,|0\rangle$$

and

$$\varphi_2 = \pm 0.145\,|\pm 4\rangle \pm 0.148\,|\pm 3\rangle \mp 0.268\,|\pm 2\rangle \mp 0.621\,|\pm 1\rangle\,.$$

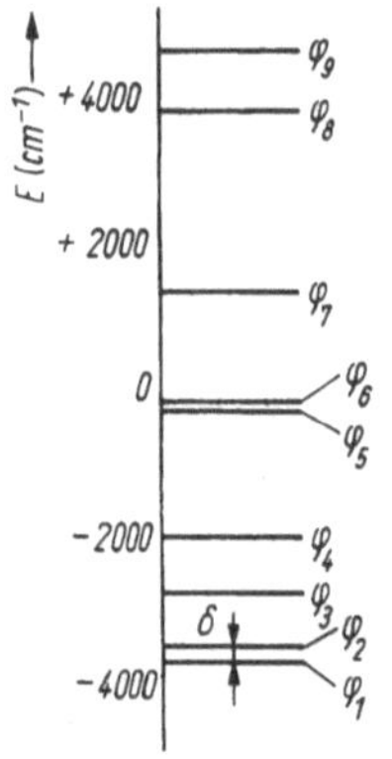

Fig. 2. Diagram of the crystal field levels for the $^3H_4$ term of the uranium ion with six-fold coordination

Thus, owing to the crystal anisotropy, the average susceptibility can be written in the form

$$\bar{\chi}_{M,6} = \frac{1}{3}(\chi_M^x + \chi_M^y + \chi_M^z) = \frac{1.0}{T} \tag{2}$$

if one neglects the high-frequency term.

The subscript 6 means the six-fold coordination of the uranium ion. Hence, the total susceptibility is equal to

$$\chi_M = \frac{1}{5}\left(\frac{0.72}{T} + \frac{3.89}{\Delta}\right) + \frac{4}{5}\frac{1.0}{T} = \frac{0.92}{T} + \frac{3.89}{5\Delta}. \tag{3}$$

The coefficients 1/5 and 4/5 correspond to the ratio of the uranium ions in both crystallographic positions. Thus, for relatively low T values the expression (3) may be approximated by $1/T$, corresponding to a magnetic moment $n_p = 2.83$ BM. This value is close enough to that given in [8, 10] as well as to the present results obtained in the temperature range up to about 300 K. However, a deviation from the linear χ^{-1} vs. T plot (see Fig. 1) above 300 K is observed, which is caused by both the contribution of the temperature-independent term and the population of the higher energy levels.

The lack of the difference in the results for $USe_{1.98}$ and $USe_{1.88}$ in the paramagnetic region suggests that the six-fold coordinated uranium atoms are mainly responsible for the magnetic behaviour of α-dichalcogenides.

2.2 Region of magnetic ordering

It has been recognized for several years [20] that the magnetic properties of the singlet ground-state systems in which the isotropic exchange interactions and the crystal field nearly balance each other, differ in a fundamental way from those of conventional magnetics. For such compounds the theory predicts a critical value for the ratio of the exchange to crystal field interaction necessary for setting up a magnetic order. Wang and Cooper [20c] have pointed out that suitable examples of such compounds are the rare earth compounds of NaCl structure with group-V and -VI anions as well as UF_4. Recently, very extensive studies on praseodymium metal and its alloys with La and Y were carried out, by measuring the magnetic susceptibility [21] and using inelastic neutron scattering [22]. One can expect a close analogy between the Pr^{3+} and U^{4+} ions which have the same ground term $^3H_4$. Unfortunately, only very recently crystal field calculations on uranium compounds with lower than cubic symmetry have been performed [17, 23]. Below, we present measurements of the magnetization on such two singlet-level systems. In the low-temperature region the magnetic properties of $USe_{1.98}$ and $USe_{1.88}$ do not differ substantially. We observe only a slight difference in the magnitude of the magnetization, but the Néel points are the same.

In Fig. 3 we present the dependence of the magnetization on the magnetic field strength, which for diselenide is linear up to about 20 kOe but begins to flatten off above this value. A similar behaviour has been observed for rare earth compounds which have been recognized as metamagnetics, with the exchange interactions being weaker than the anisotropy [24]. The temperature

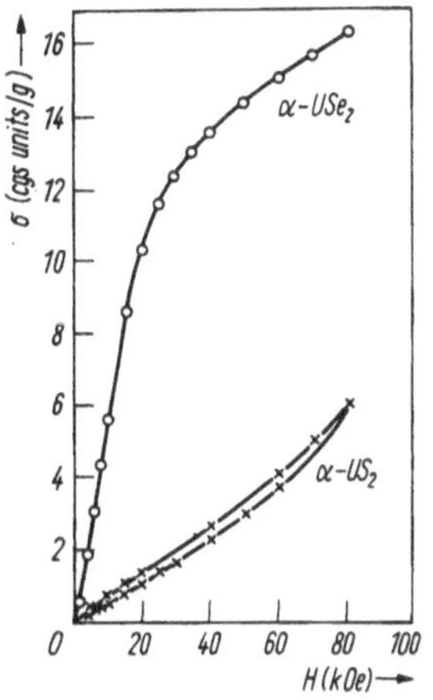

Fig. 3. The magnetization of α-US$_2$ and α-USe$_2$ as function of the magnetic field at 4.2 K

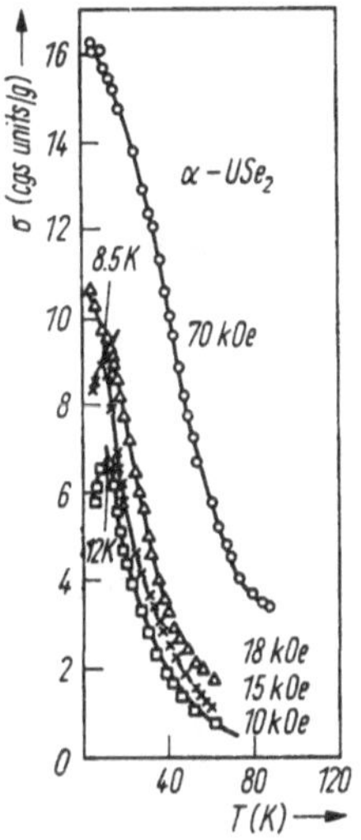

Fig. 4. The magnetization of α-USe$_2$ as function of temperature

dependence of the magnetization presented in Fig. 4 seems to provide some arguments for such an explanation of the behaviour of α-USe$_2$. The value of the critical field between 15 and 18 kOe is sufficient to suppress the antiferromagnetic order and create a "ferromagnetic" one. However, other interpretations are also possible, either based on the crystal field interactions or on the occurrence of the complicated magnetic structure realized in the presence of the two non-equivalent crystallographic positions of the uranium atoms. Moreover, the position of the Néel point of the diselenide was found to be sensitive to the magnetic field strength, changing from 12 to 8.5 K as the magnetic field increases from 10 to 15 kOe. In a magnetic field below 10 kOe no change of the Néel point is observed. The results of specific heat measurements [11] indicate that in this two singlet-level system the magnetic transition occurs in zero magnetic field.

As it follows from Fig. 3 and 5, the magnetic properties of the disulfide are very unusual. The magnetization as a function of the magnetic field is linear up to about 30 kOe, but above this value the curve becomes steeper as the magnetic field further increases. Moreover, one observes a small but well defined hysteresis. The temperature dependence of the magnetization shows a diffuse maximum in the temperature region 10 to 35 K. This maximum is observed in the lowest applied magnetic field of 1 kOe and can originate from the magnetic order of antiferromagnetic type. However, no λ-type anomaly was observed in the measurements of the low-temperature specific heat [11].

2.3 Low-temperature specific heat

It was mentioned already that the magnetic and related properties of the materials with a singlet ground state are peculiar in comparison with those of the conventional magnetic materials. Recently, Grunzweig-Genossar [25] derived an equation for calculating the magnetic entropy of a system with a single ground state, and determined graphically its dependence on the ratio X

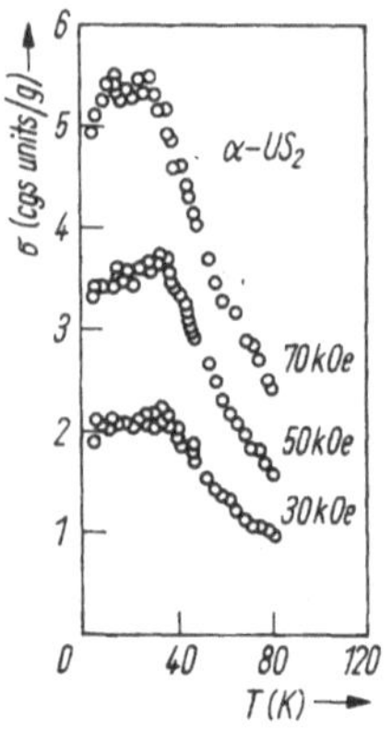

Fig. 5. The magnetization of α-US_2 as function of temperature

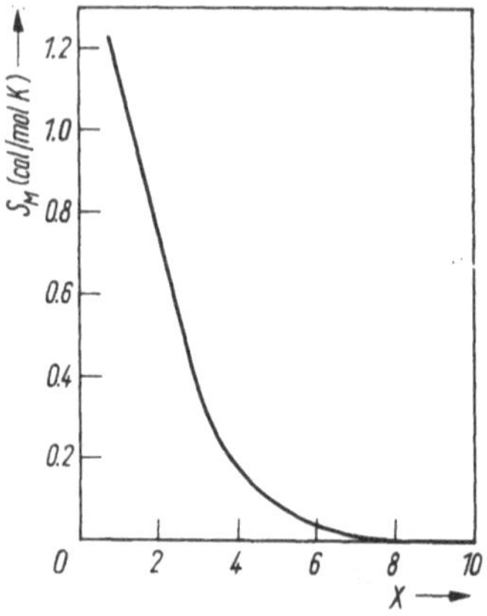

Fig. 6. The magnetic entropy as a function of the ratio X of the separation δ of the ground singlet and the first excited singlet to the transition temperature T

of the separation δ between the ground and first (excited) level to the temperature T of the magnetic transition, for the excited levels of the 2-, 3-, and 4-fold degeneracy only. Fig. 6 presents the dependence of the magnetic entropy on X for a singlet as the first excited level, calculated by using the relation given by Grunzweig-Genossar [25]. For α-USe_2 Westrum and Grønvold [11] have obtained the magnetic entropy as equal to 0.19 cal mol^{-1} K^{-1}. According to their opinion, this small value indicates that only a fraction of the uranium atoms, i.e., those with antiprismatic coordination are involved in the exchange interactions. However, it follows from the curve in Fig. 6 that a magnetic entropy equal to 0.19 cal mol^{-1} K^{-1} can also be obtained for the six-coordinated uranium atoms, for $T_N = 13$ K, assuming $\delta \approx 56$ K. The latter value is the same as given in [11]. The contribution of the eight-fold coordinated uranium atom is rather puzzling but, due to the statistical distribution of the uranium atoms in this case one can expect the splitting of the ground doublet into two singlets. Thus, the contribution of the eight-fold coordinated uranium atom to the magnetic entropy seems to be very small.

However, in the case of disulfide a λ-type anomaly has not been observed. The situation is similar to that found for praseodymium metal where the bulk susceptibility and heat capacity measurements give no indications as to the formation of a magnetic order at low temperatures. Moreover, the structure of the elemental Pr contains two crystallographically different ions in the positions of local hexagonal and cubic symmetry. Only neutron diffraction experiments [26] have revealed that at least some of the ions carry a moment in the liquid helium range and are ordered antiferromagnetically.

In the case of α-US_2 magnetic measurements provide some evidence of magnetic order. In our opinion the lack of a λ-type anomaly does not exclude the possibility of magnetic order. On examining the curve in Fig. 6 one can see that for $T_{C,N}$ of about 20 K and a separation between the ground and first excited level of about 140 K which is a reasonable value, the magnetic entropy value is well below the experimental findings, so far.

A full understanding of the magnetic properties of the uranium dichalcogenides will be possible only on the basis of neutron diffraction experiments as well as low-temperature specific heat measurements in a magnetic field. Such experiments are now in progress in our Institute.

Acknowledgements

We thank Prof. W. Trzebiatowski for his stimulating interest in this work, and Prof. B. Staliński and Dr. J. Mulak for helpful discussions.

References

[1] M. Picon and J. Flahaut, C. R. Acad. Sci. (France) **237**, 1160 (1953).
[2] W. H. Zachariasen, Acta cryst. **2**, 291 (1949).
[3] M. Picon and J. Flahaut, C. R. Acad. Sci. (France) **240**, 2150 (1955).
[4] R. C. Mooney-Slater, Z. Krist. **120**, 278 (1964).
[5] P. Khododad, C. R. Acad. Sci. (France) **245**, 934 (1957).
[6] A. J. Klein Haneveld and F. Jellinek, J. less-common Metals **21**, 45 (1970).
[7] G. V. Ellert, A. A. Elisiev, and V. K. Slovyanskich, Zh. neorg. Khim. **16**, 1451 (1971).
[8] F. Grønvold, H. Haraldsen, T. Thurmann-Moe, and T. Tufte, J. inorg. nuclear Chem. **30**, 2117 (1968).
[9] G. V. Ellert, V. K. Slovyanskich, and V. G. Sevastyanov, Zh. neorg. Khim. **16**, 1721 (1971).
[10] W. Suski and W. Trzebiatowski, Bull. Acad. Polon. Sci., Ser. Sci. chim. **12**, 277 (1964).
[11] E. F. Westrum, Jr., and F. Grønvold, J. inorg. nuclear Chem. **32**, 2169 (1970).
[12] W. Trzebiatowski and W. Suski, Bull. Acad. Polon. Sci., Ser. Sci. chim. **9**, 277 (1961).
[13] J. Leciejewicz, R. Troć, A. Murasik, and T. Palewski, phys. stat. sol. (b) **48**, 445 (1971).
[14] W. Trzebiatowski and J. Mulak, Bull. Acad. Polon. Sci., Ser. Sci. chim. **13**, 759 (1965).
[15] E. Trojnar, C. Bazan, and J. Niemiec, Bull. Acad. Polon. Sci., Ser. Sci. chim. **13**, 481 (1965).
[16] J. Grunzweig-Genossar, M. Kuznietz, and F. Friedman, Phys. Rev. **173**, 562 (1968).
[17] W. Suski, T. Gibiński, A. Wojakowski, and A. Czopnik, phys. stat. sol. (a) **9**, 653 (1972).
[18] J. Mulak and A. Czopnik, Bull. Acad. Polon. Sci., Ser. Sci. chim., in the press.
[19] W. Suski, to be published.
[20a] G. T. Trammell, Phys. Rev. **131**, 932 (1963).
[20b] B. Bleaney, Proc. Roy. Soc. **A236**, 19 (1963).
[20c] Y. L. Wang and B. R. Cooper, Phys. Rev. **172**, 539 (1968); **185**, 696 (1969).
[21] W. E. Wallace, F. Kissel, E. Segal, and R. S. Craig, J. Phys. Chem. Solids **30**, 13 (1969).
[22] R. J. Birgeneau, J. Als-Nielsen, and E. Bucher, Phys. Rev. Letters **27**, 1530 (1971).
[23] R. Troć, J. Mulak, and W. Suski, phys. stat. sol. (b) **43**, 147 (1971).
[24] G. Busch, P. Junod, P. Schwob, O. Vogt, and F. Hulliger, Phys. Letters (Netherlands) **9**, 7 (1964).
G. Busch and O. Vogt, Phys. Letters (Netherlands) **22**, 388 (1966).
G. Busch, O. Marcinček, A. Menth, and O. Vogt, Phys. Letters (Netherlands) **14**, 262 (1965).
G. Busch, O. Vogt, and P. Schwob, Proc. Coll. Internat. Champs Magnétiques Intenses, Grenoble 1966 (p. 401).
[25] J. Grunzweig-Genossar, Solid. State Commun. **8**, 1673 (1970).
[26] J. W. Cable, R. M. Moon, W. C. Koehler, and E. O. Wollan, Phys. Rev. Letters **12**, 553 (1964).

(Received April 24, 1972)

phys. stat. sol. (a) **12**, 533 (1972)

Subject classification: 14.3; 22.2.1

Institute of Radio Physics and Electronics, University of Calcutta

Low-Field Galvanomagnetic Transport in n-Type Gallium Arsenide

By

B. R. NAG, D. CHATTOPADHYAY, and G. M. DUTTA

An iterative technique for the solution of the Boltzmann equation in the presence of a magnetic field is presented. The temperature dependence of low-field Hall mobility and magnetoresistance coefficient of n-type gallium arsenide is studied in the temperature range 77 to 300 °K applying this technique. Polar-mode, deformation potential acoustic, and piezo-electric scattering processes are considered, and the results are compared with those obtained by other methods. Agreement of the Hall mobility with recent experiments is found to be satisfactory.

Es wird eine iterative Technik zur Lösung der Boltzmanngleichung bei Anwesenheit eines Magnetfeldes dargestellt. Die Temperaturabhängigkeit der Niedrigfeld-Hallbeweglichkeit und des Koeffizienten der Magnetowiderstandsänderung von n-leitendem Galliumarsenid wird im Temperaturbereich von 77 bis 300 °K mit dieser Technik untersucht. Es werden polare Moden, akustische Deformationspotential- und piezoelektrische Streuprozesse berücksichtigt und die Ergebnisse mit denen anderer Methoden verglichen. Die Übereinstimmung der Hallbeweglichkeit mit neueren Experimenten ist befriedigend.

1. Introduction

Electron mobility in polar semiconductors has in the past been calculated by the variational method [1, 2]. Lewis and Sondheimer [3] have adapted the technique for the calculation of galvanomagnetic coefficients and have given numerical values of the Hall factor to the first approximation. Devlin [4] later carried out these calculations to the twelfth approximation. It should, however, be noted that the application of the variation principle in the presence of the magnetic field is not as rigorous as it is in its absence [3]. Fortini et al. [5] have determined the transport coefficients for predominantly polar scattering by a numerical solution of the Boltzmann equation. Rode [6], on the other hand, has recently suggested a simple iteration procedure for the calculation of the drift mobility. The method is straightforward and yields accurate results without invoking excessive mathematical details contrary to previous calculations. In this paper this method is extended for the evaluation of galvanomagnetic transport coefficients in polar semiconductors. We report calculations of Hall mobility and magnetoresistance for n-type gallium arsenide using this technique. The procedure is outlined in Section 2. In Section 3 the numerical results are presented and discussed.

2. Analytical Details

The conduction band structure in the present analysis is assumed to be parabolic. Since the band gap of gallium arsenide is rather large, this simplification would not affect the results significantly in the temperature range of 77 to

300 °K considered here [6]. In this range of temperature, polar-mode scattering is known to be the predominant scattering mechanism. Though less important, scattering by acoustic modes of both the deformation potential and the piezoelectric types may also be present. We have retained all these three types of scattering processes in our calculations.

We assume an electric field F_x applied along the x-direction and a magnetic field B_z applied along the z-direction so that the Hall field F_y is developed along the y-direction. The distribution function of the carriers can then be written in the form

$$f(\boldsymbol{k}) = f_0(E) + k_x f_x(E) + k_y f_y(E) , \tag{1}$$

where $f_0(E)$ is the equilibrium Maxwell distribution function, k_x, k_y are the components of the wave vector $\boldsymbol{k}$, and f_x, f_y the perturbations caused by the fields.

Substituting (1) in the Boltzmann equation and equating the coefficients of k_x and k_y on the two sides one obtains

$$\lambda_o f_x'(E) = \lambda_{\text{ipo}}^+ f_x'^+ + \lambda_{\text{ipo}}^- f_x'^- - \frac{e\hbar}{m_e}\left\{\frac{\partial f_0}{\partial E} - \frac{\mu_H B_z^2}{\hbar} f_y'(E)\right\} \tag{2}$$

and

$$\lambda_o f_y'(E) = \lambda_{\text{ipo}}^+ f_y'^+ + \lambda_{\text{ipo}}^- f_y'^- - \frac{e\hbar}{m_e}\left\{\frac{\partial f_0}{\partial E} + \frac{f_x'(E)}{\mu_H \hbar}\right\}, \tag{3}$$

where

$$f_x' = f_x/F_x , \qquad f_y' = f_y/F_y , \qquad \text{and} \qquad F_y = \mu_H B_z F_x .$$

e, m_e are the electronic charge and effective mass, respectively, $\hbar$ is the Dirac constant, and μ_H the Hall mobility. The quantities with superscripts + and − are to be evaluated at energies $(E + k_B T_{po})$ and $(E - k_B T_{po})$, respectively, k_B being the Boltzmann constant and T_{po} the polar-phonon Debye temperature. λ_o includes the relaxation rate due to the elastic processes and the out-scattering rate due to the inelastic polar-mode scattering. λ_{ipo}^+ and λ_{ipo}^- represent the in-scattering rates due to the polar-mode scattering. For the assumed model these rates are given by the following expressions, the symbols hitherto unexplained having the same meanings as in [6].

$$\lambda_o = \lambda_{ac} + \lambda_{pe} + \lambda_{opo} , \tag{4}$$

$$\lambda_{ac} = \frac{\sqrt{2}\, E_1^2\, m_e^{3/2}\, k_B T}{\pi \varrho \hbar^4 u_a^2} E^{1/2} , \tag{5}$$

$$\lambda_{pe} = \frac{e^2 e_{14}^2 k_B T m_e^{1/2}}{2\sqrt{2}\pi \varepsilon_0^2 \hbar^2 \varrho u_p^2} E^{-1/2} , \tag{6}$$

$$\lambda_{opo} = \frac{m_e^{1/2} e^2 k_B T_{po}}{\sqrt{2}\hbar^2 E^{1/2}}\left(\frac{1}{\varepsilon_\infty} - \frac{1}{\varepsilon_0}\right)\left[N_{po} \ln\left|\frac{\sqrt{E^+} + \sqrt{E}}{\sqrt{E^+} - \sqrt{E}}\right| + \right.$$
$$\left. + (N_{po} + 1) \ln \left|\frac{\sqrt{E} + \sqrt{E^-}}{\sqrt{E} - \sqrt{E^-}}\right|\right], \tag{7}$$

$$\lambda_{\text{ipo}}^\pm = \frac{m_e^{1/2} e^2 k_B T_{po}}{\sqrt{2}\hbar^2 E^{1/2}}\left(\frac{1}{\varepsilon_\infty} - \frac{1}{\varepsilon_0}\right)\left(N_{po} + \frac{1}{2} \pm \frac{1}{2}\right) \times$$
$$\times \left[\frac{E^\pm + E}{2E} \ln\left|\frac{\sqrt{E^\pm} + \sqrt{E}}{\sqrt{E^\pm} - \sqrt{E}}\right| - \left(\frac{E^\pm}{E}\right)^{1/2}\right]. \tag{8}$$

To evaluate the Hall mobility we consider only the terms linear in B_z. In that case (2) reduces to the form valid in the absence of the magnetic field (cf. equation (42) of [6]). The perturbation f'_x can be determined by the iteration procedure described by Rode [6]. In the first step of iteration the first two terms on the right-hand side of (2) are neglected and f'_x is obtained by solving the resultant equation. In the next step $\lambda^+_{\text{ipo}} f'^+_x$ and $\lambda^-_{\text{ipo}} f'^-_x$ are evaluated using this solution and put in (2). A new f'_x is obtained by solving the equation containing $\lambda^+_{\text{ipo}} f'^+_x$ and $\lambda^-_{\text{ipo}} f'^-_x$. This procedure when repeated successively yields a convergent solution for f'_x. The rapidity of the convergence indicates that about five iterations are sufficient for an accuracy of 2 to 3%.

Once f'_x is determined it is easy to obtain f'_y from (3) using the same technique. To start with, the scattering-in terms $\lambda^+_{\text{ipo}} f'^+_y$ and $\lambda^-_{\text{ipo}} f'^-_y$ are ignored and we write as a first approximation

$$f'^{(1)}_y = -\frac{e\,\hbar}{m_e\,\lambda_o}\left\{\frac{\partial f_0}{\partial E} + \frac{f'_x}{\mu_H\,\hbar}\right\}. \tag{9}$$

Equation (9) is solved assuming a value of μ_H and this value of f'_y is substituted in the right-hand side of (3) to obtain a more accurate value of f'_y which is again used in the right-band side. As before, the process is required to be repeated only a few times to yield fairly accurate results. Solutions for f'_y are determined for a number of assumed values of μ_H. The particular value which satisfies the equation

$$\int_0^\infty f'_y\, E^{3/2}\, dE = 0 \tag{10}$$

and hence makes the current in the y-direction zero is the accepted value of Hall mobility.

The drift mobility is calculated from the relation

$$\mu = \frac{8\sqrt{2}\,\pi\, m_e^{3/2}}{3\,h^4} \int_0^\infty f'_x\, E^{3/2}\, dE\,. \tag{11}$$

The magnetoresistance effect is attributed to the second order terms in B_z, assuming $(\mu\, B_z)^2 \ll 1$. To evaluate this coefficient the value of f'_y satisfying (10) is substituted in (2) and f'_x is determined by iteration, as described above. The new mobility value is obtained from (11) and the magnetoresistance coefficient from

$$\xi = -\frac{\Delta\mu}{\mu}\,(\mu_H\, B_z)^2. \tag{12}$$

3. Numerical Results and Discussion

The numerical results were obtained with the help of the following material constants:

$$m_e = 0.072\, m\,,\ \varepsilon_0 = 13.13\,,\ \varepsilon_\infty = 11.10\,,\ T_{\text{po}} = 416\ ^\circ\text{K}\,,\ \varrho = 5.36\times 10^3\ \text{kg m}^{-3},$$
$$u_a = 5.24\times 10^3\ \text{m s}^{-1},\ E_1 = 7\ \text{eV},\ e_{14} = 0.16\ \text{Cm}^{-2},\ u_p = 4.03\times 10^3\ \text{m s}^{-1}\,.$$

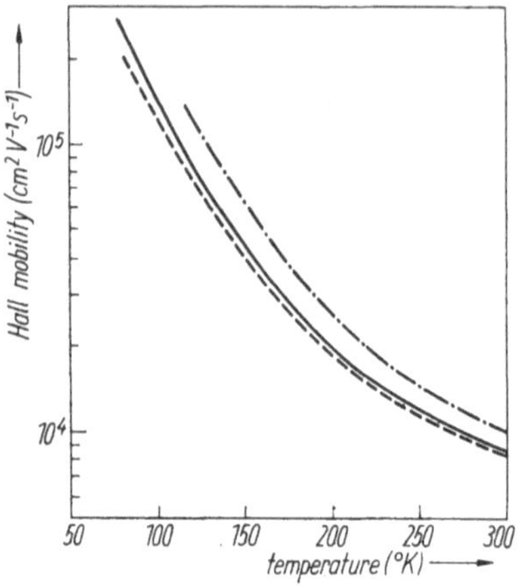

Fig. 1. Temperature dependence of Hall mobility in n-type GaAs. The solid curve is obtained in the present analysis. The dash-dotted curve represents the results of Fortini et al. [5] considering only polar-mode scattering but using identical parameter values. The dashed curve is experimentally obtained by Stillman et al. [7] for $B_z = 5$ kG

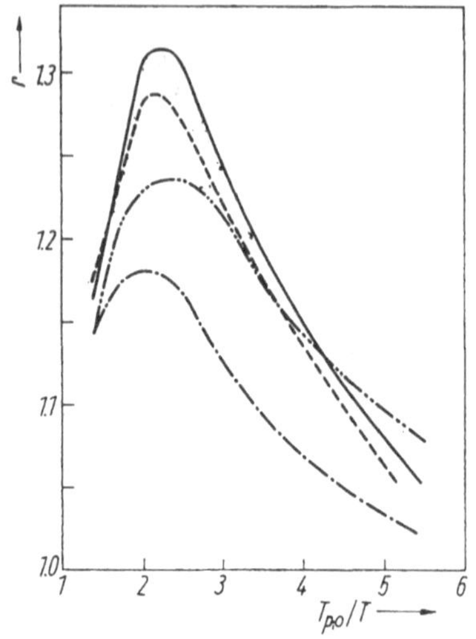

Fig. 2. Hall factor as a function of reduced reciprocal temperature. ——— present analysis; – – – – Fortini et al. [5]; –··–··– Devlin [4]; —·—·— experimentally determined by Stillman et al. [7] for $B_z = 0.5$ kG

In Fig. 1, we show the temperature dependence of Hall mobility and compare the results with experimental measurements [7] on the purest sample containing an ionized impurity concentration of 7×10^{13} cm^{-3}. One finds that the present calculations give better fit with the experimental results than those of Fortini et al. [5] considering only the polar-mode scattering with identical values of the parameters. The good agreement with experiment implies that the effect of impurity scattering is negligible in the temperature range considered.

The Hall factor r, giving the Hall to drift mobility ratio, is plotted in Fig. 2 as a function of the reduced reciprocal temperature, $Z = T_{\text{po}}/T$. An estimation of this factor r is important for a precise calculation of the carrier concentration from the measured Hall coefficient data. Furthermore, since the majority of the experiments on mobility give the Hall mobility, the Hall factor would help in reducing these data to the drift mobility. It should be noted that the variation of r predicted by the present analysis agrees closely with that obtained by Fortini et al. [5] who considered scattering by the polar modes only. The nature of the curve is also in qualitative agreement with that obtained in variational calcula-

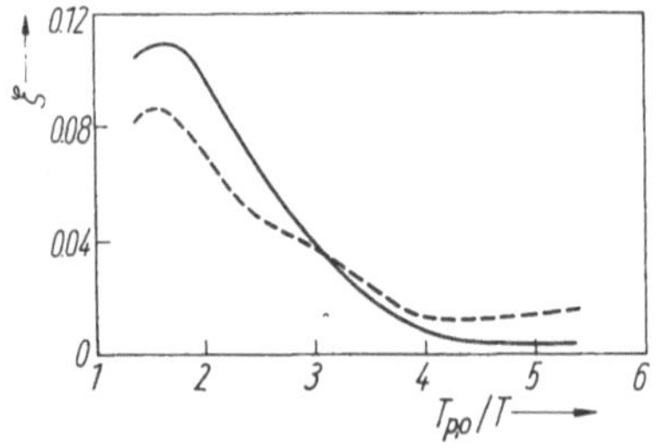

Fig. 3. Temperature variation of magnetoresistance coefficient. The solid line is derived in the present analysis. The ashed line refers to the variational calculations of Devlin [4]

tions by Devlin [4] and with the recent experiments of Stillman et al. [7], though there is a discrepancy in the exact values.

The temperature variation of the magnetoresistance coefficient is depicted in Fig. 3. It is found that the curve agrees qualitatively with that obtained by Devlin [4] in the variational calculation for pure polar-mode scattering. To the knowledge of the authors no attempt has yet been made to study experimentally the temperature dependence of ξ. Efforts in this regard would elucidate further our existing knowledge of the scattering mechanisms in the material and help in judging the correctness of the theory.

Acknowledgement

We thankfully acknowledge the computer facilities provided by the Indian Statistical Institute, Calcutta.

References

[1] D. J. Howarth and E. H. Sondheimer, Proc. Roy. Soc. (London) **A219**, 53 (1953).
[2] H. Ehrenreich, J. Phys. Chem. Solids **2**, 131 (1967).
[3] B. F. Lewis and E. H. Sondheimer, Proc. Roy. Soc. (London) **A227**, 241 (1954).
[4] S. S. Devlin, in: Physics and Chemistry of II–VI Compounds, Ed. M. Aven and J. S. Prener, North-Holland Publ. Co., Amsterdam 1967 (p. 551).
[5] A. Fortini, D. Diguet, and J. Lugand, J. appl. Phys. **41**, 3121 (1970).
[6] D. L. Rode, Phys. Rev. **B2**, 1012 (1970).
[7] G. E. Stillman, C. M. Wolfe, and J. O. Dimmock, J. Phys. Chem. Solids **31**, 1199 (1970).

(Received April 18, 1972)

phys. stat. sol. (a) **12**, 539 (1972)

Subject classification: 18.2; 21.1.1

Institute of Physics, Czechoslovak Academy of Sciences, Prague

Spatial Replica of Ferromagnetic Domains in Iron–Silicon Alloys

By

S. LIBOVICKÝ

The decomposition of some iron–silicon alloys at low temperatures is influenced by spontaneous magnetization in each ferromagnetic domain. The anisotropic decomposition structure is frozen in by cooling and in this way a spatial replica of magnetic domain structure is produced, which was present in the sample at the time of ageing (at a temperature of about 600 °C). It is then revealed at room temperature on any cut through a bulk specimen. All 90 and 180° domain walls are clearly visible. Spatial reconstruction of the internal magnetic domain structure was performed by successive polishing and etching. By the same method the interaction of domain walls with dislocation networks was observed.

Die Gefügeumwandlung von einigen Eisen–Silizium-Legierungen bei Temperaturen im Bereich von 600 °C wird durch die spontane Magnetisierung in jedem ferromagnetischen Bereich beeinflußt. Die anisotrope Umwandlungsstruktur wird durch Abkühlung eingefroren. Es entsteht dadurch ein räumlicher Abdruck der in der Probe während des Glühens existierenden, magnetischen Bereichsstruktur. Der Abdruck kann in jedem Querschnitt der massiven Probe sichtbar gemacht werden. Alle 180- und 90°-Wände sind deutlich zu unterscheiden. Der räumliche Aufbau der inneren Bereichstruktur konnte dadurch rekonstruiert werden, daß die Probe stufenweise poliert und geätzt wurde. Mit derselben Methode wurde die Wechselwirkung der Blochwände mit dem Versetzungsnetzwerk beobachtet.

1. Introduction

In metallic ferromagnets the magnetic domain structure was hitherto observed on the surface and in a thin layer below the surface (for a review see [1, 2]). The greatest thickness of samples in which the internal magnetic domain structure may be studied by X-ray topography [3] is only a few tenth of millimeter and the resolution is not very high. We have now found that it is possible to make use of the decomposition of some iron–silicon alloys influenced by spontaneous magnetization for the study of the internal magnetic domain structure of bulk samples.

The latest study of the equilibrium state diagram of the iron rich iron–silicon alloys was published by Warlimont [4]. He found an eutectoid line at 540 °C between 10 and 15 at% Si. He further found that above this eutectoid line, platelets with DO_3 superlattice grow inside the B_2 matrix. In agreement with the theory [5], the platelets are parallel to the three {001} planes.

During our study [6] of the ordering and decomposition of iron–silicon alloys by means of transmission electron microscopy and optical metallography we determined the region in the equilibrium diagram in which the DO_3 platelets are formed. They are formed in an area above the eutectoid line indicated by Warlimont and below a nearly parabolic line the top of which is at 700 °C and 13 at% Si. The spacing of the platelets grows continuously with increasing ageing temperature in a sample with given silicon content. Generally the spacing

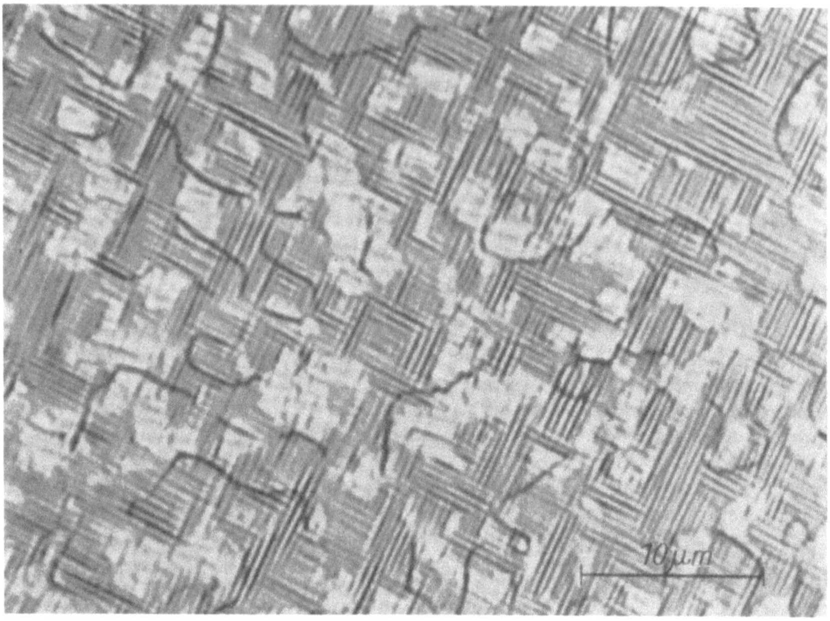

Fig. 1. Fe–13 at% Si polycrystal held 17 h at 680 °C (above Curie temperature). DO_3 platelets parallel to all three {001} planes are revealed by etching. The curved black lines are a/2 ⟨111⟩ antiphase boundaries

increases when the parabolic line is approached and no platelets are formed outside the parabolic area. The platelets revealed by etching on the surface of a sample which was aged 20 °C below the parabolic line are shown in Fig. 1 and the platelets formed by ageing just at the parabolic line are shown in Fig. 2.

The line of Curie temperatures [7 to 9] crosses the area in which the DO_3 platelets are formed. The magnetostriction constant λ_{100} is zero (at room temperature) for an alloy with silicon content of about 11.5 at% and has negative values for alloys with higher silicon content [10, 11]. We found [6] that in alloys with silicon content higher than 11.6 at % Si aged below the Curie temperature, the growth of the DO_3 platelets which are normal to the magnetization vector in each ferromagnetic domain is suppressed in some measure (Fig. 2 and 3[1])). The resulting anisotropic distribution of the platelets thus makes it possible to study the internal magnetic domain structure present at the ageing temperature, as described below.

2. Experimental Method

2.1 Sample preparation

The decomposition and the magnetic domain structure was studied in polycrystalline samples and in single crystals grown either by the Bridgman method [12] or by the floating-zone method [13]. The starting material for all samples was high-purity carbonyl iron and silicon of semiconductor grade. The samples of different shape were aged in hydrogen in a tubular resistance furnace. Before

[1]) The TEM photomicrograph in Fig. 3 was taken by Dr. A. Gemperle.

a
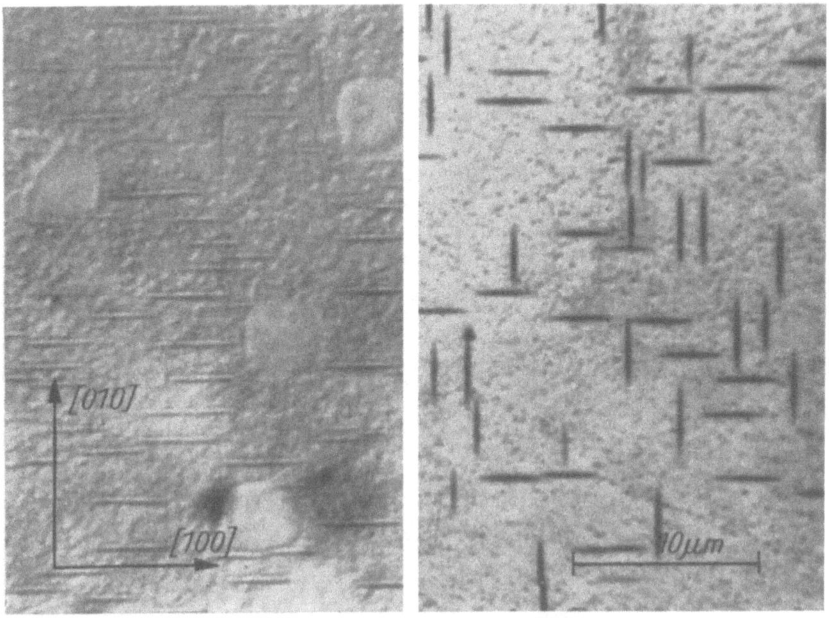

b

Fig. 2. Fe–11.7 at% Si single crystal held 15 h at 630 °C (below Curie temperature). In a) and b) two regions in one sample are shown where the formation of the (100) and (001) platelets, respectively, was enhanced during ageing. The platelets are revealed by etching

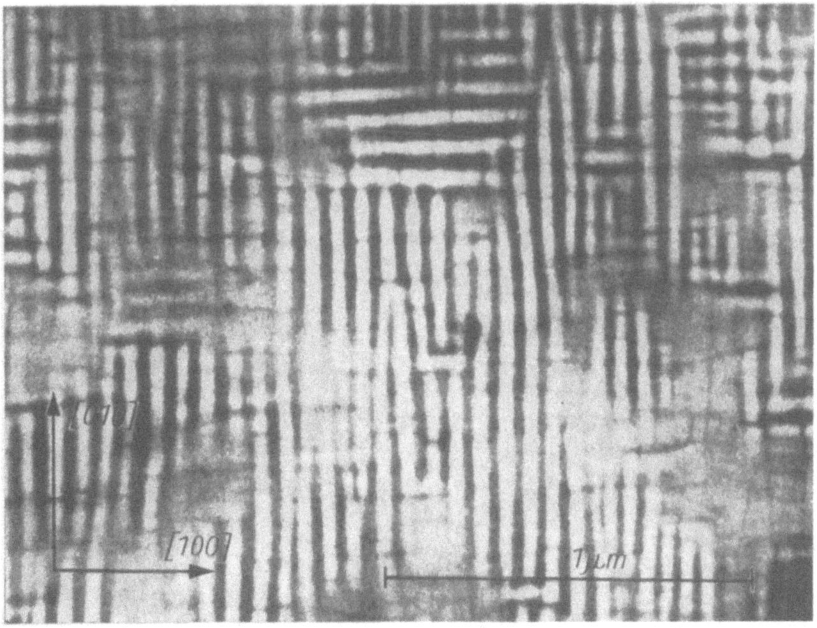

Fig. 3. TEM micrograph of Si Fe–12.5 at% a single crystal held 70 h at 580 °C. The diffuse areas are spots of DO_3 platelets parallel to (001). The growth of (010) platelets was partially suppressed in this area. Dark field 200 superlattice reflection

ageing each specimen was heated to 1000 °C and then cooled down in about 2 h to the temperature of ageing. The axial magnetic field inside the furnace was about 8 Oe during ageing. No observable difference was found in the magnetic domain structure, if the furnace was supplied by dc instead of ac current. After ageing the samples were air cooled, spark cut, ground on emery papers, and chemically polished in a mixture of $H_2O + H_2O_2 + HF$ [14]. Afterwards, they were etched anodically in an aqueous solution of CrO_3.

For the purpose of revealing the internal magnetic domain structure it is necessary to produce small spacings of the DO_3 platelets, similar to that shown in Fig. 3. We were able to produce such structure in crystals with silicon content in the range from 11.5 to 13 at% by ageing for 50 to 70 h in the temperature range 560 to 610 °C.

2.2 Domain structure at ageing temperature observed at room temperature

In this paper we shall describe only observations on {001} surfaces. The observations are actually possible on surfaces of any crystallographic orientation.

2.2.1 Domain structure revealed by etching

If the surface of a specimen is etched in a 50% aqueous solution of CrO_3 at low current density (a few $\mu A\ mm^{-2}$), the platelets are revealed (Fig. 1 and

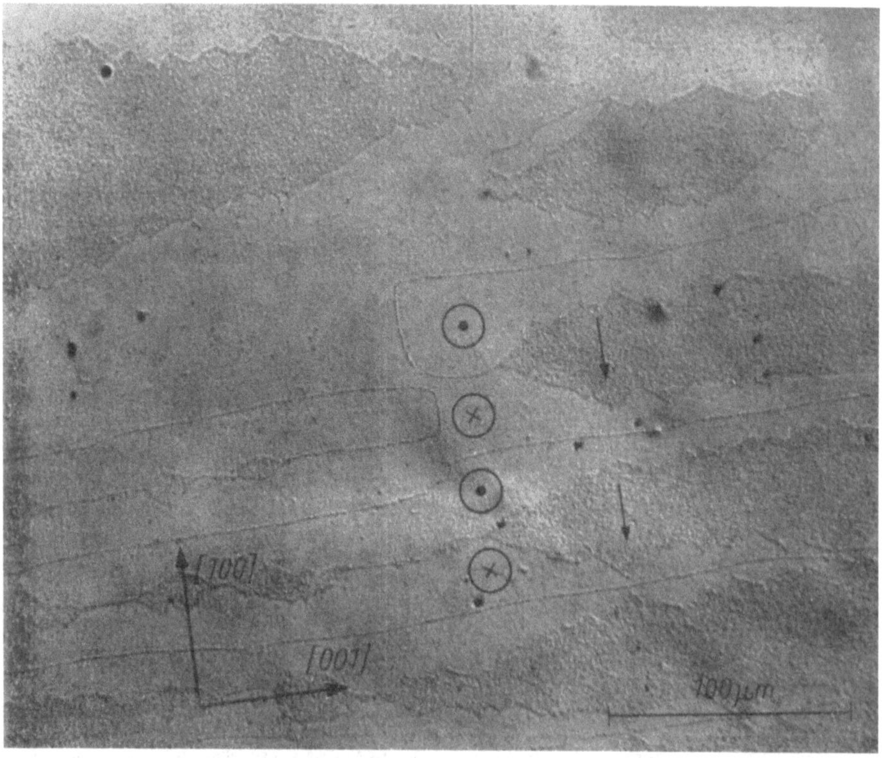

Fig. 4. Fe–12.8 at% Si single crystal held 70 h at 560 °C. Ferromagnetic domains 0.2 mm below the surface are revealed by etching. Some of magnetization vectors are indicated. Unpolarized light

Fig. 2). (Simultaneously, the antiphase boundaries are etched [15] if present in the crystal (Fig. 1).) If the spacing of the platelets is small (as in Fig. 3), the platelets perpendicular to the surface are not resolved by means of a metallographic microscope, but those parallel to the etched surface are revealed as tiny etch pits. In this way the domains in which the magnetization vector (at the time of ageing) was perpendicular to the observed surface remain smooth while those with magnetization vector parallel to the surface become rough. In the smooth areas 180° Bloch walls are seen as black (etched) lines (Fig. 4) and in the rough areas 180° walls as bright (unetched) lines (Fig. 5a).

2.2.2 Domain structure revealed by forming a birefringent layer

During the etching, an oxide layer grows epitaxially on the surface. Very thin layers formed by etching at low current density give no effect in polarized light. To produce a sufficiently thick layer without cracks, the specimens are etched for 30 s to 5 min in a saturated aqueous solution of CrO_3 at a current density of about 0.5 mA mm^{-2}. The whole surface becomes rough after such treatment (Fig. 5a). If, however, the surface is observed in polarized light with

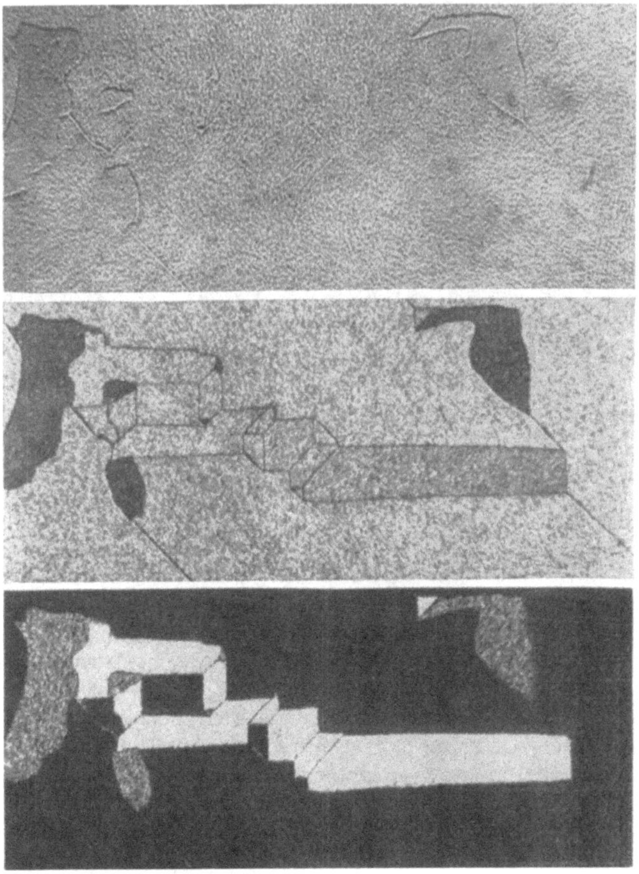

a

b

c

Fig. 5. Fe–12.5 at% Si single crystal held 70 h at 580 °C. An epitaxial oxidic layer was grown on the surface by etching;
a) unpolarized light, b) polarized light, polaroids crossed, c) polarizer rotated +3°

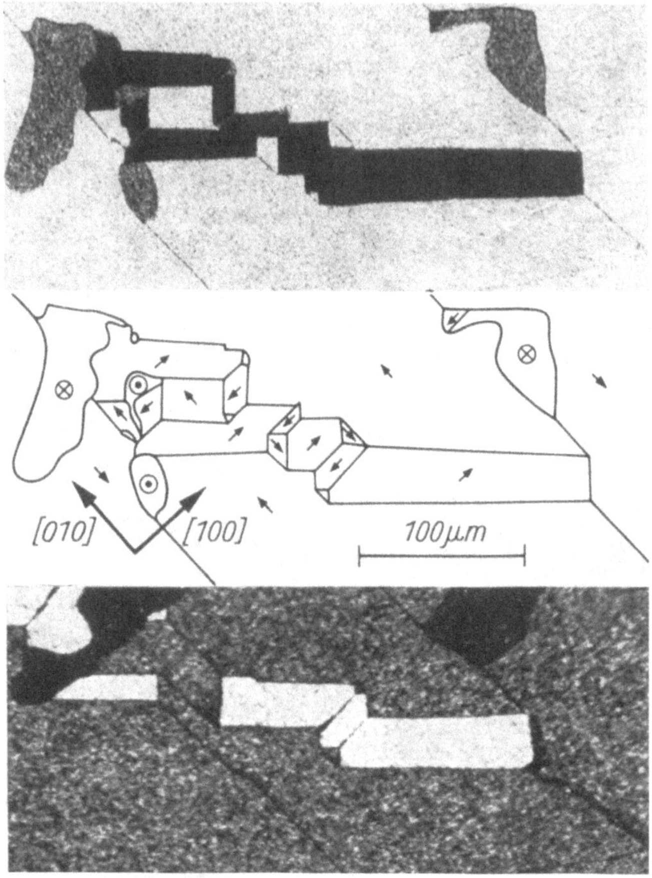

d

e

f

Fig. 5 d) polarizer rotated −3°, e) reconstruction of magnetization vectors, f) the same group of domains after polishing off 40 μm, polarizer rotated +15°

polaroids crossed and the incident light makes an angle of 45° with a ⟨001⟩ direction, all the 90 and 180° walls between domains with magnetization vectors parallel to the surface are seen as black lines in a grey area; the domains with magnetization vector perpendicular to the surface become dark (Fig. 5b). The two sets of domains with magnetization vectors parallel to the surface (and mutually perpendicular) may be clearly distinguished by turning the polarizer. When the polarizer is rotated by a few degrees, one set of domains becomes successively blue, violet, and red, the other set remains without colour contrast but becomes brighter (Fig. 5c). The colours of the two sets of domains are interchanged when the polarizer is rotated in the other direction (Fig. 5d). The colour contrast of the domains with magnetization vector perpendicular to the surface remains unchanged (cf. Fig. 5c and d). The colours may be also changed by rotating the specimen by 90°. The observations just described are possible because the oxide layer is birefringent due to planar stress, which copies the symmetry of the distribution of the DO_3 platelets in the underlying domains.

3. Internal Magnetic Domain Structure

The methods described in Section 2.2.1 and 2.2.2 reveal the magnetic domain structure which was present in the specimen at the time of ageing. (The room temperature domain structure present at the time of observation, which was made visible by the powder method, is different).

The reconstruction of the magnetization vectors in a group of small domains is accomplished in Fig. 5e.[2]) The magnetization vectors lie along the directions of easy magnetization, which are ⟨001⟩ in these crystals. The sense of the magnetization vector in one domain was chosen arbitrarily, the senses of magnetization vectors in all other domains are then uniquely determined. The senses of magnetization vectors perpendicular to the surface were determined by spatial reconstruction of the domain structure after successive polishing and etching the surface in steps of 5 μm. The domain structure after removing 40 μm is shown in Fig. 5f.

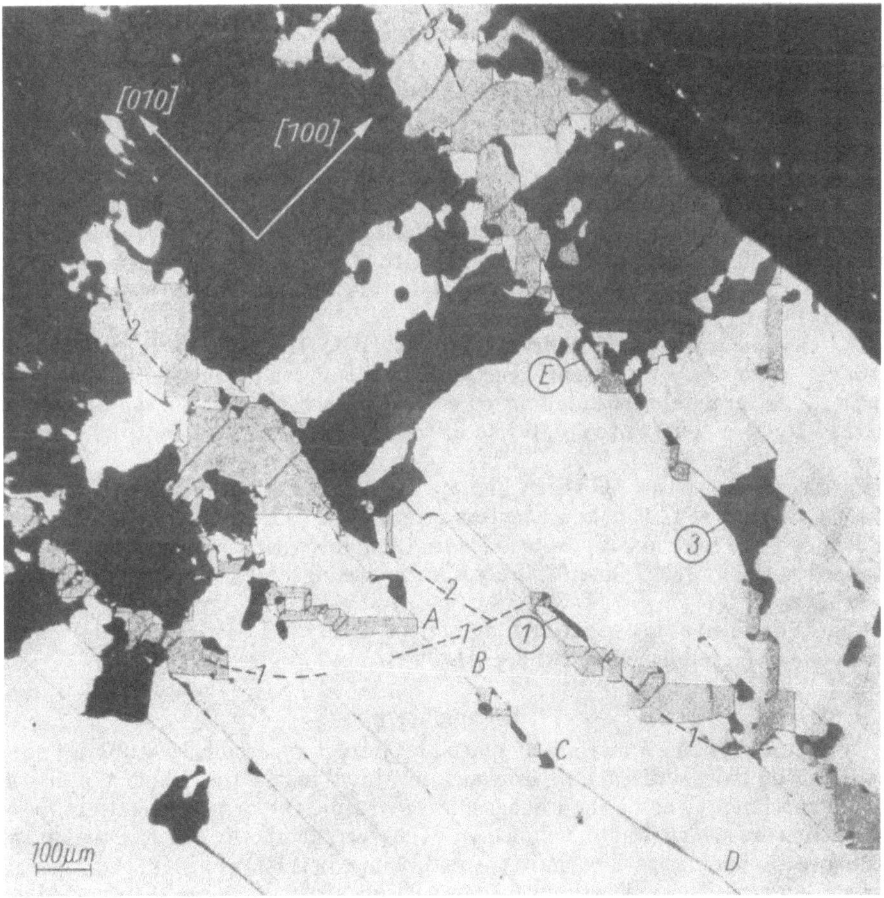

Fig. 6. Internal magnetic domain structure. Larger area round the spot shown in Fig. 5. The details indicated are referred to in the text. Polarized light, polarizer rotated −15°

---

2) In this reconstruction the author appreciates the help of Dipl.-Ing. R. Gemperle.

A larger area with magnetic domain structure on a (001) cut 5 mm from one end of a 50 mm long single crystal of semicircular cross section (∅ 15 mm) is shown in Fig. 6. The complicated domain structure is due to imperfections of the crystal. Groups of small domains are pinned to dislocation networks in low-angle boundaries of mosaic blocks which are not well shown in the microphotograph and some of them are therefore indicated by numbers. Often the 90° domain walls coincide with these low-angle boundaries. Two of such coincidences are indicated by encircled numbers. It is of interest that the Bloch walls coincide mostly with low-angle boundaries parallel or nearly parallel to {001} planes, which according to Vicena [16] have a high contribution to coercitive force.

Pinning of small domains to large precipitates was also observed.

Another common feature of the domain structure is the twisting and broadening of 180° walls. One of the twisted walls is seen between points C and D in Fig. 6. We followed the wall A–D into the depth below the surface. Marking the level shown in Fig. 6 by zero, then at the level -25 μm a small domain with magnetization vector [100] appeared at D and the wall C–D became straight and thin. On the other hand, the wall A–B, straight in Fig. 6, started to twist round point A and became broader at this level. A small portion of it is seen in Fig. 5f near point A at level -40 μm. The group of domains between points B and C diminished and disappeared completely at level -50 μm. At this level, the wall became straight and thin from A to D. The broadening of the twisted walls is due to the pinning of the ends of the wall to groups of small domains which in their turn are pinned to low-angle boundaries. The wall therefore cannot be exactly parallel to the ⟨010⟩ direction and thus should have magnetic charges. It seems that the twisted walls are rather thin domains (with magnetization vector [001] in Fig. 6).

The twisted walls broaden at some spots up to 15 μm. The broadening is, however, partially an apparent one as the walls are not perpendicular to the surface. An apparent broadening of a wall nearly parallel to the surface is marked by E in Fig. 6; the wall has diffuse borders and different colour contrast.

The domain structure shown in Fig. 4 was observed 0.2 mm below the surface of a single crystal rod of nearly cubic orientation ($6 \times 4 \times 45$ mm^3). The dark regions are the remains of surface closure domains with magnetization vectors parallel to the surface. The 90° Bloch walls are zigzag-like in agreement with the theory [17].

A more detailed study of internal domain structure in Fe–12.5% Si alloy single crystals will be published elsewhere [18].

4. Conclusion

By means of the new method of spatial replicas it is possible to study internal magnetic domain structure in iron–silicon alloys inside the whole volume of a bulk specimen. The method is based on revealing the anisotropic decomposition structure which was developed in each ferromagnetic domain under the influence of magnetization (most probably magnetostriction) in the crystals during ageing. The anisotropic decomposition structure may be revealed by etching, by forming a birefringent oxide layer on the surface, or otherwise (e.g. by surface X-ray topography [19]). All 90 and 180° Bloch walls are clearly shown up and domains as small as 2 μm are easily resolved. Pinning of groups of small domains to dislocation network in low-angle boundaries was observed. Often the 90° walls coincide with the mosaic block boundaries parallel to {001} planes.

At present, the method is confined to the study of the magnetic domain structure in alloys of iron with about 12 at% silicon and of the domain structure which was formed at temperatures of about 600 °C. The method might be employed for the investigation of domain structures in other ferromagnetic alloys which decompose in a similar manner.

Acknowledgements

The author is indebted to Dr. J. Kaczér, Dipl.-Ing. R. Gemperle, Dr. A. Gemperle, and Dr. M. Polcarová for helpful discussions during the development of the method.

References

[1] W. Andrä, J. Phys. D **1**, 1 (1968).
[2] D. J. Craik and R. S. Tebble, Ferromagnetism and Ferromagnetic Domains, North-Holland Publ. Co., Amsterdam 1965.
[3] M. Polcarová and A. Lang, Appl. Phys. Letters **1**, 13 (1962).
[4] H. Warlimont, Z. Metallk. **59**, 595 (1968).
[5] P. Schwellinger, H. J. Leamy, and H. Warlimont, Acta metall. **19**, 421 (1971).
[6] A. Gemperle and S. Libovický, unpublished.
[7] M. Fallot, Ann. Phys. (France) **6**, 305 (1936).
[8] D. Parsons, W. Sucksmith, and J. E. Thompson, Phil. Mag. **3**, 1174 (1958).
[9] D. Meinhardt and O. Krisement, Arch. Eisenhüttenwesen **36**, 293 (1965).
[10] W. J. Carr and R. Smoluchowski, Phys. Rev. **83**, 1236 (1951).
[11] R. C. Hall, J. appl. Phys. **30**, 816 (1959).
[12] B. Šesták, Czech. J. Phys. **9**, 234 (1957).
[13] S. Kadečková and B. Šesták, Kristall und Technik **4**, 243 (1969).
[14] S. Kadečková and K. Volenín, Czech. J. Phys. **B17**, 649 (1967).
[15] S. Libovický, phys. stat. sol. **20**, K85 (1967).
[16] F. Vicena, Czech. J. Phys. **B5**, 492 (1955).
[17] J. Kaczér and M. Zelený, Czech. J. Phys. **B10**, 561 (1960).
[18] R. Gemperle and S. Libovický, to be published.
[19] J. Brádler and M. Polcarová, unpublished.

(Received May 31, 1972)

phys. stat. sol. (a) **12**, 549 (1972)

Subject classification: 10.1; 21.1

Department of Physics, University of Alberta, Edmonton

Some Aspects of Stacking Fault Contrast in Thick Crystals at High Accelerating Voltages

By

S. S. SHEININ and K. Z. BOTROS

Theoretical extinction contour and stacking fault profiles were obtained for a range of orientations around the symmetry position of a systematic set and at an accelerating voltage of 1000 kV. It was found that both thickness fringe and stacking fault contrast in bright field images of thick crystals was improved by tilting away from the symmetry position. It was also found that the maximum intensity in the stacking fault image did not fall off as the crystal was tilted away from this orientation even though, under the same circumstances, there was a marked decrease in the intensity transmitted through the perfect crystal. These results therefore suggest that the symmetry position at high accelerating voltages may not be the best orientation for carrying out observations of stacking faults in thick crystals as is suggested by rocking curve data. In order to gain some insight into the mechanisms which give rise to contrast, the above results have been analyzed in terms of Bloch wave interactions. To this end, the multi-beam expressions describing the interbranch and intrabranch scattering which occurs in a faulted crystal have been derived. The results of the analysis indicate that excitation of two Bloch waves of approximately the same absorption coefficient is important in obtaining good contrast in images of thick crystals.

Es wurden der theoretische Extinktionsverlauf und die Stapelfehlerprofile für einen Orientierungsbereich um eine Symmetrielage eines systematischen Satzes bei einer Beschleunigungsspannung von 1000 kV erhalten. Es wurde gefunden, daß sowohl der Dickenstreifen als auch der Stapelfehlerkontrast bei Hellfelddiagrammen dicker Kristalle durch Kippen aus der Symmetrielage verbessert werden kann. Es wurde ebenfalls gefunden, daß die maximale Intensität in den Stapelfehlerdiagrammen nicht abfällt, wenn der Kristall aus dieser Orientierung gekippt wurde, sogar wenn unter den gleichen Umständen ein merklicher Abfall der Transmissionsintensität durch den perfekten Kristall auftrat. Diese Ergebnisse zeigen deshalb, daß die Symmetrielage bei hohen Beschleunigungsspannungen nicht die beste Orientierung für die Beobachtung von Stapelfehlern in dicken Kristallen zu sein braucht, wie es durch Rocking-Kurven-Ergebnisse nahegelegt wird. Um Einsicht in die Mechanismen, den Anlaß zu diesem Kontrast geben, zu erhalten, wurden die obigen Ergebnisse mit dem Mechanismus der Blochwellenwechselwirkung analysiert. Dazu wurden die Vielstrahlausdrücke abgeleitet, die die Interzweig- und Intrazweigstreuung beschreiben, die in einem stapelfehlergestörten Kristall auftreten. Die Ergebnisse der Analyse zeigen, daß die Anregung von zwei Blochwellen mit näherungsweise gleichem Absorptionskoeffizienten zur Erreichung von gutem Kontrast in den Diagrammen dicker Kristalle wesentlich ist.

1. Introduction

The effects of anomalous absorption at low accelerating voltages have been explained in terms of the widely differing absorption coefficients of Bloch waves 1 and 2. In the recent work of Humphreys et al. [1, 2] on rocking curves at high accelerating voltages, it has been noted that at the symmetry position Bloch waves 1 and 3 are strongly excited. Since these Bloch waves have widely differing absorption coefficients, it would be expected that the nature of the image contrast obtained at these diffraction conditions would also be characteristic of anomalous absorption.

The effects of anomalous absorption on extinction contours are characterized by the fact that fringes in thick crystals exhibit weak contrast. In a recent paper, one of the authors [3] has shown that at low accelerating voltages, tilting the crystal away from the Bragg position can result in a significant improvement in the contrast exhibited by these fringes. These results have been explained in terms of the excitation of two Bloch waves of approximately the same absorption coefficient. A question which arises from this work is how image contrast in thick crystals at high accelerating voltages would change when the crystal is tilted away from the symmetry position. The work presented in this paper was undertaken with a view to examining this question by carrying out multi-beam extinction contour calculations for a range of orientations around the symmetry position and at an accelerating voltage of 1000 kV. In order to determine how the image contrast of a lattice defect would change under the same circumstances, multi-beam calculations of images of stacking faults in thick crystals have also been carried out. The results obtained from these calculations have been analyzed with a view to gaining some insight into the mechanisms which give rise to image contrast. In the case of a faulted crystal the method of analysis used is based on the concept of scattering transitions between Bloch waves as described in the two beam case by Hirsch et al. [4]. The multi-beam expressions describing such scattering, which are required for this analysis, have not been previously presented and are therefore derived in the next section.

2. Theoretical Background

2.1 *Brief review of dynamical theory for a perfect crystal*

The wave function for the high-energy electrons in the dynamical theory can be represented as a linear combination of Bloch waves

$$\psi = \sum_i \psi^i b^i ,$$

where ψ^i is the excitation amplitude of Bloch wave b^i and

$$b^i = \sum_g C_g^i \exp 2\pi i (\boldsymbol{k}^i + \boldsymbol{g}) \cdot \boldsymbol{r} .$$

The C_g^i's in this expression are Fourier coefficients and $\boldsymbol{k}^i$ is the Bloch wave vector. The diffracted beam amplitudes at a depth z in a perfect crystal, $\widetilde{\varphi}_g(z)$, can be expressed in terms of these parameters and a scattering matrix $\widetilde{P}(z)$ as follows [4]:

$$\widetilde{\varphi}_g(z) = \widetilde{P}(z)\, \widetilde{\varphi}_g(0) , \tag{1}$$

where

$$\widetilde{P}(z) = \widetilde{C}\, \{\exp 2\pi i \gamma^i z\}\, \widetilde{C}^{-1} , \tag{2}$$

$\widetilde{\varphi}_g$ in (1) is a column vector whose elements are the diffracted beam amplitudes and $\widetilde{C}$ in (2) is a square matrix whose columns are the Fourier coefficients C_g^i of Bloch wave b^i. The curly brackets indicate a diagonal matrix whose elements are $\exp 2\pi i \gamma^i z$ and the parameter γ^i is equal to $(k_z^i - K_z)$ where $\boldsymbol{K}$ is the electron wave vector obtained after refraction by the mean lattice potential. The Bloch wave numbering system to be used in this paper is that proposed by Humphreys [5].

The values of ψ^i in a perfect crystal can be found from the application of the boundary conditions at the top surface of the crystal to be

$$\widetilde{C}^{-1}\,\widetilde{\varphi}_g(0) = \widetilde{\psi}\,, \tag{3}$$

where $\widetilde{\psi}$ is a column vector whose elements are the excitation amplitudes ψ^i. From (3) it can be seen that $\psi^i = C_0^i$. Equations (1), (2), and (3) can then be combined to give an alternate expression for the diffracted beam amplitudes, namely

$$\widetilde{\varphi}_g(z) = \widetilde{C}\,\{\exp 2\,\pi\,i\,\gamma^i\,z\}\,\widetilde{\psi}\,. \tag{4}$$

All the above equations can be modified in the usual way to take absorption into account by replacing γ^i with $\gamma^i + i\,q^i$. From (4) the expression for a particular diffracted beam amplitude can be written as

$$\varphi_g(z) = \sum_i \varphi_g^i \exp 2\,\pi\,i\,\gamma^i\,z\,, \tag{5}$$

where

$$\varphi_g^i = C_0^i\,C_g^i \exp(-2\,\pi\,q^i\,z)\,.$$

Bloch waves will be referred to as being important or otherwise in this paper on the basis of their contribution to the diffracted beam intensity. As can be seen from (5) this is determined by the value of φ_g^i.

2.2 Multi-beam theory of interbranch and intrabranch scattering for a stacking fault

The diffracted beam amplitudes for a crystal of thickness t containing a stacking fault at a depth t_1 can be written in terms of the total scattering matrix of the faulted crystal, $\widetilde{P}$, as

$$\widetilde{\varphi}_g(t) = \widetilde{P}\,\widetilde{\varphi}_g(0)\,, \tag{6}$$

where $\widetilde{P} = \widetilde{P}\,(t - t_1)\,\widetilde{P}(t_1)$. $\widetilde{P}\,(t - t_1)$ is obtained from (2) by replacing $\widetilde{C}$ with $\widetilde{Q}^{-1}\,\widetilde{C}$ where $\widetilde{Q} = \exp 2\,\pi\,i\,\boldsymbol{g}\cdot\boldsymbol{R}$ and $\boldsymbol{R}$ is the displacement of the top part of the crystal with respect to the bottom. An alternate expression for the diffracted beam amplitudes, analogous to (4) can be written as

$$\widetilde{\varphi}_g(t) = \widetilde{Q}^{-1}\,\widetilde{C}\,\{\exp 2\,\pi\,i\,\gamma^i\,t\}\,\widetilde{\psi}_1\,, \tag{7}$$

where the column matrix $\widetilde{\psi}_1$ contains the new excitation amplitudes, ψ_1^i, which are functions of the depth of the fault in the crystal. These new excitation amplitudes below the fault can be expressed in terms of the excitation amplitudes ψ^i above the fault by combining (3), (6) and (7) to give

$$\widetilde{\psi}_1 = \{\exp 2\,\pi\,i\,\gamma^i\,t\}^{-1}\,\widetilde{C}^{-1}\,\widetilde{Q}\,\widetilde{P}\,\widetilde{C}\,\widetilde{\psi} \tag{8}$$

or simply

$$\widetilde{\psi}_1 = \widetilde{B}\,\widetilde{\psi}\,.$$

The elements of the matrix $\widetilde{B}$ are given in the case of a stacking fault by

$$B^{ij} = \exp 2\,\pi\,i\,(\gamma^j - \gamma^i)\,t_1 \exp 2\,\pi\,(q^i - q^j)\,t_1 \sum_g C_g^i\,C_g^j \exp 2\,\pi\,i\,\boldsymbol{g}\cdot\boldsymbol{R}\,, \tag{9}$$

where the summation extends over all reciprocal lattice vectors $\boldsymbol{g}$.

It can be seen from (8) that the value of ψ_1^i below the fault depends on the amplitude of the same Bloch wave above the fault, ψ^i, as well as the amplitudes of the other Bloch waves above the fault ψ^j. Thus the matrix element

B^{ij} represents scattering from Bloch wave j (or the j-th branch of the dispersion surface) to Bloch wave i (or the i-th branch of the dispersion surface). In this paper we shall refer to this as interbranch scattering. On the other hand the matrix element B^{ii} represents scattering of Bloch wave i into itself and will be referred to as intrabranch scattering.

The contributions of the various intrabranch and interbranch scattering terms to the diffracted beam amplitudes can be obtained by substituting (8) into (7) giving

$$\widetilde{\varphi}_g(t) = \widetilde{Q}^{-1}\, \widetilde{C}\, \{\exp 2\pi i \gamma^i t\}\, \widetilde{B}\, \widetilde{\psi}\,. \tag{10}$$

From (10) the expression for a particular diffracted beam amplitude can be written in the form

$$\varphi_g(t) = \sum_i \sum_j \varphi_g^{ij} \exp 2\pi i \gamma^i t\,, \tag{11}$$

where

$$\varphi_g^{ij} = C_g^i\, C_0^j\, B^{ij} \exp(-2\pi i\, \boldsymbol{g} \cdot \boldsymbol{R}) \exp(-2\pi q^i t)\,.$$

The summation in (11) is from 1 to N, where N represents the number of beams taken into account in a particular calculation.

An important point to note from (9) and (11) is that fringe contrast of a stacking fault inclined to the surface of a foil must be mainly due to interbranch scattering. This can be seen from the fact that intrabranch contributions (like φ_g^{ii}) are not dependent on t_1, the depth of fault in the crystal, whereas the interbranch contributions (terms like φ_g^{ij}) are. It should be noted also that the interbranch and intrabranch contributions to the diffracted beam amplitude in (11) are complex quantities and can therefore best be illustrated by utilizing phase amplitude diagrams.

A final point to note in this section is that the dynamical theory, as outlined above, does not take into account the inelastically scattered electrons which pass through the objective aperture and contribute to the back-ground intensity. The discussion of image contrast which follows does not therefore take into account a possible reduction of contrast due to the background intensity.

3. Stacking Fault Contrast and Extinction Contours at High Accelerating Voltages

All the theoretical calculations presented in this paper were performed for an f.c.c. cobalt crystal. The (111) systematic set of reflections was considered and the reflections $\bar{5}\bar{5}\bar{5}$—555 were taken into account. The absorption parameters used in obtaining intensity profiles were those determined experimentally [6] for that material, and the Fourier coefficients of the lattice potential were Debye-Waller corrected. Bright field images only were considered and the orientations investigated correspond to values of the deviation parameter in the range from $\Delta\theta = -1.0\,\theta_B$ (symmetry position) to $\Delta\theta = 0.0$ (lowest order reflection in the exact Bragg position).

It is first of interest to consider the changes in thickness fringe contrast which occur as the crystal is tilted away from the symmetry position. The results in Fig. 1a and 1b, which show the intensity profiles obtained at values of $\Delta\theta = -1.0\,\theta_B$ and $-0.2\,\theta_B$, illustrate this change. It can be seen from Fig. 1a that at the symmetry position, thickness fringes in thick parts of the crystal exhibit poor contrast. On the other hand, the results in Fig. 1b show that when the crystal is tilted away from the symmetry position, thickness fringes in

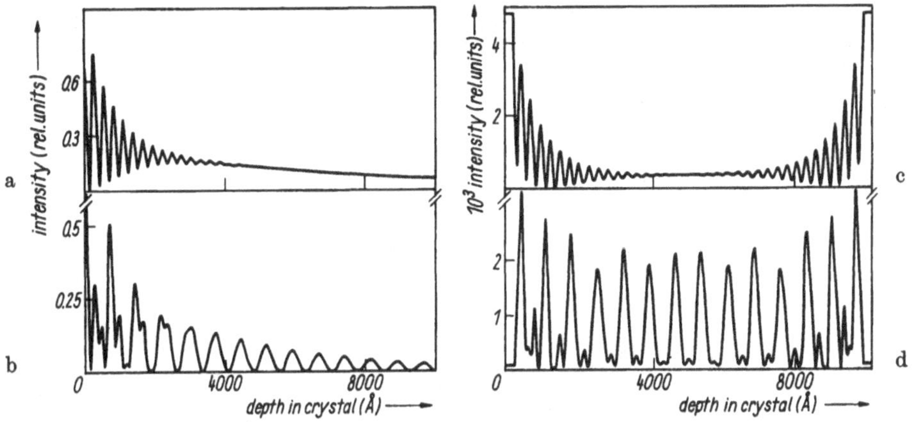

Fig. 1. Theoretical bright field intensity profiles for an f.c.c. cobalt crystal at 1000 kV with the (111) set of systematic reflections excited. a) and b) Extinction contour profiles obtained for values of $\Delta\theta$ equal to $-1.0\ \theta_B$ and $-0.2\ \theta_B$, respectively. c) and d) Profiles of an $\alpha = -2\pi/3$ stacking fault in a 1 μm thick crystal at the same values of $\Delta\theta$

thick parts of the crystal exhibit very good contrast. These results can be understood by calculating the values of φ_0^i at the top surface of the crystal in the range of orientations considered. These results, shown in Table 1, indicate that at the symmetry position thickness fringes result from the interaction of Bloch waves 1 and 3. Since these two Bloch waves have widely differing absorption coefficients, the thickness fringes would be expected to exhibit the poor contrast characteristic of anomalous absorption. At $\Delta\theta = -0.2\ \theta_B$, on the other hand, the important Bloch waves are 1, 2, and 3. Since $q^1 \gg q^2 \approx q^3$, thickness fringes in thick crystals result from the interaction of two Bloch waves of approximately the same absorption coefficient. The thickness fringes would therefore be expected to exhibit the good contrast characteristic of normal absorption.

It is next of interest to ask what stacking fault contrast is to be expected at these orientations. In order to answer this question, calculations of stacking fault profiles have been carried out for an $\alpha = -2\pi/3$ fault in an f.c.c. cobalt crystal of thickness equal to 1 μm. The profile obtained at the symmetry orientation is shown in Fig. 1c and, as can be seen, the fringes near the centre of the fault exhibit weak contrast. This behaviour is characteristic of the two beam anomalous absorption profiles obtained at low accelerating voltages [7, 8]. The

Table 1

Values of φ_0^i at the top surface of the crystal. Only Bloch waves 1 to 4 are given and the accelerating voltage was taken to be 1000 kV

$\Delta\theta$	$i = 1$	$i = 2$	$i = 3$	$i = 4$
−1.0	0.49	0.00	0.51	0.00
−0.8	0.48	0.05	0.47	0.00
−0.6	0.47	0.14	0.39	0.00
−0.4	0.45	0.23	0.32	0.00
−0.2	0.43	0.31	0.27	0.00
0.0	0.39	0.38	0.11	0.11

profile at $\Delta\theta = -0.2\,\theta_B$ is shown in Fig. 1d and, as can be seen, these fringes exhibit strong contrast in a crystal of the same thickness. It can also be noted from Fig. 1d that, except close to either surface of the crystal, the fringe spacing is equal to the effective extinction distance, $1/(\gamma^2 - \gamma^3)$ or about 900 Å.

4. Analysis of Stacking Fault Profiles in Terms of Interbranch and Intrabranch Scattering

The results obtained in Section 3 show that marked changes in the nature of stacking fault fringes near the centre of a fault occur when the crystal is tilted away from the symmetry position. The purpose of this section is to explain these results in terms of the scattering mechanisms which give rise to image contrast. These mechanisms can best be illustrated by following the Bloch wave interactions from the top to the bottom surface of the crystal. In order to follow these interactions at the two orientations of interest, calculations have been carried out of the values of φ_0^i at the top surface of the crystal and just above the fault and of values of φ_0^{ij} just below the fault and at the bottom surface of the crystal. The results which are shown in Fig. 2 only include interactions between Bloch waves 1, 2, and 3 since all other Bloch wave interactions were small and could be neglected.

The reason for the poor contrast exhibited by fringes near the centre of a fault in a crystal at the symmetry position can readily be seen by following the Bloch wave interactions in Fig. 2a. At the top surface of the crystal only Bloch waves 1 and 3 are important. Since $q^1 \gg q^3$, Bloch wave 1 will be relatively weak in comparison with Bloch wave 3 just above the fault. As a result the only significant interbranch scattering which can occur when the Bloch waves encounter the fault is from Bloch wave 3 to Bloch wave 1 (φ_0^{13}). Since q^1 is relatively high, this interbranch component of Bloch wave 1 will decrease rapidly with increasing distance from the fault with the result that it is negligible at the bottom surface of the crystal. The fringe contrast is therefore weak since, as noted in Section 2, a strong interbranch component is required in order to obtain good contrast.

The Bloch wave interactions which occur when the crystal is oriented so that $\Delta\theta = -0.2\,\theta_B$ can be seen from the values of φ_0^i and φ_0^{ij} in Fig. 2b. At the top surface of the crystal the three important Bloch waves excited are 1, 2, and 3.

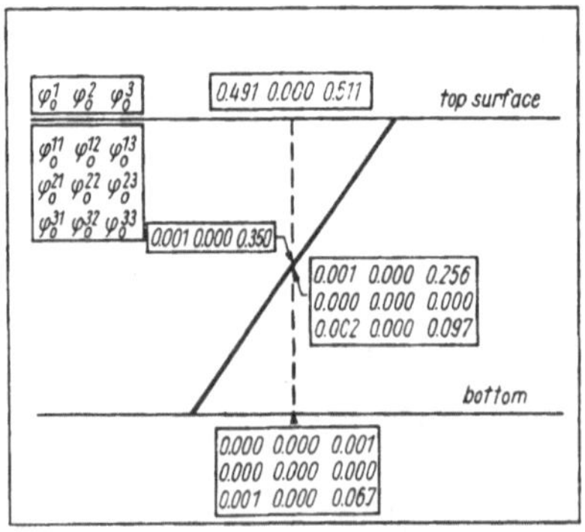

a

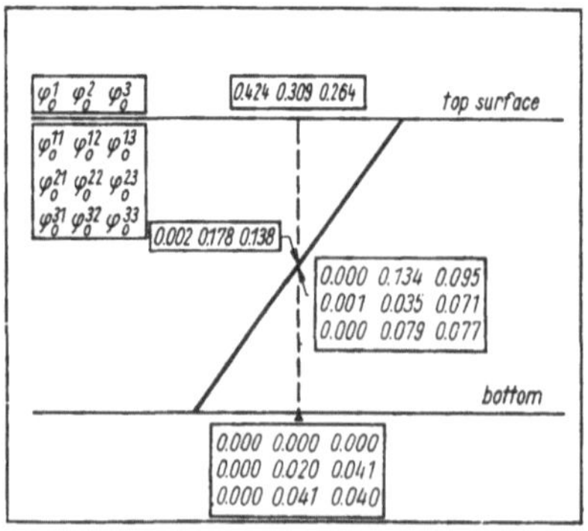

b

Fig. 2. The values of φ_0^i and φ_0^{ij} for an $\alpha = -2\pi/3$ stacking fault in f.c.c. cobalt at 1000 kV. It was assumed that the (111) set of systematic reflections was excited. a) and b) Values of $\Delta\theta$ equal to $-1.0\,\theta_B$ and $-0.2\,\theta_B$, respectively

Fig. 3. Phase amplitude diagram showing the relation between the contributions of interbranch and intrabranch scattering to the amplitude of the directly transmitted beam for an $\alpha = -2\pi/3$ stacking fault at the centre of a 1 μm thick crystal. The accelerating voltage was taken to be 1000 kV and $\Delta\theta = -0.2\,\theta_B$

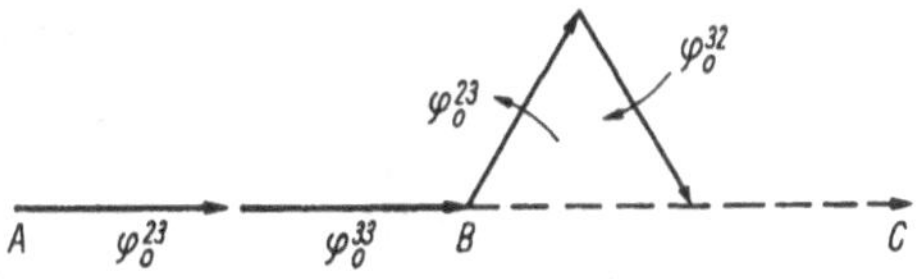

Since q^1 is relatively high, Bloch wave 1 will be weak in comparison with Bloch waves 2 and 3 just above the fault. As a result intrabranch scattering of Bloch wave 1 (φ_0^{11}) and interbranch scattering from Bloch wave 1 to Bloch waves 2 and 3 (φ_0^{21} and φ_0^{31}) will be weak while all the other interbranch and intrabranch components will be relatively high. Because of the high absorption coefficient of Bloch wave 1, the interbranch components φ_0^{12} and φ_0^{13} will weaken considerably before reaching the bottom surface of the crystal leaving only φ_0^{22}, φ_0^{33}, φ_0^{23}, and φ_0^{32} with appreciable values. The reason that strong contrast is obtained follows directly from the fact that both Bloch waves 2 and 3 have strong interbranch components (φ_0^{23} and φ_0^{32}). In order to understand why fringe spacing is equal to the effective extinction distance $1/(\gamma^2 - \gamma^3)$, the phase relationships between the interbranch and intrabranch components must be considered. A schematic representation of these phase relationships is shown in Fig. 3. This diagram can be understood by recalling that the interbranch component φ_0^{ij} will vary with the fault depth t_1 according to the term $\exp 2\pi i\,(\gamma^j - \gamma^i)\,t_1$ (see (9) and (11)). Thus at a value of t_1 equal to n extinction distances the two interbranch components φ_0^{23} and φ_0^{32} are in the same direction, giving a resultant amplitude equal to AC. As t_1 increases the vectors representing φ_0^{23} and φ_0^{32} rotate in the directions indicated with the result that when $t_1 = (n + 1/2)$ extinction distances the resultant amplitude is zero. When t_1 is equal to $(n + 1)$ extinction distances the resultant amplitude is a maximum again and equal to AC. Thus it can be seen that the amplitude of the directly transmitted beam varies from a maximum to a minimum to a maximum again if the stacking fault depth changes by one extinction distance. It might be noted that the differences in detail between the fringe profile given by the above qualitative arguments and the fringes in Fig. 1d can be understood by carrying out a more detailed analysis.

5. Discussion

5.1 Beams and Bloch waves

The first point to note from the results presented in this paper is the importance of describing image contrast in terms of Bloch wave interactions. Most calculations of image contrast based on the dynamical theory are classified according to the number of beams taken into account [9, 10]. Thus, for example, a stacking fault profile would be referred to in the literature as being two beam or multi beam. The results presented in this paper indicate that when discussing the nature of image contrast in terms of the dynamical theory, a more basic description is given in terms of the number of important Bloch waves excited and their characteristics. An example which serves to illustrate this point is the similarity between the familiar two beam stacking fault profiles obtained at low accelerating voltages and the multi-beam profiles obtained at high accelerating voltages with the crystal oriented at the symmetry position. The reason for this similarity lies in the fact that in both cases the contrast arises from the interaction of two Bloch waves of widely differing absorption coefficients.

5.2 Rocking curves and contrast

One of the primary objectives of the electron microscopist in carrying out observations of lattice defects is to choose diffraction conditions which give him high intensity, good contrast images. The crystal orientations which give maximum transmitted intensity in the perfect crystal can be determined from rocking curves. These data, however, do not provide information on such questions as what diffraction conditions must be chosen in order to obtain optimum contrast or how contrast changes with crystal orientation and accelerating voltage. As has been noted the maximum bright field intensity at low accelerating voltages occurs when the low order reflection of a systematic set is in the Bragg condition. The authors of this paper have shown [11] that in the case of bright field images of stacking faults, a decrease in contrast occurs when the crystal is tilted away from this position. Thus at low accelerating voltages the diffraction conditions for obtaining maximum bright field intensity and optimum contrast are the same. The results presented in this paper indicate that the situation at high accelerating voltages is quite different. The rocking curves obtained by Humphreys et al. [2] at high accelerating voltages indicate that the bright field intensity in the perfect crystal decreases as the crystal is tilted away from the symmetry position (note, for example, the difference in background intensities in Fig. 1c and 1d). Under the same circumstances, however, contrast in bright field images of stacking faults in thick crystals increases markedly. It is also important to note that although the intensity transmitted through the perfect crystal decreases appreciably as the crystal is tilted away from the symmetry position, intensity in the stacking fault image does not (compare, for example, the intensities at which fringe maxima occur in Fig. 1c and 1d). These results therefore suggest that at high accelerating voltages the symmetry position may not be the best orientation to carry out observations of stacking faults in thick crystals.

Acknowledgements

The authors wish to express their thanks to P. S. Turner, C. D. Cann, J. Andrew, and A. E. B. Monk for helpful discussions. They are also grateful to the National Research Council of Canada for financial support.

References

[1] C. J. HUMPHREYS and J. S. LALLY, J. appl. Phys. **41**, 232 (1970).
[2] C. J. HUMPHREYS, L. E. THOMAS, J. S. LALLY, and R. M. FISHER, Phil. Mag. **23**, 87 (1971).
[3] S. S. SHEININ, phys. stat. sol. **38**, 675 (1970).
[4] P. B. HIRSCH, A. HOWIE, R. B. NICHOLSON, D. W. PASHLEY, and M. J. WHELAN, Electron Microscopy of Thin Crystals, Butterworths, London 1965.
[5] C. J. HUMPHREYS and R. M. FISHER, Acta cryst. **A27**, 42 (1971).
[6] S. S. SHEININ and K. Z. BOTROS, phys. stat. sol. (a) **8**, 449 (1971).
[7] R. GEVERS, A. ART, and S. AMELINCKX, phys. stat. sol. **3**, 1563 (1963).
[8] H. HASHIMOTO, A. HOWIE, and M. J. WHELAN, Phil. Mag. **5**, 967 (1960).
[9] C. J. HUMPHREYS, A. HOWIE, and G. R. BOOKER, Phil. Mag. **15**, 507 (1967).
[10] R. SERNEELS and R. GEVERS, phys. stat. sol. **33**, 703 (1969).
[11] S. S. SHEININ and K. Z. BOTROS, phys. stat. sol. (a) **3**, 271 (1970).

(Received May 8, 1972)

phys. stat. sol. (a) **12**, 557 (1972)

Subject classification: 10.2; 19; 20.1; 22.5.2

Department of Physics, Indian Institute of Technology, Madras

Z-Centres in Rare-Earth Doped Alkali Halides

By

S. Radhakrishna and B. V. R. Chowdari

The peak positions and half-widths of the Z_1-, Z_2-, and Z_3-bands at LNT, in alkali halides doped with Eu^{++}, Sm^{++}, and Yb^{++} are reported. It has been shown that during the formation of the Z-centres the impurities do not directly participate but only provide the surrounding lattice interactions. In view of the present studies plausible models for the Z_1-, Z_2-, and Z_3-centres are discussed. From the peak positions of the Z_1-, Z_2-, and Z_3-bands a refined Mollwo-Ivey type of relation has been found. The half-width of an EPR line found in KCl crystals containing Yb^{++} having both the F-and Z_1-centres is 59 G. From this the half-width of the resolved Z_1-centre EPR signal is found to be 68 G. Such Z_1-centre EPR signal has been found in Sm^{++} doped KCl also.

Es werden die Lagen der Maxima und die Halbwertsbreiten der Z_1-, Z_2- und Z_3-Banden in mit Eu^{++}, Sm^{++} und Yb^{++} dotierten Alkalihalogeniden bei der Temperatur des flüssigen Stickstoffs angegeben. Es wird gezeigt, daß während der Bildung der Z-Zentren Verunreinigungen nicht direkt beteiligt sind und nur die Wechselwirkung des umgebenden Gitters liefern. Im Hinblick auf die dargestellten Untersuchungen werden plausible Modelle für die Z_1-, Z_2- und Z_3-Zentren diskutiert. Aus den Lagen der Maxima der Z_1-, Z_2- und Z_3-Banden wird eine verbesserte Mollwo-Ivey-Beziehung gefunden. Die Halbwertsbreite einer EPR-Linie, die in Yb^{++}-dotierten KCl-Kristallen mit F- und Z_1-Zentren gefunden wurde, beträgt 59 G. Daraus wird die Halbwertsbreite des aufgelösten EPR Signals des Z_1-Zentrums zu 68 G gefunden. Derartige Z_1-Zentren-EPR-Signale wurden in Sm^{++}-dotiertem KCl ebenfalls gefunden.

1. Introduction

Early work on Z-centres has been done [1 to 3] on alkali halides containing alkaline earth impurities and the data have been summarized in [4]. Under suitable optical and thermal treatment in the crystals containing these impurities Z_1-, Z_2-, Z_3-, and Z_4-centres are formed. Several other impurities like Cd^{++}, Zn^{++}, Pb^{++}, and Mn^{++} do not give Z-type centres but trap themselves electrons and get reduced in their valence state [5 to 7]. Recently it has been shown that rare earth impurities show such Z-type centres [8, 9].

A complete study of the optical properties of Z-centres in alkali halides doped with Yb^{++}, Eu^{++}, and Sm^{++} is likely to provide considerable information on the nature of the Z-centres. An important feature in the rare earth impurities is that they introduce prominent optical absorption bands [10 to 12] in the crystal, in a region well separated from the F-band, while the alkaline earth impurities do not have such characteristic bands. This would enable us to study the transformations taking place in the impurity during the process of the Z-centre formation. This was not possible in the case of the alkaline earth impurities. This paper is the result of a study of optical and electron paramagnetic resonance properties of the Z-centres in rare-earth doped alkali halides as a part of a programme studying several defect properties in these crystals.

2. Experimental Methods

The alkali halide crystals containing rare earth impurities have been grown from the melt by the Bridgman as well as by the Kyropolous technique. The respective rare earth chlorides have been added during the time of growth. Even in KBr the rare earth chlorides have been added assuming that the replacement of bromine by a small amount of chlorine has little effect. Yb^{++} doped alkali halides have also been prepared by diffusing the impurity from the vapour phase into the alkali halides at a suitable temperature in an evacuated pyrex glass tube. Optical absorption measurements at room temperature and liquid nitrogen temperature were made by using a Cary-14 spectrophotometer. Electron paramagnetic resonance spectra were recorded by using a Varian E-4 spectrometer.

3. Results and Discussion

3.1 Optical absorption studies

Fig. 1 shows the results of optical absorption studies in KBr crystals doped with Yb^{++}. Curve 1 is for an uncoloured crystal. It can be seen that there are two main bands at 2200 and 3000 Å. It is known that in rare earth impurities the principal bands arise from transitions [10 to 12] within the impurity level $4f^n \to \to 4f^{n-1}$ 5d. In cubic fields the 5d orbitals split into a triply degenerate and a doubly degenerate level. Under suitable conditions of low temperature and lowered local symmetry (due to charge compensating vacancies) these degenerate levels also split. It has been shown [11] that in Sm^{++} doped crystals the local symmetry is C_{2v} and hence only non-degenerate levels can exist in such symmetries. The two broad bands arising from $4f^n \to 4f^{n-1}$ 5d levels are now split into more bands at LNT. Such bands are most prominent in Eu^{++} and Yb^{++} doped alkali halides. In NaCl crystals containing Yb^{++} similar bands are observed at 2250 and 2950 Å. In KCl:Yb^{++} these bands are found at 2150 and 3050 Å. Curve 2 in Fig. 1 shows the spectrum of an additively coloured KBr crystal containing Yb^{++}. It can be seen that besides the characteristic impurity bands we have the well known F-band. When this crystal is bleached with

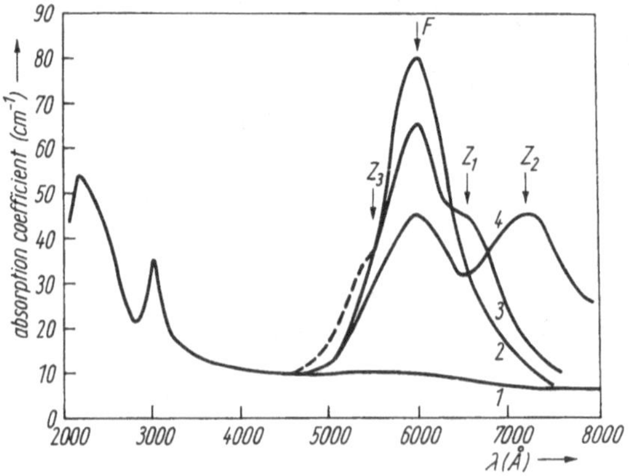

Fig. 1. Optical absorption at LNT in uncoloured and coloured KBr crystal containing Yb^{++}. Curve 1 is for an uncoloured crystal. Curve 2 is for an additively coloured crystal. Curve 3 is obtained after bleaching for 1h by F-light. Curve 4 is obtained after warming this bleached crystal to 110 °C. The dashed curve shows a part of the curve (showing only Z_3-bands) after optical bleaching by Z_2-light. The fine structure of Yb^{++} bands is not shown at LNT

F-light the F-band reduces and a new band is observed at 6550 Å. This band is the Z_1-band. The measurements for curve 2 were made at LNT. If these crystals are now warmed up to 110 °C most of the aggregate centres (R, N, and M) disappear along with the Z_1-band and another band called the Z_2-band appears at 7200 Å. This band is found to be stable up to 110 °C beyond which it starts decreasing and the F-band increases again. On long optical bleaching at LNT the Z_3-band at 5495Å and the F′-band on the longer wavelength side of the F-band are observed. It is important to note that during the formation and the destruction of the Z-centres little or no changes take place in the bands characteristic of the impurity. This suggests that the impurity does not directly trap electrons. Although such evidence has been found earlier in the case of Z-centres formed with alkali halides containing alkaline earth impurities it is important to note that such categorical evidence now comes from optical absorption measurements. This has been possible because Yb^{++} has characteristic bands in a region well separated from the F-band. Similar results are obtained in the case of NaCl and KCl crystals containing Yb^{++}. These results are summarised in Table 1.

Fig. 2 shows the optical absorption results in additively coloured europium doped NaCl crystals. Such studies have been reported earlier in the case of KCl crystals doped with Eu^{++} [9] but the Z_3-band in these crystals has not been observed. Eu^{++} also has characteristic optical absorption bands and it may be observed that these bands like in the case of Yb^{++} play no role during the trans-

Table 1

system*)	Z_1		Z_2		Z_3	
	peak position (Å)	half-width (eV)	peak position (Å)	half-width (eV)	peak position (Å)	half-width (eV)
NaCl: Ca	—	—	4715	0.30	4000	—
Sr	5060	—	5165	0.20	4000	—
Ba	—	—	—	—	—	—
Eu	5020	0.27	5165	0.35	4020	—
Sm	5050	0.26	5170	0.38	—	—
Yb	5030	0.30	5170	0.30	4050	—
KCl: Ca	5900	0.257	5985	0.28	5500	--
Sr	5900	0.271	6360	0.26	5510	
Ba	5900	0.271	6390 } 8150 }	0.35 } 0.18 }	—	—
Eu	5950	0.28	6750	0.30	5125	—
Sm	5920	0.28	6760	0.32	5200	—
Yb	5930	0.31	6780	0.30	5200	—
KBr: Ca	6700	—	7000	0.32	5560	—
Sr	6700	—	7150	0.32	5585	—
	6580	—	—	—	—	—
Ba	—	—	7150 }	0.45 }	—	—
	—	—	8490 }	0.20 }	—	—
Eu	6570	0.30	7200	0.38	5490	—
Sm	6560	0.35	7210	0.35	5495	—
Yb	6550	0.35	7200	0.35	5495	—

*) Data for Ca, Sr, and Ba are taken from [14, 9].

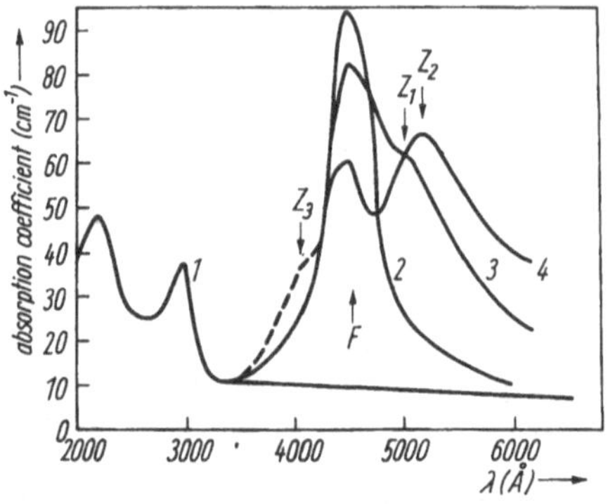

Fig. 2. Optical absorption at LNT in uncoloured and coloured NaCl crystal containing Eu^{++}. Curve 1 is for the uncoloured crystal and curve 2 is for a coloured crystal showing only the F-band. Curve 3 is obtained after optical bleaching for 1 h. Curve 4 is obtained after warming this crystal to 110 °C. The dashed line shows a part of the curve obtained after optical bleaching by Z_2-light. The fine structure of Eu^{++} bands is not shown at LNT

formation of F- to Z-centres. Such Z-centres have been observed in Eu^{++} doped KCl, KBr, and NaCl and the results along with the other parameters are summarized in Table 1.

In Sm^{2+} doped alkali halides there are two difficulties for getting correlated results before and after coloration. Uncoloured crystals containing Sm^{++} ions give [13] three prominent bands at 4200, 5400, and 6200 Å and some bands below 3000 Å. Hence it is difficult to study the effect of Z-centre formation on these bands because of the strong overlapping F-band. The second difficulty is that if the impurity concentration is large, it is exceedingly difficult to colour the crystals. Only in samples containing low concentration one can hope to get the F- and the Z-centres by suitable optical and thermal treatments. In additively coloured alkali halides we have not observed any bands which may be ascribed to Sm^+ or Sm^{+++}, i.e. there is no change of the valence state during the conversion of the F- to Z-centres. However, in irradiated crystals such conversion was observed [13]. Since the results of Z-centre formation are similar to the other impurity doped alkali halides we have only summarized the results in Table 1.

It is known that the family of impurities Ca^{++}, Ba^{++}, Sr^{++}, Sm^{++}, Yb^{++}, and Eu^{++} have a lower second ionisation potentials (Table 2) than the other impurities like Cd^{++}, Zn^{++}, Pb^{++}, Ba^{++} etc. It has been suggested [8, 9] that in view of the low second ionisation potential of the former set of impurities they have less electron affinity and thereby facilitate the formation of the Z-centre. The latter group

Table 2

impurity	first ionisation potential	second ionisation potential
Ca	6.11	11.9
Sr	5.09	11.0
Ba	5.21	10.0
Eu	5.64	11.2
Sm	5.60	11.4
Yb	6.20	12.1
Pb	7.42	15.0
Cd	8.99	16.9
Zn	9.30	18.0
Mg	7.60	15.0

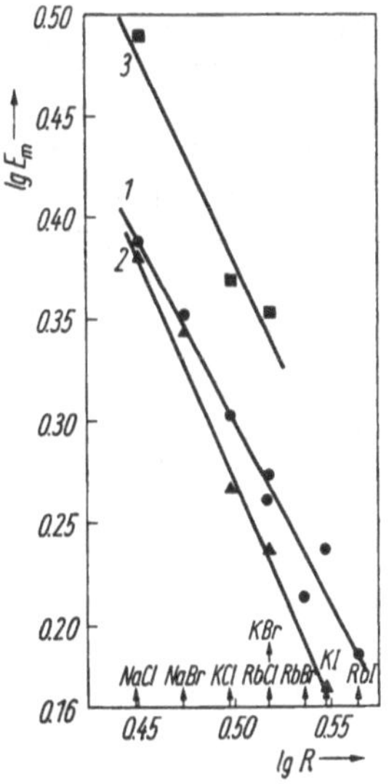

Fig. 3. Plot of lg E_m vs. lg R for the Z_1-, Z_2-, and Z_3-centres in various alkali halides. Data taken from present paper and [14, 9, 4]. The straight lines (1), (2), and (3) correspond to the Z_1-, Z_2-, and Z_3-centres

acts as electron traps as efficiently as the negative ion vacancies and are reduced in their valence state. The former impurities are not efficient electron traps and perturb only the F-centre surroundings. These studies clearly indicate that the electron centered impurity models for the Z-centres cannot be valid.

Fig. 3 shows a plot of lg E_m vs. lg R for the Z_1-, Z_2-, and Z_3-centre in various crystals. E_m is the peak position expressed in eV and R is the lattice constant. It has earlier been pointed out [14] that the Z_1- and Z_2-,peak positions depend in this manner on the lattice constants. Now we have data for these bands in more systems [9, 14]. The average peak position for the Z_1-, Z_2-, and Z_3-bands is taken in each halide doped with different impurities. This is because there is little dependence of the peak position on the impurity. From these straight lines a relation of the type $E_m = K\,R^n$ is obtained for the Z_1-, Z_2-, and Z_3-centres, where $n = -1.79$, -2.157, and -2.0 and $K = 15.6$, 22.35, and 23.82, respectively.

3.2 Electron paramagnetic resonance studies

The EPR results in Eu^{++} doped KCl have been reported recently [8]. It has been shown that there is no valence change of the Eu ion during the F → Z transformation. Eu^{++} is itself paramagnetic and this gives a strong overlap of the F and Eu^{++} signals and therefore it was difficult to draw definite conclusions. Yb^{++} and Sm^{++} are however not paramagnetic and one can look for new signals if either of these impurities changes the valence state. Fig. 4 shows results for a KCl crystal containing Yb^{++}. Curve 1 shows the F-centre EPR signal. This EPR signal has the characteristic features of the pure F-centre in KCl crystals. It has a half-width of 47 G and a g-value of 1.994. When optically bleached the line increases in width to about 59 G which is shown as curve 2 in Fig. 4. Curve 3 is obtained after heating the sample to 110 °C at which temperature the Z_1-centres disappear and Z_2-centres are found. This signal is identical to curve 1 which suggests that the Z_2-centres are diamagnetic. Assuming no change in the width of the F-centre signal during optical bleaching the Z_1- and F-centre line (curve 2) may be interpreted to be due to two different centres, one the F-centre and the

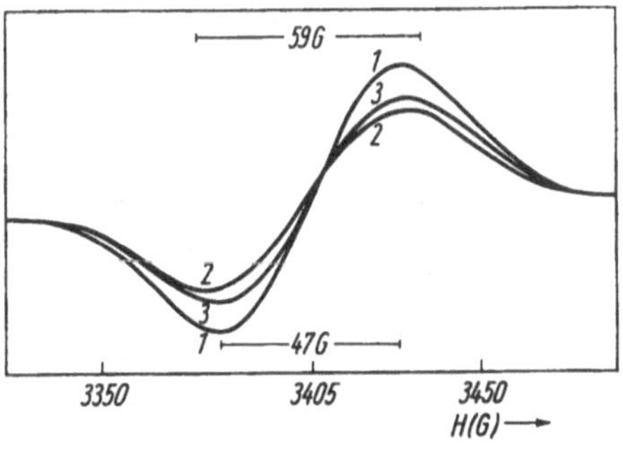

Fig. 4. EPR spectrum of coloured KCl crystals containing Yb^{++} ions. Curve 1 shows a spectrum of F-centre in KCl crystal. The half-width is 47 G and the g-value is 1.994. Curve 2 shows a combined signal of F and Z_1. The half-width is now 59 G and the g-value is 1.996. Curve 3 is obtained after heating the crystal to 110 °C when the Z_1- centres disappear

other the Z_1-centre, as we have already seen that under similar conditions the Z_1-band is formed in the optical absorption spectrum. The width of the separated Z_1-band is 69 G while the g-value is 1.996. These results are similar to the EPR line for the Z_1-centre reported earlier in Ca- and Sr- doped alkali halides [15].

From our experimental results the following conclusions can be drawn.

a) The impurities do not directly trap electrons as we do not observe any change in the optical absorption bands characteristic of the divalent impurity. Further, we do not observe any evidence of the formation of monovalent states of the impurity either from optical absorption studies or from electron paramagnetic resonance studies.

b) Since the peak position of the Z_1-band in a particular alkali halide crystal has little dependence on the impurity it can be concluded that the impurity acts only as a perturbing agent near the F-centre, i.e. the impurity itself is of secondary importance. Such a result has been obtained even from EPR studies. The width of the Z_1-centre signal is the same as that of the Z_1-centre formed in KCl crystals containing Ca^{++}, Sr^{++}, or Yb^{++}.

c) Our results support the ionized Z_2-centre models for the Z_3-centre since we observe the Z_3-centre and F'-centres when the Z_2-centres are bleached at LNT.

These conclusions can be best explained by the following models.

An F-centre with a cation vacancy as nearest neighbour and the divalent impurity at a nearby cation site seems to be the most plausible model for the Z_1-centre [16]. An F'-centre in the neighbourhood of the divalent impurity seems to explain all the experimental results for the Z_2-centre [8, 17, 18]. An ionized Z_2-centre is considered to be the Z_3-centre [17].

Acknowledgements

The authors would like to thank Dr. S. Subramaniam, Department of Chemistry, Indian Institute of Technology, Madras, for help in recording the EPR spectra and Prof. C. Ramasastry for his interest in this work.

References

[1] H. PICK, Z. Phys. **114**, 127 (1939).
[2] F. SEITZ, Phys. Rev. **83**, 134 (1951).
[3] J. H. SCHULMAN and W. D. COMPTON, Color Centers in Solids, Pergamon Press, 1963.
[4] S. C. JAIN and S. RADHAKRISHNA, J. Sci. Indian Res. **26**, 324 (1967).
[5] S. C. JAIN and S. RADHAKRISHNA, Phys. Rev. **172**, 972 (1968).
[6] S. C. JAIN, S. RADHAKRISHNA, and K. S. K. SAI, J. Phys. Soc. Japan **27**, 1179 (1967).
[7] M. IKEYA and N. ITOH, Solid State Commun. **7**, 355 (1969).
[8] B. V. R. CHOWDARI, Solid State Commun. **9**, 1659 (1971).
[9] K. H. UMBACH and H. J. PAUS, Internat. Conf. Color Centers, Abstract No. 77, Reading (UK) 1971.
G. KENNTNER and H. J. PAUS, ibid., Abstract No. D55.
[10] D. S. MCCLURE and Z. KISS, J. chem. Phys. **39**, 3251 (1963).
[11] W. B. BRON and M. WAGNER, Phys. Rev. **145**, 689 (1966).
[12] A. K. MEHRA, J. Opt. Soc. Amer. **58**, 853 (1968).
[13] F. K. FONG, J. A. CAPE, and E. Y. WONG, Phys. Rev. **151**, 299 (1966).
[14] J. HINGSAMMER and H. JODL, Phys. Letters (Netherlands) **25A**, 131 (1967).
[15] H. KAWAMURA and K. ISHIWATARI, J. Phys. Soc. Japan **13**, 574 (1958).
[16] F. ROSENBERGER and F. LÜTY, Solid State Commun. **7**, 249 (1969).
[17] H. OHKURA, Phys. Rev. **136**, 446 (1964).
[18] W. ADLHART, G. GEHRER, and E. LUSCHER, Solid State Commun. **9**, 117 (1971).

(Received April 3, 1972)

phys. stat. sol. (a) **12**, 563 (1972)

Subject classification: 14.3.3; 17; 22.6

Department of Physics, Brunel University, Uxbridge, Middlesex

Electrical Forming in and Electron Emission from Thin Film Aluminium–Borosilicate Glass–Aluminium Sandwiches

By

E. H. Z. TAHERI, C. A. HOGARTH, and R. D. GOULD

Some electrical conduction properties of thin co-evaporated films of SiO_x(70%)–B_2O_3 (30%) are described and a region of differential negative resistance in the current–voltage characteristic is observed for films of thickness approximately 3×10^{-5} cm. The films, deposited between aluminium electrodes, show evidence of electroforming and of electron emission into a vacuum when a bias voltage is applied. At voltages of order 30 V at which the negative resistance occurs, electron emission into a vacuum increases very rapidly with voltage. The emission current density is of order 10^{-5} Acm^{-2} and since the circulating current through the sandwich is lower than in similar films of SiO, the transmission efficiency is high.

On décrit quelques propriétés électriques des couches minces de SiO_x (70%)–B_2O_3 (30%), préparées par co-évaporation dans le vide et on observe une région de résistance négative dans la caractéristique courant–tension pour les couches d'une epaisseur de 3×10^{-5} cm environ. Les couches, formées entre les électrodes d'aluminium, montrent l'évidence de formation électrique et de l'émission d'électrons lorsqu'on applique une tension directe à la couche dans le vide. Pour les tensions appliquées d'environ 30 V où on observe la résistance négative, l'émission électronique dans le vide s'augmente rapidement avec la tension. La densité du courant d'émission est environ 10^{-5} Acm^{-2} et puisque le courant qui circulent dans cette couche est plus petit que celui qui est dans les pareilles couches de SiO, l'efficacité de transmission est grande.

1. Introduction

Amorphous materials are of current topical interest and one class of such materials consists of thin layers of mixed oxides which possess a glassy structure. Hogarth and Wright [1] developed a co-evaporation technique for depositing mixed layers having controllable compositions. Timson and Hogarth [2] have used this technique to prepare co-evaporated films of silicon oxide and boric oxide with thicknesses in the range (1 to 2) $\times 10^{-4}$ cm and have reported on the electrical conduction characteristics and also the results of electron-spin resonance measurements, which show that one effect of admixtures of B_2O_3 to the SiO is to reduce the density of dangling bonds, and also the electrical conductivity characteristic of the SiO without additives. In general the current–voltage characteristic was found to be of the form $I \propto U^n$ where n was approximately four but varied somewhat with film composition. The ac conductance σ_{ac} at frequency f was of the form $\sigma_{ac} \propto f^n$ for $n \approx 0.9$ to 1.0.

Recently these measurements have been extended by Hogarth and Taheri [3] to cover dc measurements on SiO_x(70%)–B_2O_3 (30%) films having thicknesses in the range (3 to 6) $\times 10^{-5}$ cm. The resistivity of such films is very high (10^{15} to 10^{16} Ωcm) but the current levels are significantly higher at a given voltage than for those of the thicker films described by Timson and Hogarth [2]. A decrease in current with increasing voltage was observed at voltages somewhat below the onset of complete electrical breakdown. This behaviour was ascribed to the rupture of field-assisted formed filaments of the type discussed by Dearnaley et al. [4] and by Sutherland [5]. The essential postulate is that conducting filaments of metal atoms from the electrodes, alternating with anion vacancies, can be established between the metal contacts when a sufficiently high voltage is applied. The process is known as forming and the voltage at which it occurs is the forming voltage (U_F). The filaments tend to rupture when the current levels through them lead to high local power dissipation and then the current will fall with increasing voltage. The filaments may re-form and the current will then rise again. This type of behaviour has been observed with different assemblies including sandwiches of Al–SiO–Au, [6], Al–Al_2O_3–Au [7], and Au–SiO–Au [8]. Associated with the negative resistance region is a memory effect and electron emission from the sandwich into a vacuum. Early work suggested that forming with aluminium electrodes only would not occur but more recently Pivot et al. [9] and Doucas and Walsh [10] have formed devices with aluminium positive electrodes. Although aluminium is a less favourable material for forming than, for example gold, forming can occur with aluminium but sometimes only in special cases. Hickmott [11] has formed Al–Al_2O_3–Al devices which were bisulphate anodized but was unable to form pure samples. He attributed the increased ability to form to the impurities present and this explanation may hold for the samples of Doucas and Walsh [10] which had an additional CdS layer and of Hogarth and Taheri [3] which contained 30% of B_2O_3 as an additive to the SiO.

The observation of forming phenomena and the appearance of a negative resistance region suggested that the SiO–B_2O_3 films might act as electron emitters. Since the resistivity of the films is high it would be expected that the transmission ratio α (emission current/circulating current, I_e/I_c) in these samples would be higher than the best values quoted for cold cathodes using SiO.

2. Experimental Methods

The technique for the preparation of mixed oxide films has been described by Hogarth and Wright [1] and this technique was followed. SiO_x–B_2O_3 films were prepared by thermal evaporation from separate sources in a Balzer BA 510 coating unit. The lower vapour pressure of B_2O_3 requires a higher source temperature ($\approx$ 2000 °C) than SiO. Tantalum and molybdenum boats were used respectively for SiO and B_2O_3. Since SiO_x–B_2O_3 films show a high sensitivity to water vapour for concentrations of B_2O_3 greater than 50 mol%, a 30% concentration of B_2O_3 was used in this work. Aluminium was used for both electrodes each of thickness approximately 5×10^{-6} cm and deposition was on to cleaned Corning 7059 borosilicate substrates. The pressure during deposition was $\approx 4 \times 10^{-6}$ Torr and substrate temperatures were stable at $\approx$ 180 °C. The source-substrate distance was 30 cm and the specimen area was $\approx 4 \times 10^{-6}$ m². The deposition rate (0.8 at. units/s) was constantly monitored during evaporation, using conventional quartz crystal techniques. SiO_x–B_2O_3 layers were deposited with thickness $\approx$3000 at. units in order to ascertain whether the negative resistance and electron emission were more pronounced for the lower thicknesses.

All voltage–current characteristics and electron emission measurements were performed in a subsidiary vacuum system at a pressure of $\approx 10^{-6}$ Torr. Voltage bias was provided from a Solartron stabilized dc power supply. An aluminium anode was positioned approximately 3 cm above the sandwich and was biassed so that an accelerating voltage of 100 V was always applied. The emission current collected at the anode was measured with a Keithley 610 C electrometer. A Comark electronic thermometer using a chromel–alumel thermocouple attached to the top of the samples was used to monitor the sample temperature during electrical measurements.

3. Results

The general circulating current I_c against voltage bias U_b curve was of the form $I \propto U^n$, where $n \approx 4$ to 6 and is directly comparable with the earlier results of Hogarth and Taheri for 3000 to 6000 at. units films. On increasing the bias above a certain voltage (U_F, the forming voltage) a substantial increase in I_c was observed. Typical increases were from 10^{-5} to 10^{-3} A. Current levels were substantially higher thereafter even at voltages below U_F and electron emission into vacuum could now be detected.

All samples were subjected to an increasing dc bias until the forming voltage was reached. This was somewhat variable between samples but was usually in the range 12 to 20 V. When forming was complete the bias voltage was reduced to zero and then reapplied to give the formed characteristic. In some cases the devices would spontaneously un-form and then re-form and general stability was not as good as for Al or Au–SiO–Au devices. However the devices did show remarkable dielectric properties and some sample could withstand voltages up to 400 V.

Results typical of a substantial number of samples investigated are given below.

Fig. 1 shows a typical characteristic of circulating current I_c against bias voltage U_b. A peak current is exhibited at 30 V followed by a negative re-

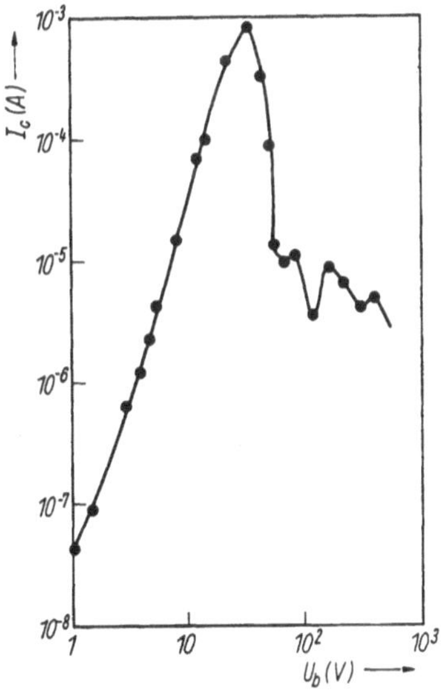

Fig. 1. Variation of circulating current with bias voltage

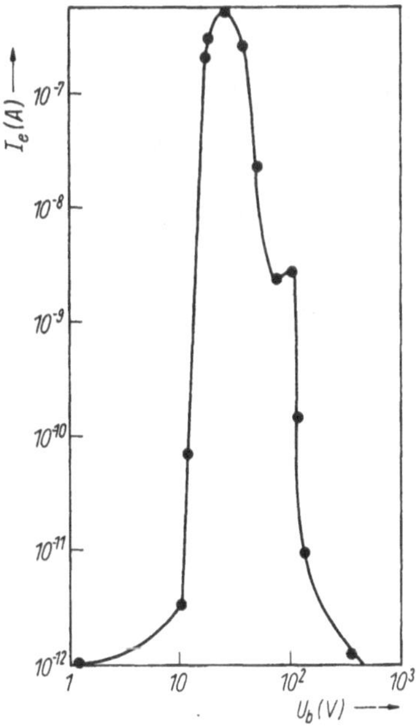

Fig. 2. Variation of emission current with bias voltage

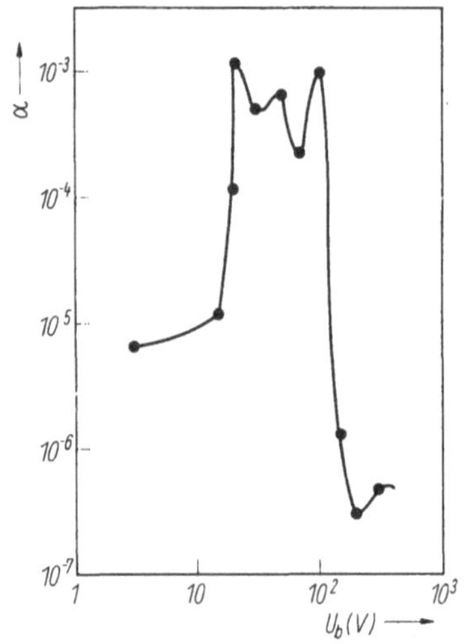

Fig. 3. Variation of transmission ratio ($\alpha = I_e/I_c$) with bias voltage

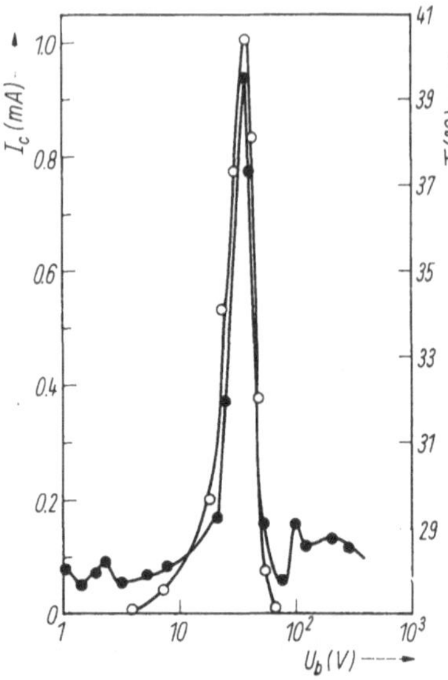

Fig. 4. Variation of temperature (●) with bias voltage. (I_c vs. U_b (○) shown for comparison)

sistance region. Above the maximum value the general level of current falls although there are sometimes small increases in current at voltages greater than the voltage for current maximum.

Fig. 2 shows the emission current I_e variation with U_b. This is apparent at low voltages and rises very steeply at about 10 V to a maximum which occurs at the same value of voltage as the maximum in I_c. I_e then slowly decreases until at some voltage usually ≈ 100 V the emission current falls very steeply and sometimes disappears altogether. Subsequent retracing of the characteristic show the same general form and current levels of the same order, thus indicating that dielectric breakdown has not begun to destroy the samples.

Fig. 3 shows the variation of transmission ratio ($\alpha = I_e/I_c$) with U_b. The maximum value of α is 10^{-3} which is comparable with that for other thin film cold cathodes. The large rises and falls of current at applied voltages of order 10 and 100 V respectively, correspond to similar changes in I_e.

In order to test whether any appreciable quantity of heat was dissipated in the device especially at very high voltage levels a thermocouple was attached to the top of some devices and the temperature monitored.

Fig. 4 shows the variation of temperature T with U_b. The variation of I_c with U_b is also shown for comparison. The T–U_b curve in general follows the I_c–U_b curve and the maximum temperature occurs when the current flow is highest. At higher voltages the temperature relaxes to the ambient temperature and no significant heat dissipation is detectable.

4. Discussion

The use of co-evaporated borosilicate glass in a metal–insulator–metal (MIM) structure has shown some new results. It was found possible to form the devices electrically even though aluminium electrodes were employed and the circulating currents through the devices were quite low and certainly not so high as for simple oxide devices previously reported. This result is consistent with the improved dielectric properties of these glasses previously reported [1]. The maximum value of transmission efficiency observed, 10^{-3} was on films 3×10^{-5} cm thick with electrodes of thickness $\approx 5 \times 10^{-6}$ cm. This value is comparable with the maximum estimated for SiO_x films [8] of thickness 8×10^{-6} cm and assuming a top electrode having zero thickness and thus zero attenuation. We may expect to obtain even higher transmission ratios on thinner films and with a thinner top electrode.

The very high dielectric strength of the mixture allows high voltages to be applied across the insulator without breakdown occurring. Although currents are low the high applied voltage means that considerable power is being dissipated in the device. For example a value of $U_b = 300$ V and a circulating current of 0.33 mA corresponds to a power dissipation of order 0.1 W. Although this is of the same order as power dissipated in the simple oxide devices, the lower circulating currents would be expected to give more stable operation.

The model of Dearnaley et al. [4] is supported by the temperature measurements which show that the temperature is at a maximum before the onset of negative resistance. Beyond this region the temperature falls and there is no evidence of thermally-induced breakdown. This poses the question of the failure of emission at voltages of order 100 V. There is now substantial evidence [12, 13] to show that in MIM structures there is a barrier layer region at one electrode so that the local electric field in this region is very high indeed. For example a potential drop of 40 V across a layer of thickness 3000 at. units corresponds to an applied field across the specimen of 1.3×10^6 V cm^{-1}, but for the lower range of applied voltages, if most of the potential drop occurs a space-charge layer of thickness, say, 300 at. units, then the effective field which is exciting electrons to an energy sufficient to be emitted is 1.3×10^7 cm^{-1} and it may well be greater than this value in some experiments. As the applied voltage is increased to higher values the potential fall across the specimen becomes more uniform and the *effective* field exciting the electrons is lower and thus the probability of emission will decrease. We believe that this redistribution of potential accounts for the gradual decrease in emission current from a U_b-value of 100 V as shown in Fig. 2. In the sandwich structures using SiO_x [8, 13] which have been carefully studied in relation to this effect, a significant diminution of electron emission current is not observed and breakdown of the device occurs while emission is still taking place. This is probably due to the larger circulating currents in the SiO_x devices.

References

[1] C. A. Hogarth and L. A. Wright, Proc. Internat. Conf. Phys. Semicond., Moscow 1968 (p. 1274).
[2] P. A. Timson and C. A. Hogarth, Thin Solid Films **10**, 321 (1972).
[3] C. A. Hogarth and E. H. Z. Taheri, Thin Solid Films **10**, 455 (1972).
[4] G. Dearnaley, D. V. Morgan, and A. M. Stoneham, J. non-cryst. Solids **4**, 593 (1970).
[5] R. R. Sutherland, J. Phys. D (Appl. Phys.) **4**, 468 (1971).
[6] R. R. Verderber and J. G. Simmons, Radio Electr. Engng. **33**, 347 (1967).
[7] T. W. Hickmott, J. appl. Phys. **36**, 1885 (1965).
[8] R. A. Collins and R. D. Gould, Solid State Electronics **14**, 805 (1971).

[9] J. PIVOT, M. BOUDEUILLE, A. CACHARD, and C. H. S. DUPUY, phys. stat. sol. (a) **2**, 319 (1970).
[10] G. DOUCAS and D. WALSH, Thin Solid Films **9**, 25 (1971).
[11] T. W. HICKMOTT, Thin Solid Films **9**, 431 (1971).
[12] T. W. HICKMOTT, J. appl. Phys. **35**, 2679 (1964).
[13] R. D. GOULD and C. A. HOGARTH, Thin Solid Films, to be published.

(Received May 10, 1972)

phys. stat sol. (a) **12**, 569 (1972)

Subject classification: 19; 22.8.2

Institut d'Electronique Fondamentale, Laboratoire associé au C.N.R.S., Faculté des Sciences, Orsay (a), and Laboratoire de Magnétisme, C.N.R.S., Bellevue (b)

Hyperfine Interaction of Cr^{3+} and Cu^{2+} Ions in Cr- and Cu-Doped YIG

By

LE DANG KHOI (a), M. ROTTER[1]) (a), and R. KRISHNAN (b)

The hyperfine fields of Cr^{3+} and Cu^{2+} ions in substituted YIG were measured by NMR at 4.2 °K to be about 327 and 195 kOe, respectively. These values are rather high compared to those observed in other host lattices. Hyperfine interaction was also evaluated for Cr^{3+} and Cu^{2+} in the octahedral sites. The ionic Hamiltonian includes the crystal field energy, the spin–orbit coupling, and the exchange energy. The hyperfine fields thus estimated are in agreement with the experimental results.

Es wurden die Hyperfeinfelder von in YIG substituierten Cr^{3+}- und Cu^{2+}-Ionen mit NMR bei 4,2 °K zu 327 bzw. 195 kOe gemessen. Diese Werte sind ziemlich hoch im Vergleich zu den in anderen Wirtskristallen beobachteten. Für Cr^{3+} und Cu^{2+} an Oktaederplätzen wurde die Hyperfeinwechselwirkung berechnet. Der ionische Hamiltonian enthält die Kristallfeldenergie, die Spin–Bahnkopplung und die Austauschenergie. Die so berechneten Hyperfeinfelder stimmen mit den experimentellen Ergebnissen überein.

1. Introduction

In this paper we report our investigations using spin echo techniques of hyperfine fields at Cu^{2+} and Cr^{3+} nuclei in the yttrium–iron garnet host lattice. The samples are aggregates of crystals oriented at random.

A few words about the crystals are necessary to characterize them and to better understand the results.

YIG:Cu^{2+} — The crystals of $Y_3Fe_{5-2x}Cu_xGe_xO_{12}$ were prepared by the flux method by one of us. A description of it has been published earlier [1]. The magnetic properties were also investigated by ferrimagnetic resonance techniques [2]. It was shown that Cu^{2+} ions were present mostly on the octahedral (16a) sites. From there they contribute to the anisotropy (K_1), the single-ion contribution at 4.2 °K being $\approx 0.10\ cm^{-1}$. It has also been demonstrated from line width measurements that these crystals have a negligible amount of Fe^{2+} ions [2]. In the samples investigated $x = 0.09$.

YIG:Cr^{3+} — The crystals of $Y_3Fe_{5-x}Cr_xO_{12}$ with $0 < x < 0.43$ were prepared using an appropriate flux system which yields good homogeneity of the Cr_2O_3 distribution in the crystal [3]. Ferrimagnetic resonance studies indicate that the contribution to K_1 from Cr^{3+} ions present on octahedral sites (16a) is negative. The crystals studied here had $x = 0.43$.

NMR can be a convenient probe for studying the behaviour of paramagnetic impurity ions in magnetic materials. Often one must ascertain whether the observed signals arise from domains or domain walls. For a steady-state method

[1]) On leave from the Faculty of Mathematics and Physics, Charles University, Prague.

the nuclear signals in ferrimagnetic samples are proportional to the square of the enhancement factor η [4]. If the hyperfine fields are isotropic the signals will come mainly from nuclei in domain walls. The situation is quite different for the spin echo method since the signal is simply proportional to η.

Let us compare the spin echo signals S_D and S_W in domains and domain walls, respectively. The ratio S_D/S_W is equal to $\eta_D D/\eta\delta$ where D is domain width and δ wall thickness. By using the expression for η_D and η_W [5], this ratio is found to be of the order of unity. This result can also be estimated otherwise. Since η is proportional to the electronic susceptibility this ratio precisely equals χ_D/χ_W where χ_D and χ_W are the contributions to radiofrequency susceptibility from domain rotations and wall motion, respectively. They were known to be of the same order of magnitude [6]. Now if the hyperfine fields are anisotropic the resonance line will be broadened in domain walls. Moreover the nuclear relaxation times are much shorter in domain walls than in domains [7]. Under these circumstances the spin echo spectrum is expected to arise chiefly from nuclei in domains. There are thus evidences that only domain signals in Er IG [8], Dy IG [9], MnP [10], and MnBi [11] are observable by the spin echo method.

2. Experimental Results and Interpretation

2.1 Cr^{3+} ($3d^3$, $L = 3$, $S = 3/2$)

The resonance frequencies in MHz of ^{53}Cr and ^{57}Fe nuclei at a-sites in zero external field are given in Table 1 and their respective resonance spectra at 4.2 °K are shown in Fig. 1. The corresponding hyperfine fields are compared with those of Fe^{3+} and Cr^{3+} in the octahedral environment in MgO [12] in Table 2.

Table 1
Resonance frequencies of Fe and Cr at a-sites in Cr-doped YIG

	^{57}Fe	^{53}Cr
4.2 °K (MHz)	75.8	≈78.8
77 °K (MHz)	75	≈78

The difference in the fields for Cr^{3+} arises from orbital and dipolar contributions to the hyperfine interaction, the octahedral symmetry in YIG being much lower than in the cubic MgO. In YIG the Cr^{3+} ions occupy the a-sites of C_{3v} symmetry

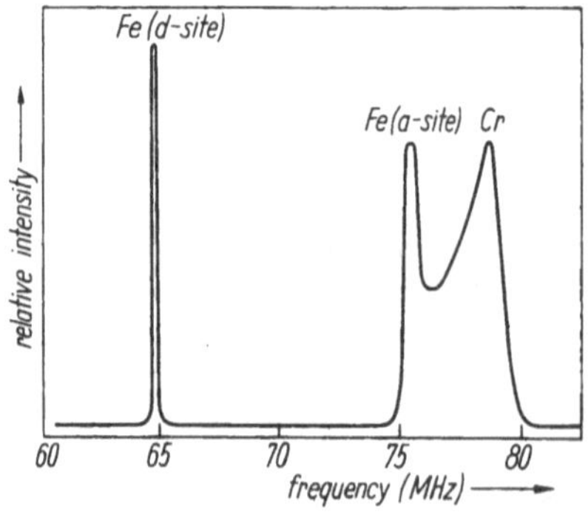

Fig. 1. Resonance spectrum at 4.2 °K of ^{57}Fe nuclei at a- and d-sites and of ^{53}Cr at a-sites

Table 2

Hyperfine fields of Fe^{3+} and Cr^{3+} in YIG and in MgO

	H_n (kOe) for Fe^{3+}	H_n (kOe) for Cr^{3+}
YIG	550	327.4
MgO	565	303

with the c-axis along [111] which is also the direction of easy magnetization. Therefore in zero external field there will be two magnetically inequivalent a-sites with statistical weights 1:3. To explain the hyperfine field value and the form of the ^{53}Cr spectrum it was necessary to consider Cr^{3+} in the actual crystal field symmetry. For the C_{3v} symmetry the ionic Hamiltonian [13] can be written with Steven's operators O_l^m and with the simple forms of spin–orbit coupling and exchange energy as

$$\mathcal{H} = B_2^0\, O_2^0 + B_4^0\, O_4^0 + B_4^3\, O_4^3 + \lambda\, L\, S + 2\, \mu_B\, S\, H_{ex},$$

where

$$B_2^0 = \langle L || \alpha || L \rangle\, A_2^0\, \langle r^2 \rangle\,,$$

$$B_4^m = \langle L || \beta || L \rangle\, A_4^m \langle r^4 \rangle\,.$$

The quantities $A_l^m\, \langle r^l \rangle$ are referred to as crystal field parameters [13]. The coefficients $\langle L || \alpha || L \rangle$, $\langle L || \beta || L \rangle$ and the O_l^m matrix elements are tabulated [13].

On choosing the exchange field H_{ex} direction as the quantized direction the Hamiltonian for the second site whose local c-axis is along another diagonal was obtained by rotating the coordinate system with Euler angles ($\pi/6$, arc cos $\frac{1}{3}$, 0). The O_l^m transformations were performed using tabulated values given by Baker and Williams [14].

It is well known that for the iron group the calculated value of the crystal field parameter from the point charge model is always several times (about 3 for Ni^{2+} in $KNiF_3$ [13]) smaller than that deduced from experimental data. We assumed here a multiplicative factor 3 for the terms A_4^0, A_4^3, after having calculated them by summation over a sphere radius 3 a_0 (lattice parameter). The coefficient A_2^0 was estimated from the upper limit of quadrupole splitting in connection with the spectrum width. This value is only 1.37 times higher than the calculated one. Using the $\langle r^l \rangle$ values calculated by Freeman and Watson [12] we deduced the following parameters:

$$B_2^0 = 83\ \text{cm}^{-1}\,, \quad B_4^0 = -13.7\ \text{cm}^{-1}\,, \quad B_4^3 = -340\ \text{cm}^{-1}\,.$$

A reasonable value of λ for Cr^{3+} in the crystal is about 70 cm^{-1}, instead of the free ion value of 91 cm^{-1} [13]. Since the thermal variations of ^{53}Cr and ^{57}Fe resonance frequencies are similar, the exchange field H_{ex} at Cr^{3+} could be roughly evaluated from the Curie temperature T_c as $H_{ex} = kT_c/2\, \mu_B\, S \approx 2$ MOe.

The ion ground state is a mixture of $|M_L, M_S\rangle$ states and is obtained by diagonalizing the ionic Hamiltonian within the L, S multiplicities of the free ion ground term. The orbital H_{orb} and dipolar H_d terms are deduced from the hyperfine interaction Hamiltonian [13] with usual notations as

$$H_{\rm orb} = -2\,\mu_{\rm B}\langle r^{-3}\rangle\,\langle L\rangle\,,$$

$$(H_{\rm d})_z = -2\,\mu_{\rm B}\,\xi\langle r^{-3}\rangle\,[\langle L\,(L+1) - 3\,L_z^2\rangle\,\langle S_z\rangle - {}$$
$$- \tfrac{3}{2}\langle L_x\,L_z + L_z\,L_x\rangle\,\langle S_x\rangle - \tfrac{3}{2}\langle L_y\,L_z + L_z\,L_y\rangle\,\langle S_y\rangle]\,.$$

The other components are obtained by circular permutations. Taking $\langle r^{-3}\rangle =$ $= 3.5$ at. units (10% lower than the calculated free ion value [12]) we calculated these terms, neglecting orbital reduction. It is noted that the Fe^{3+} ion is actually larger [15] than the Cr^{3+} ion on octahedral sites and shows no 4s-like character [16]. The contact term H_c was evaluated by Geschwind and cited in [12] to be roughly -285 kOe. The calculated values of the hyperfine fields given in Table 2 are slightly lower than the observed value (327.4 kOe), indicating that the orbital contribution is underestimated. Since the $\langle r^{-3}\rangle$ effective value is necessarily smaller than the free ion value and the orbital reduction factor k less than unity, this would mean that the ratio λ/B_4^3 must be slightly larger. In order to derive the complete spectrum we evaluated the electric field gradients (EFG) arising from the lattice and the 3d shell [17]. They can be written as

$$e(q_{zz})_{\rm lat} = \left(\frac{\partial^2 V_{\rm axial}}{\partial z^2}\right)_{r=0} = 4\,A_2^0\,,$$

$$e(q_{zz})_{\rm 3d} = e\,\langle L||\,\alpha\,||L\rangle\,\langle r_q^{-3}\rangle\,\langle L\,(L+1) - 3\,L_z^2\rangle\,.$$

Assuming for the lattice EFG an antishielding factor γ_∞ of -9 as for Fe^{3+} [18] and $\langle r_q^{-3}\rangle = 3.5$ at. units ($\langle r_q^{-3}\rangle$ equals $\langle r^{-3}\rangle$ approximatively) the total quadrupole coupling $P = e_q^2\,Q/4\,I\,(2\,I-1)\,h$ was calculated. The calculated spectrum shown in Fig. 2 is similar to that observed taking into account inhomogeneous broadening.

2.2 Cu^{2+} ($3d^9$, $L = 2$, $S = 1/2$)

The ^{63}Cu, ^{65}Cu resonance spectrum at 4.2 and 2.1 °K is shown in Fig. 3. The signal of this very broad spectrum was too weak to be observed at 77 °K. At 4.2 °K the signal to noise ratio was about five. It should be pointed out that for a ferrimagnetic insulator the nuclear signal generally increases more rapidly than $1/T$ if the anisotropic energy remains nearly the same in this temperature range. Indeed in this case the susceptibility χ'' of the sample which determines the factor Q is expected to decrease with temperature. From the spectrum we deduced an average value for the hyperfine field of about 195 kOe. This value

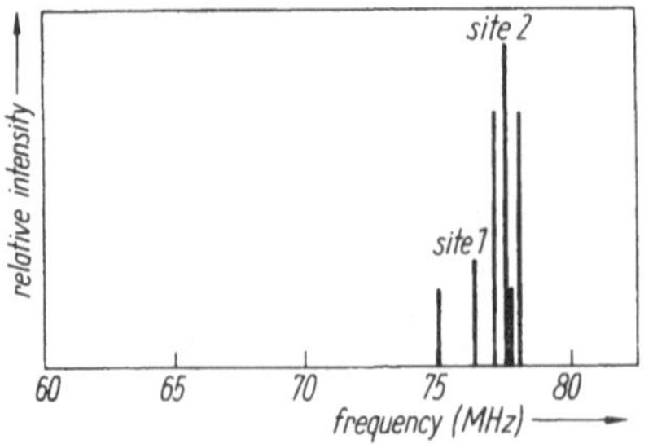

Fig. 2. Calculated spectrum of ^{53}Cr at two inequivalent a-sites

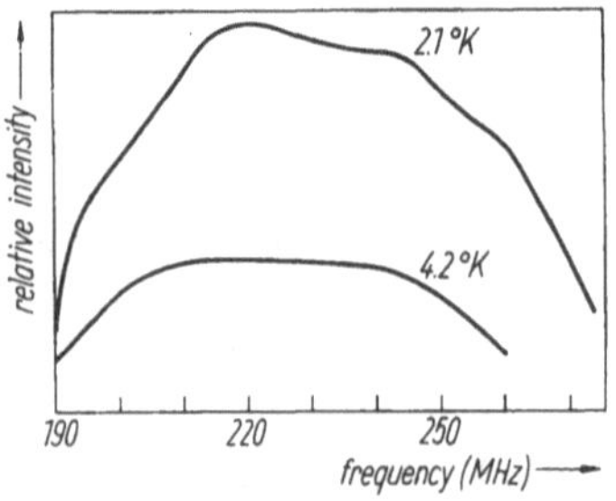

Fig. 3. Resonance spectrum at 4.2 and 2.1 °K of ^{63}Cu and ^{65}Cu nuclei in Cu-doped YIG

Table 3

Calculated hyperfine fields of Cr^{3+} at two inequivalent a-sites

	H_{orb}(kOe)	H_d (kOe)	H_c (kOe)	H_n (kOe)
site 1	−25.5	−7	−285	−317.5
site 2	−40	+2.4	−285	−322.6

Table 4

Calculated hyperfine fields of Cu^{2+} at two inequivalent a-sites

	H_{orb} (kOe)	H_d (kOe)	H_c (kOe) $= 122 \times 2\langle S_i \rangle_i$ $i = x, y, z$	H_n (kOe)
site 1 z-component	293	41	−122	212
site 2 z-component	117	−11.8	− 85.5	19.7
x-component	262	42	− 85.2	218.8

corresponds to a hyperfine constant A that is a weighted mean value for the two isotopes ^{63}Cu, ^{65}Cu of 0.015 cm^{-1} while the reported A values for Cu^{2+} in the octahedral environment in different salts lie between 0.010 and 0.013 cm^{-1} [19].

However, here, unlike in the case of Cr^{3+}, Fe^{3+} ions are replaced by divalent copper and tetravalent germanium ions which induce fluctuations in the crystal fields. We assumed in a first approximation that the local symmetry of the a-site is still C_{3v} neglecting also the possible local Jahn-Teller distortions due to the Cu^{2+} ion in the octahedral site. Further for Cu^{2+} the exchange energy is uncertain and the spin–orbit coupling is no longer a small perturbation with regard to the crystal field so that we could only make an attempt to evaluate the hyperfine interaction. The process then becomes similar to that described for Cr^{3+} in Section 2.1. In YIG Cu^{2+} (0.70 Å) is expected to be less ionic than Fe^{3+} (0.67 Å) due to its larger ionic radius. We used here the experimental value of the cubic term D_q (about 1200 cm^{-1}) for Cu^{2+} in octahedral complex [13]. From the diagram [13] of the splitting of a D-state in a cubic field, D_q and B_4^3 are related by $D_q = (9/10\sqrt{2})\, B_4^3$. This gives a multiplicative factor of 5.7 which is 1.9 times higher than the factor for Cr^{3+}. We assumed that A_0^2 is also 1.9 times higher. The parameters B_l^m were then defined by the following values:

$$B_2^0 = 560\ \text{cm}^{-1}\,,\quad B_4^0 = +76\ \text{cm}^{-1}\,,\quad B_4^3 = +1885\ \text{cm}^{-1}\,.$$

λ was taken as 700 cm^{-1}, a plausible value for Cu^{2+} in the octahedral field [13]. H_{ex} was assumed to be 2 MOe. Using the experimental value $\langle r^{-3} \rangle$ of 6.3 at. units [13], an orbital reduction factor k of 0.7, and a contact term H_c of −122 kOe [20] we deduced the hyperfine fields at two sites. These results explain qualitatively the rather high observed value in YIG. Assuming the same antishielding factor γ_∞ as for Cr^{3+} we calculated the nuclear spectrum (Fig. 4).

In fact the effect of Jahn-Teller distortion and of the simultaneous presence of Cu^{2+} and Ge^{4+} ions in the proximity of Cu^{2+} ion decreases the local symmetry and, therefore, increases the number of magnetically inequivalent sites. The inhomogeneous width arises then from the dispersion of the hyperfine fields and electric field gradients due both to these inequivalent sites and to the fluctua-

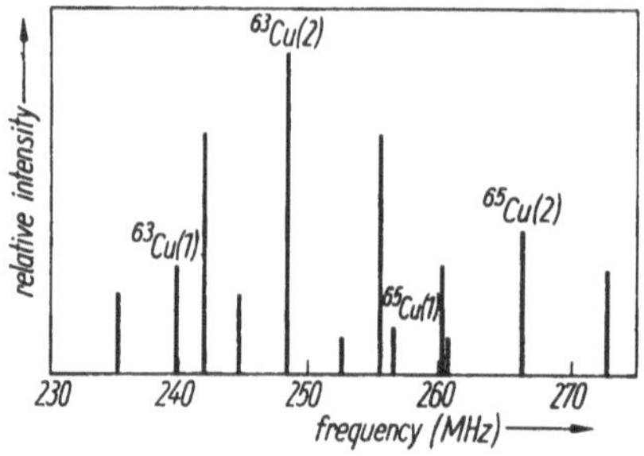

Fig. 4. Calculated spectrum of ^{63}Cu and ^{65}Cu at two hypothetical inequivalent a-sites

tions in crystal fields. Under these circumstances the copper resonance could be expected to show a totally unresolved spectrum.

3. Conclusion

The agreement between the observed spectrum and the semi-empirically calculated one is satisfactory for the Cr^{3+} ion in Cr-doped YIG. This follows from the fact that the orbital ground state of Cr^{3+} in octahedral environment is only slightly perturbed by the spin-orbit coupling and the exchange interaction. As for the Cu^{2+} ion there are some uncertainties about the exact octahedral symmetry, the exchange field, and covalent effect. Nevertheless, from the experimental and estimated results it is considered that Cu^{2+} ions are actually present at the octahedral sites, which is consistent with the conclusion of [2].

Acknowledgement

One of us (M. R.) would like to acknowledge Prof. P. Grivet for the kind hospitality during his stay in Orsay.

References

[1] R. KRISHNAN, phys. stat. sol. **26**, K47 (1968).
[2] R. KRISHNAN, J. appl. Phys. **40**, 1557 (1969).
[3] R. KRISHNAN, J. Crystal Growth **13/14**, 582 (1972).
[4] A. M. PORTIS and A. C. GOSSARD, J. appl. Phys., Suppl. **31**, 205S (1960).
[5] F. HARTMANN-BOUTRON, Thesis, Paris 1963.
[6] W. P. WOLF, Ferrimagnetism, Rep. Progr. Phys. **24**, 212 (1961).
[7] M. WEGER, Phys. Rev. **128**, 1505 (1962).
[8] LE DANG KHOI, Internat. Conf. Magnetism, Grenoble, 1970; J. Physique **32**, C1-1146, (1971).
[9] R. L. STREEVER and P. J. CAPLAN, Phys. Letters (Netherlands) **35A**, 8 (1971).
[10] H. NAGAI, T. HIHARA, and E. HIRAHARA, J. Phys. Soc. Japan **29**, 622 (1970).
[11] T. HIHARA and Y. KOI, J. Phys. Soc. Japan **29**, **343** (1970).
[12] A. J. FREEMAN and R. E. WATSON, in: Magnetism, Vol. II A, Eds. G. T. RADO and H. SUHL, Academic Press, 1963.
[13] A. ABRAGAM and B. BLEANEY, Electron Paramagnetic Resonance of Transition Ions, Chap. 7, Clarendon Press, Oxford 1970.
[14] J. M. BAKER and F. I. B. WILLIAMS, Proc. Phys. Soc. **78**, 1340 (1961).
[15] S. GELLER, J. appl. Phys., Suppl. **31**, 30S (1960).
[16] L. R. WALKER, G. K. WERTHEIM, and V. JACCARINO, Phys. Rev. Letters **6**, 98 (1961).
[17] B. BLEANEY, Hyperfine Interactions, Academic Press, 1967.
[18] R. M. STERNHEIMER, Phys. Rev. **146**, 140 (1966).
[19] G. E. PAKE, Paramagnetic Resonance, W. A. Benjamin, Inc., New York 1962.
[20] A. ABRAGAM, J. HOROWITZ, and M. H. L. PRYCE, Proc. Roy. Soc. **A230**, 169 (1955).

(Received June 9, 1972)

phys. stat. sol. (a) **12**, 575 (1972)

Subject classification: 3; 22.9

Department of Physics, Thermionic Emission Laboratory, The Pennsylvania State University, University Park

Diffusion Zone Process
A New Method for Growing Crystals

By

A. H. MADJID, A. R. VAALA, J. PEDULLA, JR., and W. F. ANDERSON, JR.

The diffusion zone process, which is a new method for growing crystals from solutions, is described. This method essentially consists in the preparation of a solution of the solute from which crystals are to be grown in a suitable solvent and in the diffusion of a second solvent (in which solute is less soluble than in the first mentioned solvent), into the original solution through a semipermeable membrane. Aqueous, organic solvent, and melted salt solutions may be used for the purpose. The method is versatile and often leads to results where ordinary methods fail to produce crystals of appreciable size. Cases in point are carotenoid pigments and the transition metal complex amines which may be grown to centimeter sizes using the diffusion zone process.

Der Diffusionszonenprozeß, eine neue Methode zur Züchtung von Kristallen aus Lösungen, wird beschrieben. Diese Methode besteht in der Hauptsache in der Vorbereitung einer Lösung der Substanz, von der Großkristalle erzeugt werden sollen, in einem geeigneten ersten Lösungsmittel. In einem weiteren Schritt wird ein zweites Lösungsmittel, in dem die Substanz weniger löslich ist als im ersten, in die ursprüngliche Lösung durch eine halbdurchlässige Membran eindiffundiert, wobei der Beginn der gewünschten Kristallbildung meist von Keimen an der Membran ausgeht. Wässerige, organische und Salzschmelzenlösungen können für den Zweck Verwendung finden. Die Methode ist vielseitig und führt oft zu Resultaten in Fällen, wo gewöhnliche Methoden keine Kristalle von experimentell nutzbarer Größe ergeben. Als Beispiel seien die Karotenoidfarbstoffe und die Übergangsmetallaminkomplexe erwähnt, die sich mittels der Diffusionszonenmethode zu zentimetergroßen Kristallen züchten lassen.

1. Description of Process

We wish to report upon a new method for growing crystals which has recently been developed in our laboratory.

The process requires a minimum amount of equipment and it is applicable to grow crystals of a wide variety of materials. Crystals that are difficult or impossible to grow by conventional methods are often easily grown to large sizes by using the diffusion zone process. The method essentially consists of choosing a solvent and saturating it with the solute from which the crystals are to be grown. This may be accomplished by dissolving the solute in the solvent; by dissolving one or more substances in the solvent that react with each other (or react with the solvent), to yield the required solute; or by any other effective means that produces a saturated solution of the desired solute in the solvent. (This solvent will be referred to as the *loading solvent* when pure, as the *loaded solvent* when saturated with the solute, and as the *over-loaded solvent* when an excess of solute is present.)

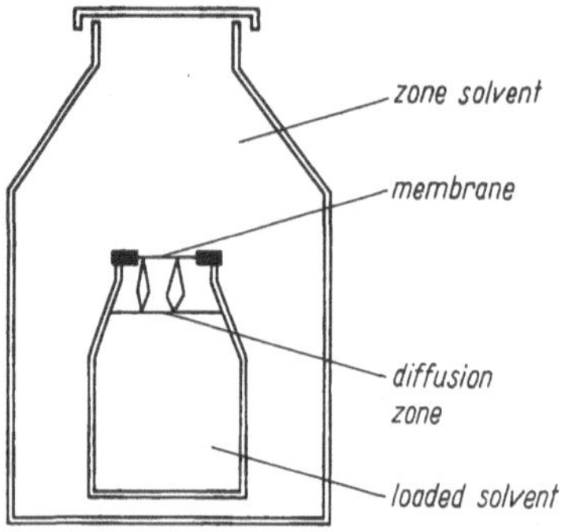

Fig. 1. Schematic of the diffusion zone set-up for growing crystals

In a second step a second solvent, or *zone solvent* is chosen according to the following two major criteria. The solute from which the crystals are to be grown should be less soluble (or insoluble), in the zone solvent; and the zone solvent should preferably, but not necessarily, be miscible with the loading solvent.

In the final step, the loaded or over-loaded solvent is separated from the zone solvent by a semipermeable membrane. A convenient way to accomplish this is shown in Fig. 1. The loaded solvent is placed in a vessel such as a wide-mouth bottle. The bottle is then sealed with the semipermeable membrane and it is then plunged into a larger vessel which is filled with the zone solvent. The choice of membrane is critical for good results and criteria for this choice can only be estimated empirically. Glass frits, porous discs of plastic or ceramic materials, mammalian membranes, cellophane, cheesecloth, felt, and even the purely virtual membranes represented by holes in impermeable discs all were used by us successfully for specific purposes.

Crystal growing by the diffusion zone method is best carried out in a constant temperature ambient that is free of major mechanical vibrations and disturbances. Depending upon the substance, solvent, and membrane used, crystals will start growing from the membrane within hours, days, or weeks. A major factor in the growing process is the establishment of a sharp concentration gradient zone between the two solvents through diffusion of the various constituents into the growing vessel through the semipermeable membrane. This zone, starting at the membrane, slowly moves away from the membrane down the vessel containing the loaded solvent. The moving zone tends to lead the growth of the

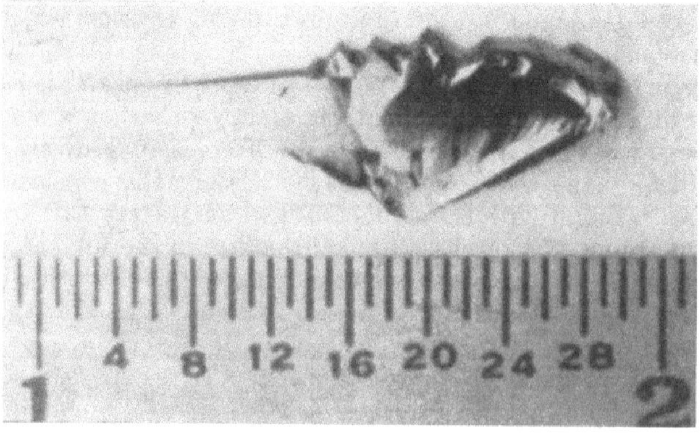

Fig. 2. Large β-carotene crystal grown by the diffusion zone method

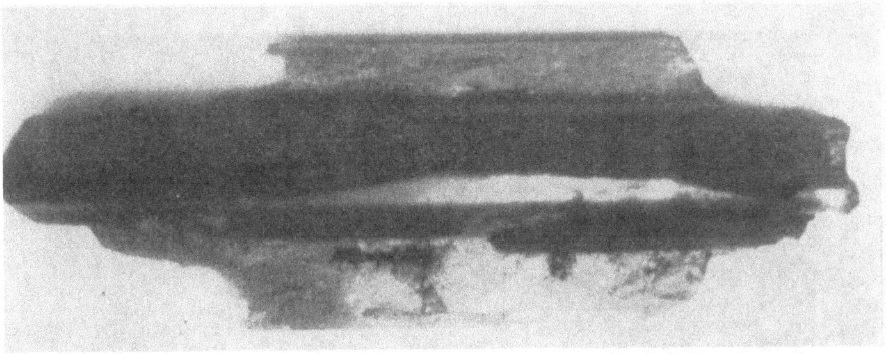

Fig. 3. 9 cm long crystal of copper tetramine sulfate monohydrate grown by the diffusion zone method

crystals from initial nucleation sites on, or in the membrane. The membranes usually provide a sufficient, but not an excessive, number of nucleation sites and this maximizes the harvest of large crystals.

2. Applicability of Process

The diffusion zone process is applicable at low and high temperatures for water and organic solvents and recent tests indicated that this method can also be used for growing refractory crystals using molten salts and fluxes as solvents. This application is particularly promising because the choice of appropriate oxidizing or reducing melts as solvents may offer the possibility of effectively controlling the valence of the ions when growing crystals of transition metal compounds.

The process is evidently potentially very versatile being applicable to the production of a large variety of crystals and the growing of crystals of the C_{40}

Table 1

Crystals successfully grown by the diffusion zone process

crystals grown by diffusion zone method	length	remarks
low temperatures		
β-carotene	up to 3.8 cm	high quality crystals
copper tetramine sulfate	up to 9 cm	high quality crystals
nickel hexamine sulfate	up to 0.6 cm	high quality crystals
nickel hexamine nitrate	up to 0.6 cm	high quality crystals
copper sulfate	up to 2.5 cm	fair quality crystals
nickel sulfate	up to 0.4 cm	high quality crystals
ammonium chloride	—	very poor crystals
selenium	microscopic	high quality crystals
sodium sulfate	up to 1.0 cm	whiskers
unidentified cobalt complex	up to 2.0 cm	whiskers
high temperatures		
unidentified iron oxide	microscopic	possibly $Li_{0.5}Fe_{2.5}O_4$
unidentified boride	up to 1.0 cm	high quality crystals

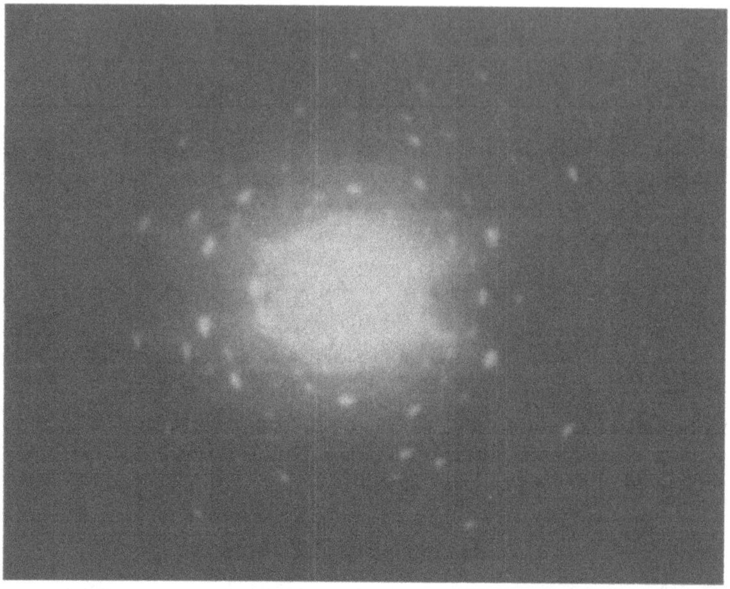

a

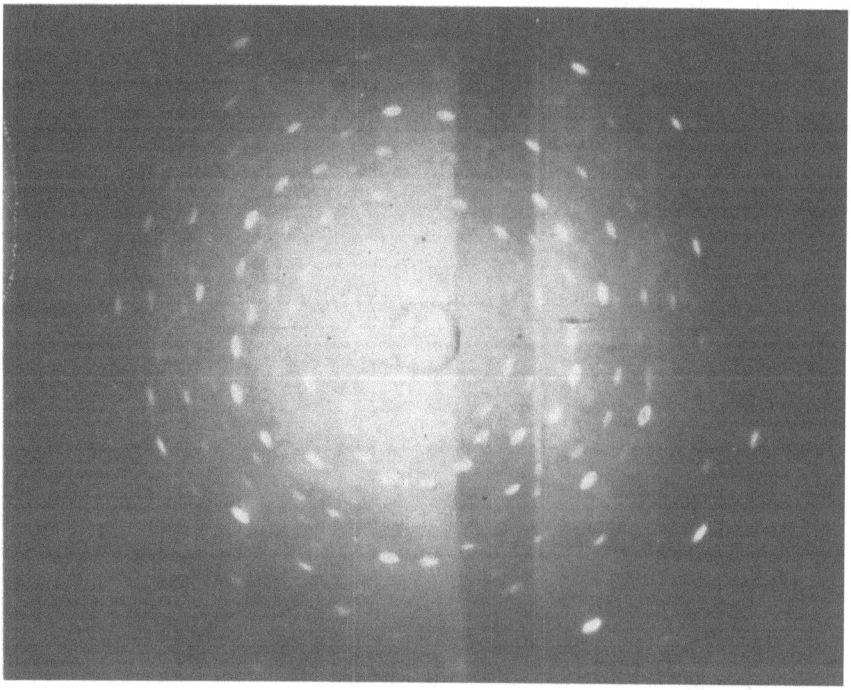

b

Fig. 4. Laue transmission X-ray diffraction pattern of diffusion zone grown crystals. a) β-carotene; b) copper tetramine sulfate monohydrate. The shape and the arrangement of the individual spots clearly indicate that the crystals grown are of a high quality carotenoid pigment β-carotene — that accompanies chlorophyll in all autotrophic plants, and that is a precursor of vitamin A and the visual pigment retinene — may serve as an example of the applicability of the diffusion zone method to the growing of crystals that are not easily grown by conventional methods.

β-carotene has previously been grown by Chapman and Cherry [1] in millimeter sized crystals and in the form of thin platelets. With the diffusion zone process, large crystals of β-carotene can be grown using organic solvents such as benzene and methyl alcohol as the loading and the zone solvents, and mammalian membranes as the semipermeable separators. The largest crystal grown by us measured $3.8 \times 0.7 \times 0.2\ cm^3$ after a growing time of several weeks. But much larger crystals might, no doubt, be produced by extending the growing time beyond this period. A large β-carotene crystal is shown in Fig. 2.

Other crystals that have successfully been grown by the diffusion zone process are shown in Table 1. Particularly gratifying are the results in the case of the unstable and hard to grow crystals of the transition metal complex amines. Fig. 3 shows a good quality crystal of copper tetramine sulfate monohydrate that is over 9 cm long. The quality of the crystals thus produced is in many instances very good. Fig. 4 shows X-ray diffraction patterns of β-carotene and copper tetramine sulfate monohydrate. The shape of the individual spots and the pattern of the transmission Laue pictures clearly indicate that the crystals grown are not only large but that they also are characterized by a high degree of crystalline perfection.

Acknowledgement

The authors wish to express their appreciation to Miss S. Wallace for preparing the manuscript.

References

[1] D. Chapman and R. J. Cherry, Nature (London) **203**, 641 (1964).

(Received February 15, 1972)

phys. stat. sol. (a) **12**, 581 (1972)

Subject classification: 10.1; 21

Institute for Solid State Physics, University of Tokyo, Minato-ku Tokyo

Motion of Screw Dislocations under Stress in a Model B.C.C. Lattice

By

F. MINAMI, E. KURAMOTO, and S. TAKEUCHI

Using a model b.c.c. lattice, the behaviour of the screw dislocation under various types of the shear stress at 0 °K has been investigated by computer simulation. The dislocation core moved in a zigzag way changing its slip plane alternately on two {110} planes so that the resultant slip trace was along the twinning {112} plane for every type of the shear stress. Calculated Peierls stresses were 0.0025, 0.0015, and 0.0027 μ for the shear on {110} plane, on {112} plane in the twinning direction, and on {112} plane in the anti-twinning direction, respectively, where μ is the shear modulus. These results agree semi-quantitatively with the experimental results on refractory b.c.c. metals at low temperatures, which indicates strongly that the plastic anisotropy in b.c.c. metals is the result of the characteristic behaviour of the screw dislocations in overcoming the Peierls potential.

Mit einem k.r.z. Modellgitter wurde die Bewegung von Schraubenversetzungen unter verschiedenen Schubspannungsarten bei 0 °K durch Rechnersimulation untersucht. Der Versetzungskern bewegt sich in Zickzackform abwechselnd auf zwei {110}-Ebenen; deshalb ist die resultierende Gleitspur die {112}-Zwillingsebene für alle Schubspannungsarten. Die berechneten Peierlsspannungen für die Scherungen auf {110}-Ebenen, auf {112}-Ebenen in Zwillingsrichtung und auf {112}-Ebenen in Gegenzwillingsrichtung betragen 0,0025, 0,0015 bzw. 0,0027 μ, wobei μ der Schubmodul ist. Diese Ergebnisse stimmen halbquantitativ mit den experimentellen Ergebnissen an hochschmelzenden k.r.z. Metallen bei tiefen Temperaturen überein. Dies zeigt, daß die plastische Anisotropie von k.r.z. Metallen vom charakteristischen Verhalten der Schraubenversetzungen beim Überschreiten des Peierlspotentials herrührt.

1. Introduction

For the purpose of understanding the plastic behaviour characteristic of b.c.c. metals [1], a number of computer works have been performed on the atomic configuration of the screw dislocation core in a b.c.c. lattice based upon assumed interatomic potentials [2 to 8]. Among them, Suzuki calculated the two-dimensional Peierls potential map for the motion of the screw dislocation using a simple interaction potential between the two neighbouring [111] atomic rows, obtaining a Peierls stress about an order of magnitude larger than that estimated experimentally [4]. He attributed the discrepancy between the calculation and the experiment to the contribution of the zero-point energy of the screw dislocation to overcome the Peierls potential [4, 9].

Recently, Heinrich and Schellenberger calculated non-equilibrium configuration and energy of a screw dislocation in a b.c.c. lattice using guide parameters representing the position of the dislocation [10]. They obtained a selection rule for the translation of the screw dislocation from one equilibrium position to the other. Basinski et al. computed the behaviour of the screw dislocation core under the shear stress for the b.c.c. and h.c.p. phases of a model sodium lattice [11], using an interatomic potential determined on a fundamental basis [12]. According to their results, dislocation movement in b.c.c. lattice takes place by unit translation on {110} planes with the selection rule that two consecutive translations cannot take place on the same {110} plane. The Peierls stress was calculated to be 0.0105 to 0.021 μ, where μ is the shear modulus, which is larger than the experimentally estimated values of the usual b.c.c. metals by more than a factor of two. On the other hand, independently of the above calculations two of the present authors (S. T. and E. K.) and Suzuki proposed a model of dislocation motion with selection rules concerning the conservation or alteration of the type of the core configuration, in an attempt to explain the slip behaviours of tantalum single crystals at liquid helium temperature [13].

In this paper, using a model b.c.c. lattice the behaviour of a screw dislocation under stress at absolute zero temperature has been investigated by computer simulation similar to the work of Basinski et al. [11]. The results on the orientation dependence of the Peierls stress and the traces of dislocation motion have shown fairly good agreement, in contrast to the results of Basinski et al., with the experimental results on Ta deformed at liquid helium temperature [13], and Mo [14] and W [15] at low temperatures.

2. Computation Procedures

The model b.c.c. lattice used was the same as that proposed by Suzuki, where the atomic configuration around a screw dislocation was determined by the interaction potential between the nearest atomic rows parallel to the Burgers vector. Taking the z-axis parallel to the Burgers vector, the interaction potential per unit length between i-th and j-th atomic rows, Φ_{ij}, may have a maximum when the coordinates of the atomic positions of the two rows, z_i and z_j, are equal and have the periodicity b, where b is the strength of the Burgers vector. Thus, Φ_{ij} can be expressed by a Fourier series as

$$\Phi_{ij} = \sum_{k=1}^{\infty} A_k \cos \frac{2\,k\,\pi\,(z_i - z_j)}{b}\,. \tag{1}$$

To a first approximation, only the first term will be used in the present calculation. The amplitude A_1 is related to the shear modulus μ by the equation $A_1 = \mu\,b^2/(2\sqrt{3}\,\pi^2)$. Then

$$\Phi_{ij} = \frac{\mu\,b^2}{2\sqrt{3}\,\pi^2} \cos \frac{2\,\pi\,(z_i - z_j)}{b}\,. \tag{2}$$

The special feature of employing (2) is that the equilibrium value of z of any atomic row with respect to the surroundings can be calculated exactly by the following equation:

$$z = \frac{b}{2\pi} \tan^{-1} \frac{\sum\limits_k \sin \dfrac{2\,\pi\,z_k}{b}}{\sum\limits_k \cos \dfrac{2\,\pi\,z_k}{b}}\,, \tag{3}$$

where the summation is taken over six nearest atomic rows. For this reason, computation time is much reduced compared with the usual interatomic potential model in which the equilibrium atom position cannot be obtained without iterative method. This model, however, has the defect that the effect of the

Fig. 1. The crystal block used in the present calculation viewed along the Burgers vector. Circles indicate atomic rows

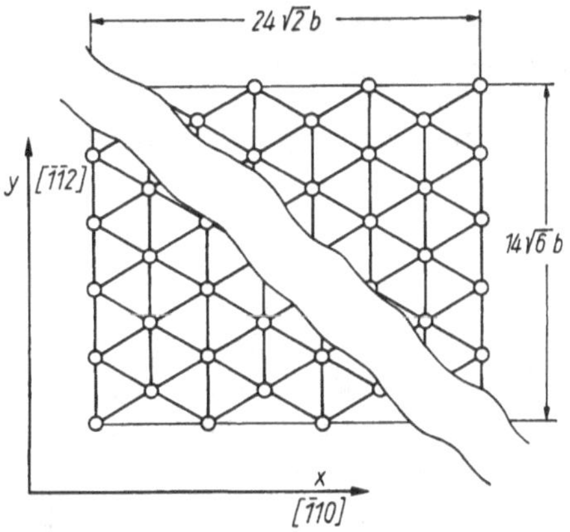

dilatational strain cannot properly be taken into consideration, because the coordinates of the atomic rows in the xy plane are fixed.

The computation process is as follows:

(i) We took a rectangular crystal block with dimensions $24\sqrt{2}\,b \times 14\sqrt{6}\,b$, which consisted of 1570 atomic rows, as shown in Fig. 1.

(ii) At the centre of the block, a screw dislocation was put in based on linear elasticity, and the whole crystal was relaxed with a free boundary condition.

(iii) The forces $\pm f_z$ were acting on each atomic row on the two opposite surfaces of the block giving a definite shear stress τ in the crystal.

(iv) The whole crystal was relaxed keeping the forces at the boundary atomic rows constant. In this process we adopted the initial condition that the crystal was homogeneously sheared by τ/μ for the sake of computation time.

(v) By successively changing the value of τ the behaviour of the dislocation was examined.

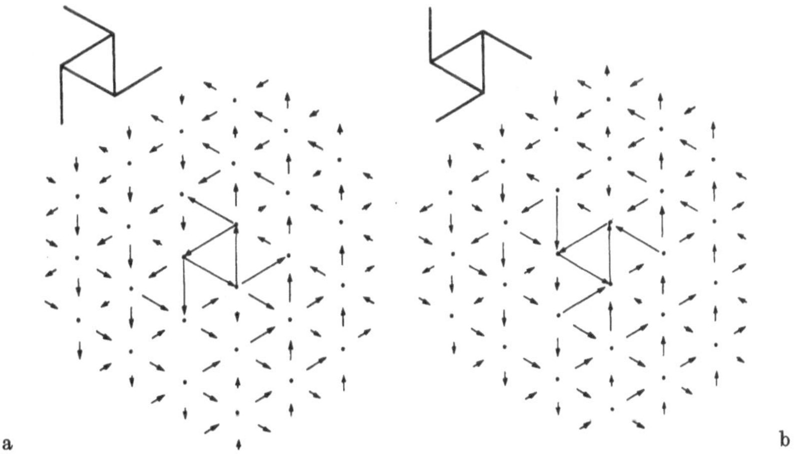

Fig. 2. Two types of the equilibrium core configurations without applied stress viewed along the Burgers vector. Arrows indicate relative displacements of atomic rows in the direction of the Burgers vector between two neighbouring atomic rows with respect to the status of the perfect crystal. The length of the arrow joining the two atomic rows corresponds to one third of the strength of the Burgers vector. Symbolized marks for the two configurations are also presented

In these calculations, the convergence condition was adopted such that the displacements of any atomic rows were less than $1 \times 10^{-4}\, b$ in successive iterations. Shear stresses were applied in two opposite directions in the $(\bar{1}10)$ plane, and in the twinning and anti-twinning directions in the $(\bar{1}\bar{1}2)$ plane.

3. Results

The core configuration with no applied stress is shown in Fig. 2a by the differential displacement representation employed by Vítek et al. [5], where the relative displacements along the z-axis between the nearest atomic rows with respect to the status of the perfect crystal are indicated by arrows. As pointed out by Vítek [16], because of the $\langle 110 \rangle$ diad symmetry of the lattice there should be another equilibrium configuration with equivalent energy, which is obtained by x-axis diad operation of the configuration of Fig. 2a as shown in Fig. 2b. These two configurations will be denoted by C_1 and C_2, and are symbolized by the marks presented in the figure. It should be noted that the equilibrium configurations obtained from various crystal models including the present one are qualitatively the same [5 to 7].

When the stress was applied in the $(\bar{1}10)$ plane so that the force $\tau \boldsymbol{b}$ was acting in the $[\bar{1}\bar{1}2]$ direction on the dislocation, the dislocation with C_1 configuration in site 1 first began to move to site 2 converting the configuration into C_2 at a low stress, and then at a stress of $0.0025\,\mu$ it moved to site 3 reversing the configuration, as shown in Fig. 3. Thus, the 1–2–3 process in the figure is a identity translation. When $\tau\, \boldsymbol{b}$ was acting in the $[11\bar{2}]$ direction, a 1–4–5 . . . movement began to take place at the same stress as above.

When the shear stress was applied in the $(\bar{1}\bar{1}2)$ plane in the twinning direction so that $\tau\, \boldsymbol{b}$ was acting in the $[1\bar{1}0]$ direction, the dislocation core began to move from 1 to 2 and then to 3 at a stress of $0.0015\,\mu$ alternately changing the configuration, as shown in Fig. 4.

The most interesting case was that of the stress applied in the $(\bar{1}\bar{1}2)$ plane in the anti-twinning direction. The dislocation in site 1 with C_1 configuration began to move unexpectedly towards site 2 at a stress of $0.0027\,\mu$ constituting a split configuration as shown in Fig. 5b and subsequently it moved to site 3 and 4 at the same stress, as shown in Fig. 5.

The displacement of the core and the change in configuration for the three types of shear stress are schematically summarized in Fig. 6 using the symbolized

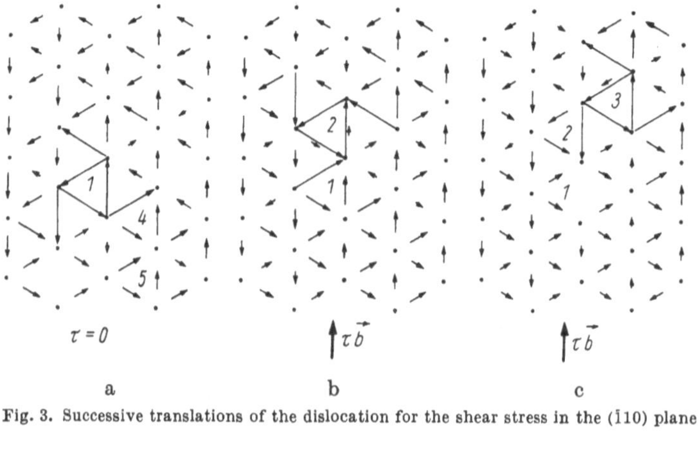

Fig. 3. Successive translations of the dislocation for the shear stress in the $(\bar{1}10)$ plane

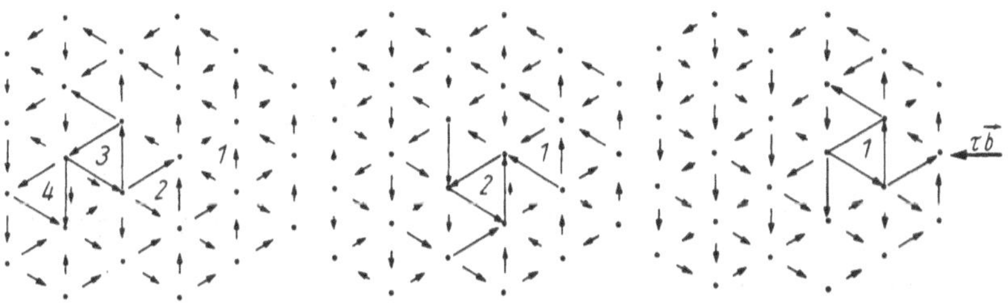

Fig. 4. Successive translations of the dislocation for the shear stress in the ($\bar{1}\bar{1}2$) plane in the twinning direction

Fig. 5. Successive translations of the dislocation for the shear stress in the ($\bar{1}\bar{1}2$) plane in the anti-twinning direction

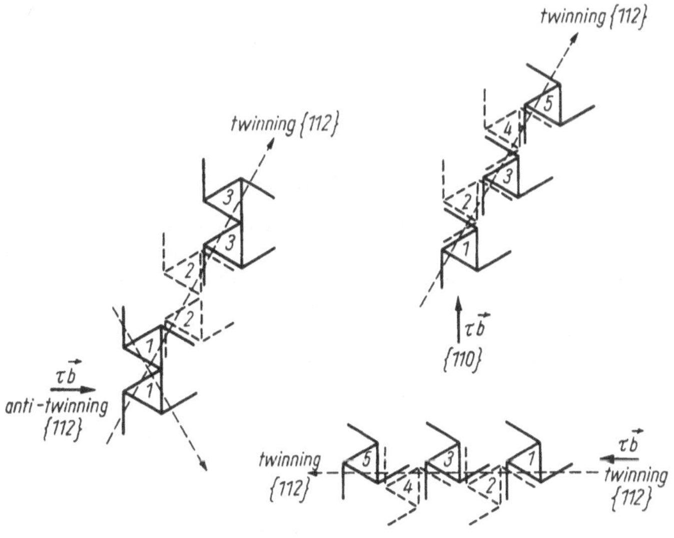

Fig. 6. Schematic representation of successive changes in the core configuration for the three types of shear stresses using symbolized marks from Fig. 2

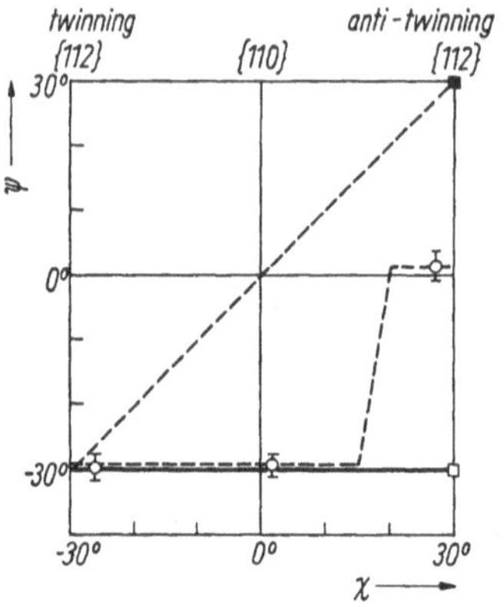

Fig. 7. χ–ψ relation obtained in the present calculation (solid line) and experimental results on Ta at 4.2 °K [13] (dashed curve with measuring points), where χ is the angle between the maximum shear stress plane and {110}, and ψ is that between the slip plane and {110}

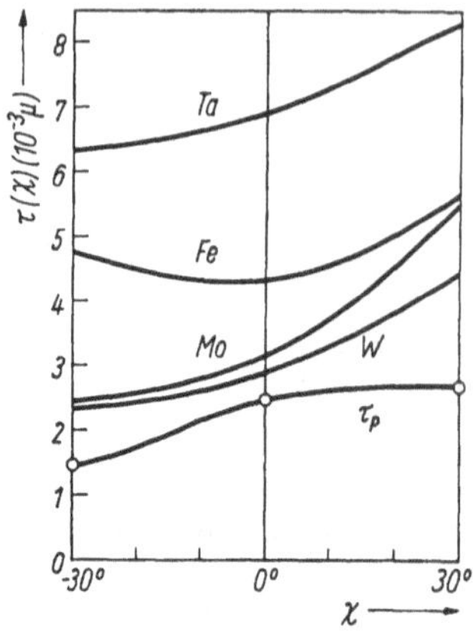

Fig. 8. τ_p–χ relation in the present results and $\tau_0(\chi)$–χ relations for Ta [13], Fe [17], Mo [14], and W [15], where $\tau_0(\chi)$ is the extrapolated shear stress to 0 °K resolved on the maximum shear stress plane. τ_0 values were obtained from τ_y–T curves in the temperature ranges above 4.2, 77, 20, and 77 °K for Ta, Fe, Mo, and W, respectively

marks from Fig. 2. It is found that in all cases the resultant slip traces are along the twinning {112} plane. In the case of anti-twinning {112} shear, the dislocation moves along only one of the two symmetric twinning {112} planes with respect to the direction of the force. This may have been caused by a deviation of the dislocation position from the initial symmetry plane of the crystal. In a real crystal, however, it is expected that the zigzag twinning {112} identity translations will take place, giving rise to a macroscopic anti-twinning {112} slip. In Fig. 7 the conventional χ–ψ relation of the present results is shown together with experimental results on Ta at 4.2 °K [13], where χ is the angle between the maximum shear stress plane and {110} and ψ is that between the slip plane and {110}.

The orientation dependence of the calculated Peierls stress τ_p is plotted in Fig. 8 in the conventional form. In this figure, $\tau_0(\chi)/\mu$–χ relations for Ta [13], Mo [14], W [15], and Fe [17] are also shown, where $\tau_0(\chi)$ is the shear stress extrapolated to 0 °K resolved on the maximum shear stress plane.

4. Discussion

The macroscopic yield stress of b.c.c. metals at low temperatures is considered to be determined by the mobility of screw dislocations, supported by the direct observations of dislocations by many workers. Thus, the yield stresses at 0 °K as extrapolated from τ_y–T relations can be taken to correspond to the Peierls stress of the screw dislocation if the solution hardening effect due to impurities is disregarded. It is found from Fig. 8 that the orientation dependence of τ_p agrees qualitatively with the experiments except for Fe, and moreover the absolute level of τ_p may be regarded as in satisfactory agreement with the observed values especially of VIa metals. The χ–ψ relation also agrees with the experiments on Ta except for the orientation $\chi = 30°$ as is shown in Fig. 7. Thus, it may be concluded that the present results are in semi-quantitative agreement with the plastic behaviour of b.c.c. metals. Consequently, specific orientation dependences of the mechanical properties in b.c.c. metals are considered

Fig. 9. Interaction potential between atomic rows per atomic distance in the present model (solid curve) and that deduced from the interatomic potential of sodium according to Basinski et al. [12] (dashed curve). The origin of the ordinate is taken at the perfect lattice position

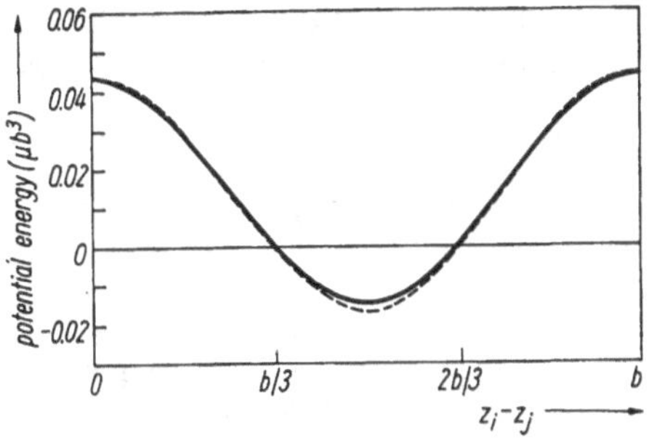

to be the results of the characteristic behaviour of screw dislocations overcoming the Peierls barrier.

The present τ_p values are by an order of magnitude smaller than that of the original calculation by Suzuki [4] in spite of the same crystal model. The discrepancy may result from the difference in the process of calculation. Suzuki calculated the Peierls potential energy only by displacing rigidly the strain field based on the linear elasticity without allowing relaxation of the lattice, and consequently τ_p values much higher than the present results must result.

The results on a model b.c.c. sodium lattice by Basinski et al. [11] are different from the present results not only in the stress level of τ_p values but also in the orientation dependence of τ_p and the slip trace, despite the similar computation process. The τ_p values of Basinski et al. ranged between 0.0105 and 0.021 μ, the maximum value being at $\chi = 0°$ in contrast to $\chi = 30°$ in the present results, and identity translations occurred not only along the twinning {112} plane but also along the anti-twinning {112} plane. The main differences in the model and the computation process between those of Basinski et al. and the present authors are as follows: (i) Difference in the model lattice, i.e., the former used an interatomic potential determined from first principles while the latter used a simple interaction potential between atomic rows. To compare the potentials the interatomic potential used by Basinski et al. was reduced to the interaction potential between the atomic rows and is shown in Fig. 9 together with the present potential. (ii) Difference in the process of shearing of the crystal, i.e., Basinski et al. applied shear stress by shearing the crystal initially by an amount τ/μ and fixed the boundary atoms of the crystal block during the relaxation process, whereas the present work applied a constant shear stress during relaxation without fixing the boundary. (iii) No dilatational relaxation around the core was allowed in the present work while the three-dimensional relaxation was allowed by Basinski et al. (iv) Other possible differences in the computation process, e.g. different convergence conditions. It is not known which of these factors significantly influenced the results, but the difference in the potential does not seem to be significant because the two potential curves in Fig. 9 are in surprisingly good agreement with each other.

Heinrich and Schellenberger showed a selection rule for the type of dislocation configuration and translation site [10]. According to them, the identity translation along {110} plane can take place as well as that along {112} plane contradictory to the present results and those of Basinski et al. It should be noted, however, that Heinrich and Schellenberger treated dislocation motion processes different from those by Basinski et al. and the present authors, i.e., the former treated the thermally activated motion of the dislocation while the latter groups treated the dislocation motion at 0 °K.

5. Conclusion

The orientation dependence of the Peierls stress and the slip trace of screw dislocations in b.c.c. crystals were studied by computer simulation using a model b.c.c. lattice. The results show fairly good agreement with the experimental facts on b.c.c metals at very low temperatures. Thus, the plastic anisotropy in b.c.c. transition metals is concluded to result from the characteristic behaviour of the screw dislocations overcoming the Peierls potential.

Acknowledgements

The authors would like to thank Prof. T. Suzuki and Prof. T. Sugeno of The University of Tokyo, and members of their research groups for useful discussions.

References

[1] J. W. CHRISTIAN, Proc. 2nd Internat. Conf. Strength of Metals and Alloys, Asilomer 1970, Amer. Soc. Metals, Cleveland 1970 (p. 31).
[2] R. CHANG, Phil. Mag. **16**, 1021 (1967).
[3] R. BULLOUGH and R. C. PERRIN, in: Dislocation Dynamics, Ed. A. ROSENFIELD, G. T. HAHN, A. L. BEMENT, and R. I. JAFFEE, McGraw-Hill Publ. Co., New York 1968 (p. 175).
[4] H. SUZUKI, in: Dislocation Dynamics, Ed. A. ROSENFIELD, G. T. HAHN, A. L. BEMENT, and R. I. JAFFEE, McGraw-Hill Publ. Co., New York 1968 (p. 679).
[5] V. VÍTEK, R. C. PERRIN, and D. K. BOWEN, Phil. Mag. **21**, 1049 (1970).
[6] Z. S. BASINSKI, M. S. DUESBERY, and R. TAYLOR, Phil. Mag. **21**, 1201 (1970).
[7] G. DIENER, R. HEINRICH, and W. SCHELLENBERGER, phys. stat. sol. (b) **44**, 403 (1971).
[8] L. L. BOYER and J. R. HARDY, Phil. Mag. **24**, 647 (1971).
[9] H. SUZUKI, Fundamental Aspects of Dislocation Theory, Ed. J. A. SIMMONS, R. DE WITT, and R. BULLOUGH, Nat. Bur. Stand. (U.S.), Spec. Publ. 317, Vol. I, 1970 (p. 253).
[10] R. HEINRICH and W. SCHELLENBERGER, phys. stat. sol. (b) **47**, 81 (1971).
[11] Z. S. BASINSKI, M. S. DUESBERY, and R. TAYLOR, Canad. J. Phys. **49**, 2160 (1971).
[12] Z. S. BASINSKI, M. S. DUESBERY, A. P. POGANY, and R. TAYLOR, Canad. J. Phys. **48**, 1480 (1970).
[13] S. TAKEUCHI, E. KURAMOTO, and T. SUZUKI, Acta metall. **20**, 909 (1972).
[14] S. S. LAU and J. E. DORN, phys. stat. sol. (a) **2**, 825 (1970).
[15] A. S. ARGON and S. R. MALOOF, Acta metall. **14**, 1449 (1966).
[16] V. VÍTEK, Proc. 2nd Internat. Conf. Strength of Metals and Alloys, Asilomer 1970, Amer. Soc. Metals, Cleveland 1970 (p. 389).
[17] W. A. SPITZIG and A. S. KEH, Acta metall. **18**, 611 (1970).

(Received May 5, 1972)

phys. stat. sol. (a) **12**, 589 (1972)

Subject classification: 15; 21.1

Physics Department, Harcourt Butler Technological Institute, Kanpur

Thermo-Electric Behaviour of Hydrogenated Nickel

By

Amar Nath Nigam and Vinod Saran Garg

Nickel wire subjected to hydrogen diffusion by electrolysis loses hydrogen when left to itself. The changing phase shows a time variation in its thermo-electric power. The phenomenon has been studied with nickel wires annealed at different temperatures. The thermo-e.m.f. variations have been used to derive the shape of the 3d band of nickel and certain new structures in the band shape have been detected.

Nickeldraht, der einer Wasserstoffdiffusion durch Elektrolyse ausgesetzt ist, verliert Wasserstoff, wenn er sich selbst überlassen wird. Die Änderungsphase zeigt eine Zeitvariation in ihrer Thermospannung. Diese Erscheinung wurde an Nickeldraht, getempert bei verschiedenen Temperaturen, untersucht. Die Thermo-EMK-Änderungen wurden benutzt, um die Form des 3d-Bandes von Nickel abzuleiten, wobei gewisse neue Strukturen im Bandverlauf gefunden wurden.

1. Introduction

The change in the physical properties of metals on hydrogen diffusion is a well known fact [1]. The thermo-electric effects are important from the point of view of solid-state theory. Harrison [2] investigated Fe–Cu thermocouples after heating them in hydrogen atmosphere and found that the thermo-e.m.f. is lowered with respect to the same couple preheated in vacuum. Franzini and Gazzaniga [3] employed electrolysis as a mean to diffuse hydrogen into wires of Fe, Ni, Pt, and Pd, the latter being used as cathodes. Couples were made of hydrogenated and unhydrogenated wires of the same metal by soldering. The results are similar in nature to those of Harrison [2].

The present paper describes the thermo-electric behaviour of hydrogenated nickel. Hydrogen was diffused by electrolysis following Janko's method [4] which enables the introduction of large amounts of hydrogen temporarily in the metal. When left to itself the wire starts losing hydrogen. The time variation of thermo-e.m.f. of this changing phase was recorded. It is found possible to interpret the changed phase in terms of the filling of the 3d band of nickel by the electrons contributed by hydrogen. The subsequent escape of hydrogen empties the newly filled part of the 3d band. Thermo-e.m.f. variation therefore gives information about the shape of the unfilled part of the 3d band. Certain structures in this part of the band have been detected. It was further found that annealing has a considerable influence on the thermo-e.m.f. of the metal and also on hydrogen diffusion.

2. Experimental

2.1 Annealing

Nickel wire samples (diameter 0.25 mm) were enclosed in glass or quartz capsules, evacuated to 10^{-5} Torr pressure and then sealed. Each capsule was separately heated to the desired temperature in a furnace, the temperature maintained for three hours and then the sample was allowed to be furnace-cooled. After annealing the samples were chemically polished by a solution recommended by deLang [5].

2.2 *Electrolysis*

The solution was acidulated water containing traces of thio-urea. Nickel wire sample was attached to the platinum cathode. The anode was also a platinum wire. Thio-urea enhances hydrogen diffusion. Each sample was subjected to hydrogen diffusion for 216 h at 30 mA, 15 V.

Effects of thio-urea on hydrogen diffusion were described by Janko [4] and subsequently studied by other investigators in connection with the magnetic properties [6] and X-ray K-absorption edge and its fine structure [7, 8]. Both properties show only a temporary change on hydrogen diffusion. Magnetic properties are regained after 7 to 8 h if the sample is left to itself. The X-ray absorption edge shape is regained after half an hour.

2.3 *Thermo-e.m.f. measurements*

A ten wire potentiometer was used to measure the thermo-e.m.f. by the method described by Starling [9]. Ni–Cu couples were made by mechanical twisting of the ends of the two wires and the junction dipped in mercury to ensure good electrical contact. The hot junction was at room temperature while the cold maintained at the temperature of melting ice. A thermometer dipped in mercury recorded the temperature of the junction. If a fresh couple was dipped in pre-cooled mercury, the null point showed constancy after only one minute. This is, therefore, the time needed by the cold junction to attain the steady state. Each time first a measurement was carried out with an unannealed pure nickel wire. This being complete, the electrolytic diffusion of hydrogen that had been going on for 216 h was switched off and simultaneously a stopwatch started. The wire subjected to hydrogen diffusion was taken out at once, dried by filter paper and put in position for thermo-electric measurements. The changes in the null point position were measured after allowing the cold junction to attain the steady state i.e. after one minute. After each minute the null point position was measured and the observations continued till the sample approached a steady state.

The least count of our ten wire potentiometer was 6 μV/cm. The null point remained insensitive within 1/2 cm. Thus a safe estimate of the accuracy in the measurement is ± 3 μV. For calculating the thermo-power this has to be divided by the room temperature, 30 °C, the cold junction being at 0 °C. The thermo-power is therefore correct up to ± 0.1 μV/deg. The changes we describe below are much greater than this in magnitude.

The absolute value of thermo-electric power of copper is $+3$ μV/deg. This was subtracted from the measured thermo-power to get the thermo-power of nickel.

3. Results and Discussion

The variation in thermo-power with time for annealed samples is shown in Fig. 1. To discuss these curves we recall that according to the band theory of metals, the thermo-power is given by [10]

$$\frac{\mathrm{d}\theta}{\mathrm{d}t} = -\frac{\sigma}{\varepsilon}, \tag{1}$$

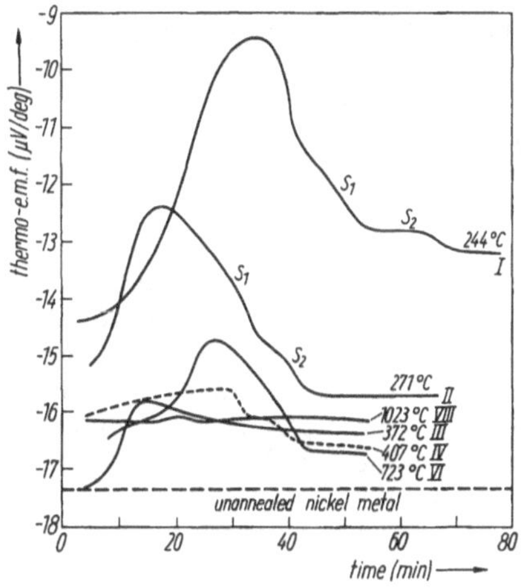

Fig. 1. Time variation in thermo-electric power of hydrogenated nickel wires pre-annealed at different temperatures. Annealing temperatures are mentioned at the curves. Sample V was pre-annealed at 937 °C, on hydrogenation it became too brittle to be handled for any further experiment

where

$$\sigma = \frac{1}{3}\pi^2 k^2 T \left(\frac{\partial}{\partial E} \ln \sum_i n_i \tau_i v_i^2\right)_{E=\zeta}. \tag{2}$$

ζ is the Fermi energy, ε is the electronic charge, τ the relaxation time for conduction electrons due to lattice vibrations and is proportional to $E^{3/2}$, and n_i is the number of electrons in the i-th band proportional to $E^{1/2}$. In nickel 3d and 4s electrons form the conduction band. Equation (2) can be expanded as

$$\sigma = \frac{1}{3}\pi^2 k^2 T \left[\frac{\partial}{\partial E} \ln (n_s \tau_s v_s^2 + n_d \tau_d v_d^2)\right]_{E=\zeta} =$$

$$= \frac{1}{3}\pi^2 k^2 T \frac{1}{n_s \tau_s v_s^2 + n_d \tau_d v_d^2} \left[n_s \frac{\partial}{\partial E}(\tau_s v_s^2) + \tau_s v_s^2 \frac{\partial n_s}{\partial E} + \right.$$

$$\left. + n_d \frac{\partial}{\partial E}(\tau_d v_d^2) + \tau_d v_d^2 \frac{\partial n_d}{\partial E}\right]_{E=\zeta}.$$

With $\tau \sim E^{3/2}$ and $n \sim E^{1/2}$ and omitting the constants we get

$$\sigma = \frac{\pi^2 k^2 T}{\zeta} + \frac{1}{3} \frac{\pi^2 k^2 T}{\zeta^{1/2}} \left(\frac{\partial n_s}{\partial E} + \frac{\partial n_d}{\partial E}\right). \tag{3}$$

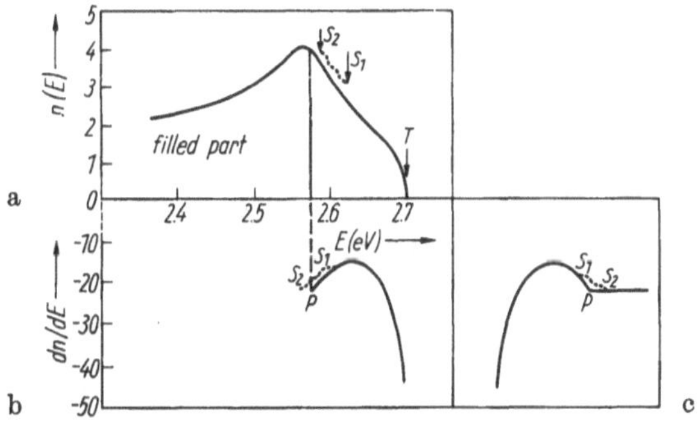

Fig. 2. a) $n(E)$ versus E curve for the d-band in nickel as drawn by Flechter [11]. b) dn/dE vs. E curve as derived from Fig. 2a. c) mirror image of Fig. 2b about the vertical axis

We can therefore regard σ to be proportional to $\partial n_s/\partial E + \partial n_d/\partial E$. The $n(E)$ vs. E curve for the 3d band of nickel as calculated by Flechter [11] is shown in Fig. 2a. The dn/dE vs. E curve was calculated by numerical differentiation and is shown in Fig. 2b. This curve is shown only for the unfilled part of the band which in Fig. 2a lies on the right-hand side of the vertical line. The diffused hydrogen contributes its electron to the unfilled part of the 3d band [8]. When hydrogen starts leaving the system the 3d band supposed to be filled will begin emptying from the top (point T in Fig. 2a). This process will continue till hydrogen vacates the system and the original state is restored. Thereafter thermopower will remain constant with time. Since the emptying process of the band is just reverse of the filling in we draw a mirror image of Fig. 2b in 2c. This is drawn up to the relevant point P and thereafter the constancy of thermopower is indicated by the horizontal line. The shape of the curve in Fig. 2c is similar to the curves shown in Fig. 1, but for the fact that these experimental curves show structures S_1, S_2. The corresponding structures in the theoretical curve of Fig. 2a are shown by the dotted line separated from the main curve but above it at the expected position.

A comparison of the curves in Fig. 1 and 2c is justified only if the emptying process of the temporarily filled part of the 3d band is linear in time. A trial showed that only curve 1 of Fig. 1 is the closest fit with that in Fig. 2c. Therefore, the emptying process is linear in time only for the sample pre-annealed at 244 °C but not for others. The exact reason for this behaviour is not clear but might be due to the different dislocation densities created by annealings at different temperatures in different samples.

Fig. 3 shows the behaviour of an unannealed hydrogenated sample. The curve is quite different from those in Fig. 1. One looks like the mirror image of the other with respect to the time axis. The sign of the charge carriers seems to have been reversed. Again, the exact reason for this behaviour is not clear.

Fig. 4 shows the steady state thermo-electric power of annealed wires both before and after hydrogen diffusion, the steady state being ensured a few months after the hydrogen effusion was over. Wires annealed at 407 °C show the maximum change in thermo-electric power before and after hydrogen diffusion. The X-ray orientation factors of this sample show a maximum change on hydrogen diffusion.

Thermo-electric studies with 8 μm nickel foils were also carried out. When saturated with hydrogen, the foil showed a change in the sign of the thermo-e.m.f. However due to extreme embrittlement prolonged observations with foils could not be undertaken and the phenomenon is still under study.

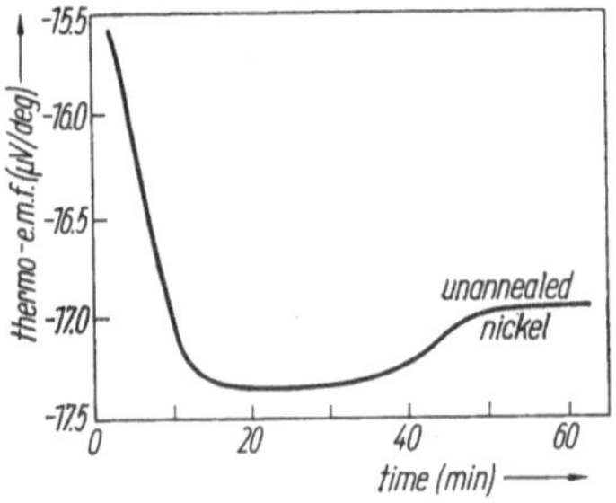

Fig. 3. Time variation in thermo-electric power of unannealed hydrogenated nickel wire

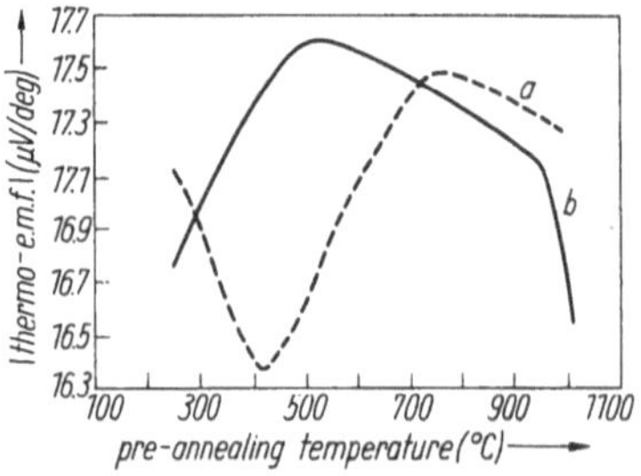

Fig. 4. Steady state thermo-electric powers of wires pre-annealed at different temperatures. Curve (a) represents hydrogenated and (b) unhydrogenated wires

Acknowledgements

The authors thank Mr. R. N. Kapoor for lending his D'Arsonval galvanometer, Dr. T. M. Srinivasan for vacuum facilities he provided to seal the specimen in glass and quartz capsules and to Mrs. Skillen for translating an Italian paper [3] into English. One of us (VSG) is grateful to the University Grants Commission for the award of a research fellowship.

References

[1] D. P. Smith, Hydrogen in Metals, University of Chicago Press, 1958.
[2] E. P. Harrison, Phil. Mag. **3**, 177 (1902).
[3] T. Franzini and G. Gazzaniga, Reale Ist. lombardo, Sci. Lettere, Rend. (Milan) **66**, 115 (1933).
[4] A. Janko, Naturwissenschaften **10**, 255 (1960).
[5] J. J. de Lang, Metalen **9**, 2 (1954).
[6] H. Bauer and E. Schmidt-Bauer, Z. Phys. **164**, 367 (1961).
[7] A. N. Nigam, Current Sci. **33**, 76 (1964).
[8] A. Faessler and R. Schmidt, Z. Phys. **190**, 10 (1966).
[9] S. G. Starling, Electricity and Magnetism, Chap. 7, Longmans Green & Co., London 1941 (p. 219).
[10] A. H. Wilson, Theory of Metals, Cambridge University Press, Cambridge 1958 (p. 205).
[11] G. C. Flechter, Proc. Phys. Soc. **A65**, 192 (1952).

(Received May 12, 1972)

phys. stat. sol. (a) **12**, 595 (1972)

Subject classification: 1.2 and 14.1; 21; 21.1.1

The Research Institute for Iron, Steel, and Other Metals, Tôhoku University, Sendai

Thermo-Hysteresis Phenomenon of the Electrical Resistivity in Fe_2Ti Suggesting Its Martensitic Transformation

By

K. Ikeda, T. Nakamichi, and M. Yamamoto

Some indirect evidence is found of the occurrence of a martensitic transformation below room temperature in the Laves phase compounds in the Fe–Ti alloy system. The detailed measurements of the electrical resistivity from 4.2 to 1200 °K show a large thermo-hysteresis, suggesting that the low temperature transformation develops just below room temperature and the high temperature one begins at about 560 °K in the stoichiometric compound. Although the X-ray powder photographs show no clear evidence of a new phase, it may be interpreted that the new phase consists of a very small volume with a fine structure.

Ein indirekter Hinweis über das Auftreten der Martensit-Transformation bei Raumtemperatur in den Laves-Phasen-Verbindungen des Fe–Ti-Legierungssystems wurde gefunden. Die ausführlichen Messungen des elektrischen Widerstands von 4,2 bis 1200 °K zeigen eine große Thermo-Hysteresis, die darauf hinweist, daß die Tieftemperatur-Transformation sich gerade unterhalb Raumtemperatur entwickelt und die Hochtemperatur-Transformation bei 560 °K in der stöchiometrischen Probe beginnt. Obwohl die Röntgenstrahl-Pulverphotographien keinen klaren Beweis einer neuen Phase zeigen, kann gefolgert werden, daß die neue Phase aus einem sehr kleinen Volumen mit feiner Struktur besteht.

1. Introduction

The intermetallic compound Fe_2Ti is the Laves phase compound with $MgZn_2$-type crystal structure (see Fig. 1), which was first reported by Witte and Wallbaum [1]. Murakami et al. [2] have reported, based on X-ray measurements, that its single phase appears in a rather wide composition range (26.9 to 35.4 at% Ti). On the other hand, Nakamichi [3] has reported, from magnetic measurements that it ranges from 30.5 to about 35.2 at% Ti.

Concerning the magnetic properties of this compound, Kocher and Brown [4] have first reported based on their neutron diffraction and Mössbauer effect measurements that it is ferromagnetic even at room temperature and has magnetic moments of (0.10 ± 0.03) and (0.20 ± 0.02) μ_B per iron atom in the 2a and 6h crystal sites, respectively, with a hyperfine field of some 5 kOe. Piegger and Craig [5], however, have concluded from their static magnetic measurements that it is strongly paramagnetic with $\chi_{mol} = 12.7 \times 10^{-3}$ e.m.u./mol. On the other hand, Wallace [6] has pointed out that it is antiferromagnetic with the Néel temperature of 297 °K and obtained fairly good Mössbauer absorption lines of ^{57}Fe with a hyperfine field of 71 kOe as well as some quadrupole effect at liquid air temperature. Under such situations, Nakamichi [3] has reported from the magnetization measurements that its magnetism changes abruptly at about stoichiometric composition from ferromagnetism in the Fe-rich composition range to antiferromagnetism in the Ti-rich one. He has also suggested that excess iron atoms which are assumed to occupy the normal titanium atomic sites substitutionally, seem to play an important role in the occurrence of ferromagnet-

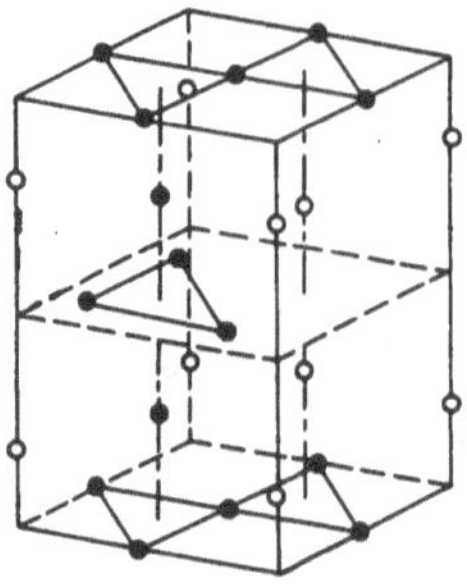

Fig. 1. Crystal structure of the $MgZn_2$-type Laves phase compound Fe_2Ti ● Fe; ○ Ti

ism in the Fe-rich composition range. Mössbauer [7] and magnetic measurements at high fields [8] proved recently that fact in greater detail.

In connection with magnetic properties of this compound, it was expected that the measurement of electrical resistivity might give some useful information for this compound, such as the abrupt change in the temperature dependence of resistivity at the stoichiometric composition. However, the large abnormality in the temperature variation of resistivity and, moreover, its thermo-hysteresis behaviour which were not first expected were discovered [9, 10]. These facts suggested the possibility of a new phase at lower temperatures. This paper reports on these facts in detail.

2. Specimens and Experimental Procedures

Raw materials used for alloying are electrolytic iron and sponge titanium, the compositions of which are tabulated in Table 1. Nine batton ingots, with compositions from 28.4 to 35.6 at % Ti, were prepared by arc melting the raw materials three times in argon gas atmosphere at a pressure of about 700 Torr. Then, they were cooled as slowly as possible in order to avoid any cracking, because the Laves phase compounds are brittle. Since the surfaces of the batton ingots were slightly oxidized, the oxide films were removed by etching with 1 HF + 1 HNO_3 + 2 glycerin. After that, they were homogenized at 1000 °C in a 10^{-7} Torr vacuum for 48 h to avoid oxidation [11]. From these batton ingots, specimens for the resistivity measurements were cut to about $2.0 \times 2.0 \times \times 18\ mm^3$ with a diamond cutter, those for magnetic measurement are obtained as small block of about 500 mg by crushing, and those for the X-ray diffraction were further crushed into powders with an agate mortar. These specimens were

Table 1

Purities and impurities of raw materials

material	purity (%)	impurity (%)
Fe	99.9	C 0.011, S 0.010, Si 0.006, Mn 0.001, P 0.005, Ni 0.025, Cr trace, Mo 0.001, W trace, V 0.001, N 0.004, O 0.073, Co 0.010, Ti 0.003
Ti	>99.5	Fe 0.07, Cl 0.04, Mn 0.003, Mg 0.03, Si 0.01, N 0.008, C 0.01, H 0.002

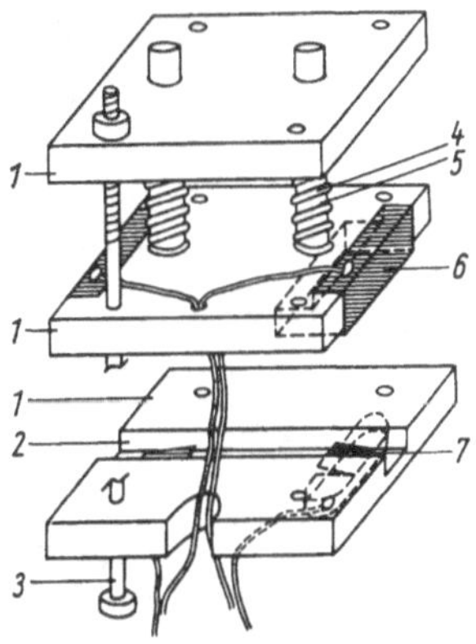

Fig. 2. The special tool for the disposition of the specimens with four terminals, by which the electrical resistivity was measured below room temperature. (1) insulated plates; (2) specimen set groove; (3) holding bolt; (4) spring; (5) spring holder; (6) current terminal; (7) voltage terminal

again heat-treated under the same conditions as before for 10 h to remove strains induced during cutting and crushing. The initial balanced compositions and the final chemically analysed ones of the ingots are shown in Table 2.

The electrical resistivity was measured by a conventional potentiometric method in the temperature range from 4.2 to about 1150 °K. Since the specimens were brittle and easily oxidizable, we made a special tool for the disposition of specimens (see Fig. 2). For measurements below room temperature, twospiral springs made specimen contact on one side with knife-edge-type voltage terminals which were made of stainless steel, and simultaneously on the other side with the plate-type current terminals which were made of copper. At higher temperatures, both the voltage and current terminals were made of 0.3 mm diameter platinum wire. The plate fixing these terminals was made of an acrylic plate below room temperature and of a steatitic plate above room temperature. The apparatus was maintained in a 10^{-7} Torr vacuum for measurements.

Magnetic measurements were made from 4.2 °K to room temperature using a Bozorth-type magnetometer [12, 13]. The temperatures of the specimens were measured with a 2% Co–Au and Cu thermocouple at lower temperatures, and with a 13% Rh–Pt and Pt thermocouple at higher temperatures. X-ray powder photographs were taken using FeK_α radiation emitted under the voltage of 30 kV, current of 7 mA and filtered by a manganese filter of 0.2 mm thickness.

Table 2
Initial and chemically analyzed composition of Fe–Ti alloy specimens

specimen no.	balanced compositions (at% Ti)	chemically analyzed compositions (at% Ti)
29	29.0	28.5
30	30.0	30.0
31	31.0	30.9
32	32.0	31.6
33	33.0	32.6
$33^1/_3$	33.33	33.2
34	34.0	33.5
35	35.0	34.6
36	36.0	35.7

3. Experimental Results

3.1 The increment of the electrical resistivity due to cooling below room temperature

The electrical resistivity was measured on nine specimens with compositions ranging from 28.5 to 35.7 at% Ti, in order to examine the temperature variation below room temperature. Fig. 3 shows the cooling and heating curves of the

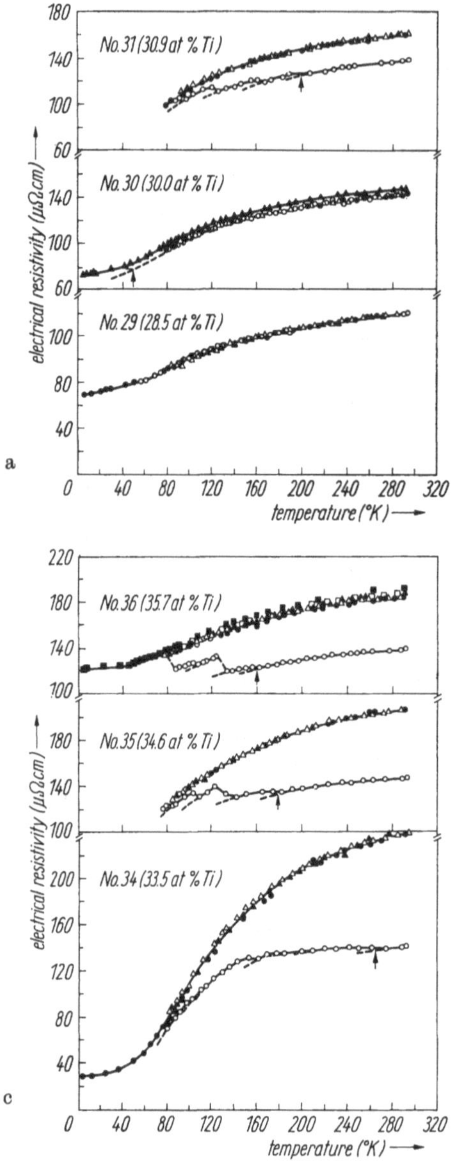

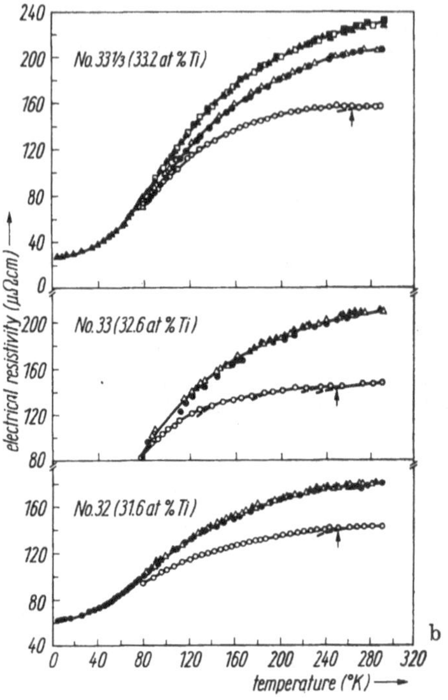

Fig. 3. The cooling and heating curves of the electrical resistivity of the Laves phase compounds, $Fe_{2+x}Ti_{1-x}$. Open symbols represent cooling runs, full symbols heating runs; circles 1st, triangles 2nd; squares 3rd runs

electrical resistivity of these specimens. As can be seen no difference between both curves is found in the specimen containing 28.5 at% Ti and scarcely any in the specimen containing 30.0 at % Ti. However, a substantial difference appears in the specimens with compositions beyond 30.9 at% Ti. Namely, the cooling curve of the 30.9 at % Ti specimen shows a zigzag and step-wise increasing tendency of the electrical resistivity, notwithstanding a decrease in temperature.

However, the heating curve does not show such an abnormal zigzag curve, but a smooth one. Moreover, it was confirmed that after one cycle of first cooling below room temperature and then heating to room temperature, both the cooling and heating curves of the resistivity in the second and third runs almost coincide with the first heating one. The other interesting phenomenon is found in the 33.2 at% Ti specimen, in which the resistivity at room temperature increases by about 31% compared with the initial value after once cooling it down to 77 °K and further, after cooling down to 4.2 °K the final increase amounts to about 45%. The third interesting phenomenon is found in comparison between the stoichiometric specimen and others, in which it must be pointed out that the smoothness in the cooling curve of resistivity below room temperature in the former is in striking contrast to the zigzag process in the latter. Fourthly, in the composition dependence of this increment of the resistivity, it is largest near the stoichiometric composition. This suggests that this phenomenon is essentially the problem in the Laves phase compound of Fe_2Ti.

3.2 Thermo-hysteresis curves of the electrical resistivity

Fig. 4 shows the temperature variation of the electrical resistivity of three specimens containing 31.6, 33.2, and 34.6 at% Ti, respectively. In all specimens the values of the resistivity showing large increments at room temperature once after cooling them down to 4.2 °K begin to decrease near 560 °K as they are heated at a rate of about 200 deg/h above room temperature and approach nearly constant values at about 1150 °K. After that, when they are cooled down to room temperature again, the specimen near the stoichiometric composition (33.2 at% Ti–Fe) recovers nearly the initial low resistivity at room temperature, while the specimen in the Ti-rich composition (34.6 at% Ti–Fe) does not recover perfectly its initial one. For the latter, however, it was found that a prolonged annealing at about 900 °C for 20 h is effective for the recovery of its resistivity. Therefore, one migh tinfer that its recovery may be completed in all specimens, if the measurement could be done up to higher temperatures than 1150 °K.

It is characteristic of this thermo-hysteresis phenomenon of the resistivity that it occurs below room temperature and exists over a fairly wide temperature range. It is natural to consider that this thermo-hysteresis is due to some phase transformation. In order to confirm, what sort of phase transformations occur, we made magnetic measurements and an analysis of X-ray powder photographs. Hereinafter, for convenience in defining the state of the specimen, the state with low resistivity after the heat-treatments at high temperature is called the high

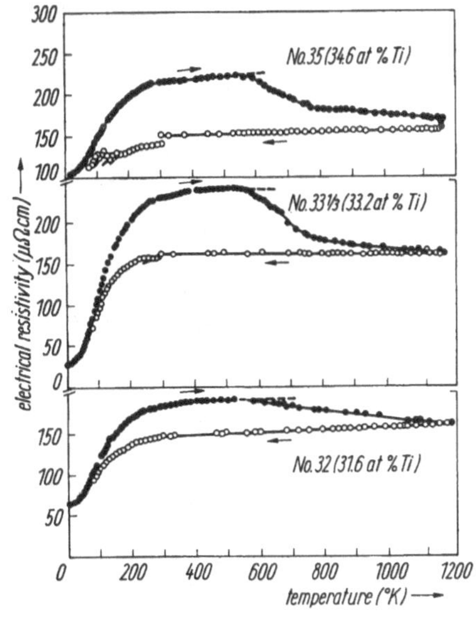

Fig. 4. Thermo-hysteresis curves of the electrical resistivity in three Laves phase compounds in the Fe–Ti alloy system with compositions of 31.6, 33.2, and 34.6 at% Ti, respectively. ○ cooling curves; ● heating curves

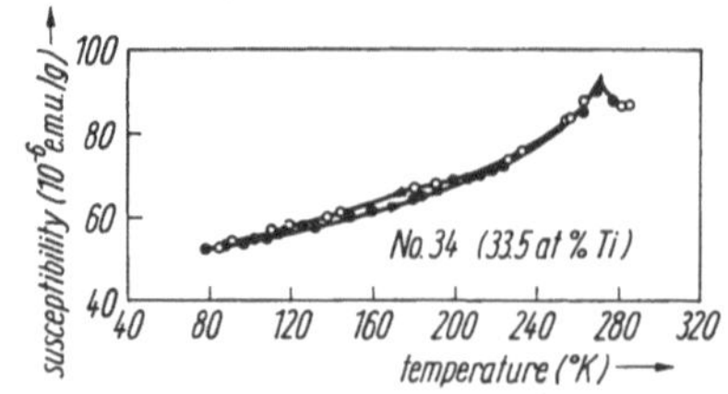

Fig. 5. Temperature variations of susceptibility of the 33.5 at% Ti–Fe alloy. ○ cooling curve; ● heating curve

temperature phase and that with higher resistivity after cooling down below room temperature is called the low temperature phase.

3.3 Magnetic measurements

Magnetic measurements were made on the specimen containing 33.5 at% Ti, in which thermo-hysteresis of the resistivity is remarkable, in the temperature range from 77 °K to room temperature in order to deduce the magnetic state of this transformation at lower temperatures. As shown in Fig. 5, the magnetic measurement on cooling the specimen which is annealed at 1000 °C for 10 h shows an antiferromagnetic peak characteristic of this compound in a descending curve, and also the heating curve shows almost the same temperature dependence of the magnetic susceptibility as the descending one. According to this experimental result, the Néel temperature is 270 °K and the maximum susceptibility is 93 e.m.u./g, which are nearly consistent with Nakamichi's data [3]. This ex-

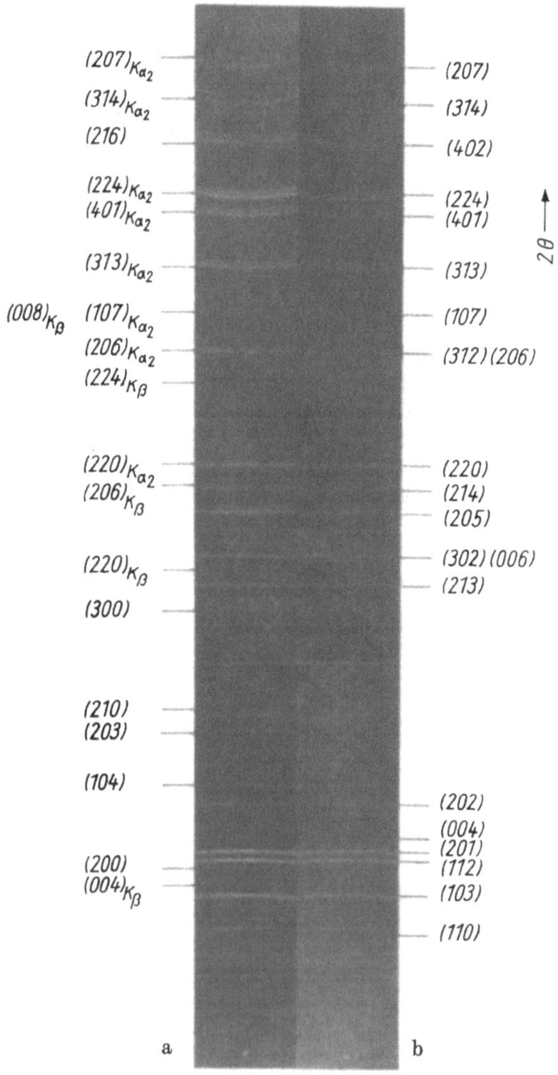

Fig. 6. X-ray powder diffraction patterns of 33.2 at% Ti–Fe specimen annealed at 1000 °C (a) and afterone cooling treatment down to 77 °K (b)

perimental result shows that the thermo-hysteresis of the resistivity has no direct connection with the magnetic transformation temperature, because the transformation which causes the thermo-hysteresis of the resistivity is in progress at far lower temperatures than the Néel temperature.

3.4 X-ray measurement results

X-ray powder photographs of the specimens were taken in order to distinguish the crystal structures of the low-temperature phase from that of the high temperature phase of the specimens employed in the measurement of resistivity. Since it is considered that the low-temperature phase is stable even at room temperature according to the results of the measurement of the electrical resistivity, both crystal structures of the powder specimen annealed at high temperature and once cooled down to 77 °K, were examined simultaneously at room temperature by the Debye-Scherrer method. Both show none of the reflected lines without those of the $MgZn_2$ type structure. Fig. 6 shows X-ray powder diffraction patterns of 33.2 at% Ti specimen after annealing and that after once cooling down to 77 °K. Since the photograph was taken in back reflection in order to find the lattice parameter exactly, it contained many Fe-$K_{\alpha 2}$ lines and some K_β lines, which were not filtered perfectly by a manganese filter, besides Fe-$K_{\alpha 1}$ lines. As can be seen in this photograph, both are hardly different and all diffraction lines are in good agreement with the position and the relative intensity of lines which are calculated assuming the $MgZn_2$ type structure, except some broad diffraction lines of the specimen after the cooling treatment.

Fig. 7 shows the composition dependence of the lattice parameter of the $MnZn_2$ type Laves phase compounds, which was determined from two high-angle diffraction lines, (205) and (220). As shown in the figure, the lattice parameters seem to be smaller after cooling in the composition range from 30.0 to 34.6 at% Ti, and that of the specimen containing 33.2 at% Ti contracts by about 0.15% compared with the initial value, though it is not far over the limit of error.

4. Discussions

Here, we summarize the above-mentioned experimental results and discuss the cause of the thermo-hysteresis of the resistivity in Fe_2Ti. Naturally, one should consider this thermo-hysteresis suggests something like a phase transformation at low temperatures for this Fe_2Ti specimen. The following phase transformation-like mechanisms may be considered as a cause of thermo-hysteresis of resistivity; 1. the magnetic transformation in connection with the ferromagnetism or antiferromagnetism of these compounds, 2. the occurrence of microcracks due to the brittle character of this compound at low temperature, and 3. diffusionless transformation. Concerning the first, it must be mentioned

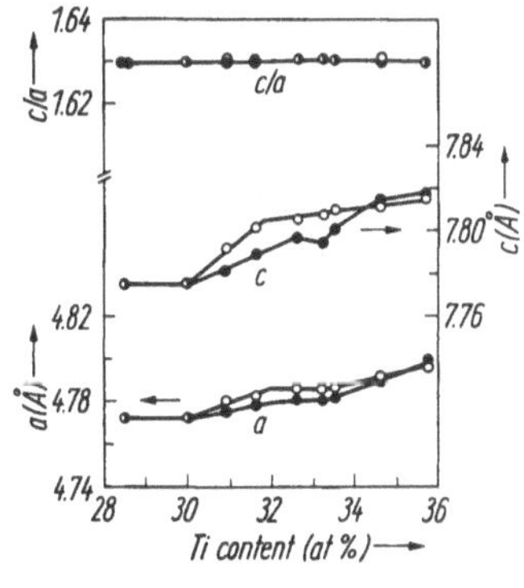

Fig. 7. The composition dependence of the lattice parameter of the $MgZn_2$-type Laves phase compounds, $Fe_{2+x}Ti_{1-x}$. ○ as-annealed; ● after cooling to liquid N_2 temperature; ◑ both values coincide

that the magnetic transformation is independent of the resistivity thermo-hysteresis, as described in 3.1. Concerning the second, as will be reported later, such a thermo-hysteresis is not found in the electrical resistivity of the similar Laves phase compound Fe_2Zr with a stoichiometric composition. Therefore, it is difficult to consider this hysteresis phenomenon in the resistivity of Fe_2Ti as a result of microcracks owing to the simple brittle character of the Laves phase compound.

On the other hand, in order to interpret this phenomenon as a diffusionless transformation it may be useful to understand the following four features observed in this transformation. First, it occurs below room temperature and over a considerably wide temperature range. For instance, in the 33.2 at% Ti specimen, the starting temperature for the transformation from the high temperature phase to the low-temperature phase is 265 °K and the transformation from the low-temperature phase to the high-temperature phase begins at 560 °K and then the width of the hysteresis ranges over about 300 deg. It compares with the width of the temperature range of the martensitic transformation in the Fe–Ni alloy system, which amounts to 400 deg in the 20 at% Ni–Fe alloy [14]. The second reason is that the transformation quantity is determined mainly by the lowest temperature in the cooling process, as mentioned in detail in the previous Section 3.1). The third reason is interpreted with the help of Fig. 8. This figure shows the composition dependence of the starting temperature of the low-temperature transformation of Fe_2Ti, which was determined from the step-wise variation of the electrical resistivity in the cooling curve. As shown in this figure, it lies just below room temperature in the stoichiometric specimen and falls rapidly in the Fe-rich and slowly in the Ti-rich specimens. This fall of the starting temperature of the transformation in the non-stoichiometric specimens, which is especially rapid on the Fe-rich side, may be understood from the fact [9] that excess iron atoms dissolved substitutionally in the Fe-rich composition range and excess titanium atoms in the Ti-rich composition range play respectively the role to interrupt the motion of the transformation dislocations in the martensitic transformation, and this interruption increases with increasing deviation from stoichiometry. The zigzag increase of resistivity observed in the first cooling curve of the electrical resistivity in the non-stoichiometric specimens might be considered to be due to excess iron or titanium atoms which obstruct the transformation dislocations as barriers and cannot be overcome by them until internal stresses have been accumulated to some degree due to further reduction of temperature.

Concerning the crystal structure of the compound, it was concluded that specimens, both treated and not treated in the cooling process, show the same diffracted lines of the $MgZn_2$ type structure, but it seems that these are slightly broader and the lattice parameter also is slightly smaller in the former specimen. If we assume

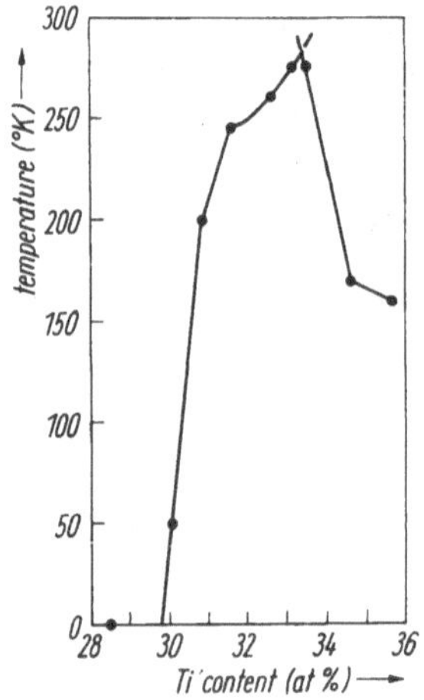

Fig. 8. The composition dependence of the starting temperature of the low-temperature transformation in $Fe_{2+x}Ti_{1-x}$

that the broadness of the diffracted lines is due to stacking faults accompanied by the motion of dislocations, and the occurrence of vacancies owing to such stacking faults, the phenomena of the increment of the resistivity due to the cooling treatment may be explained as the result of small second phases and other crystal defects dispersed in the whole crystal.

This mechanism is also advantageous to the magnetization measurements, because small differences between the cooling and heating curves may suggest that the low-temperature phase occupies only a small portion of the crystal. Whereas, the increment of the resistivity is possible owing to a small but dispersed irregularity in the whole crystal. However, at the present stage, the low-temperature phase could not yet be clarified from the analysis of X-ray diffraction. The first reason might be the possibility that the powder specimen used for the X-ray diffraction was so fine that the diffusionless transformation could not occur [15].

Secondly the low-temperature phase may be one of the modified Laves phases as described by Komura [16] who reported that there are the other Laves phases which have 5-, 9-, and 10-layer structures in addition to the principal three structures of the 2-layer ($MgZn_2$-type), 3-layer ($MgCu_2$-type), and 4-layer structure ($MgNi_2$-type) or the μ phase (Fe_7W_6-type) which has a very similar crystal structure to that of the Laves phase. And, it may also be expected that the diffraction lines of the $MgCu_2$-type structure existing slightly in the low-temperature phase could not be confirmed, because the diffraction lines of the $MgCu_2$-type structure are in good agreement with the position of that of the $MgZn_2$-type one. At present, expecting the appearance of such complex Laves-phase structure, we are now making further investigations by preparing a single crystal specimen in order to confirm the crystal structure of the low-temperature phase of this compound more precisely.

Acknowledgements

The authors wish to express their sincere thanks to Mr. T. Yamada, Mr. T. Sato, and Mr. T. Ono for the preparation of the alloy specimens.

References

[1] H. Witte and H. J. Wallbaum, Z. Metallk. **30**, 100 (1938).
[2] Y. Murakami, K. Kimura, and Y. Nishimura, Rep. Nat. Res. Inst. Metals **1**, 32 (1958) (in Japanese).
[3] T. Nakamichi, J. Phys. Soc. Japan **25**, 1189 (1968).
[4] C. W. Kocher and P. J. Brown, J. appl. Phys., Suppl. **33**, 1091 (1962).
[5] E. Piegger and R. S. Craig, J. chem. Phys. **39**, 137 (1963).
[6] W. E. Wallace, J. chem. Phys. **41**, 3857 (1964).
[7] G. K. Wertheim, J. H. Wernick, and R. C. Sherwood, Solid State Commun. **7**, 1399 (1969); J. appl. Phys. **41**, 1325 (1970).
[8] M. Okazaki, C.R. Acad. Sci. (France) **270B**, 254 (1970).
[9] K. Ikeda, T. Nakamichi, and M. Yamamoto, J. Phys. Soc. Japan **27**, 1361 (1969).
[10] K. Ikeda, T. Nakamichi, and M. Yamamoto, J. Phys. Soc. Japan **30**, 1504 (1971).
[11] M. Okazaki, K. Kai, T. Nakamichi, and M. Yamamoto, J. Phys. Soc. Japan **25**, 1509 (1968).
[12] R. M. Bozorth, H. J. Williams, and D. E. Walsh, Phys. Rev. **103**, 572 (1956).
[13] M. Iwata, Oyo Butsuri **34**, 812 (1965) (in Japanese).
[14] Y. Shirakawa, Sci. Rep. Res. Inst. Tôhoku Univ. **27**, 528 (1939).
[15] S. Kachi, Y. Bando, and S. Higuchi, Japan. J. appl. Phys. **1**, 307 (1962).
[16] Y. Komura, J. Phys. Soc. Japan **23**, 398 (1967).

(Received May 17, 1972)

phys. stat. sol. (a) **12**, 605 (1972)

Subject classification: 5; 18.1; 18.4; 22.6; 22.8

Department of Physics (a) and Department of Chemistry (b), Indian Institute of Technology, Kanpur

Mössbauer Studies of the Thermal Effects on β-FeOOH, $(Cr_{0.8}Fe_{0.2})(OH)_3$, and Corresponding Oxides

By

V. G. JADHAO (a), R. M. SINGRU (a), and C. N. R. RAO (b)

Mössbauer parameters such as isomer shift, quadrupole splitting, internal magnetic field, and line-width show marked variation during the thermal conversion of paramagnetic β-FeOOH to antiferromagnetic α-Fe_2O_3. The quadrupole splitting and the line-width show significant changes accompanying the decomposition of paramagnetic $(Cr_{0.8}Fe_{0.2})(OH)_3$ to paramagnetic $(Cr_{0.8}Fe_{0.2})_2O_3$. The line-width progressively decreases with the growth of the crystallites of the oxides in both the systems.

Mößbauerparameter, wie isomere Verschiebung, Quadrupolaufspaltung, inneres magnetisches Feld und Linienbreite, zeigen eine merkliche Änderung während der thermischen Umwandlung des paramagnetischen β-FeOOH zu antiferromagnetischem α-Fe_2O_3. Die Quadrupolaufspaltung und Linienbreite zeigen bedeutende Änderungen, die den Zerfall des paramagnetischen $(Cr_{0,8}Fe_{0,2})(OH)_3$ zum paramagnetischen $(Cr_{0,8}Fe_{0,2})_2O_3$ begleiten. Die Linienbreite nimmt progressiv mit dem Kristallwachstum in beiden Systemen ab.

1. Introduction

The thermal conversion of some iron oxyhydroxides to ferric oxide has been studied by Mössbauer spectroscopy by a few workers [1 to 3]. Thus, Dézsi et al. [1] have examined the conversion of β-FeO · OH and δ-FeO · OH into α-Fe_2O_3 by recording the Mössbauer spectra of the oxyhydroxides at various temperatures in the range 80 to 1000 K. However, in such studies, the Mössbauer spectra would indicate combined effects due to thermal conversion as well as relaxation phenomena. On the other hand, Vlasov et al. [2] have studied the conversion of δ-FeO · OH into α-Fe_2O_3 by recording the room temperature Mössbauer spectra of samples preheated to different temperatures in the range 393 to 923 K. This method brings out the effects associated with the thermal conversion alone.

We have adopted the second method to investigate the thermal conversion of β-FeO · OH into α-Fe_2O_3 by annealing the oxyhydroxide samples at different temperatures in the range 393 to 923 K and studying their Mössbauer spectra at room temperature. Our interest was to study the variation of the various Mössbauer parameters such as isomer shift, quadrupole splitting, and internal magnetic field accompanying the thermal decomposition as well as the thermal effects on these parameters after the complete conversion to the oxide. The oxyhydroxide β-FeO · OH is paramagnetic [1] above 295 ± 4 K while α-Fe_2O_3 is antiferromagnetic [4] below 950 K.

In order to study the thermal effects on the Mössbauer parameters in a simpler system (containing Fe_2O_3) where there is no magnetic transition, we have chosen to examine the decomposition of the hydroxide $(Cr_{0.8}Fe_{0.2})(OH)_3$ to $(Cr_{0.8}Fe_{0.2})_2O_3$. This oxide has a very low (155 K) Néel temperature [4, 5].

2. Experimental

The samples of FeO·OH were prepared by ammoniacal hydrolysis of the nitrate and they were subsequently heated in a furnace at temperatures 393, 453, 553, 593, 623, 653, 723, and 923 K for 3 h. The decomposition of FeO·OH begins [6] at 533 K and it crystallizes into α-Fe_2O_3 at 608 K. Similarly samples of $(Cr_{0.8}Fe_{0.2})(OH)_3$ were prepared by mixing proper amounts of standardized $Fe(NO_3)_3$ and $Cr(NO_3)_3$ and treating the mixture with excess of NH_4OH. After filtering, the residue was annealed at temperatures of 523, 623, 663, 723, and 923 K for a period of 3 to 3.5 h. X-ray diffraction studies showed that in both cases, the amorphous oxyhydroxides transformed into crystalline oxides above the known crystallization temperatures. The lattice parameters for α-Fe_2O_3 and $(Cr_{0.8}Fe_{0.2})_2O_3$ as found from our X-ray studies were in good agreement with the literature values [2, 7]. All these samples previously fired at different temperatures were studied by Mössbauer spectroscopy at room temperature.

The absorbers were prepared by taking the required amount of a dried and well-powdered sample and pressing it between aluminium foils.

The Mössbauer spectra were recorded with the help of a constant acceleration drive similar to the Bell Laboratory drive [8]. Our Mössbauer spectrometer employed a Philips loudspeaker to drive the source and a Hewlett-Packard velocity transducer to sense the motion. A triangular signal (frequency 10.5 Hz) obtained from Hewlett-Packard Model 202A function generator was used as the drive signal. The spectra were stored in a 512-channel (Nuclear Data, Series 1100) analyzer operated in the multi-scaling mode. The radioactive source used was ^{57}Co embedded in a 0.001" thick copper foil. The 14.4 keV γ-rays were detected with an argon-filled proportional counter having good energy resolution. The Mössbauer spectrometer was calibrated using absorbers like stainless steel 310, sodium nitroprusside, α-Fe_2O_3, and enriched iron foil. The intrinsic line-width (FWHM) for stainless steel (SS 310) absorber was observ edto be 0.42 mm/s. All the spectra reported here were measured at room temperature.

The Néel temperatures of the solid solution samples of $(Cr_{0.8}Fe_{0.2})_2O_3$ previously annealed at 723 and 923 K were determined as follows. The absorbers were mounted inside a cryostat [5] using liquid nitrogen for cooling. The temperature of the absorber was varied in the range 120 to 180 K by a small nichrome heater. The temperatures were read with the help of a resistance thermometer coupled to a Wheatstone bridge. The absorber and the ^{57}Co-source were maintained at zero relative velocity. Our measurements of the Néel temperature are in agreement with the reported value of Cox et al. [4].

Differential thermal analysis was carried out on Aminco or du Pont thermoanalyzer and the X-ray diffraction patterns were recorded on a GE X-ray diffractometer.

3. Results and Discussion

The parameters of the Mössbauer spectra describing the thermal conversion of β-FeO·OH into α-Fe_2O_3 are given in Table 1. The profiles of Mössbauer

Table 1

Mössbauer parameters for the conversion of β-FeO · OH into α-Fe_2O_3

annealing temperature (K)	isomer shift*) δ (mm/s)	quadrupole splitting 2ε (mm/s)	internal magnetic field H_{int} (kOe)	Mössbauer line-width 2 G (mm/s)	remarks
393	0.58 ± 0.06	0.63 ± 0.06	—	0.74 ± 0.06	β-FeOOH
453	0.65 ± 0.03	0.74 ± 0.03	—	0.90 ± 0.05	β-FeOOH
553 p	0.52 ± 0.02	0.87 ± 0.03	—	1.10 ± 0.05	β-FeOOH
af	0.58 ± 0.06	0.30 ± 0.06	525 ± 5	0.72 ± 0.05	α-Fe_2O_3
593 p	0.53 ± 0.03	1.05 ± 0.03	—	0.61 ± 0.05	β-FeOOH
af	0.68 ± 0.06	0.22 ± 0.04	502 ± 5	0.54 ± 0.05	α-Fe_2O_3
623	0.68 ± 0.06	0.22 ± 0.05	509 ± 5	0.61 ± 0.05	α-Fe_2O_3
653	0.71 ± 0.07	0.22 ± 0.10	509 ± 5	0.70 ± 0.07	α-Fe_2O_3
723	0.53 ± 0.03	0.24 ± 0.06	515 ± 3	0.50 ± 0.05	α-Fe_2O_3
923	0.53 ± 0.03	0.24 ± 0.06	515 ± 3	0.40 ± 0.05	α-Fe_2O_3

*) measured with respect to sodium nitroprusside.
p paramagnetic β-FeOOH; af antiferromagnetic α-Fe_2O_3.

spectra corresponding to the conversion of β-FeO · OH to α-Fe_2O_3, are shown in Fig. 1.

At 393 K the sample of β-FeO · OH is known [1] to be paramagnetic and the Mössbauer spectrum shows the characteristic doublet structure (Fig. 1). As the annealing temperature is raised, the thermal conversion into α-Fe_2O_3 sets in as shown by the spectrum of the sample annealed at 553 K. This spectrum is composed of both the paramagnetic oxyhydroxide and antiferromagnetic oxide showing the characteristic six-finger pattern. Obviously, the thermal

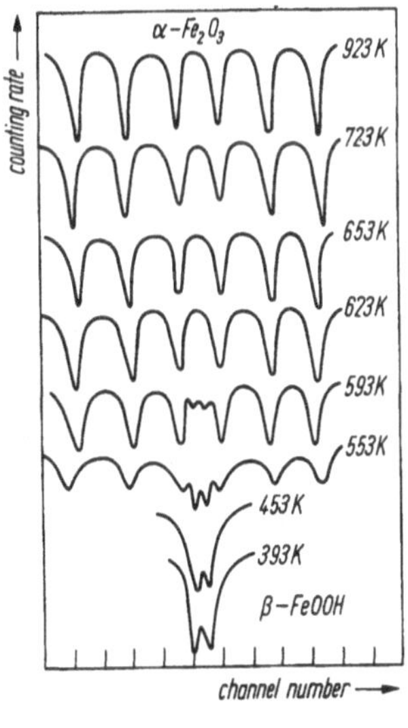

Fig. 1. Mössbauer spectra of β-FeOOH samples (annealed at various temperatures) used as absorbers at room temperature. The source used was ^{57}Co in copper matrix

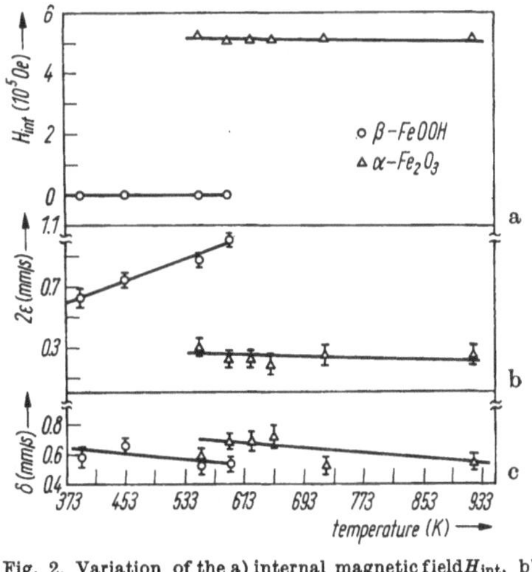

Fig. 2. Variation of the a) internal magnetic field H_{int}, b) electric quadrupole splitting 2ε, and c) the isomer shift δ during the thermal conversion β-FeOOH → α-Fe_2O_3

decomposition into α-Fe_2O_3 is only partial. The decomposition apparently takes place over a wide range of temperature; further the crystallization of α-Fe_2O_3 takes place slowly continuing up to 593 K (Fig. 1). The thermal conversion appears to be nearly complete around 653 K and complete at 923 K. At this temperature, the Mössbauer spectrum is truly characteristic of α-Fe_2O_3. Among the Mössbauer parameters, the isomer shift does not vary much with the annealing temperature and retains the value characteristic of Fe^{3+}. The quadrupole splitting shows a marked discontinuity in the temperature range where the internal magnetic field also develops accompanying the thermal conversion.

Mössbauer parameters for samples annealed at 533 and 593 K have been calculated for the paramagnetic as well as the antiferromagnetic contribution to the observed composite spectra. The thermal conversion is not complete at these temperatures, and the paramagnetic lines characteristic of β-FeO · OH give a higher value of quadrupole splitting than that calculated from the six finger spectrum of the antiferromagnetic α-Fe_2O_3 (Table 1 and Fig. 2). Our values for the Mössbauer parameters for the paramagnetic and antiferromagnetic states agree with those of Dézsi et al. [1]. The behaviour of the line-width in the Mössbauer spectra (Fig. 1) is shown in Fig. 3. It appears that the line-width for the paramagnetic lines shows a sharp decrease after thermal decomposition sets in. The line-width in the antiferromagnetic oxide also shows a sharp decrease during the thermal conversion, but stabilizes once the crystallization into α-Fe_2O_3 is complete. This behaviour in line-width is similar to the results of Vlasov et al. [2] on the thermal conversion of δ-FeO · OH into α-Fe_2O_3. The line-width of α-Fe_2O_3 continues to decrease slightly with the annealing temperature up to 923 K (Fig. 3) probably due to the progressive increase in crystallite size.

Mössbauer spectra of the $(Cr_{0.8}Fe_{0.2})(OH)_3$–$(Cr_{0.8}Fe_{0.2})_2O_3$ system are shown in Fig. 4. The thermal decomposition as shown by DTA is over at about 650 K

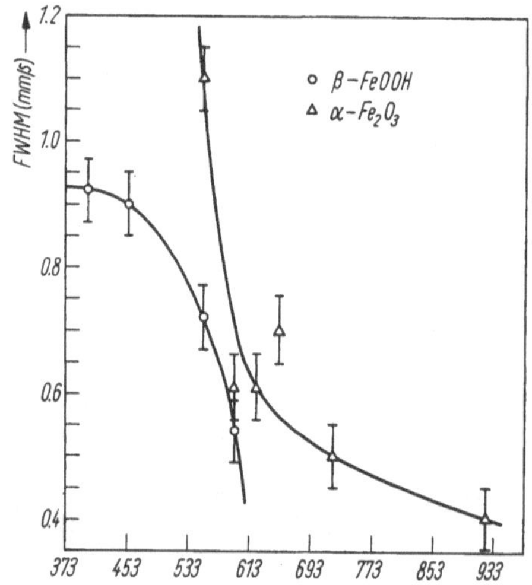

Fig. 3. Variation of the line-width of the β-FeOOH–α-Fe_2O_3 system with annealing temperature

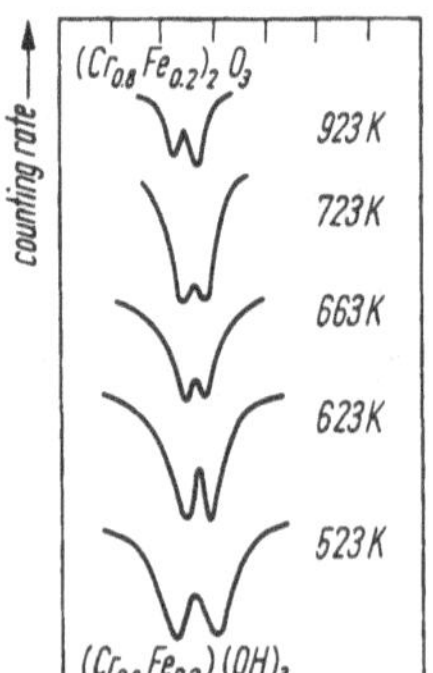

Fig. 4. Mössbauer spectra of $(Cr_{0.8}Fe_{0.2})(OH)_3$ samples (annealed at various temperatures) used as absorbers at room temperature. The source used was ^{57}Co in copper matrix

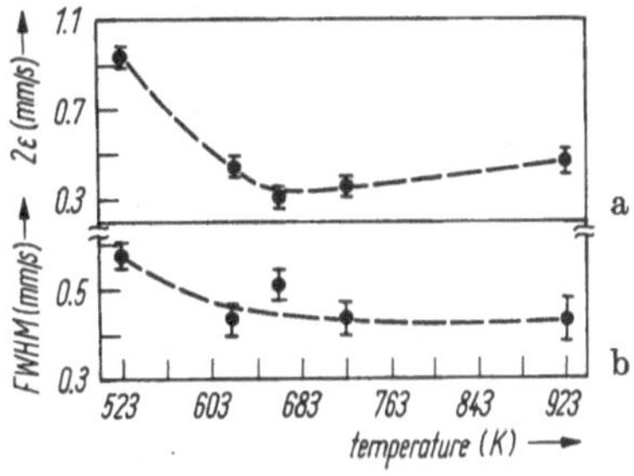

Fig. 5. Variation of the Mössbauer a) quadrupole splitting, and b) linewidth of the $(Cr_{0.8}Fe_{0.2})(OH)_3 - (Cr_{0.8}Fe_{0.2})_2O_3$ system with annealing temperature

Table 2

Mössbauer parameters for the conversion of $(Cr_{0.8}Fe_{0.2})(OH)_3$*) into $(Cr_{0.8}Fe_{0.2})_2O_3$

annealing temperature (K)	isomer shift**) δ (mm/s)	quadrupole splitting $2\,\varepsilon$ (mm/s)	Mössbauer line-width 2 G (mm/s)	remarks
523	0.64 ± 0.03	0.93 ± 0.03	0.57 ± 0.03	$(Cr_{0.8}Fe_{0.2})(OH)_3$
623	0.67 ± 0.03	0.45 ± 0.03	0.43 ± 0.03	$(Cr_{0.8}Fe_{0.2})_2O_3$
663	0.61 ± 0.03	0.31 ± 0.03	0.50 ± 0.03	$(Cr_{0.8}Fe_{0.2})_2O_3$
723	0.68 ± 0.03	0.36 ± 0.03	0.43 ± 0.03	$(Cr_{0.8}Fe_{0.2})_2O_3$
923	0.53 ± 0.05	0.46 ± 0.05	0.41 ± 0.05	$(Cr_{0.8}Fe_{0.2})_2O_3$

*) Formula based on chemical analysis; the exact structure is not known.
**) measured with respect to sodium nitroprusside.

and this is evident in the Mössbauer spectra as well. As this thermal conversion involves transition from one paramagnetic state to another, the Mössbauer spectra for all the five samples are two-finger patterns. The isomer shift shows a small variation through the thermal conversion (Table 2), but the quadrupole splitting decreases (Fig. 5b) sharply up to 650 K, at which temperature the thermal conversion is also complete. The line-width shows a decrease up to the decomposition temperature (Fig. 5) and there is only a slight decrease due to crystallite size effect at higher temperatures.

We see the Mössbauer spectrum of the oxide fired to 923 K shows asymmetry (Fig. 4). The asymmetry can arise from one of the three causes [9]: (i) Goldanski-Karyagin effect; (ii) magnetic relaxation, and (iii) particle orientation or crystallite effect. If (i) is to be taken as responsible for the asymmetry, under the conditions of our sample preparation this would imply freezing in of the anisotropic Debye-Waller factor [10] in the polycrystalline sample of this non-cubic system even after the sample is cooled to room temperature. We cannot, however, rule out (iii) since crystallites appear to grow up to the highest temperature (923 K) studied by us.

Acknowledgements

We are grateful to Dr. K. S. Raju for preparing the various samples. Thanks are due to Dr. J. K. Srivastava of the Tata Institute of Fundamental Research for his help in the measurement of Néel temperature for $(Cr_{0.8}Fe_{0.2})_2O_3$.

References

[1] I. Dézsi, L. Keszthelyi, D. Kulgawczuk, B. Molnár, and N. A. Eissa, phys. stat. sol. **22**, 617 (1967).
[2] A. Ya. Vlasov, G. V. Loseva, E. F. Makarov, N. V. Murashko, E. P. Petukhov, and V. A. Povitskii, Soviet Phys.—Solid State **12**, 1177 (1970).
[3] J. Danon, in: Chemical Applications of Mössbauer Spectroscopy, Ed. V. I. Goldanskii and R. H. Herber, Academic Press, Inc., New York 1968 (p. 159).
[4] D. E. Cox, G. Shirane, and S. L. Ruby, Phys. Rev. **125**, 1163 (1962).
[5] J. K. Srivastava and R. P. Sharma, phys. stat. sol. **35**, 491 (1969).
[6] M. Natarajan, G. V. Chandrashekhar, and C. N. R. Rao, J. chem. Engng. Data **13**, 235 (1968).
[7] R. K. Cerbo and A. V. Seybolt, J. Amer. Ceram. Soc. **42**, 430 (1959).
[8] R. L. Cohen, P. G. McMullin, and G. K. Wertheim, Rev. sci. Instrum. **34**, 671 (1963).
[9] N. Malathi and S. P. Puri, Chem. Phys. Letters **10**, 154 (1971).
[10] V. I. Goldanskii, E. F. Makarov, and V. V. Khrapov, Phys. Letters (Netherlands) **3**, 344 (1963).

(Received April 18, 1972)

phys. stat. sol. (a) **12**, 611 (1972)

Subject classification: 10.1; 21

National Research Institute for Metals, Nakameguro, Tokyo

Behavior of Dislocations in Niobium under Stress

By

S. IKENO and E. FURUBAYASHI

Single crystal foils of pure niobium were stretched at room temperature within a transmission electron microscope operating at 500 kV and formation mechanisms of the dislocation structure were studied. Under a stressed condition, dislocation lines of screw character showed curvatures much larger than those observed in the other b.c.c. metals with high frictional stresses. The large curvatures were responsible for the dislocation behavior observed in the present study, e.g. the conservative motion of jogs or the strong dependence of debris formation on the foil orientation. It is also concluded that dislocation dipole clusters so far reported in pure b.c.c. metals were originated from jogs, rather than from the trapping of dislocations with opposite signs.

Einkristallschichten von hochreinem Niob wurden bei Raumtemperatur innerhalb eines bei 500 kV arbeitenden Elektronenmikroskops zugverformt und die Bildungsmechanismen der Versetzungsstrukturen untersucht. Unter Spannung zeigten die Linien der Schraubenversetzungen eine größere Krümmung als die gewöhnlichen k.r.z.-Metalle, welche die höhere Reibungsspannung haben. Die großen Krümmungen sind für die in der vorliegenden Untersuchung beobachteten Eigenschaften der Versetzungen, wie konservative Wanderung der Sprünge und die starke Abhängigkeit der Debrisbildung von der Schichtorientierung, verantwortlich. Es wird geschlossen, daß Versetzungsdipolcluster, die für reine k.r.z.-Metalle berichtet wurden, mehr auf Sprünge als auf Einfang durch Versetzungen mit entgegengesetzten Vorzeichen zurückzuführen sind.

1. Introduction

The dislocation structures observed in an early stage of deformation of b.c.c. transition metals have been characterized by a lattice friction stress [1 to 3]. Isolated dislocations of predominantly screw orientation and many debris loops are observed under the condition of large frictional stress at low temperatures [4 to 10], or owing to solution hardening [2, 11 to 14]. On the other hand, edge dislocation dipoles and their clusters are predominantly observed in pure b.c.c. metals deformed at relatively high temperatures [4 to 10, 15 to 19].

These two extreme structures will reflect some differences in the deformation mechanisms, depending on the frictional stress. It is not a priori evident, however, that these dislocation structures surely represent those during deformation. Particularly in the low-frictional-stress condition, as anticipated from works on f.c.c. metals [20 to 28], the dislocation structure was considered to rearrange while unloading and thinning the specimens. Direct observation of the dynamic behavior of dislocations under stress is then a useful method to clarify this point.

As a typical case of the high-frictional-stress condition, the dislocation structure in Fe–3% Si has been examined [29, 30] along this line, using a high-voltage electron microscope. The dislocation structure in unloaded and thinned specimens of the alloy was similar to those during deformation. In the present work, the dislocation structure and its behavior under stress in Nb were studied at room temperature, as a typical case of the low-frictional-stress condition [6, 7, 17].

2. Experimental Procedure

2.1 Preparation of specimens

Starting material was electron-beam-melted 99.91% Nb rods of 8 mm diameter, obtained from Kawecki Inc. After sufficient pre-heating for degassing right below the melting point, single crystals were made with a high-frequency floating zone furnace under vacuum $< 10^{-5}$ Torr.

Slices about 1 mm thick were cut from the single crystals with a modified Servomet spark cutting machine. Copper wire was used for the electrode in order to avoid mechanical stress due to cutting. Foil specimens for the tensile study in the microscope were prepared from the slices, by chemical, jet, and finally electrolytic polishing successively at 25 °C (Table 1). They became thin foils having a small hole in the centre, as described previously [30].

Table 1

Procedure of preparing foil specimens

order of procedure	solution (vol%)		temperature (°C)	voltage (V)	electrode
1. chemical polishing	HF	50	25	—	—
	HNO_3	50			
2. jet polishing	HF	10	25	140	platinum wire
	H_2SO_4	20			
	CH_3OH	70			
3. final polishing	HF	20	25	10	platinum plates
	H_2SO_4	20		(1 A/cm²)	
	H_2O	60			

The tensile axes were previously chosen, but expected slip systems were not always operative because of the complicated stress conditions in the foils. However, the geometry of the systems were important with respect to the foil surfaces, as shown schematically in Fig. 1; the extreme cases are *type 1* and *type 2*, in which the foil planes were parallel to the screw and edge component respectively.

A possibility remained in electropolishing of Nb that a certain amount of hydrogen was absorbed in Nb. The hydrogen content in the final thin foils however was < 4 ppm in weight, which was confirmed from the lack of hydrides

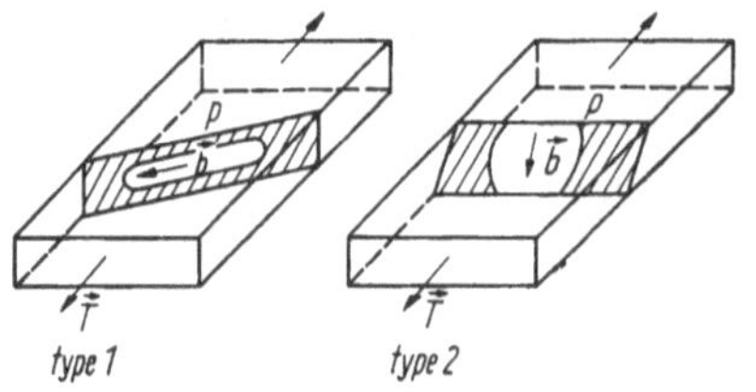

Fig. 1. Geometry of slip systems with respect to the foil surfaces. The ***b***, *p*, and ***T*** represent Burgers vector, slip plane, and tensile direction, respectively. ***b*** is parallel to the surfaces in *type 1*, and makes about 45° with the surface in *type 2*

Table 2

Contents of impurities (wt. ppm)

stage	O	N	C	H	Ta	W	B	Fe
as received	100	20	25	10	200	100	100	30
single cryst.	70	27	20	—	—	—	—	—
foils	—	—	—	<4*)	—	—	—	—

*) Estimation from the invisibility of hydrides at 158 °K [31].

on cooling the foils down to 150 °K in the microscope [31]. Table 2 shows the content of impurities. The amount of impurities suggested that the dislocation behavior in the present Nb would not be affected by the impurities [2, 32 to 35].

2.2 Observation of deforming specimens

The deformation of the foils was carried out with a combined stage of tensile and tilting operation [36] in a Shimadzu SMH-5B electron microscope operating at 500 kV.

The orientation of the foils was determined with selected area diffraction patterns containing Kikuchi lines and Laue zones, and the accuracy was better than 2°. The thickness of the foils was measured from the length of screw dislocation segments. Most observations were made in specimen areas between 0.7 and 0.9 μm thickness.

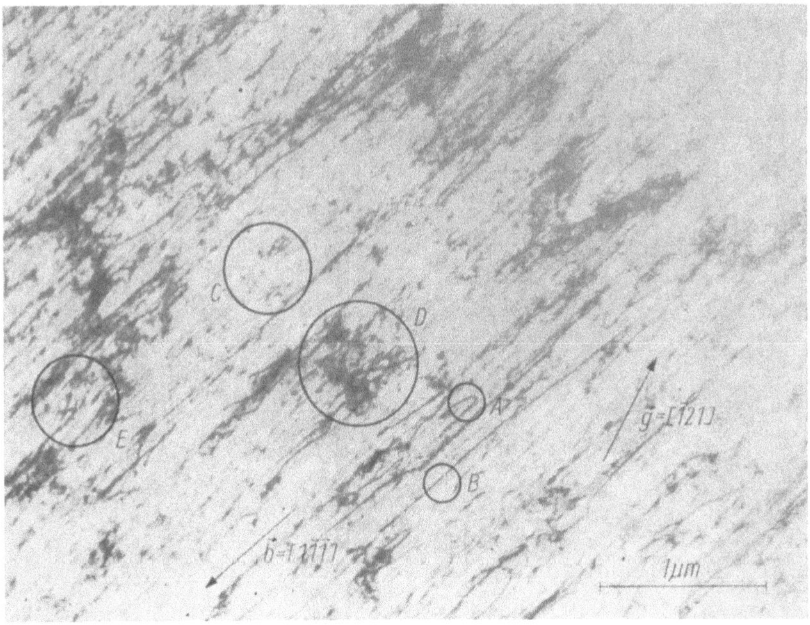

Fig. 2. Dislocation structure in a specimen of *type 1*. ***g*** represents reflection vector. Foil surfaces are parallel to (101)

The successive position of the dislocations was recorded on photographic films while the applied stress was decreasing very slowly but almost constant throughout a series of micrographs.

3. Results

3.1 Observation of dislocation structure and its activity

Fig. 2 shows the dislocation structure in a specimen deforming with a slip system of *type 1*. Sharply curved edge components (A), long screw components (B), many small debris loops (C), and dislocation clusters (D) are seen. The dislocation clusters are forming as a result of the interaction between the dislocations and the debris loops. At (E), several edge dislocations are piling up against a cluster.

Fig. 3 shows the dislocations in a specimen deforming with a slip system of *type 2*. Many dislocation segments are screw dislocations of one family cut by the foil surfaces. Edge components and the debris loops are scarcely seen. The large tendency for screw orientation is associated with the smaller mobility of screw components than that of edge ones.

Fig. 4 shows the dislocation structure by the double-slip deformation with the systems of $\boldsymbol{b}_1$ and $\boldsymbol{b}_2$. Both systems are of *type 1*. Besides edge components (A), screw components (B), small debris loops (C), and dislocation clusters (D), the interaction among dislocations of different systems is seen (E, E′). Furthermore, at (F, F′) a debris loop is forming by double cross slip.

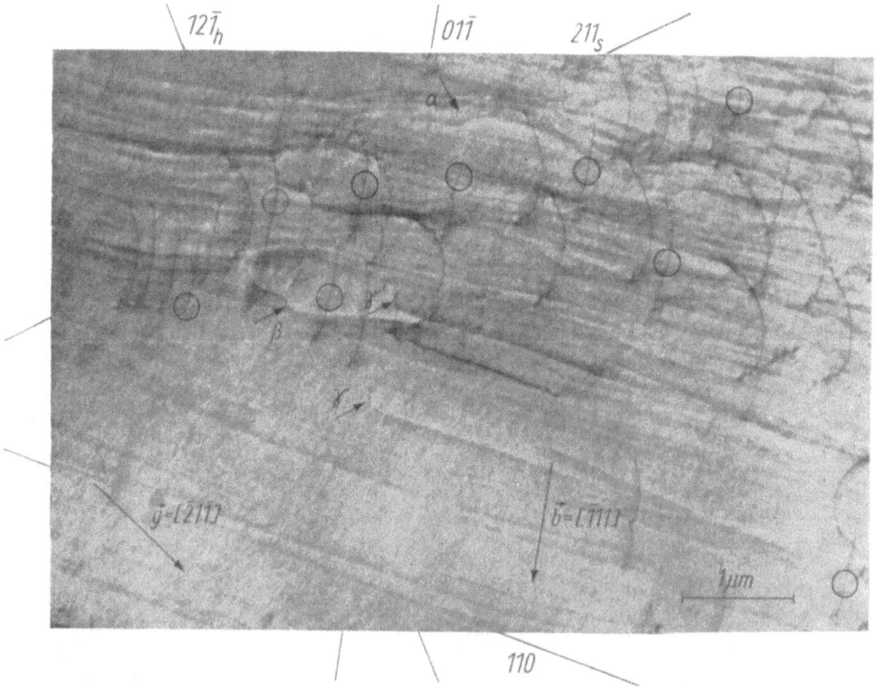

Fig. 3. Dislocation structure in a specimen of *type 2* under stress. Lines denoted by (110), (01$\bar{1}$), (12$\bar{1}$)$_h$, and (211)$_s$ are the intersections of each crystallographic plane with foil surfaces. Foil surfaces are nearly parallel to (132)

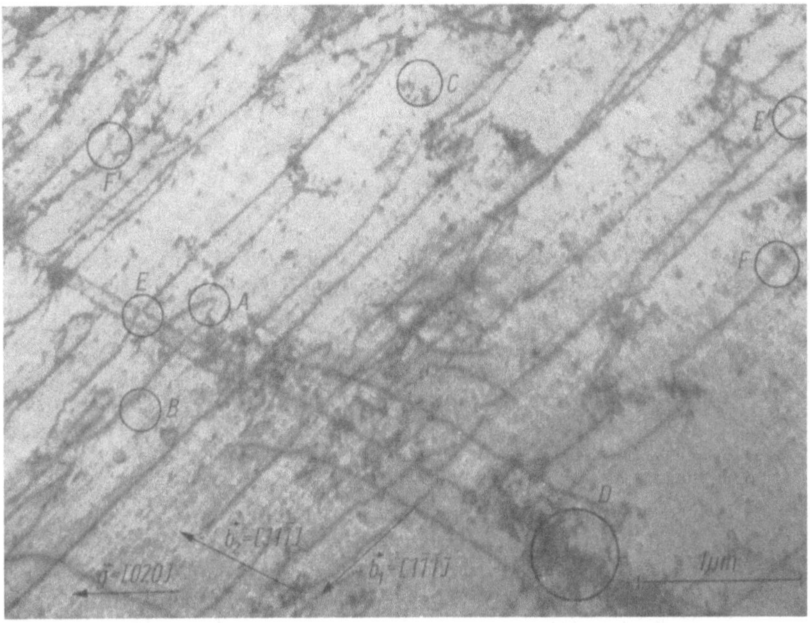

Fig. 4. Dislocation structure due to deformation by double-slip systems of b_1 and b_2. Foil surfaces are (101)

The activity of the dislocations was, in general, not homogeneous but enhanced in some band regions, e.g. in the middle of Fig. 3. This arised probably from the preferred operation of the dislocation sources along the bands. The deformation was restricted virtually in the bands, before a certain number of the dislocations, e.g. 20, passed over. Then the dislocation activity diminished noticeably and, in turn, that of the surrounding regions was enhanced. A cause of the phenomena was considered a kind of 'inertia' effect in which moving dislocations were more operative than the stationary ones. The alternative operation of the slip was surely due to the reverse stress resulting from the local strain in the operating bands.

On the other hand, even in the double-slip orientation, the activation of each slip system was not simultaneous, as seen in Fig. 4. It is recognized that the system of $\boldsymbol{b}_1$ is now in operation. Once a system had operated, the deformation tended to continue mostly with the same system, in general, until the orientation changed to reduce the activity virtually zero.

3.2 Detailed observation of dislocation migration

Slip traces of individual dislocations were easily observed on a microscopic scale, if a low-index reflection was used. This made possible to discuss the crystallography of slip.

The slip traces were not always parallel to low-index planes, {011}, {112}, or {123}. As shown in Fig. 3, the traces are rather curved continuously, though they are almost straight and parallel to (110) in the lower part, where the densities of the dislocations and their traces are extremely low. Such wavy traces seem consistent with the idea that the slip planes are corrugated on an atomistic

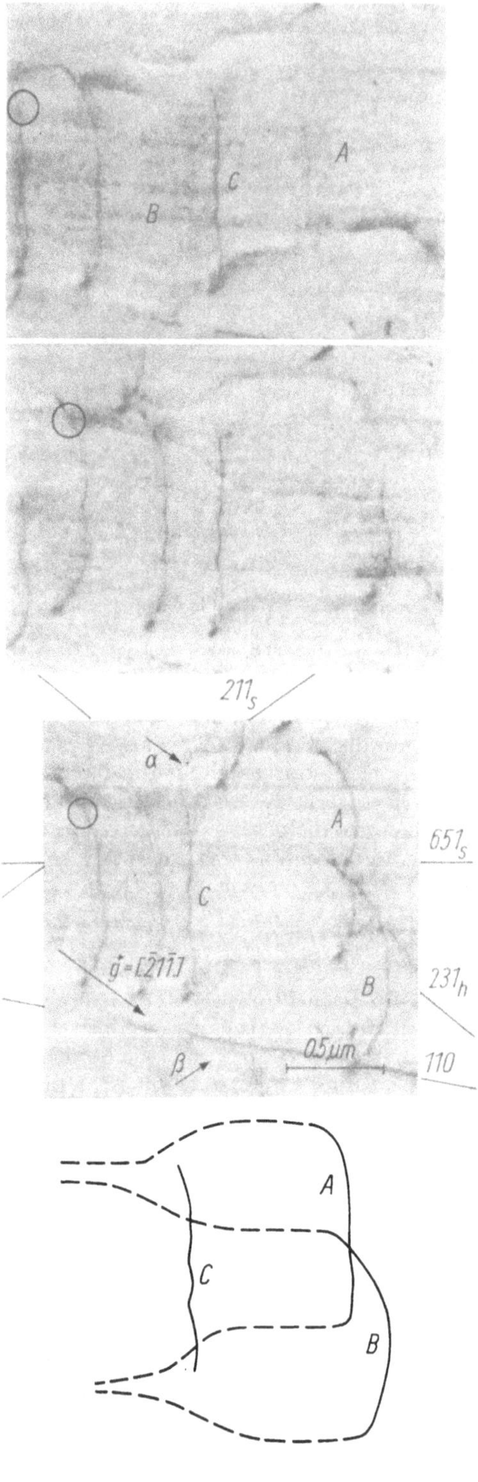

a
b
c
d

Fig. 5. Cross slip of screw dislocations (A) and (B) under the elastic stress field of a stationary dislocation (C). Non-conservative motion of jog can be seen at the dislocation marked by a ring

scale [3, 37]. In some cases, crystallographically segmented traces were often observed. But they could not be identified as any low-index planes even with precise orientation analysis.

Fig. 3 further exemplifies that several dislocations of the same sign are piling up against each stationary dislocation. These dislocations often made a detour around the obstacle by cross slip. In Fig. 5a, dislocations A and B are in the course of cross slip in the elastic stress field of a stationary dislocation C, as discussed by Li [38]. Multiple images of the (A) and the (B) are due to the migration during photographic exposure. (A) and (B) cross-slip upward and downward, respectively, depending on their original routes being above or below (C). After ceasing the cross slip, they glide along a {651} type plane rather than (110). The traces for the cross slip are smoothly curved up to the steepest cross slip planes α and β, close to $(211)_s$ and $(231)_h$, respectively. Here s and h are notations for the slip asymmetry, indicating the sense of the deviation from {110} towards soft {211} (twinning shear direction) and hard $\{12\bar{1}\}$ (anti-twinning direction), respectively [3]. Similar phenomena are also seen in Fig. 3, where the steepest planes of cross slip seem to be $(211)_s$ at α and $(12\bar{1})_h$ at β.

Fig. 6 shows annihilation between the opposite sign of screw dislocations. Dislocations (A) and (B), or (C) and (D) annihilate in c), leaving slip traces shown by

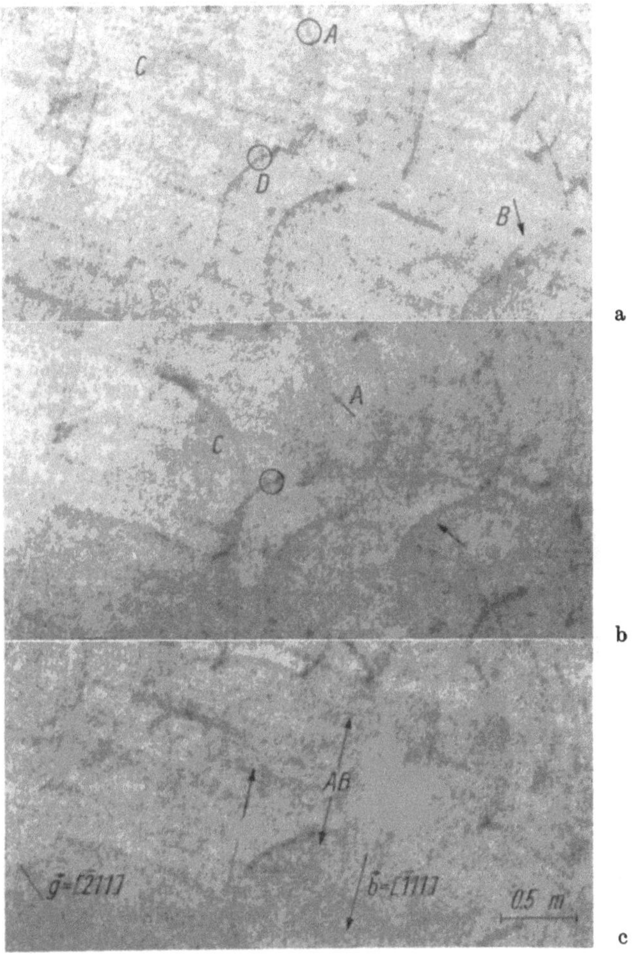

Fig. 6. Annihilation between opposite sign of screw dislocations by cross slip. (A) and (B), (C) and (D) respectively, annihilate each other

(AB) or (CD), respectively. Each annihilation leaves a pair of traces characterized by sharp deflections where the annihilation occurs. Similar traces are also seen in Fig. 3 as indicated by γ and γ'. The traces are mostly parallel to $(01\bar{1})$ which have an inclination of 60° with the original slip plane (110).

3.3 Flexibility of edge and screw dislocations

The dislocations had curvatures during migration as seen in the previous micrographs. The curvatures increased with the applied stress, as seen in Fig. 7. This suggested a sizable dragging effect by the foil surfaces. The curvatures of edge dislocations, however, were much larger than those of screw ones[1]), as

[1]) 'Curvature of edge or screw dislocations' may sound ambiguous because curved dislocations on slip plane cannot be edge or screw as a whole. Here 'edge' or 'screw' are used for dislocation segments not parallel or nearly parallel to the Burgers vector, respectively.

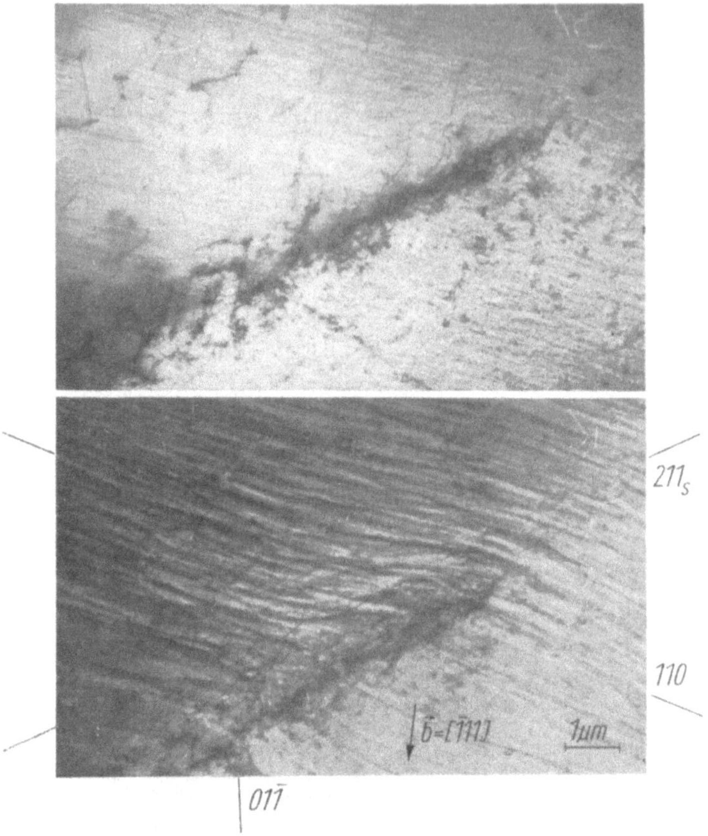

a

b

Fig. 7. Effect of applied stress on the dislocation curvature, as viewed around pre-existent dislocation tangle. Foil surface parallel (132). a) Before applying stress; b) after some deformation (under stress)

shown in Fig. 2. The difference is closely understood by looking at the intersection with the foil surfaces and cannot, therefore, be ascribed wholly to a surface effect. The curvatures are considered in this way as indexes showing the easiness in the deviation of the dislocation line from edge or screw orientation.

3.4 Behaviour of jogs and mobility of screw dislocations

Screw dislocations were often accompanied with cusps. The dislocations indicated by small rings in Fig. 3 did not migrate while the others moved over a large distance. These stationary dislocations were characterized either by small curvatures or cusps. However, further continuous observation showed that the stationary dislocations began to migrate after appropriate increase in the applied stress. The immobility of the dislocations was probably due to the indetectable size of jogs, and their transition to mobile ones was ascribed to annihilation of the jogs with the aid of cross slip, as described below.

In order to investigate the nature of the cusps, their dynamic behaviour was closely examined. The dislocation indicated by a ring in Fig. 5 moved to the right keeping a cusp in the center. From such observation, it was concluded

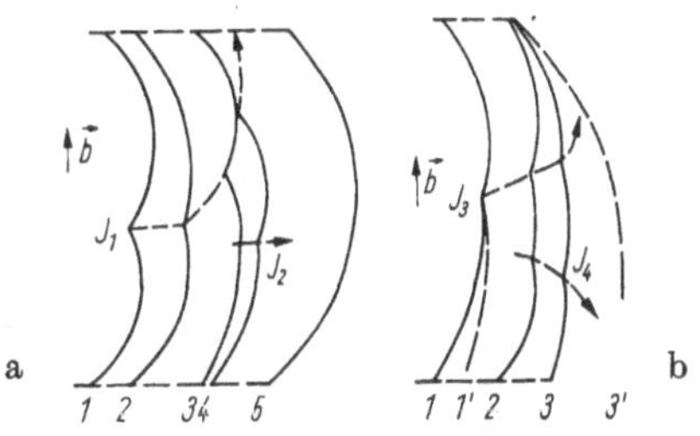

Fig. 8. Conservative and non-conservative motion of jogs as duplicated from a series of micrographs. J_1, J_2, J_3, and J_4 are jogs and curves 1, 2, ... indicate successive positions of a curved screw dislocation. a) and b) show different dislocations. 1 and 1' or 3 and 3' denote double images during exposure

that the cusps were due to jogs rather than the pinning by impurities, and the jogs moved non-conservatively.

Furthermore, conservative motion of jogs also occurred extensively when they situated at portions of the dislocations slightly deviated from screw orientation, as seen in Fig. 8. Fig. 8 shows successive positions of a dislocation as duplicated from a series of micrographs similar to Fig. 5. Conservative and non-conservative motion of jogs J_1, J_3 and J_4, and sudden formation and disappearence of J_2 are seen.

Fig. 9 presents a diagram showing the motion of twelve screw dislocations indexed as (A), (B), ... (L). The position of each dislocation was measured on each micrograph taken successively at every 6 s. The abscissa shows the order of micrographs, or therefore the time elapsed. The ordinate indicates the distance from a reference point and therefore the migration distance. The length of small horizontal bars shows the exposure time of 1.5 s. Double images due to the migration during the exposure are shown by the width of the bars. Relatively low probability of finding the double images was an evidence for the dislocation motion being jerky rather than continuous. Furthermore, stop of dislocation migration was associated with the formation of new jogs with a high probability, as pointed by arrows. This phenomenon has also been observed in Fe [40] and Fe–3% Si [30]. This, together with the low mobility of joggy dislocations, suggested that the motion of screw dislocations was strongly influenced by the formation and annihilation of the jogs.

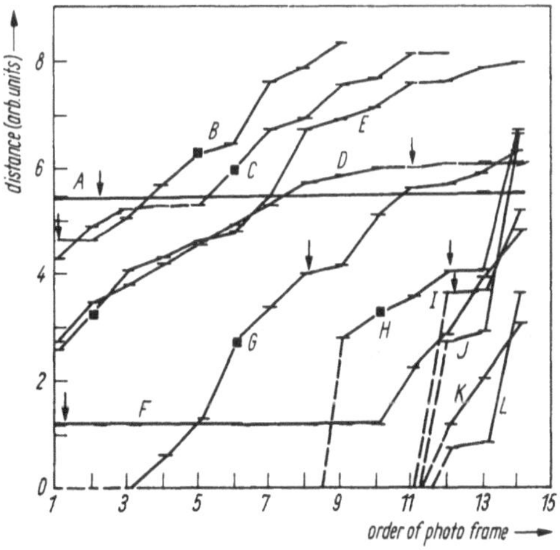

Fig. 9. Distance versus time curve showing the motion of screw dislocations, as replotted from micrographs similar to Fig. 6. The letters (A), (B), ..., (L) represent twelve screw dislocations

4. Discussion

4.1 Correlation between the dislocation curvature and the jog behavior

In contrast to the present Nb, screw dislocations under stress had smaller curvatures in Fe–3% Si [30]. Screw dislocations of pure Fe [39] and Nb [40] deforming at low temperatures have also smaller curvatures. The difference is considered to be a direct consequence of the difference in the frictional stress rather than the surface effect.

In the Nb specimens of *type 2*, debris loops were seldom formed while they were produced rather frequently in *type 1* specimens. This orientation effect can be readily understood if one assumes that the motion of screw dislocations is governed by a double-kink mechanism, and the kink motion or conservative jog motion occurs for a longer distance than the foil thickness. In *type 2* specimens the jogs can readily escape to the foil surfaces, whereas in *type 1* specimens, the conservative motion makes the jogs or the kinks combine with others, resulting in kink complexes [41] or superjogs as sources of dislocation multiplication and pinch-off debris [30].

However, in Fe–3% Si, many debris were formed irrespective of the orientation [30]. This means that the migration distance along the screw dislocation in the alloy is much shorter than the foil thickness. The average spacing between neighboring jogs was as large as 1.1 μm in the Nb, as compared with 0.17 μm in Fe–3% Si (Table 3). These values are considered to be an indication of the migration distance of jogs along the screw dislocations.

Table 3

Jog densities on screw dislocations in metals

metals	deformation temperature (°K)	jog density (10^4 cm^{-1}) (mean spacing (μm))	deformation state	references
Nb	296	0.9 (1.1)	foil	present result
Nb– 5 at% Mo	295	1.4 (0.7)	bulk	FOXALL and
Nb–16 at% Mo	295	3 (0.3)	bulk	STATHAM [2]
Fe	296	3 (0.3)	foil	IKEDA [5]
	197	1 (1)	bulk	
Fe–5.4 at% Si	296	6 (0.17)	foil	FURUBAYASHI [30]
Mo	373	4 (0.25)	bulk	LOESCH and
	78	17 (0.06)		BROTZEN [9]
	293	4 (0.25)	bulk	LUFT and KAUN [10]

In order for a curved screw dislocation to move, the foremost part F, shown in Fig. 10a, will be the most favourable site of nucleation of double kinks, as long as the Peierls potential is considered. Therefore, the probability of forming jogs is expected low in the curved dislocations in the Nb, because a jog is caused by a collision between two kinks from different nucleation sites [30, 41]. This gives rise also to the low production rates of the debris in the Nb. The jogs in the configuration of Fig. 10c are mobile conservatively with ease. In the curved

Fig. 10. Three possible submicroscopic configurations of screw dislocation segment in *type 2* specimen under stress. a) A curved dislocation without jogs. Double-kink formation occurs at the foremost part F. b) A jog locates in the middle. c) A jog locates at a sidewise position

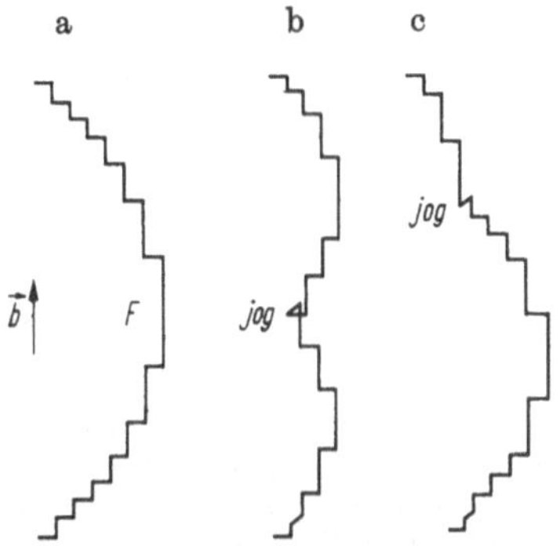

dislocations, therefore, the conservative motion of jogs is enhanced. The observed differences between the Nb and Fe–3% Si can thus be ascribed, in these respects, to the dislocation curvature.

4.2 Causes of edge dislocation clusters

In the previous investigations, it has been observed that edge dislocation clusters are formed in bulk-deformed and thinned Nb crystals deformed at room temperature [6, 7, 17], while in the present study, the dislocations under stress had a considerable tendency for screw orientation, and many debris loops were formed according to the migration of the screw dislocations. Clusters of dislocations were developed at the debris loops by collecting dislocations. Therefore, it can be that edge dislocation clusters so far reported were formed by rearrangement while unloading and thinning, rather than the trapping of edge dislocations from different dislocation sources [7, 17, 42]. The trapping mechanism seems unlikely further because of the low probability of the simultaneous operation of the sources, as discussed in Section 3.1.

5. Summary and Conclusion

Nb single crystal foils, which were prepared by floating zone melting in high vacuum and thinning with electrolytic polishing, were deformed during observation in the 500 kV electron microscope. The results are summerized as follows:

1. The dislocations in Nb deforming at room temperature were predominantly of screw character and debris loops were observed to form. It is suggested that the dislocation structure so far observed in pure b.c.c. metals can arise from the rearrangement while unloading and thinning.
2. Operation of dislocation sources was essentially inhomogeneous and occurred alternatively due to an 'inertia' effect, in which operating sources were more operative until they were work-hardened to operate new sources.
3. Cross slip and annihilation of screw dislocations were analysed in detail with the aid of slip traces.
4. Many jogs were formed and disappeared during the migration of screw dislocations. The formation mode was consistent with the double-cross-slip mechanism. The dislocation mobility was strongly affected by the jog distribution. The dislocation clusters came from these jogs, rather than the trapping of opposite sign of dislocations.
5. By comparing with the previous observation on Fe–3% Si, effects of frictional stress on dislocation behavior were discussed qualitatively. Particularly, it was demonstrated that the larger curvatures of the moving screw dislocations in

Nb were responsible for the conservative motion of jogs and the orientation dependence of the debris formation observed in the present study.

Acknowledgements

The authors would like to express their sincere thanks to Dr. H. Yoshida, Dr. R. Yoda, and Dr. T. Takeuchi for their encouragement in the course of this work. They also are very grateful to Dr. S. Ohba and Dr. Y. Sasaki, Dr. Y. Tamura and Dr. S. Ikeda for helpful discussion.

References

[1] P. B. Hirsch, Trans. Japan Inst. Metals, Suppl. **9**, 30 (1968).
[2] R. A. Foxall and C. D. Statham, Acta metall. **18**, 1147 (1970).
[3] J. W. Christian, Proc. II. Internat. Conf. Strength Metals and Alloys, Asilomar 1970, Vol. 1, Amer. Soc. Metals, 1971 (p. 31).
[4] T. Yamashita and Y. Taneda, Japan. J. appl. Phys. **5**, 266 (1963).
[5] S. Ikeda, J. Phys. Soc. Japan **27**, 1564 (1969).
[6] D. K. Bowen, J. W. Christian, and G. Taylor, Canad. J. Phys. **45**, 903 (1967).
[7] G. Taylor and J. W. Christian, Phil. Mag. **15**, 893 (1967).
[8] A. Lawley and H. L. Gaigher, Phil. Mag. **10**, 15 (1964).
[9] H. W. Loesch and F. R. Brotzen, J. less-common Metals **13**, 565 (1967).
[10] A. Luft and L. Kaun, phys. stat. sol. **37**, 781 (1970).
[11] J. R. Low, Jr. and A. M. Turkalo, Acta metall. **10**, 215 (1962).
[12] N. S. Stoloff, R. G. Davies, and R. C. Ku, Trans. MS AIME **233**, 1500 (1965).
[13] S. Takeuchi, E. Furubayashi, and T. Taoka, Acta metall. **15**, 1179 (1967).
[14] R. E. Ruffo and T. E. Mitchell, Trans. MS AIME **242**, 907 (1968).
[15] A. S. Keh, Phil. Mag. **12**, 9 (1965).
[16] J. W. Edington and R. E. Smallman, Acta metall. **13**, 1313 (1964).
[17] A. Foxall, M. S. Duesbery, and P. B. Hirsch, Canad. J. Phys. **45**, 607 (1967).
[18] W. A. Spitzig and T. E. Mitchell, Acta metall. **14**, 1311 (1966).
[19] R. J. Arsenault and A. Lawley, Phil. Mag. **15**, 549 (1967).
[20] V. Essmann, phys. stat. sol. **3**, 932 (1963).
[21] V. Essmann, Acta metall. **12**, 1468 (1964).
[22] V. Essmann, phys. stat. sol. **12**, 707 (1965); **12**, 723 (1965).
[23] J. W. Steeds, Proc. Phys. Soc. **A292**, 343 (1966).
[24] F. W. Young and F. A. Scherill, Canad. J. Phys. **45**, 757 (1967).
[25] H. Fujita, J. Phys. Soc. Japan **23**, 1349 (1967).
[26] J. C. Crump, III and F. W. Young, Phil. Mag. **17**, 381 (1968).
[27] H. Mughrabi, phys. stat. sol. **39**, 317 (1970).
[28] H. Mughrabi, Phil. Mag. **23**, 869 (1971); **23**, 897 (1971).
[29] E. Furubayashi, Trans. Japan Inst. Metals, Suppl. **9**, 939 (1968).
[30] E. Furubayashi, J. Phys. Soc. Japan **27**, 130 (1969).
[31] D. G. Westlake, Trans. MS AIME **245**, 287 (1969).
[32] D. G. Westlake, Trans. MS AIME **245**, 1969 (1969).
[33] K. V. Ravi and R. Gibala, Acta metall. **18**, 623 (1970).
[34] H. C. Kim and P. L. Pratt, phys. stat. sol. **41**, 797 (1970).
[35] B. Harris, phys. stat. sol. **18**, 715 (1966).
[36] H. Fujita, T. Taoka, M. Iwanaga, Y. Sakamoto, and S. Shibata, Trans. Nat. Res. Inst. Metals **9**, 54 (1967).
[37] H. Suzuki, Dislocation Dynamics, McGraw-Hill Publ. Co, 1968 (p. 679).
[38] J. C. M. Li, J. appl. Phys. **32**, 593 (1961).
[39] S. Ikeda, to be published.
[40] S. Ikeno and E. Furubayashi, to be published.
[41] H. Suzuki, Nachr. Akad. Wiss. Göttingen, math.-phys. Kl. **6**, 1 (1971).
[42] A. S. Tetelman, Acta metall. **10**, 813 (1962).

(Received May 18, 1972)

phys. stat. sol. (a) 12, 623 (1972)

Subject classification: 18.3; 5; 20.1; 22.8.2;

Physico-Technical Institute of Low Temperatures, Academy of Sciences of the Ukrainian SSR, Kharkov

On the Cation Distribution in $Ni_{1-x-y}Fe^{2+}_xZn_yFe^{3+}_2O_4$ Spinel Ferrites

By

V. A. POTAKOVA, N. D. ZVEREV, and V. P. ROMANOV

Infrared and Mössbauer measurements on spinel ferrites of the $Ni_{1-x-y}Fe^{2+}_xZn_yFe^{3+}_2O_4$ system have been made. As a result from the analysis, it is supposed that Fe^{2+} ions are present in both the octahedral and tetrahedral lattice sites of the compounds investigated.

Infrarot- und Mößbauermessungen wurden an Spinell-Ferriten des $Ni_{1-x-y}Fe^{2+}_xZn_yFe^{3+}_2O_4$-Systems durchgeführt. Als Ergebnis der Analyse wird angenommen, daß Fe^{2+}-Ionen sich sowohl auf Oktaeder- als auch auf Tetraederplätzen der untersuchten Verbindung befinden.

1. Introduction

To explain the behaviour of magnetocrystalline anisotropy in the ferrites $Ni_{1-x-y}Fe^{2+}_xZn_yFe^{3+}_2O_4$ the possible location of Fe^{2+} in tetrahedral sites in these solid solutions has been assumed in [1, 2]. To verify this assumptions the vibrational spectra (700 to 300 cm^{-1}) of ferrous- and zinc-doped Ni ferrites and the Fe^{57} Mössbauer spectrum of $Ni_{0.53}Fe^{2+}_{0.36}Fe^{3+}_{2.07}O_4$ ferrite have been measured.

2. Experimental

The vibrational spectra were measured by means of a double-beam grating spectrophotometer „DS-402G jasko" (with optical resolution $\pm 1.5\ cm^{-1}$) and the Mössbauer spectrum was obtained by means of constant acceleration NGR spectrometer using an AI-256-1 analyzer (Co^{57} in Pd as source and a proportional $Xe\text{-}CH_4$ counter as detector). The perfection of the synthetic single- and polycrystalline samples was checked by X-ray analysis. The crystal-chemical formulas and the lattice parameters of the compounds studied are summarized in Table 1.

3. Results and Discussion

Fig. 1 presents the Mössbauer spectra of the highly Fe^{2+} doped ferrite $Ni_{0.53}Fe^{2+}_{0.36}Fe^{3+}_{2.07}O_4$ at $T = 285$ and 87 °K. As seen from the the figure, the spectra arise from the overlap of at least three Zeeman patterns. The weakest sextiplet with the lowest splitting is naturally ascribed to ferrous ions. The ion pattern of octahedral sites is known to have a higher splitting than that of tetrahedral sites, which is due to the covalence effects. This implies that in this case the sextiplet with a higher splitting has to be ascribed to Fe^{3+}_{oct} ions and that with somewhat lower splitting to Fe^{3+}_{tetr} ions. However, in the spectrum obtained the intensity of the Fe^{3+}_{oct} pattern is lower than that of the sextiplet with a smaller splitting. According to the cation distribution of spinels $Me^{2+}Fe^{3+}_2O_4$, the number of Fe^{3+} ions in both octahedral and tetrahedral positions must be the same: (Fe^{3+}) $[Me^{2+}Fe^{3+}]O_4$. As a result, the most intensive sextiplet of the spectrum in Fig. 1 has contributions both from Fe^{3+} ions of the tetrahedral sublattice and Fe^{3+} ions on octahedral sites. Thus in the octahedral sublattice Fe^{3+} ions have at least two non-equivalent positions $(Fe^{3+}_{oct})'$ and $(Fe^{3+}_{oct})''$ as in $Fe^{2+}Fe^{3+}_2O_4$ [3]. The intensity of the observed sextiplet from Fe^{2+} as compared to the patterns due to $(Fe^{3+}_{oct})'' + Fe^{3+}_{tetr}$ and $(Fe^{3+}_{oct})'$ is essentially lower than expected from the crystal-chemical formula of the given sample. This leads to the conclusion that the low-intensity sextiplet is due only to a part of the $(Fe^{2+})'$ ions. The remaining part of $(Fe^{2+})''$ ions contributes to one of the above sextiplets. The groups of six vertical primes in Fig. 1 show the position of the lines due to the contributions of various iron ions, and Table 2 presents the parameters of the corresponding patterns.

Table 1

Crystal-chemical formulas and some parameters of the samples studied*)

sample, formula unit ±0.01	preparation technique	lattice constant ±0.005 (Å)	resistivity (Ωcm)	absorption line positions (cm⁻¹) ν_h	ν_l
$Ni_{0.99}Fe^{2+}_{0.02}Fe^{3+}_2O_4$	solution crystallization in melting	8.343	10^5	667.5 605 537.5	400
$Ni_{0.78}Fe^{2+}_{0.19}Fe^{3+}_{2.01}O_4$	Verneuil technique	8.349	4.6	650 595 545	400
$Ni_{0.53}Fe^{2+}_{0.36}Fe^{3+}_{2.07}O_4$	Verneuil technique	8.357	1.05	660 580 530	380
$Ni_{0.48}Zn_{0.14}Fe^{2+}_{0.33}Fe^{3+}_{2.04}O$	Verneuil technique	8.360	—	580 530	380
$Ni_{0.88}Zn_{0.14}Fe^{2+}_{0.03}Fe^{3+}_{1.97}O_4$	solution crystallization in melting	8.340	10^9	587.5	395
$Ni_{0.95}Fe^{2+}_{0.02}Fe^{3+}_{2.01}O_4$	ceramic technique, T_{sint}=1250°C	8.344	141	657 597.5 535	380
$Ni_{0.87}Fe^{2+}_{0.04}Fe^{3+}_{2.06}O_4$	ceramic technique, T_{sint}=1250°C	8.343	81.2	662.5 590 530	375
$Ni_{0.86}Fe^{2+}_{0.1}Fe^{3+}_{2.03}O_4$	ceramic technique. T_{sint}=1250°C	8.346	37.6	665 587.5 525	375

*) The resistance of the single crystals was measured along [100] direction.

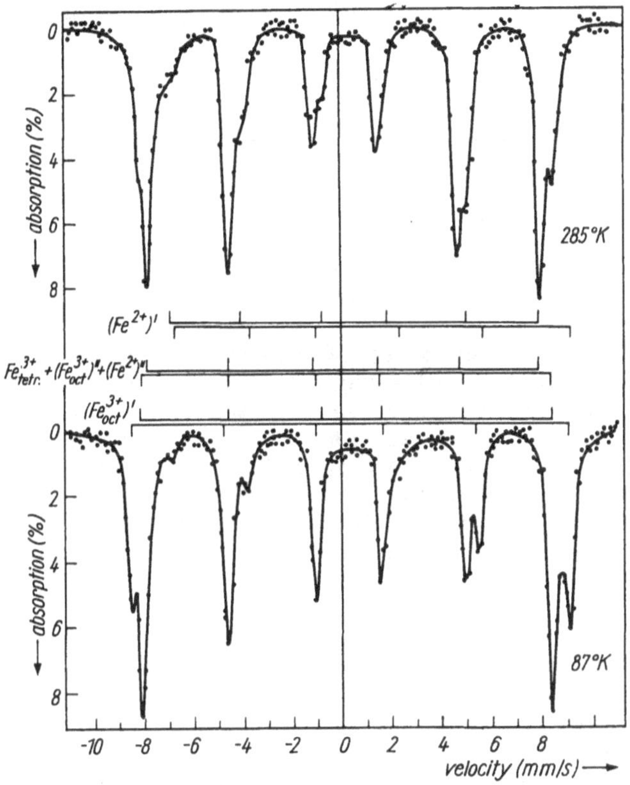

Fig. 1. Mössbauer spectra for $Ni_{0.53}Fe^{2+}_{0.36}Fe^{+3}_{2.07}O_4$ single crystal (the direction of γ-quanta is perpendicular to (110) plane). The background is 340000 impulses

Unfortunately, the Mössbauer data obtained are not sufficient for the unambiguous assumption that the low-intensity sextiplet of the spectrum in Fig. 1 is due to Fe^{2+} ions occupying tetrahedral sites in the lattice, because two versions

Table 2

Parameters for Mössbauer spectra of the $Ni_{0.53}Fe^{2+}_{0.36}Fe^{3+}_{2.07}O_4$ single crystal

line system	H_{eff} (kOe)		$Q\ S$ (mm/s) *)		δ (mm/s) **)	
	T=285°K	T=87 °K	T = 285 °K	T = 87 °K	T = 285 °K	T = 87 °K
$(Fe^{3+}_{oct})''$	520 ± 5	549 ± 3	0.00 ± 0.05	–0.04 ± 0.05	0.32 ± 0.06	0.41 ± 0.06
$(Fe^{3+}_{oct})'' + Fe^{3+}_{tetr} + (Fe^{2+})''$	496 ± 4	516 ± 3	0.08 ± 0.05	0.04 ± 0.05	0.18 ± 0.05	0.27 ± 0.05
$(Fe^{2+})'$	470 ± 10	500 ± 10	0.16 ± 0.08	0.42 ± 0.08	0.55 ± 0.10	0.94 ± 0.10

*) $Q\ S = (I_6 - I_5) - (|I_1| - |I_2|)$, I_i is the line position.
**) Referred to Fe metal absorber.

of cation distribution and line identification are possible, which agree with the experimental results

$$\left(\underline{\mathrm{Fe}_{x}^{2+\prime}\,\mathrm{Fe}_{1-\frac{2}{3}x}^{3+}\right)\left[\mathrm{Fe}_{0.36-x}^{2+\prime\prime}\,\mathrm{Fe}_{\frac{2}{3}x+y}^{3+\prime\prime}}\,\mathrm{Fe}_{1.07-y}^{3+\prime}\,\mathrm{Ni}_{0.53}\right]\mathrm{O}_4$$

and

$$\underline{(\mathrm{Fe}^{3+})\,[\mathrm{Fe}_{0.36-x}^{2+\prime\prime}\,\mathrm{Fe}_{y}^{3+\prime\prime}}\,\mathrm{Fe}_{1.07-y}^{3+\prime}\,\mathrm{Fe}_{x}^{2+\prime}\,\mathrm{Ni}_{0.53}]\,\mathrm{O}_4\,.$$

To prove more strictly the assumption of Fe^{2+} ions in tetrahedral sites of the $Ni_{1-x-y}Fe_x^{2+}Zn_yFe_2^{3+}O_4$ system of ferrite spinel we measured infrared spectra of these compounds.

The infrared transmission spectra of single crystals are shown in Fig. 2. The vibrational spectra of spinels are known to consist of two absorption bands in the range 700 to 300 cm^{-1}: a high-frequency band ν_h due to the intrinsic vibrations of tetrahedral complexes and a low-frequency band ν_l due to the intrinsic vibrations of octahedral complexes [4]. It is seen from Fig. 1 that besides the main high-frequency band ν_h there are two sidebands ν_h' and ν_h'' (curve 1) even for the slightly doped Ni ferrite. When the ferrous doping of the ferrite was increased the primary band ν_h is seen to move towards the long-wavelength range and the intensities of the high-frequency sidebands ν_h' and ν_h'' were increased (Fig. 2, curves 2, 3).

The displacement of the primary band indicates the possible presence of Fe^{2+} ions in tetrahedral sites. If this is the case, the increase of the total amount of Fe^{2+} in a sample will by all means result in increasing its amount in tetrahedral sites as well, that is the effective charges of these sites are decreased [5], leading to the observed decrease in frequency of the intrinsic vibrations of tetrahedral complexes. The existence of the sidebands can be considered as an evidence of Fe^{2+} being present in tetrahedral sites as well. The local lattice deformations can occur due to the Jahn-Teller effect in Fe^{2+} [5] which can lead to a non-cubic component in a crystal-field potential and to the splitting of the ν_h-band which is known to be connected to threefold degenerate intrinsic vibrations of tetrahedral complexes.

Similar data have been obtained on polycrystalline samples (Table 1).

When Zn ions are added to a highly ferrous-doped ferrite the intensities of the high-frequency sidebands decrease (Fig. 2, curve 4). In our opinion this fact can be due to a replacement of Fe^{2+} ions on tetrahedral sites by Zn ions.

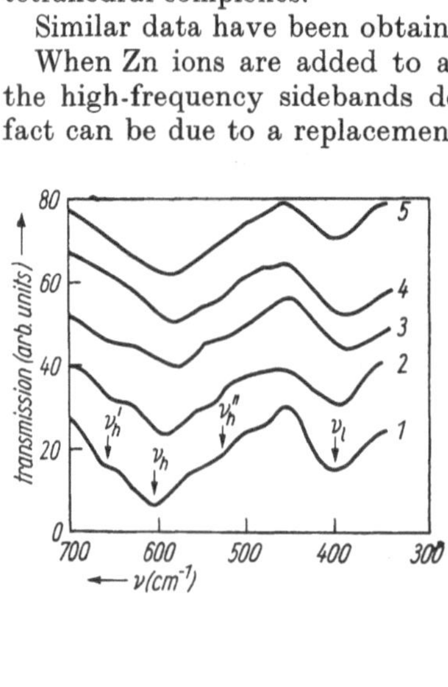

Fig. 2. Infrared transmission spectra for single crystals:
1. $Ni_{0.99}Fe_{0.02}^{2+}Fe_2^{3+}O_4$,
2. $Ni_{0.78}Fe_{0.19}^{2+}Fe_{2.01}^{3+}O_4$,
3. $Ni_{0.53}Fe_{0.36}^{2+}Fe_{2.07}^{3+}O_4$,
4. $Ni_{0.48}Zn_{0.14}Fe_{0.33}^{2+}Fe_{2.04}^{3+}O_4$,
5. $Ni_{0.88}Zn_{0.14}Fe_{0.03}^{2+}Fe_{1.97}^{3+}O_4$.
The spectra are shifted along to the transmission scale

The spectrum from a $Ni_{0.88}Zn_{0.14}Fe^{2+}_{0.03}Fe^{3+}_{1.97}O_4$ sample (curve 5) is the evidence for this; one can see no sidebands in this spectrum although the Fe^{2+} fraction is the same as in the $Ni_{0.99}Fe^{2+}_{0.02}Fe^{3+}_2O_4$ sample (curve 1). The shift of the high-frequency band towards the long-wavelength range observed for Zn-doped ferrites in comparison with the low-doped $Ni_{0.99}Fe^{2+}_{0.02}Fe^{3+}_2O_4$ arose quite naturally from the fact that Zn [6] together with Fe^{2+} displacing Fe^{3+} ions from tetrahedral sites increased the effective masses of these complexes, i.e. the values of the elastic bond constants increased and the frequency of the intrinsic vibrations of these complexes decreased.

4. Conclusions

Thus in terms of the proposed assumptions the given analysis of infrared and Mössbauer spectra for solid solutions of the $Ni_{1-x-y}Fe^{2+}_xZn_yFe^{3+}_2O_4$ system suggests as in earlier papers [1, 2] to conclude about the possible positions of Fe^{2+} not only in the octahedral sites of this system but also in the tetrahedral ones.

Acknowledgements

The authors are indebted to Prof. V. V. Eremenko for valuable and helpful advice; to Dr. A. V. Antonova for useful discussions and to O. P. Gromovenko for supplying the $Ni_{0.53}Fe^{2+}_{0.36}Fe^{3+}_{2.07}O_4$ single crystal.

References

[1] A. A. Schwartz and V. A. Potakova, Relaxation Phenomena in Solid States, Metallurgizdat, Moscow 1968 (p. 673).
[2] V. A. Potakova, V. P. Romanov, N. D. Zverev, O. I. Gromovenko, and E. V. Rubalskaya, phys. stat. sol. (a) **4**, 327 (1971).
[3] R. Hargrove and W. Kündig, Solid State Commun. **8**, 303 (1970).
[4] R. Waldren, Phys. Rev. **99**, 1727 (1955).
[5] M. Goodenough, Magnetism and the Chemical Bond, Wiley, New York/London, 1963.
[6] T. Nagamiya, K. Yosida, and R. Kubo, Adv. Phys. **4**, 1 (1955).

(Received June 9, 1972)

phys. stat. sol. (a) **12**, 629 (1972)

Subject classification: 20.1; 1; 22.8.2

Department of Physics, University of Technology, Eindhoven

Ionic Ordering and Infrared Spectra of Some II–IV Spinels

By

V. A. M. Brabers

The infrared absorption spectra in the range from 200 to 700 cm^{-1} are reported for the spinels $NiMn_2O_4$, $NiMnZnO_4$, $LiZnNbO_4$, and M_2TiO_4 with M=Mg, Mn, Fe, Co, and Zn. The compounds Mn_2TiO_4 and Zn_2TiO_4 could be obtained with 1:1 order on octahedral sites. This ordering gives a superstructure in the IR spectrum, which was not observed in the case of the cobalt and iron compound because of the normal-inverse disorder occurring in the cobalt titanate and the oxygen nonstoichiometry of the iron titanate. The infrared spectra of $NiMn_2O_4$ and $NiMnZnO_4$ give evidence of the short-range order in these materials from which the cation valencies in $NiMn_2O_4$ can be deduced to be $Ni_\nu^{2+}Mn_{(1-\nu)}^{2+}(Ni_\nu^{2+}Mn_{2-2\nu}^{3+}Mn_\nu^{4+})O_4$.

Die Ultrarotabsorption der Spinelle $NiMn_2O_4$, $NiMnZnO_4$, $LiZnNbO_4$ und M_2TiO_4 mit M=Mg, Mn, Fe, Co und Zn wurde im Bereich von 200 bis 700 cm^{-1} gemessen. Die Verbindungen Mn_2TiO_4 und Zn_2TiO_4 wurden durch langsames Abkühlen in einen tetragonalen Spinell mit 1:1-Ordnung auf Oktaederplätzen überführt. Diese Ordnung gibt eine Überstruktur im Ultrarotspektrum, die im Falle der Kobalt- und Eisen-Verbindung nicht beobachtet wurde. Die Ultrarotabsorption von $NiMn_2O_4$ und $NiMnZnO_4$ weist auf eine Nahordnung hin, woraus geschlossen werden kann, daß die Kationenwertigkeiten in $NiMn_2O_4$ durch die Formel $Ni_\nu^{2+}Mn_{(1-\nu)}^{2+}(Ni_\nu^{2+}Mn_{2-2\nu}^{3+}Mn_\nu^{4+})O_4$ dargestellt werden.

1. Introduction

The first simplified model to interpret the vibrational spectra of ionic spinels was reported in 1955 by Waldron [1]. This model is based on the fact that the primitive cell of the spinel structure is a rhombohedron, constructed exclusively from ion groups with T_d symmetry, namely, two AO_4 tetrahedrons, consisting of four oxygen ions with a tetrahedral cation A in the centre, and one B_4 tetrahedron, which is composed of four octahedral cations B. This rhombohedral unit cell permits the classification of the vibrations into the species of the point group T_d, with the result that only four modes are infrared active. Waldron supposed that the two modes with the highest frequencies ν_1 and ν_2 could be attributed to the motion of the oxygen ions with respect to the cations and especially that the high-frequency mode ν_1 could be ascribed to the stretching of the tetrahedral group and ν_2 to the stretching of the octahedral group. The low-frequency modes ν_3 and ν_4 were assigned to oscillations of the cations. These statements imply that interactions between the four vibration modes are ignored.

Lutz [2] concluded on the basis of a normal coordinate analysis that the number of IR active modes is really limited to four, but that isolated vibrations of the polyhedrons are excluded too, although more or less independent vibrations of the ions in octahedral configuration may occur. However, this theoretical inference is in contradiction with the extended experimental work of Preudhomme and Tarte [3, 4] on spinels with a tetrahedral high-valency cation. These authors made in a series of papers [4, 5] a systematic experimental approach to the assignment problem for the spinel structure, using the methods of isotopic and isomorphic substitution outlined in an earlier paper [6]. Their main conclusions are that the high-frequency absorption ν_1 has to be assigned to an isolated vibration of the polyhedron which contains the highest-valency cation and that the lower frequency bands are related to complex vibrations. This implies that the generally accepted systematic assignment of the high-frequency absorption to the tetrahedral group is only justified if the tetrahedral cation has the highest valency and that the statement of Lutz [2] about the impossibility of a characteristic tetrahedral vibration, is not correct.

White and de Angelis [7] treated the vibrational spectrum of the spinel structure by a factor group analysis and arrived at the same number if IR active modes as Waldron. For the assignment of the vibrational spectrum of the spinel structure however, they used again the site group analysis proposed by Waldron, because the factor group analysis gives no clear insight into the assignment of the normal modes to specific atomic motions.

Seen all the problems encountered with the assignments, the question rises if IR spectroscopy anyway can be used in the study of the crystal chemistry of spinels. A positive answer to this question was given by Preudhomme [8], who distinguished three applications namely (i) the determination of the coordination of the cations in the spinel structure, (ii) the study of the cation ordering, (iii) the study of the deformation of the cubic spinel stucture. Topic (i) can be considered as a special case of topic (ii) the ionic ordering, i.e. the distribution of cations between octahedral and tetrahedral sublattice. A theoretical basis of topic (ii) is given by White and de Angelis [7], who examined group-theoretically the influence of a few types of ordering on the phonon spectrum of the spinel structure. Generally, ordering of cations results in a lowering of the overall symmetry, which means an increasing number of IR active vibrations. So, the infrared spectrum of the ordered spinels will show a fine structure compared with the disordered structure, but on the other hand the appearance of extra absorptions in the spectra of solid solutions does not indicate any kind of ordering [5]. One of the first reported examples of the influence of cation ordering on the IR spectrum is $LiAl_5O_8$, where 1 : 3 order on octahedral sites occurs [9]. De Angelis et al. [10] concluded from the vibrational spectra of spinels with tetrahedral 1 : 1 order, that ordering only partially decouples the motions of the tetrahedrons. A compilation of the literature on spectra of ordered spinels is found in [8].

Ionic ordering is in some cases also detectable with X-ray diffraction techniques by means of the appearance of extra reflections or distortions of the cubic structure. However, if only superstructure reflections may occur and the X-ray scattering power of the ordered cations is very close to each other, this ordering can only be detected with IR spectrocopy. An excellent example of this is given by Preudhomme [11] for the manganate spinels of the type $Mn_3X^{2+}Li_2O_8$.

The third topic, which can be studied with IR spectroscopy is the distortion of the cubic spinel structure. This can be a local one, limited to a few atomic distances, or a cooperative one, resulting in a lowering of the symmetry of the crystal structure. The cause of the distortions may be ionic ordering processes mentioned already under topic (ii), and distortions originating from an electronic nature like the Jahn-Teller effect.

The change of the crystal structure is easily quantitatively measured with X-ray diffraction; the IR spectrum gives only qualitatively an indication by means of a splitting or a variation in the shape of the absorption bands and if the distortions are small only a line broadening or a line shift may be observed [12]. Local distortions, however, are difficult to observe with X-ray techniques. A local distortion, which is very common is the displacement of the oxygen ions along the ⟨111⟩ axis as a consequence of the large tetrahedral ions. This distortion is described by the oxygen parameter u. With X-ray techniques this parameter can be calculated from the intensities of certain diffraction lines [13]. Nothing is known about the influence of the oxygen parameter on the IR spectrum. Červinka [14] introduced a second oxygen parameter for the cubic solid solutions $Mn_xFe_{3-x}O_4$, which was assumed to have a close connection with local Jahn-Teller distortions of $Mn^{3+}O_6^{2-}$ octahedrons. Indications of these local tetragonal distortions were also found by the author with IR spectroscopy, since the absorption bands of these materials were split [15], which points out that under certain circumstances the IR spectrum is very sensitive to local distortions.

In this article some applications of IR spectroscopy on spinels will be given, i.e. the application on ordering phenomena in titanate spinels and the determination of the cation valencies in $NiMn_2O_4$.

2. Experimental

Polycrystalline specimens of $NiMn_2O_4$, $NiMnZnO_4$, $LiZnNbO_4$, and M_2TiO_4 with M=Co, Mg, Mn, Fe, and Zn were prepared with the usual ceramic techniques. The details of the preparation are given in Table 1. The preparation temperatures of $NiMn_2O_4$ and $NiMnZnO_4$ are chosen at relatively low values because a segregation of NiO goes on in these materials above 940 and 700 °C respectively. The crystal structure was determined with X-ray diffraction using CuKα or FeKα radiation. The IR absorption measurements were performed in the range of 200 to 700 cm^{-1} by means of the KBr pellet technique. The spectra were recorded on a Hitachi double-beam IR-spectrometer, type EPI-L.

Table 1

compound	sinter temperature	time	atmosphere	starting materials	lattice parameters
Mg_2TiO_4	1350 °C	48 h	air	MgO; TiO_2	8.45 Å
Mn_2TiO_4	1200 °C	24 h	H_2-H_2O mixture 5:1	$MnCO_3$; TiO_2	8.68 Å
Fe_2TiO_4	1200 °C	24 h	H_2-H_2O mixture 3:1		8.54 Å
Co_2TiO_4	1300 °C	6 h	CO_2	Co-oxalate; TiO_2	8.46 Å
Zn_2TiO_4	1200 °C	3 h	air	ZnO; TiO_2	8.47 Å
$LiZnNbO_4$	1000 °C	5 h	O_2-CO_2 mixture 1:1	Nb_2O_5; Li_2CO_3; ZnO	8.61; 8.41 Å
$NiMn_2O_4$	940 °C	48 h	oxygen	Ni-oxalate; Mn-oxalate	8.39 Å
$NiMnZnO_4$	680 °C	150 h	oxygen	Ni-oxalate; Mn-oxalate; Zn-oxalate	8.31 Å

3. Results and Discussion

3.1 Disordered titanates

In Fig. 1 the infrared spectra are given of some titanate spinels prepared with the sinter procedure and cooled to room temperature within half an hour. An interpretation of the spectra in terms of localized vibrations of polyhedrons, which are not coupled to each other, is not very significant. The experimental evidence, given by Preudhomme and Tarte [5], that in normal spinels the polyhedron with the highest-valency cation is responsible for the highest-frequency absorption cannot be adopted for the titanate spinels, because the octahedrons are occupied by Ti^{4+} ions and Mn^{2+} ions as well. However, the general feature, which follows from the group-theoretical treatment [1, 2, 7], i.e. the number of infrared active modes is limited to four, is still valid for the titanate spinels. All of the spectra, except that of Mg_2TiO_4, show in the measured range from 200 to 700 cm^{-1} three absorptions; in addition to this triplet, another absorption was reported by Preudhomme [8] below 200 cm^{-1} for Co_2TiO_4 and Zn_2TiO_4, which completes the number to four. These absorption bands are rather broad compared with those reported in literature for normal spinels [5, 16]. This is in agreement with the fact that Ti^{4+} ions have a strong preference for the octahedral environment, which means that titanate spinels are mainly inverse. The local symmetry is now lowered by the presence of two kinds of cations on octahedral sites and the triply degenerate vibration will split into three vibrations of different frequency. Because of a statistical distribution of the cations within the octahedral sublattice, there will be a spread in these frequencies, which results in the above-mentioned broadening of the absorptions.

Mg_2TiO_4 shows a different behaviour; the number of absorptions in the 200 to 700 cm^{-1} range is already six, which could be an indication of ionic ordering. From the theoretical possibilities of ordering in the spinel structure, only the normal-inverse order and the 1:1 order on octahedral sites can occur in this case [17]. However, with X-ray diffraction no superstructure reflections nor a transition to the tetragonal structure was observed if a specimen was cooled slowly from 1350 °C to room temperature with a cooling velocity of 7 deg/h. Also the IR spectrum did not change, which is the inference that the larger number of IR active modes is not caused by ordering by the presence of the two kinds of cations on one sublattice. The frequency of a vibration is determined by various parameters like the cationic mass, the bonding forces, and the cation–anion dis-

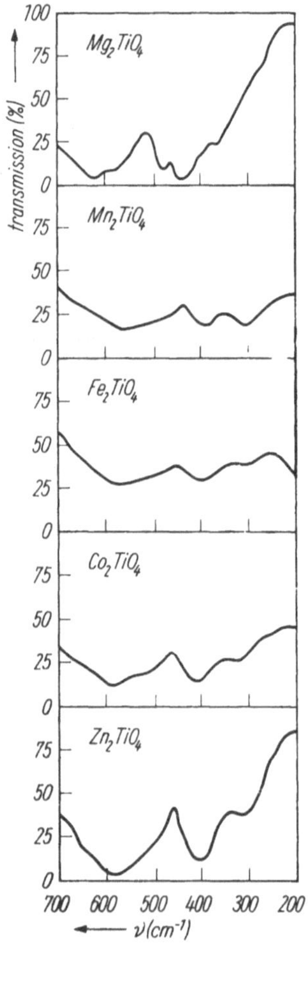

Fig. 1. Infrared spectra of disordered titanate spinels

tances. There is now one distinct difference between Mg_2TiO_4 and the other reported titanates and that is the cationic mass of the bivalent cations. The mass of the Mg^{2+} ion is about a factor two smaller than the mass of the titanium ion, whereas the bivalent cations in the other titanates have a mass which deviates maximally only 20%. This has the consequence that the variation in frequency caused by mass effects have to be greater in the case of Mg titanate, which can be the reason for the splitting of the absorptions in Mg_2TiO_4 while the absorption bands in the other titanates are only broadened. So the occurrence of extra lines in the IR spectra is not automatically the proof of an ordering process [7, 5].

3.2 Ordered titanates

Because of the preference of the Ti^{4+} ion for octahedral environment, two ordering processes can go on in the titanate spinels, namely, the normal-inverse ordering and the 1:1 ordering on octahedral sites. The 1:1 order on octahedrons can give two modifications, one belongs to the space group D_4^3-P 4_1 22 with a primitive tetragonal unit cell and the other to D_{2h}^{28}-Immb with a body centered orthorhombic unit-cell. Both unit-cells are two times smaller than the f.c.c. unit-cell of the spinel structure. A precondition that the 1:1 ordering can occur is that the structure is inverse, i.e. there have to be two kinds of cations on octahedral sites in the ratio 1:1. From thermal expansion measurements [18] and electrical conductivity measurements [19] on manganese-iron spinels, it was concluded that at relatively low temperatures (300 to 450 °C) a cation exchange between octahedral and tetrahedral sites is possible within a few minutes. Nothing is known about the kinetics of this exchange process, but the incidence of a temperature-dependent cation distribution in spinels is very common. Also about the kinetics of the 1:1 ordering not much is known, only that in the case of Zn_2TiO_4, the time to get a total conversion to the ordered structure is at 480 °C in the order of a few hours and in the transition range of 500 to 550 °C still larger [20]. Billiet and Poix [20] found also that a Zn_2TiO_4 specimen, quenched from 1200 °C could not be transformed to the ordered structure by annealing below 550 °C, which was ascribed to the cation exchange between the sublattices, which would not take place below that temperature. But, it is also possible that this phenomenon is caused by the disturbance of the short range ordering on octahedral sites, which goes on during slow cooling from 1200 to 600 °C; in

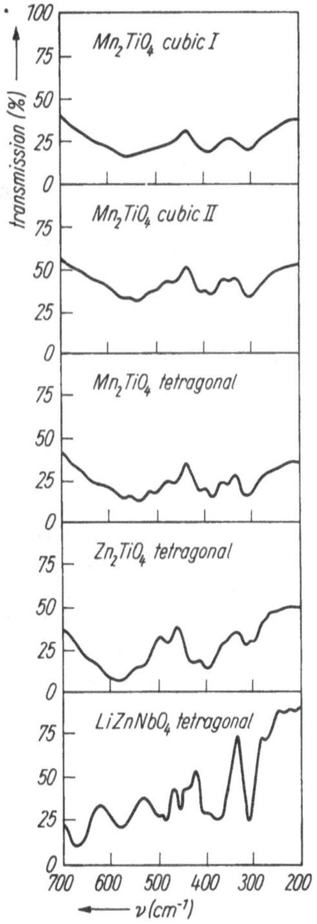

Fig. 2. Infrared spectra of spinels with 1:1 order on octahedral sites

the quenched specimens the ions have to diffuse over larger distances to reach the complete ordered state, which occurs very slowly at low temperature. Also the identical values of the lattice parameter observed for quenched and slowly cooled cubic specimens are indicative that the cation exchange does not play an important role in this ordering process. It is not clear which diffusion determines the kinetics of the octahedral 1:1 ordering process.

To get ordered titanate spinels, specimens of the five titanates were slowly cooled with 7 deg/h from 1000 °C to room temperature. The annealing of Mn_2TiO_4, Fe_2TiO_4, and Co_2TiO_4 was carried out in vacuum (10^{-5} Torr) sealed silica tubes, to avoid oxidation. The results of the X-ray measurements on sintered and annealed specimens are given in Table 2.

Table 2

compound	sintered specimens	annealed
Mg_2TiO_4	$a = 8.45$ Å	$a = 8.45$ Å
Mn_2TiO_4	$a = 8.68$ Å	$c = 8.58$ Å; $a = 8.74$ Å
Fe_2TiO_4	$a = 8.54$ Å	$a = 8.54$ Å
Co_2TiO_4	$a = 8.46$ Å	$a = 8.45$ Å
Zn_2TiO_4	$a = 8.47$ Å	$c = 8.42$ Å; $a = 8.49$ Å

Mn_2TiO_4 and Zn_2TiO_4 are the only substances the structure of which changes from cubic into tetragonal ($P4_1$ 22). The transition temperatures are 770 and 550 °C for Mn_2TiO_4 and Zn_2TiO_4, respectively [20, 21]. A specimen of Mn_2TiO_4 (II), which was cooled from 1200 °C to room temperature within 2 h remained cubic, although line broadening in the X-ray diagram was observed. The infrared spectra of these annealed specimens are given in Fig. 2, where for comparison the spectra of cubic Mn_2TiO_4 (I) and tetragonal $LiZnNbO_4$ (1:1 order of octahedral Li^+ and Nb^{5+} ions) are given, too. From these spectra it can be inferred that the X-ray line broadening observed in Mn_2TiO_4 (II) originates from the beginning of the octahedral 1:1 order, since the fine structure, which appears in the IR spectrum has the same features as the spectrum of the completely ordered tetragonal spinel, is only less pronounced.

According to White and de Angelis [7] the number of IR active modes for this ordered structure is 31. Because of the complex nature of the vibrations, which was shown experimentally by Preudhomme [8] by means of isotopic substitutions in Zn_2TiO_4, it is impossible to interpret the spectra in terms of localized vibrations. The only conclusion that can be drawn from these spectra is that there is an ordering process. The kind of ordering and the crystal structure has to be determined with X-ray diffraction, although by a tentative comparison of the spectra, one can draw a conclusion about the kind of ordering as is shown for the cubic Mn_2TiO_4 with broadened X-ray lines.

The appearance of the spectrum of ordered Mn_2TiO_4 is very close to that of Zn_2TiO_4, obviously because the atomic masses and the valencies of the cations are equivalent. The valencies and atomic masses in $LiZnNbO_4$, however, are quite different, which results in a very different spectrum. In the case of the manganite spinels in Section 3.3 we shall use the reversed reasoning that IR spectra with the same features, i.e. location, number, and shape of absorption bands, originate from spinels with equivalent atomic masses and cation valencies.

The annealed Fe_2TiO_4 specimens did not show any variations in the IR spectrum. The absence of ordering can be explained by the non-stoichiometry of the specimen containing a small excess of oxygen, by which cation vacancies on octahedral sites are created and the 1:1 ratio is disturbed.

In Co_2TiO_4 the 1:1 order is prevented by the normal-inverse disorder. Near 400 °C, an anomaly in the thermal expansion analogous to those in $MnFe_2O_4$ [18], indicated the exchange of cations between the octahedral and tetrahedral sublattices.

3.3 Nickel manganite

$NiMn_2O_4$ crystallizes in the cubic spinel structure, in contradistinction to most of the other manganites [22] which are tetragonal spinels deformed by the Jahn-Teller effect of the octahedral Mn^{3+} ion. A neutron diffraction study by Boucher et al. [23] proved the temperature dependence of the distribution of Ni and Mn ions among the two sublattices; between 350 and 940 °C the inversion parameter ν, determined by the formula

$$Mn_\nu\, Ni_{1-\nu}\, (Ni_\nu\, Mn_{2-\nu})\, O_4\ ,$$

varied from 0.94 for low temperature to 0.74 for a specimen quenched from 940°C. The ions before the brackets are located on tetrahedral sites, those between the brackets on octahedral sites. On account of the cubic structure, Bongers [24] excluded the presence of Mn^{3+} ions and proposed the cation distribution $Mn^{2+}(Mn^{4+}\,Ni^{2+})O_4$.

Boucher et al. [25] proposed the distribution $Mn_\nu^{3+}Ni_{1-\nu}^{2+}(Ni_\nu^{2+}Mn_{2-\nu}^{3+})O_4$, based on a particular model for the magnetic structure, Because of the sensitivity for cation valencies, the infrared spectrum offers now an opportunity to decide in this controversy about the cation valencies by comparing the spectrum of $NiMn_2O_4$ with that of a proper reference material, namely $NiMnZnO_4$. $NiMnZnO_4$ is a cubic spinel, and owing to the large stability of the Ni^{2+} ion and the preference of Zn^{2+} for tetrahedral environment, the cation distribution of this compound is $Zn^{2+}(Ni^{2+}Mn^{4+})\,O_4$. Fig. 3 shows the IR spectra of the quenched specimens and those cooled slowly with 7 deg/h. The bands in the spectra of the quenched specimens are not so distinguished as those of the slowly cooled samples, which is in agreement with the increasing normal-inverse disorder at higher temperatures. For $NiMn_2O_4$ even a small shifting of the absorption bands is observed indicating that relatively large quantities of cations migrate. In addition to the increasing ordering of cations between the sublattices by means of slow cooling, short-range

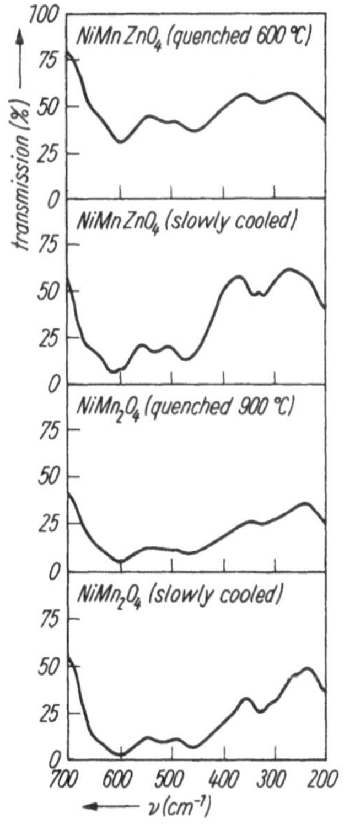

Fig. 3. Infrared spectra of $NiMn_2O_4$ and $NiMnZnO_4$, quenched and slowly cooled

ordering on octahedral sites occurs, which was verified by Boucher et al. [23] for $NiMn_2O_4$ by neutron diffraction.

It turned out that the octahedral Mn ions have a tendency to surround themselves with Ni ions as nearest cation neighbours and vice versa the Ni ions with Mn ions. This short-range ordering can be considered as the beginning of the 1:1 order on octahedral sites, which favours the idea that $NiMn_2O_4$ is a II–IV spinel; octahedral Mn^{3+} ions have a tendency to form clusters, which would prevent the short-range ordering. On closer inspection of the IR spectra of slowly cooled $NiMn_2O_4$ and $NiMnZnO_4$ the excellent similarity of the spectra has led to the conclusion that both spinels have the same ionic structure, i.e. the tetrahedral sites are occupied by bivalent cations and the octahedral sites mainly by bi- and quadrivalent cations. So the structure of $NiMn_2O_4$ has to be presented by the formula

$$Mn_\nu^{II}\, Ni_{1-\nu}^{II}\, (Ni_\nu^{2+}\, Mn_{2-2\nu}^{3+},\, Mn_\nu^{4+})\, O_4\,.$$

Further, an investigation of the semiconducting properties of $NiMn_2O_4$ which is in progress, has also proved the simultaneous presence of Mn^{4+} and Mn^{3+} ions on octahedral sites [26].

Acknowledgements

The author expresses his thanks to Mr. P. Hummeling and Mrs. M. Kuyer for recording the IR spectra, and Mr. M. J. Heijligers for this help with the preparation of the materials.

References

[1] R. D. WALDRON, Phys. Rev. **99**, 1727 (1955).
[2] H. D. LUTZ, Z. Naturf. **24a**, 1417 (1969).
[3] P. TARTE, Spectrochim. Acta **19**, 49 (1963).
[4] J. PREUDHOMME and P. TARTE, Spectrochim. Acta **28A**, 69 (1972).
[5] J. PREUDHOMME and P. TARTE, Spectrochim. Acta **27A**, 961 (1971); **27A**, 845 (1971); **27A**, 1817 (1971).
[6] J. PREUDHOMME and P. TARTE, Spectrochim. Acta **26A**, 2207 (1970).
[7] W. B. WHITE and B. A. DE ANGELIS, Spectrochim. Acta **23A**, 985 (1967).
[8] J. PREUDHOMME, Thesis, University of Liege, 1970.
[9] P. TARTE, Silicates Ind. **28**, 345 (1963).
[10] B. A. DE ANGELIS, V. G. KERAMIDAS, and W. B. WHITE, J. Solid State Chem. **3**, 358 (1971).
[11] J. PREUDHOMME, C. R. Acad. Sci. (France) **267C**, 1632 (1968).
[12] K. SIRATORI, J. Phys. Soc. Japan **23**, 948 (1967).
[13] L. P. S. KOLNICK, S. KONDO, and L. R. LAVINE, J. appl. Phys. **29**, 198 (1958).
[14] L. ČERVINKA, J. Phys. Chem. Solids **26**, 1917 (1965).
[15] V. A. M. BRABERS, phys. stat. sol. **33**, 563 (1969).
[16] S. HAFNER, Z. Krist. **115**, 113 (1961).
[17] C. HAAS, J. Phys. Chem. Solids **26**, 1225 (1965).
[18] V. A. M. BRABERS, J. Phys. Chem. Solids **32**, 2181 (1971).
[19] V. A. M. BRABERS, Ber. dtsch. keram. Ges. **47**, 648 (1970).
[20] Y. BILLIET and P. POIX, Bull. Soc. chim. (France) **3**, 477 (1963).
[21] H. VINCENT, J. C. JOUBERT, and A. DURIF, Bull. Soc. chim. (France) **1**, 246 (1966).
[22] R. BUHL, J. Phys. Chem. Solids **30**, 805 (1969).
[23] B. BOUCHER, R. BUHL, and M. PERRIN, Acta cryst. **B25**, 2326 (1969).
[24] P. BONGERS, Thesis, Leiden 1957.
[25] B. BOUCHER, R. BUHL, and M. PERRIN, J. Phys. Chem. Solids **31**, 363 (1970).
[26] V. A. M. BRABERS, J. C. J. M. TERHELL, and R. LIETH, to be published.

(Received May 15, 1972)

phys. stat. sol. (a) 12, 637 (1972)

Subject classification: 14.4; 2; 22.9

Northwestern University, Department of Materials Science, Evanston, Illinois

Dielectric Properties of Cholesteric Liquid Crystals[1]

By

D. G. SHAW[2] and J. W. KAUFFMAN

The dielectric constant and loss of the three cholesteric liquid crystals, cholesteryl acetate (CA), propionate (SP), and stearate (CS) in the solid, liquid crystal, and isotropic liquid state are presented as a function of temperature and frequency. These data provide evidence for the existence of two different cholesteric liquid crystal states for CA and CS, one on heating and one on cooling. The liquid crystal state on heating behaves dielectrically like a solid and on cooling behaves dielectrically like a liquid. The dielectric constant on cooling is much greater than that on heating. The liquid crystal state of CP on the other hand behaves dielectrically like a liquid on both heating and cooling. The permanent and induced dipole moments and the orientational and instantaneous polarizations are calculated for the three esters. The instantaneous polarizations are strongly dependent on the ester chain length and are of much greater magnitude than the orientational polarizations. Both the permanent dipole and the orientational polarizations are independent of ester chain length.

Die Dielektrizitätskonstante und der Verlustwinkel von drei Cholesterin-Flüssigkristallen, Cholesteryl-azetat (CA), -propionat (CP) und -stearat (CS), werden im festen Zustand, als Flüssigkristall und im isotropen, flüssigen Zustand als Funktion der Temperatur und Frequenz dargestellt. Diese Werte ergeben den Beweis für die Existenz von zwei verschiedenen Cholesterin-Flüssigkristallzuständen für CA und CS, einer nach dem Aufheizen und einer nach dem Abkühlen. Die Flüssigkristalle verhalten sich nach dem Aufheizen dielektrisch wie ein Festkörper und nach dem Abkühlen wie eine Flüssigkeit. Die Dielektrizitätskonstante ist nach dem Abkühlen viel größer als nach dem Aufheizen. Der Flüssigkristallzustand von CP verhält sich andererseits dielektrisch wie eine Flüssigkeit sowohl beim Aufheizen als auch beim Abkühlen. Die permanenten und induzierten Dipolmomente und die Orientierungs- und Instantanpolarisationen werden für die drei Ester berechnet. Die instantanen Polarisationen hängen stark von der Kettenlänge der Ester ab und sind größer als die der Orientierungspolarisationen. Sowohl die permanenten Dipolmomente als auch die Orientierungspolarisationen sind von der Kettenlänge der Ester unabhängig.

1. Introduction

It is well known that the cholesteric liquid crystals exhibit many unusual electro-optical phenomena arising from the crystalline-like structure of the fluid state. Some of the phenomena are: 1. ac and dc field induced transitions [1, 2] from the cholesteric to the nematic liquid crystal state resulting in a change from optically negative to optically positive refringence, 2. ac and dc field dependence [3, 4] of the light-scattering properties, and 3. field and

[1]) This research was supported by the Advanced Research Projects Agency of the Department of Defense through Northwestern University. This paper is submitted in partial fulfilment of the requirements for a Ph.D. degree in Materials Science.

[2]) Permanent address: General Electric Industrial & Power Capacitor Products Department, Advance Development Laboratory, John St., Hudson Falls, New York 12839.

temperature [5] dependence of the pitch of the helical cholesteric structure. The materials in which these effects are observed are mostly mixtures of a cholesterylhalide and a cholesteryl ester. Individually these components exhibit the same phenomena only at higher fields than in the mixture.

Several recent papers and reports [2, 5, 6] attempt to determine the dielectric properties of the halide–ester mixture and the role of induced and permanent dipole moments in the electro-optical effects. Most of this work has concentrated on the cholesteryl chloride–cholesteryl myristate mixture. Further study of the dielectric properties of the individual components of the halide–ester mixture is necessary before a complete explanation of these effects and a dielectric characterization of the cholesteric liquid crystal state can be obtained.

Previous studies of the electrical properties of cholesteric liquid crystals have been concerned with the current–voltage [7] and conductivity [8] characteristics. The present work on the dielectric properties is a continuation of this study.

2. Dielectric Theory

When an electric field is applied to a dielectric the electrical charges in the material are displaced from their equilibrium positions giving rise to a net polarization. The net polarization consists of an orientational polarization, P_M, an electronic polarization, P_E, an atomic polarization, P_A, and often an interfacial polarization, P_I, i.e. $P_{net} = P_M + P_A + P_I + P_E$.

P_M arises from the orientation of the permanent molecular dipole moment in the direction of the applied field. The allignment of the molecular dipole moment with the electric field is opposed by thermal fluctuations and near neighbor frictional forces. Onsager [9] has derived a relation between P_M, the permanent dipole moment, μ_0, and temperature, T:

$$P_M = \frac{3\,\varepsilon'\,(\varepsilon_0 + 2)\,4\,\pi\,N\,\mu_0^2}{(2\,\varepsilon' + \varepsilon_0)\,(\varepsilon' + 2)\,9\,kT}\,, \tag{1}$$

where ε' is the low-frequency dielectric constant, ε_0 is the very high-frequency dielectric constant (often identical to the solid-state dielectric constant in molecular crystals), and N and k are Avogadro's number and Boltzmann's constant, respectively. Maxwell [10] has shown that the dielectric constant measured at a particular frequency is related to the index of refraction, n, measured at the same frequency by the equation

$$\varepsilon' = n^2\,. \tag{2}$$

The electric field also induces into the dielectric an electronic polarization, P_E, by displacing the electrons relative to their positive nuclei and an atomic polarization, P_A, by displacing atoms relative to one another. Debye [11] has shown that this induced or instantaneous polarization can be written in terms of the induced dipole moment, α_0, and the sum of the electronic and atomic dielectric constants, ε_0, as,

$$P_{ind} = P_E + P_A = \frac{4\,\pi\,N\,\alpha_0}{3} = \frac{(\varepsilon_0 - 1)\,M}{(\varepsilon_0 + 2)\,d}\,, \tag{3}$$

where M is the molecular weight and d is the density. The observed dielectric constant, ε', can be related to the induced and permanent dipole moments by the Onsager equation [9],

$$\frac{(\varepsilon' - 1)\, M}{(\varepsilon' + 2)\, d} = \frac{3\,\varepsilon'\,(\varepsilon_0 + 2)}{(2\,\varepsilon' + \varepsilon_0)\,(\varepsilon' + 2)} \left(\frac{4\,\pi\,N\,\mu_0^2}{9\,kT}\right) + \frac{\varepsilon_0 - 1}{\varepsilon_0 + 2}\left(\frac{M}{d}\right). \tag{4}$$

The equations relating the various polarizations, the induced and permanent dipole moments, and the experimentally obtained dielectric constants have now been presented. A fourth type of polarization often found in heterogeneous materials is the interfacial polarization which is due to the charging of interfaces at boundaries within the dielectric. This polarization often takes seconds or minutes to form.

With each of the polarizations discussed above can be associated a dielectric constant and the observed dielectric constant, ε', may be written as a sum of the orientational, electronic, atomic, and interfacial dielectric constants as $\varepsilon' = \varepsilon_M' + \varepsilon_E' + \varepsilon_A' + \varepsilon_I'$. A schematic diagram of ε' versus frequency in Fig. 1 demonstrates the different components of the dielectric constant and how they disappear as frequency increases. At increasing frequencies the polarizations requiring long relaxation times don't have enough time to form and the dielectric constant consequently decreases. An increase in the dielectric loss, ε'', or dissipation factor, $\tan\delta = \varepsilon''/\varepsilon'$, is observed at the frequencies at which this dielectric dispersion takes place. This dispersion is due to a dissipation of part of the electric field energy in heat. The dielectric loss versus frequency is also demonstrated in Fig. 1.

The dielectric constant components, ε_M', ε_E', ε_A', and ε_I' can be obtained by capacitance measurements as a function of frequency, temperature and state of matter, and by the index of refraction measurements. The dielectric constant is computed from the specimen capacitance, C, and the vacuum capacitance, C_0, according to the equation $\varepsilon' = C/C_0$. In the solid state the permanent molecular dipoles are frozen in the lattice and $\varepsilon' = \varepsilon_E' + \varepsilon_A'$. This relation assumes that the dielectric constant is being measured at frequencies sufficiently high for interfacial polarizations to be neglected. The value of ε_E' is obtained from index of refraction and equation (2). In polar molecules the dielectric constant is expected to increase on melting by the amount ε_M' since the molecular dipoles are much less restricted than in the solid state. The various polarizations and the induced and permanent dipole moments are computed from this dielectric constant according to equations (1) to (4).

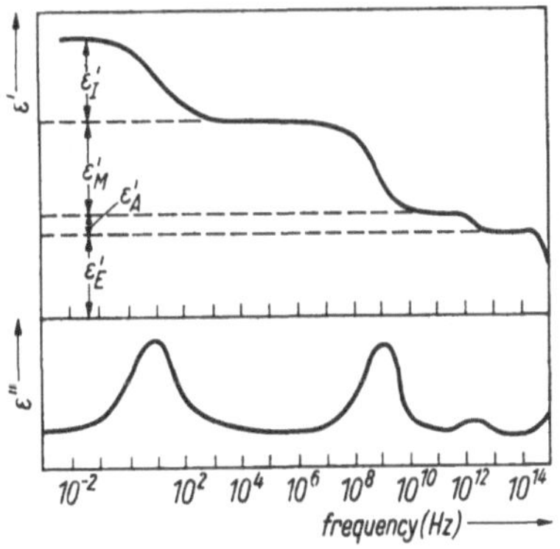

Fig. 1. Schematic diagram of dielectric constant, ε', and loss, ε'', as a function of electric field frequency

3. Experimental

The purity of the cholesteryl esters, the specimen chamber, and the electrode configuration has been previously discussed [7] however a brief description of each will be given here for completeness. The cholesteryl esters were obtained from General Biochemicals Co. in chromatographically pure form as determined by TLC. Carbon–hydrogen analysis and melting point determinations were made before and after each experiment to estimate the degradation due to heating and cooling cycles. The specimens were stored in the dark under dry nitrogen at 0 °C and all measurements were made in the dark under dry nitrogen atmosphere.

A hermetically sealed three terminal micrometer adjustable electrode cell was used for all capacitance measurements. The lower electrode consists of a cup into which the ester was melted. The guard and upper electrodes were then immersed in the melt. The electrodes were gold plated to prevent the specimen degradation and the electrode pitting observed with stainless steel electrodes. The capacitance measurements were ascertained to be due to volume capacitance rather than electrode or circuit effects by verifying the $C = \varepsilon A/d$ relationship for volume capacitance of a parallel plate condenser. The $C \sim 1/d$ relation was observed for cholesteryl, acetate, propionate, and stearate. The active electrode area, A, was 0.75 cm^2.

The capacitance measurements were made using a ratio-arm bridge circuit, accuracy of which was within 0.01%. The specimen temperature was measured to within 0.1 °C using a thermocouple on the upper and lower electrodes and a potentiometric measuring circuit.

The index of refraction was measured on a Bausch and Lomb Abbe refractometer at variable temperatures. The refractometer was first calibrated for distilled water and then the index of refraction measured for the three esters. The accuracy of the measurement was about 0.1% in the liquid state and decreases to about 1% in the solid state. The density was measured using a U-shaped glass tube. One end was sealed and a capillary was mounted on the other end. A cholesteryl ester fused pellet and quartz chips were inserted into the tube. The tube was evacuated and liquid mercury was then introduced into the tube through the capillary. As the apparatus is heated in the oil bath by heat lamps the specimen and mercury expands. This specimen expansion can be measured by weighing the mercury that is collected from the top of the capillary every 5 °C and then correcting for the expansion of mercury.

4. Results and Discussion

The variation of dielectric constant and dissipation factor on increasing and decreasing temperature for cholesteryl acetate, propionate, and stearate is shown in Fig. 2 to 4. Little change in ε' is observed on increasing temperature for CA and CP until we reach a critical temperature at which ε' increases abruptly. This discontinuous increase in ε' corresponds to a sudden loosening or relaxation of the molecular lattice which causes the motion of the permanent molecular dipoles to become less restricted. In polar molecules this phenomenon is usually observed at the melting point. In the case of CP (Fig. 3) this abrupt increase in ε' is observed at the melting point ($\approx$ 100 °C) and the liquid crystal state of CP is seen to behave dielectrically like a true liquid in which the molecular dipoles are relatively free to respond to the electric field. On further

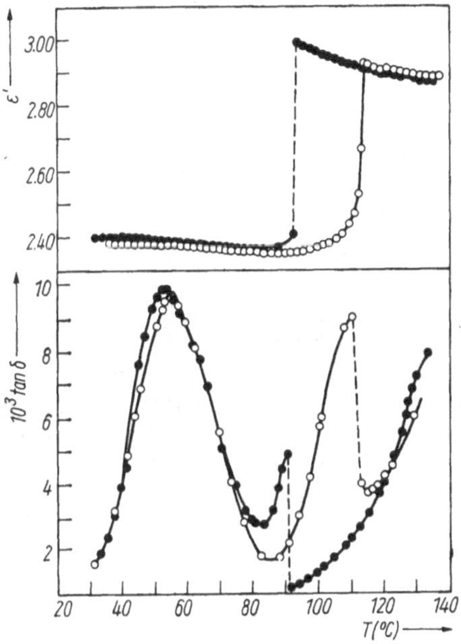

Fig. 2. Dielectric constant ε', and dissipation factor, tan δ, plotted as a function of temperature for cholesteryl acetate. Heating and cooling rates 15 deg/h, applied field 10 V/cm peak-to-peak at 10 kHz. (●) cooling, (○) heating

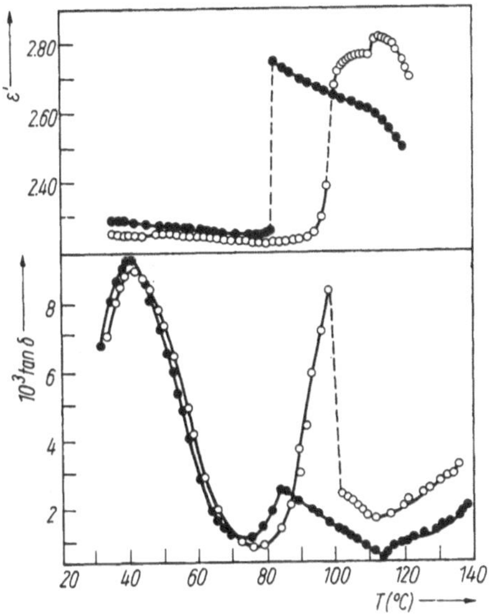

Fig. 3. Dielectric constant, ε', and dissipation factor, tan δ, plotted as a function of temperature for cholesteryl propionate. Heating and cooling rates 15 deg/h, applied field 10 V/cm peak-to-peak at 10 kHz. (●) cooling, (○) heating

heating CP through the liquid crystal state a slight increase in ε' is observed at the clearing point (liquid crystal–isotropic liquid transition temperature) at about 115 °C.

The dielectric behavior of CA and CS on heating is quite different from the behavior of CP. The large discontinuous increase in ε' which is observed at the melting point of CP is not observed at the melting point of CA (≈ 95 °C) and CS (≈ 75 °C) in Fig. 2 and 4. On heating the dielectric constant of the liquid crystal state of CA and CS is about the same as that of the solid state and the motion of the permanent dipoles are restricted as in the solid state. A large increase in ε' occurs not at the melting point but at the clearing point. This indicates a basic difference in the liquid crystal structure

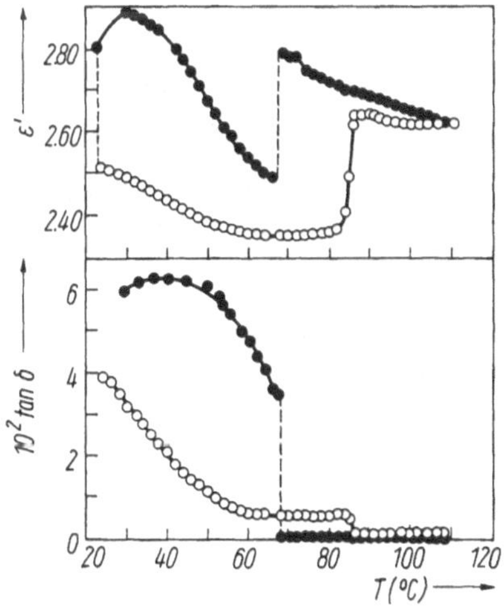

Fig. 4. Dielectric constant ε', and dissipation factor tan δ, plotted as a function of temperature for cholesteryl stearate. Heating and cooling rates 15 deg/h, applied field 10 V/cm peak-to-peak at 10 kHz. ● cooling, ○ heating

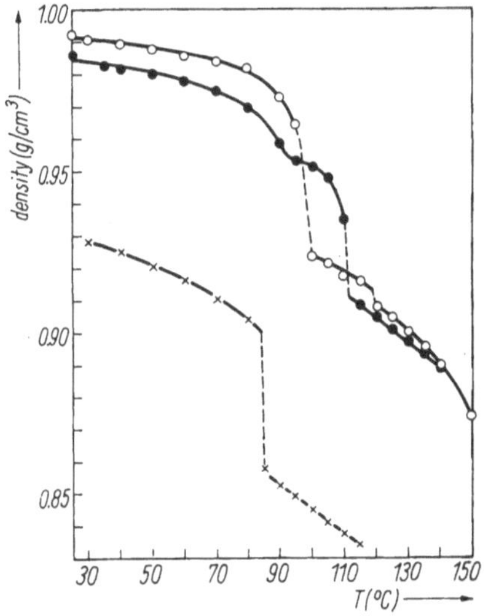

Fig. 5. Density plotted versus temperature for cholesteryl acetate (●), propionate (○), and stearate (×) (curves obtained on heating)

between the acetate and stearate esters and the propionate ester. It is a well known fact that the cholesteric liquid crystal state exhibits a helical structure in which the lathe shaped molecules lie in planes with the long axis of the molecules in each plane pointed in the same direction. The molecular direction in each successive plane is rotated slightly from the plane beneath it. This rotation of the molecular direction in successive planes results in a helical structure with a characteristic pitch and interplanar spacing. The constrained motion of the molecular dipoles in the liquid crystal state of CA and CS on heating which is evident from the low dielectric constant is probably a function of pitch or interplanar spacing. However, to fully explain the dielectric properties of the liquid crystal state in terms of the helical pitch or interplanar spacing would require detailed structural information which is not available at the present time.

If the dielectric characteristics discussed above are due to a contracted or a relaxed condition of the molecular lattice then these characteristics should also be reflected in any density change at the melting and clearing points. The variation of density with temperature on heating for the three esters is shown in Fig. 5. Here we see a large decrease in density at the melting point for CP and at the clearing point for CA and CS. Thus, the freeing of molecular dipoles, as evident from the increase in ε', corresponds to a relaxing or expansion of the molecular structure evident from the decrease in the density.

On cooling the large magnitude of ε' in both the isotropic liquid and liquid crystal state indicates that the dipoles are free to move and the intermolecular forces remain in a relaxed condition in both states. There appears to be two dielectrically different liquid crystal states for CA and CS; the liquid crystal state on heating seems to "remember" the solid state structure from which it came and behaves dielectrically like a solid while the liquid crystal state on cooling "remembers" the isotropic liquid from which it came and behaves dielectrically like a liquid. A transition from the low ε' liquid crystal structure to the high ε' liquid crystal structure could not be made without first proceeding to the isotropic liquid and the reverse transition could only be made by proceeding through the solid state. Barrall and Sweeney [12] have reported a difference in the mesophase textures of dicholesteryl adipate on heating and cooling. The present work provides quantitative evidence for the existence of two cholesteric liquid crystal states, one with a low ε' and the other with a much larger ε'. These results indicate the need for a detailed investigation of the structure of these materials in order to understand the cholesteric liquid crystal state.

The transition temperatures mentioned above were determined by observing the ester under a polarizing microscope using a heating slide and a thermocouple to measure and control temperature. The melting and clearing points were determined optically to within 1.5 °C and are listed in Table 1 for the three esters. These transition temperatures do not differ greatly from ones previously reported [13].

Table 1

Table of solid–liquid crystal and liquid crystal–isotropic liquid phase transition temperatures (±1.5 °C) for cholesteryl acetate, propionate, and stearate on heating and cooling

	S.S.–L.C. (°C)		L.C.–I.L. (°C)	
specimen	heating	cooling	heating	cooling
cholesteryl acetate	≈95	85	≈115	≈115
cholesteryl propionate	100	80	116	110
cholesteryl stearate	75	68	85	72

The abrupt decrease in ε' on cooling in Fig. 2 to 4 occurs at the liquid crystal–solid crystal transition and corresponds to a freezing of the molecular dipoles in the molecular lattice.

A very large variation in ε' and $\tan \delta$ as a function of T and a considerable amount of hysteresis between heating and cooling for the long chain CS in the solid state is observed in Fig. 4. This same general result was observed in specimens from two separate lots with heating and cooling rates of 15 deg/h. However, the change in ε' and the degree of hysteresis was much lower in a specimen from a third lot with a slower cooling rate of 1 deg/h. This suggests that the solid state structure of CS is strongly dependent on the cooling rate and various forms of the solid state can be obtained at varying cooling rates. It is unlikely that the large change in ε' in the CS solid state is due to either the atomic, electronic, or orientational polarization since P_E and P_A are not expected to be temperature-dependent and P_M is negligible in the solid state. It is probable however, that the variation in ε' can be attributed to a change in the interfacial polarization, either due to a change in the P_I relaxation frequency with temperature or to a change in the magnitude of P_I due to a structural change as a function of temperature. A change in P_1 could arise either from a change in structure or some other mechanism changing the interfacial electrical environment such as a contraction of the lattice on cooling resulting in a slight change in the grain boundaries. A change in the solid state structure would give rise to the large dissipation factor and its variation with temperature in Fig. 4. Similar interfacial effects in CA and CP could also account for the peaks in the dissipation factor in Fig. 2 and 3. However, the dissipation factor here is much smaller than in CS. The DTA measurements of Barall et al. [14] indicate the possibility of a solid state transition in CA at 44 °C. It has been shown [15] that solid state structures of varying grain sizes can be produced by varying the cooling rate of CA. A similar grain size dependence on cooling rate in CS could explain the dependence of ε' vs. T curves on the cooling rate and the grain boundaries could be responsible for the interfacial polarization.

The induced dipole moment can be calculated from the solid state dielectric constant, ε_0, and equation (3). The density in this expression was obtained from the graph in Fig. 5. The induced moments are listed in Table 2 and vary from

Table 2

Table of induced dipole moments and instantaneous polarizations for cholesteryl acetate, priopionate, and stearate

specimen	induced dipole moment, α_0 (10^{-23} e.s.u. cm)	induced polarization (C/cm²)
cholesteryl acetate (CA)	5.41	136.4
cholesteryl propionate (CP)	5.30	133.5
cholesteryl stearate (CS)	9.17	231.3

about 5.3×10^{-23} e.s.u. cm for CA and CP to 9.2×10^{-23} e.s.u. cm for CS. The instantaneous polarization can also be calculated from equation (3) and is also listed in Table 2. The induced polarization is about 135 C/cm² for the acetate and propionate ester and about 231 C/cm² for the stearate ester. Since the induced dipole moment and instantaneous polarization are proportional to the molecular weight, it is not surprising that α_0 and P_{ind} are much greater for the long chain stearate molecule than for the short chain acetate and propionate molecules.

The component of the dielectric constant due to the polarization of electrons in the molecules can be obtained by index of refraction measurements at optical

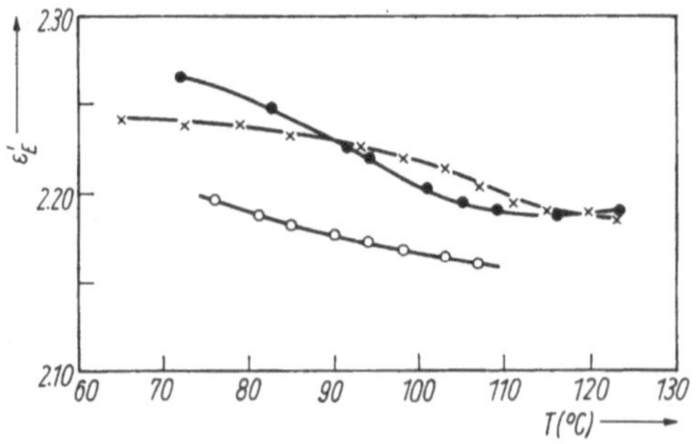

Fig. 6. Dielectric constant at optical frequencies, ε'_E, plotted versus temperature for cholesteryl acetate (●), propionate (×), and stearate (○). Dielectric constant was computed from the index of refraction obtained at optical frequencies

frequencies and equation (2). This component of the dielectric constant, ε_E', is plotted in Fig. 6 as a function of temperature. The slight temperature dependence may be due to a decrease in density at increasing temperature. The magnitude of ε_E' varies between 2.15 and 2.27 and is almost equal to the solid state dielectric constant, ε_0. The atomic component of the dielectric constant, ε_A', is the difference between ε_0 and ε_E' and is about 0.11 for CA, 0.02 for CP, and 0.20 for CS. It is to be expected that ε_A' is only a small fraction of the total solid state dielectric constant in the molecular crystals.

The permanent dipole moment, μ_0, for the three esters can be calculated from the values of ε' and ε_0 obtained from Fig. 2 to 4, the density obtained from Fig. 5 and equation (4). The values of μ_0 are listed in Table 3 for CA, CP, and CS and are about 1.5×10^{-18} e.s.u. cm. These permanent dipole moments

Table 3

Table of permanent dipole moments and orientational polarizations for cholesteryl acetate, propionate, and stearate

specimen	permanent dipole moment, $\mu_0(10^{-18}$ e.s.u.cm)	orientational polarization (C/cm²)	temperature (°C)
cholesteryl acetate	1.501	32.953	120
		31.997	140
cholesteryl propionate	1.450	32.346	100
		32.521	110
cholesteryl stearate	1.472	34.395	100
		33.578	110

are five orders of magnitude larger than the induced dipole moments although ε_M' is only about a fourth of ε_0. The orientational polarizations, calculated from equation (1) and listed in Table 3 for various temperatures, are about 32 C/cm². The orientational polarization is much smaller than the induced polarization. The dipole polarization exhibits little variation with increasing chain length. There has been some discussion as to whether the electro-optical effects are due to permanent or induced dipole orientation. If the electro-optical effects are due to permanent dipole polarization, increasing the chain length should produce little change in the observed effect. If, however, the observed effect is very dependent on the length of the ester chain, it is probably due to the induced dipole moment. The electro-optical effects of the liquid crystal state in the CA and CS esters should be more pronounced on cooling than on heating since the motion of the molecular dipoles is much more hindered on heating than on cooling.

The frequency dependence of ε' and $\tan\delta$ for the three esters is illustrated in Fig. 7 to 9. Little frequency dependence of ε' is observed except for a

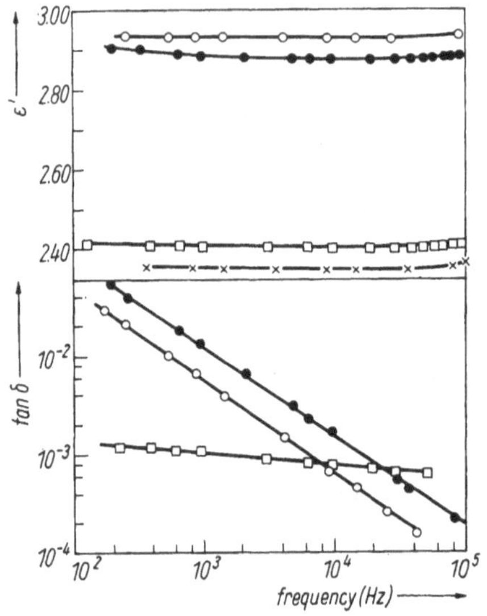

Fig. 7. Dielectric constant, ε', and dissipation factor, tan δ, plotted versus frequency for cholesteryl acetate in the solid (90 (×) and 28 °C (□)), liquid crystal (108 °C (○)), and isotropic liquid (120 °C (●)) state

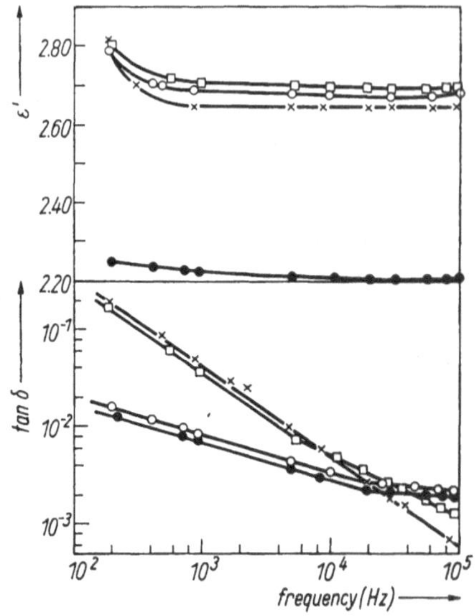

Fig. 8. Dielectric constant, ε', and dissipation factor tan δ, plotted versus frequency for cholesteryl propionate in the solid (25 °C (●)), liquid crystal (102 (□) and 108 °C (○)), and isotropic liquid (120 °C (×)) state

large increase in the solid state ε' for CS between 10^2 and 10^5 Hz and a slight rise in ε' in the liquid state of CP with decreasing frequency. The increase in ε' and tan δ with decreasing frequencies in Fig. 7 to 9 suggests that an interfacial polarization is present at low frequencies below 10^3 Hz. This dielectric dispersion and absorption is interpreted as an interfacial phenomenon rather than a Debye relaxation phenomenon (involving molecular dipoles) since the large increase in ε'_M on melting at 10^4 Hz indicates that the Debye relaxation frequency is greater than 10^4 Hz. Fig. 7 to 9 also suggest the possibility of an interfacial polarization in the liquid states of CA, CP, and CS, however, at a much lower frequency than in the solid state of CS. The relaxation frequencies for the interfacial polarization in the liquid states are shifted to higher frequencies with increasing temperatures. This effect is a general characteristic of the effect of temperature on the relaxation frequencies of various polari-

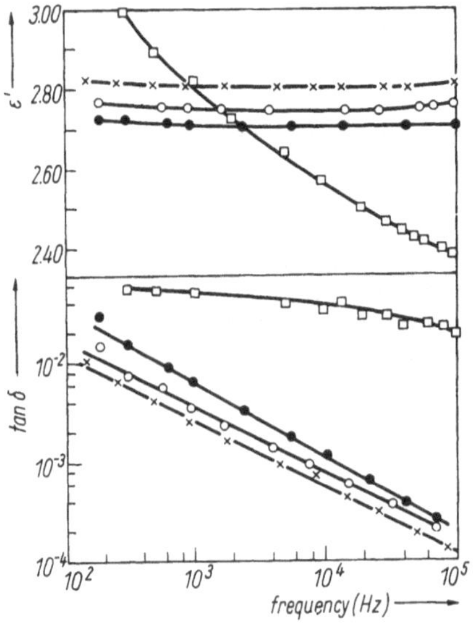

Fig. 9. Dielectric constant, ε', and dissipation factor, tan δ, plotted versus frequency for cholesteryl stearate in the solid (54 °C (□)) and isotropic liquid (104 (●), 90 (○), and 75 °C (×)) state

zations [16]. The net effect of the shift of the relaxation frequencies in the $\varepsilon(T)$ curves would be to cause ε' and $\tan\delta$ to increase with temperature. The increase of $\tan\delta$ with T is observed in the liquid states of CA, CP, and CS in Fig. 2 to 4; however, little change in ε' is observed. This could be due to a competing effect of temperature causing fluctuations in dipole alignment and a lowering of the orientational polarization predicted by equation (1) or to a lack of change of ε' with frequency.

5. Conclusion

The dependence of dielectric constant and dissipation factor on temperature, frequency, and state of matter has been investigated here for the two short chain cholesteryl esters, acetate, and propionate, and the long chain stearate ester. The induced and permanent dipole moments and orientational and instantaneous polarizations have been calculated. Some of the results of these measurements and calculations are:

1. Evidence was presented for the existence of two different cholesteric liquid crystal states for CA and CS, one on heating and one on cooling. The liquid crystal state on heating behaves dielectrically like the solid state and on cooling behaves dielectrically like a true liquid with a much larger ε'. The dielectric behavior on heating corresponds to the behavior of the density versus temperature curves on heating.
2. The CP ester liquid crystal state on the other hand behaves dielectrically like a liquid on both heating and cooling.
3. A dependence of the solid state dielectric properties on the rate of cooling suggests that the quenching rate plays a role in the determination of the solid state structure. This was interpreted as being due to a variation in the interfacial polarization with varying cooling rates (possibly a grain size effect).
4. Evidence for an interfacial polarization at frequencies below 10^2 Hz in the solid state of CS and the liquid states of CA, CP, and CS was presented.
5. The induced dipole moments of 5.4, 5.3, and 9.2×10^{-23} e.s.u. cm and the instantaneous polarizations of 136.4, 133.5, and 231.3 C/cm^2 for CA, CP, and CS respectively were calculated. The induced moments and instantaneous polarizations increased with longer chain cholesteryl esters.
6. The permanent dipole moment of about 1.5×10^{-18} e.s.u.cm and an orientational polarization of about 32 C/cm^2 was calculated for all three esters. The length of the ester chain has little effect on the magnitude of the orientational polarization or permanent dipole moment.
7. There was no evidence of a Debye relaxation phenomena up to 150 kHz.
8. Little change in ε' or $\tan\delta$ was observed on increasing applied ac field up to 1000 V/cm.

The great difference between the dielectric properties of the liquid crystal state of CA and CS on heating and cooling may suggest completely different structures. These results indicate the need for structural studies for the cholesteryl esters on heating and cooling and at varying heating and cooling rates.

Acknowledgement

The authors wish to thank Mr. P. Forsyth for his able technical assistance in making the dielectric measurements.

References

[1] J. J. Wysocki, J. Adams, and W. Haas, Phys. Rev. Letters **20**, 1024 (1968).
[2] H. Baessler and M. M. Labes, J. chem. Phys. **51**, 5397 (1969).
[3] J. H. Muller, Mol. Cryst. **2**, 167 (1966).
[4] W. J. Harper, Mol. Cryst. **1**, 325 (1966).
[5] H. Baessler, T. M. Laronge, and M. M. Labes, J. chem. Phys. **51**, 3213 (1969).
[6] J. J. Wysocki and J. Adams, Bull. Amer. Phys. Soc. **14**, 739 (1969).
[7] D. G. Shaw and J. W. Kauffman, phys. stat. sol. (a) **4**, 467 (1971).
[8] D. G. Shaw and J. W. Kauffman, J. chem. Phys. **54**, 2424 (1971).
[9] L. Onsager, J. Amer. Chem. Soc. **58**, 1486 (1936).
[10] J. C. Maxwell, in: Treatise on Electricity, Vol. II, Oxford/London (1881).
[11] P. Debye, Z. Phys. **13**, 97 (1912).
[12] E. M. Barrall, II and M. A. Sweeney, Mol. Cryst. **5**, 257 (1969).
[13] G. W. Gray, J. chem. Soc. 3733 (1956).
[14] E. M. Barrall, II, R. S. Porter, and J. F. Johnson, J. chem. Phys. **71**, 895 (1967).
[15] M. B. Rhodes, R. S. Porter, W. Chu, and R. S. Stein, Mol. Cryst. and liquid Cryst. **10**, 731 (1970).
[16] J. B. Birks, in: Modern Dielectric Materials, Academic Press Inc., New York 1960 (p. 20).

(Received March 29, 1972)

phys. stat. sol. (a) 12, 649 (1972)

Subject classification: 18.2; 22.8

International Laboratory for High Magnetic Fields and Low Temperatures (a) and Institute for Low Temperatures and Structure Research, Polish Academy of Sciences, Wroclaw (b)

Magnetic Properties of UAsY-Type Compounds (Y = S, Se, Te) in Ordered State

By

C. Bazan (a) and A. Zygmunt (b)

The dependence of the magnetization for the ternary compounds UAsS, UAsSe, and UAsTe on temperature and magnetic field has been investigated in the temperature range between 3 K and the respective Curie point in fields up to 80 kOe. Measurements were performed for both polycrystalline and single crystal samples. In the latter case the magnetization along the c-axis, being the easy one, was measured only. All these compounds exhibit large anisotropy of the order 10^7 erg/cm^3. The average values of the magnetic moments of the uranium ion (1.1 to 1.4 BM) along the c-axis are smaller than those reported for other uranium compounds having similar crystal structure. Both UAsS and UAsSe have rectangular hysteresis loops in helium temperature region, with BH product of the order 10^7 erg. They also display unusual temperature variation of the magnetization, resulting probably from the temperature dependence of the magnetocrystalline anisotropy.

Исследовано зависимость намагниченности соединении UAsS, UAsSe, UAsTe от температуры и магнитного поля в области полей до 80 кэ, и в температурной области от 3 К до температуры Кюри. Измерения проводились для порошкообразных образцов и для монокристаллов в направлении оси c. Температура Кюри указанных соединений составляет 124, 113 и 66 К соответственно. Соединения обладают исключительно большой магнитной анизотропией (эквивалент поле 10^6 э), вызванной по всей вероятности действием кристаллическкго поля. Средние магнитные моменты на ион урана вдоль оси c (1,1 до 1,4 μ_B) меньше чем у других соединений с подобной кристаллографической структурой. У UAsS и UAsSe при гелиевых температурах обнаружена прямоугольная петля гистерезиса, а при более высоких температурах своеобразный температурный ход намагниченности. Оба явления обусловлены повидимому температурной зависимостью магнитной анизотропии.

1. Introduction

Previous investigations of the UAsY compounds (Y = S, Se, Te) performed in the temperature range 77 to 1000 K [1] have shown them to display a magnetic transition. For UAsS and UAsSe, which crystallize in the PbFCl-type structure (P4/nmm), the Curie temperatures are 128 and 118 K, respectively. In the case of UAsTe, for which the UGeTe-type structure (I4/mmm) was assumed, the Curie temperature was expected below 77 K. In the paramagnetic state all these compounds show a curvilinear χ^{-1} versus T dependence. Such behaviour could be qualitatively explained in terms of the effect of the crystal on the U^{4+} ion [1].

In this paper data obtained from magnetic measurements at low temperatures and high magnetic fields are presented.

2. Experiment

Magnetic measurements were made on both single-crystal and polycrystalline samples, obtained as described in [1]. A number of thin single crystal plates perpendicular to the c-axis were glued together to form a cylinder of 4 mm in diameter and 8 mm in length. The c-axis of all crystals was parallel to the cylinder axis with an accuracy of 6°. Measurements of the magnetization were made in a moving-sample magnetometer. The sample placed in the cryostat was allowed to move between two pick-up coils by means of a small iron-core magnet located outside the cryostat. A digital voltmeter (V-524 type) was used to measure the peak value of the induced voltage. The magnetometer was calibrated against nickel.

The static magnetic fields up to 80 kOe were generated by a water-cooled iron-free electromagnet of Bitter-type [2].

The temperature of the sample in the helium temperature region was determined by helium vapour pressure measurements. The higher temperatures were measured with a copper–constantan thermocouple calibrated against platinum, and a germanium thermometer.

Magnetic measurements were carried out either at fixed magnetic fields and temperatures ranging from 3 K to slightly above the Curie points, or at constant temperatures but in changing magnetic fields.

3. Results and Discussion

In Fig. 1 and 3 the isothermal magnetization versus field curves $\sigma_g = f(H)_T$ are given for single crystals of UAsS and UAsSe along the c-axis. As it is seen the rapid increase in the magnetization takes place at some critical field strength H_c, above which the magnetization practically does not change. The remanent magnetization reaches about 98% of the high-field value. The hysteresis loop

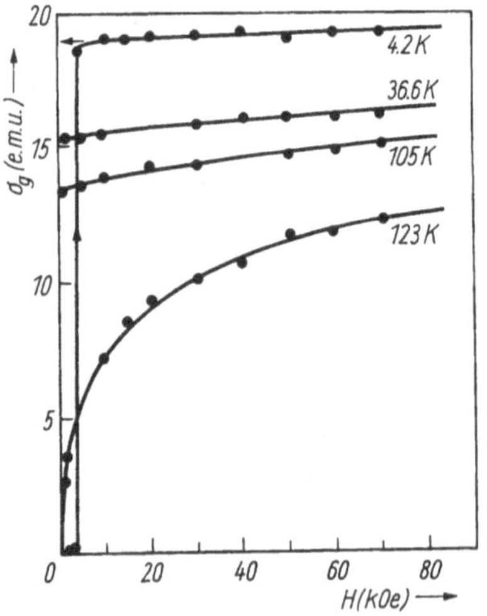

Fig. 1. The dependence of the magnetization (per gramme) on the field for UAsS single crystal along the c-axis

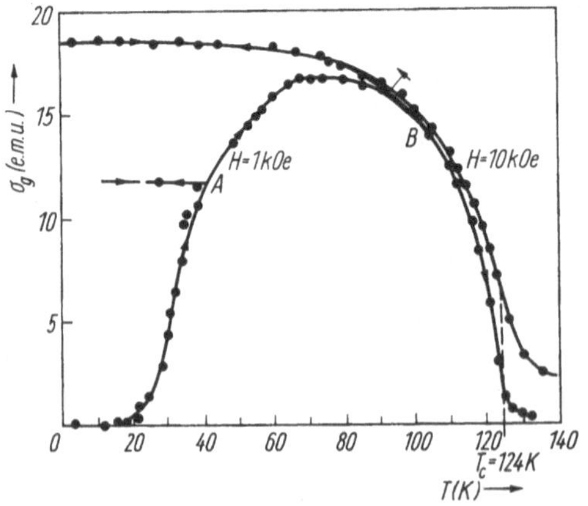

Fig. 2. The dependence of the magnetization per gramme for UAsS single crystals on the magnetic field (**H** ∥ **c**). The arrows denote the direction of the temperature change. $H_c(4) = 4$ kOe

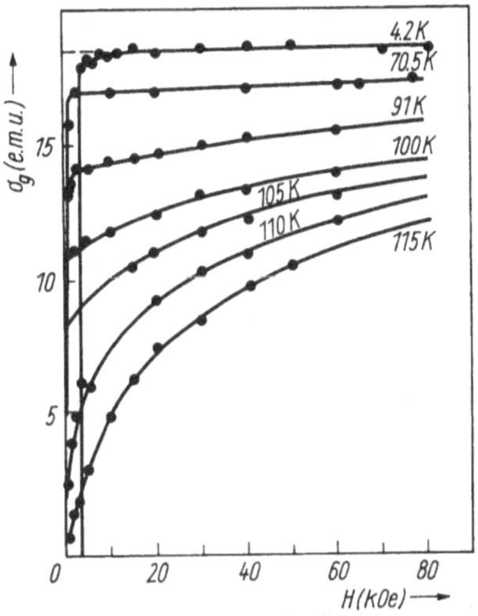

Fig. 3. The dependence of the magnetization (per gramme) on the field for UAsSe single crystal along the c-axis

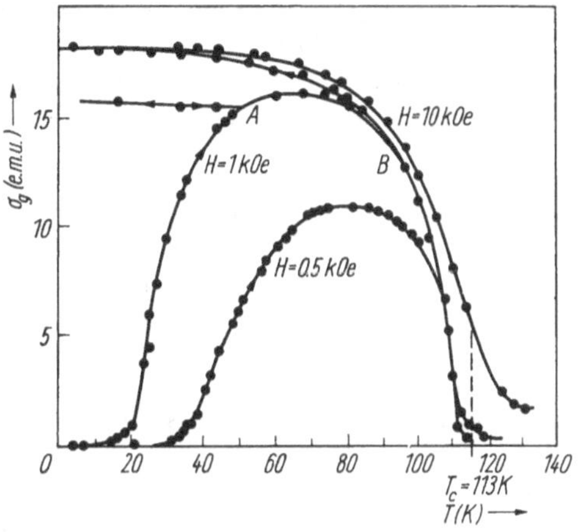

Fig. 4. The dependence of the magnetization per gramme for UAsSe single crystal on the magnetic field ($\boldsymbol{H} \parallel \boldsymbol{c}$). The arrows denote the direction of the temperature change. $H_c(4) = 3.8$ kOe

(Fig. 5) shows nearly rectangular form. The coercive field H_c diminishes with increasing temperature.

The magnetization versus temperature curves $\sigma_g = f(T)_H$ (Fig. 2 and 4) show also unusual features. When the sample is heated in fields smaller than H_c the magnetization appears only at some limiting temperature, then increases, goes through a maximum, and falls down when reaching the Curie temperature.

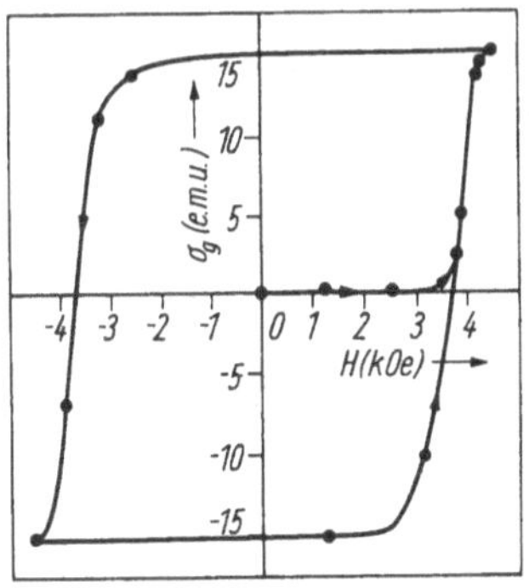

Fig. 5. The hysteresis loop for UAsSe single crystal, $\boldsymbol{H} \parallel \boldsymbol{c}$; $T = 4.2$ K

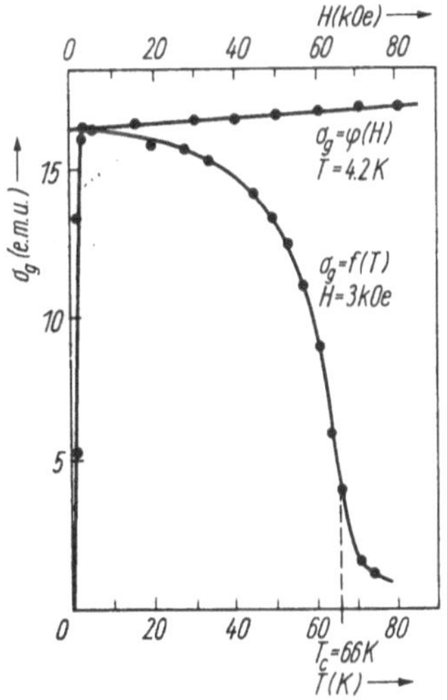

Fig. 6. The dependence of the gramme magnetization for UAsTe on the field and temperature

The value of the magnetization in some point A below the maximum reached on heating is not further changed if the sample was allowed to cool down to 4 K again. On the contrary, the magnetization in some point B above the maximum increases on subsequent cooling and runs along the curve obtained for fields $H > H_c$.

Similarly, the temperature dependence of the electrical resistivity of UAsS and UAsSe in the magnetically ordered state [3] shows also anomalous behaviour in comparison to other simple ferromagnets. Contrary to the first two compounds, UAsTe does not show any irregularities in the magnetization process down to 4 K, and behaves like an ordinary ferromagnet. The results obtained on powder samples indicate, in general, similar but more complicated relations.

The values of the magnetization, the magnetic moments, and other data obtained in this work are presented in Table 1.

Table 1

Magnetization, Curie temperatures, and magnetic moments of UAsY compounds

compound	crystal structure		magnetization		magnetic moment μ_{cr} (BM)	Curie temp. (K)	coercive field at 4 K H_c (kOe)
	type	space group	powder σ_p (e.m.u./g)	single crystal σ_{cr} (e.m.u./g)			
UAsS	PbFCl	P4/nmm	8.3	18.4	1.14	124	4.0
UAsSe	PbFCl	P4/nmm	7.5	18.1	1.36	113	3.8
UAsTe	UGeTe	I4/mmm	10.4	16.4	1.29	66	—

Previous magnetic measurements on single crystals of these three compounds at 77 K [1] have shown them to have high magnetic anisotropy. There was stated that the magnetic moments are aligned parallel to the c-axis. Measurements on powder samples at 4.2 K show that in fields up to 80 kOe the magnetization is practically independent of the field strength except for the low-field region. This fact confirms the existence of a high anisotropy and gives the value of 10^5 Oe as the lower limit of the anisotropy field. This value is confirmed also by magnetostriction measurements [4] which have revealed the lack of saturation in fields as high as 140 kOe.

The high magnetic anisotropy may result from the strongly axial character of the crystalline field [1] which is at least by an order of magnitude higher than that of the magnetic exchange interactions.

Thus the rapid increase of the magnetization may be explained by the formation of a compensated structure of 180° domains. Above some critical field H_c necessary for domain wall movement the domain structure suddenly rebuilds into a single domain, which manifests itself in a jump in the magnetization curve (Fig. 1 and 3). The diminution of the magnetic anisotropy with temperature results in lowering H_c.

Preliminary neutron diffraction studies made for UAsSe [5] at 4 K seem to confirm the existence of a simple ferro- or ferrimagnetic structure. No additional magnetic peaks were observed which could contradict the parallel alignment of the magnetic moments along the c-axis.

The absence of the magnetization jump in UAsTe may arise from its different crystallographic structure.

The values of the magnetic moments μ_{cr} given in Table 1 refer to the average moment per uranium ion determined for single crystals along the c-axis. The magnetic moments μ_{cr} obtained in this work are smaller than those in other uranium compounds with similar crystal structure [6, 7].

On the assumption that only U^{4+} ions are present in the lattice the above fact may be explained as the result of the influence of the strong crystal field which splits the lowest level $^3H_4$ into a number of Stark levels. The subsequent splitting occurs under the influence of the molecular field (much smaller than the crystal field) and the lowest level $|M\rangle$ at low temperatures gives rise to a small value of the observed magnetic moment $\langle M| \, 4/5 \, \hat{J}_z \, |M\rangle$.

It should be noted that UAsY compounds will be investigated by neutron diffraction in order to account for their interesting magnetic behaviour.

Acknowledgements

The authors would like to express their gratitude to Prof. W. Trzebiatowski for his kind interest in this investigation, to Dr. W. Ziętek for helpul discussions as well as to Mr. T. Mydlarz for his aid in performing measurements.

References

[1] A. Zygmunt and M. Duczmal, phys. stat. sol. (a) **9**, 659 (1972).
[2] K. Trojnar and C. Bazan, Prib. i Tekh. Eksper., in the press.
[3] A. Wojakowski, private communication.
[4] R. Z. Levitin, private communication.
[5] J. Leciejewicz, private communication.
[6] A. Oleś, J. Phys. Radium **26**, 561 (1965).
[7] J. Leciejewicz, R. Troć, A. Murasik, and A. Zygmunt, phys. stat. sol. **22**, 517 (1967).

(Received May 10, 1972)

phys. stat. sol. (a) **12**, 655 (1972)

Subject classification: 14.3.3; 8; 22

Zentralinstitut für Festkörperphysik und Werkstofforschung der Deutschen Akademie der Wissenschaften zu Berlin, Institutsteil für magnetische Werkstoffe, Jena

On the Theory of Thermal Switching in Semiconducting Devices

II. Discussion of the Static Current–Voltage Characteristic and the Temperature, Electric Field, and Current Density Distribution in the Case of Cylindrical Symmetry

By

W. BRODKORB and W. HAUBENREISSER

Starting from the general expressions given in [1] the temperature, electric field, and current density distribution and the static current–voltage characteristic are calculated numerically for both the special cases heat flow in axial direction, parallel or antiparallel to the current (axial case), and heat flow in radial direction, perpendicular to the current (radial case). In the parameter range $b \equiv (1/2)\,\Delta E/k_B T_0 > 2$ S-shaped behaviour of the current–voltage characteristic was obtained, but no well defined current and temperature channel which are connected with an inflection point in the radial case. Additionally the dependence of the critical voltage U_z on several geometry, material, and cooling parameters is discussed.

Ausgehend von den in [1] angegebenen allgemeinen Ausdrücken, werden die Temperatur-, Feld- und Stromdichteverteilung sowie die statische Strom–Spannungskennlinie für die beiden Spezialfälle Wärmefluß in axialer Richtung, parallel oder antiparallel zum Strom (axialer Fall) und Wärmefluß in radialer Richtung, senkrecht zum Strom (radialer Fall) numerisch berechnet. In dem Parameterbereich $b \equiv (1/2)\,\Delta E/k_B T_0 > 2$ werden S-förmige Strom–Spannungskennlinien erhalten, jedoch resultieren im radialen Fall keine Strom- und Temperaturkanäle mit einem Wendepunkt. Zusätzlich wird die Abhängigkeit der kritischen Spannung U_z von verschiedenen Geometrie-, Material- und Abkühlungsparametern diskutiert.

1. Introduction

In Part I [1] we have calculated the temperature, electric field, and current density distribution and the static current–voltage characteristic of a cylinder with top and bottom contacts in dependence on several geometry and material parameters. It was pointed out that assuming the quasi-homogeneity of temperature distribution the heat transport equation completed by the term of Joule's heating could be solved in an analytical form. However, the static Maxwell equations are satisfied exactly only for temperature distributions dependent on the radial or axial coordinate alone. In the intermediate case, the obtained electric field distribution is a good approximation if the relation $\partial E_z/\partial r \approx 0$ is fulfilled.

In the following we shall discuss such temperature, electric field, and current density distributions and current–voltage characteristics for some special cooling conditions on the surface of a cylindrical semiconducting device.

2. The Axial Case

For $a_L = \infty$, i.e. no heat flow in the radial direction (a_L and a_E are cooling parameters characterizing the heat flux into the surroundings of the device, see Part I) we get from I (10) for the temperature distribution $v(z) \equiv T(z)/T_0$:

$$v(z) = \vartheta + \frac{1}{\beta}\left[1 - \frac{\cosh d_U \frac{2z}{D}}{\frac{\sinh d_U}{d_U}}\right] \tag{1}$$

with

$$d_U = \frac{U}{U_c} e^{\gamma \frac{U}{U_c}} \frac{e^{-b/\vartheta}}{\vartheta}, \qquad U_c = \sqrt{\frac{4\varkappa k_B T_0^2}{\sigma_0 \Delta E}},$$

$$\gamma = \frac{1}{2}\frac{U_c}{U_0}, \qquad b = \frac{1}{2}\frac{\Delta E}{k_B T_0}.$$

(T_0 ambient temperature, $\varkappa$ thermal conductivity, σ_0 electrical conductivity for $v = \infty$ and $\gamma = 0$, ΔE energy gap parameter, U_0 voltage parameter, k_B Boltzmann constant, D thickness of the device.) The relative mean temperature ϑ is defined by (see I (7, 13)

$$\frac{1}{\beta} + \left(1 + \frac{1}{\beta} - \vartheta\right) \frac{\frac{\sinh d_U}{d_U}}{\cosh d_U - 2\frac{\sinh d_U}{d_U} + a_E d_U \sinh d_U} = 0\,.$$

Finally, it follows for the electrical field distribution

$$\frac{E_z(z)}{(U_c/D)} = \frac{U}{U_c}\frac{1}{\beta}\frac{\cosh d_U \frac{2z}{D}}{\frac{\sinh d_U}{d_U}}\,. \tag{2}$$

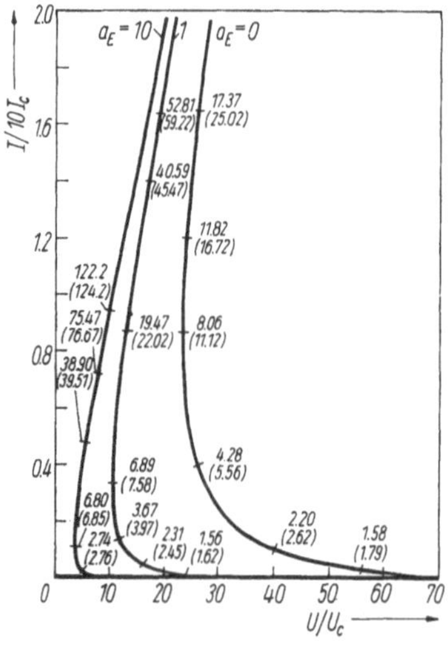

Fig. 1. Current–voltage characteristics for $b \equiv (1/2)\,\Delta E/(k_B T_0) = 4$, $\gamma \equiv (1/2)\;U_c/U_0 = 10^{-4}$, $a_L \equiv (2\varkappa)/(Dh_L) = \infty$ (heat flow parallel or antiparallel to the current — axial case) and different $a_E \equiv (2\varkappa)/(Dh_E)$ values ($I_c = (\pi R^2 \sigma_0 U_c)/D$, $U_c = \sqrt{(4\varkappa k_B T_0^2)/\sigma_0 \Delta E}$). The numbers on the characteristics denote the relative temperature ($T(z)/T_0$ in the central plane of the cylinder (values in brackets) and the relative mean temperature $\vartheta \equiv 1/D \int_{-D/2}^{+D/2} dz\, T(z)/T_0$ of the device

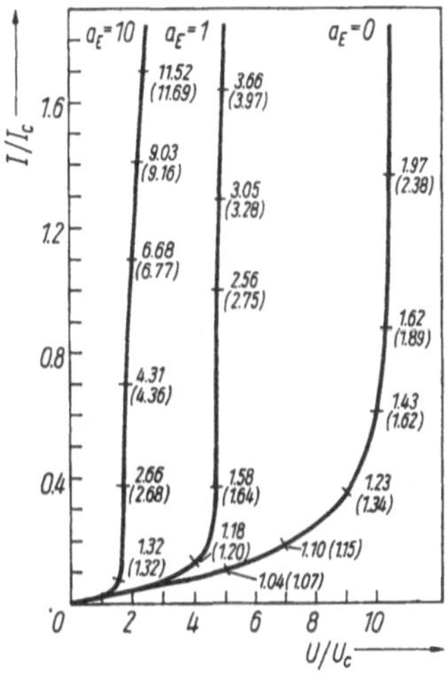

Fig. 2. Current–voltage characteristics for $b = 2$, $\gamma = 10^{-4}$, $a_L = \infty$, and different a_E-values (axial case).

The current is constant in this case and parallel or antiparallel to the heat flow in the axial direction.

In Fig. 1 we have plotted the current–voltage characteristic resulting from (1, 2) and I (15) for different cooling parameters a_E. From these results it must be concluded that, opposite to Böer [2], in our approximation the current–voltage characteristic shows an S-shaped behaviour also in an one-dimensional model with a z-dependent temperature and field distribution. We see that with decreasing a_E (i.e. with increasing heat flow into the electrodes) the characteristic passes through an increasing negative part at which the threshold voltage shows even the tendency to increase. However, it must be emphasized that the form of curves shown in Fig. 1 exists only for $b > 2$, for $b \leqq 2$ one obtains a characteristic without negative differential resistance (Fig. 2). Additionally on the characteristics here and in the following figures the relative temperature $T(r, z)/T_0$ in the centre of the cylinder (values in brackets) and the mean temperature ϑ are marked. In Fig. 3 and 4 the dependence of the threshold voltage U_z is plotted (U_z is the voltage for the beginning of the negative differential resistance range) on the cooling parameter a_E for different b-values and the parameter b for different a_E-values, respectively ($\gamma = 10^{-4}$, $a_L = \infty$). As it could be expected the critical voltage (for fixed b- and γ-values) increases with increasing heat flow into the electrodes and (for fixed a_E parameters) decreases with decreasing b-values (for instance increasing ambient temperature if ΔE is constant).

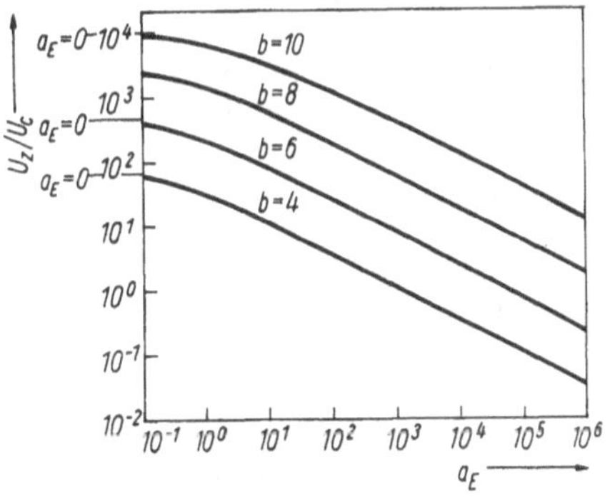

Fig. 3. Dependence of the normalized threshold voltage U_z/U_c on the cooling parameter a_E in the axial case ($\gamma = 10^{-4}$, $a_L = \infty$)

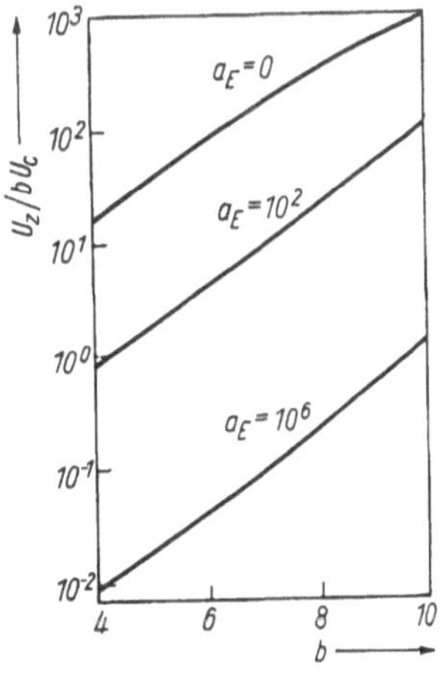

Fig. 4. Dependence of the normalized threshold voltage U_z/U_c on the parameter b for different cooling parameters a_E in the axial case ($\gamma = 10^{-4}$, $a_L = \infty$).

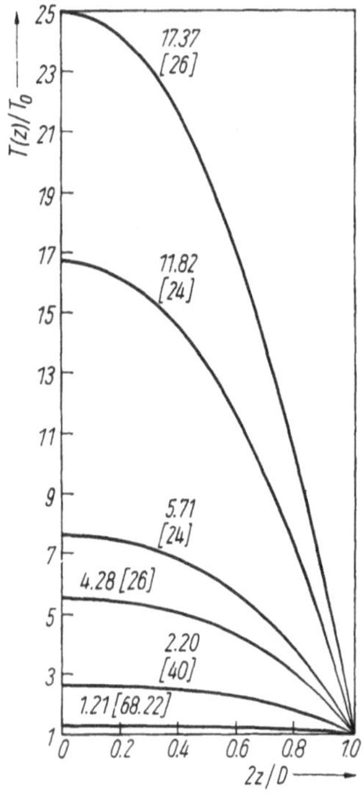

Fig. 5. Normalized temperature distribution as function of the axial distance from the central plane of the device ($a_E = 0$, $a_L = \infty$, $\gamma = 10^{-4}$, $b = 4$). The values in brackets correspond to the applied voltage U/U_c, the other figures denote again the mean temperature ϑ

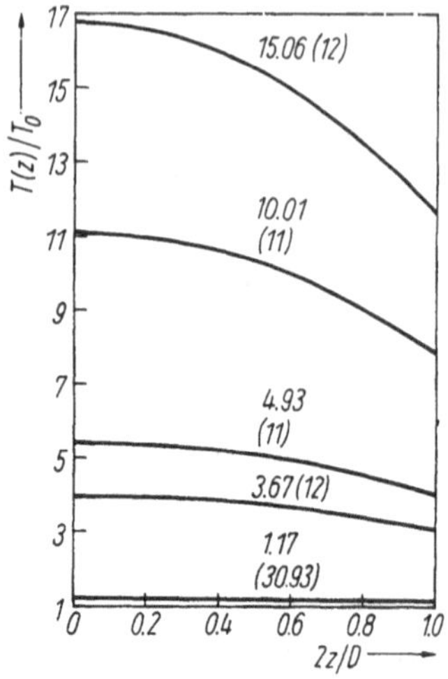

Fig. 6. Normalized temperature distribution as function of the axial distance from the central plane of the device ($a_E = 1$, $a_L = \infty$, $\gamma = 10^{-4}$, $b = 4$). Parameters as in Fig. 5

Typical temperature and field profiles are shown in Fig. 5 to 8. Fig. 5 and 6 indicate that in the high ohmic branch of the characteristic up to the threshold voltage ($\vartheta \leqq 1.21$ and 1.17, respectively) the temperature is nearly constant within the cylindrical semiconductor, but with increasing supplied electric power (i.e. increasing mean temperature ϑ, see I (14)) the temperature distribution becomes more and more inhomogeneous. The upper physical limit of the temperature in the centre is given by the melting point of the material. Contrary, the field profile is smoothed visibly with increasing power (Fig. 7).

3. The Radial Case

With $a_E = \infty$ in I (12) we obtain for the eigenvalues λ the solutions $\lambda_1 = i\sqrt{2}\,d_U$ and $\lambda_m = (m-1)\,\pi$ $(m = 2, 3, \ldots)$. With these values the following temperature and current density distributions result:

$$v(r) \equiv \frac{T(r)}{T_0} = \vartheta - \frac{1}{\beta} + \frac{1}{2\beta}\frac{2R}{D}d_U\,\frac{J_0\left(\frac{2R}{D}d_U\frac{r}{R}\right)}{J_1\left(\frac{2R}{D}d_U\right)}, \tag{3}$$

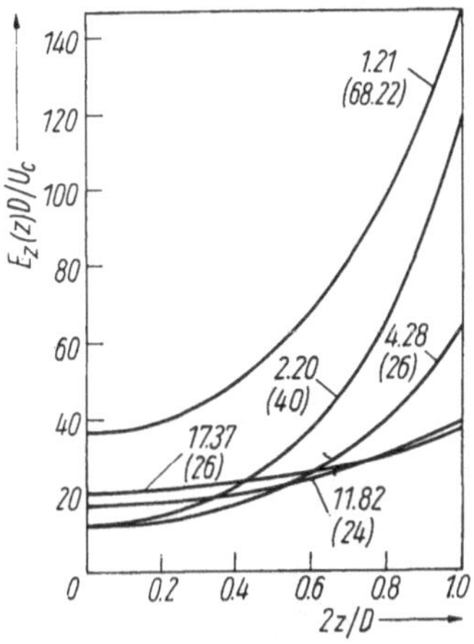

Fig. 7. Normalized field distribution as function of the axial distance from the central plane of the cylinder ($a_E = 0$, $a_L = \infty$, $\gamma = 10^{-4}$, $b = 4$)

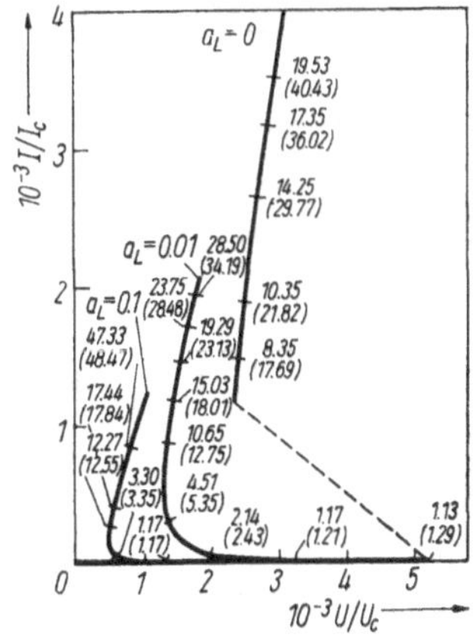

Fig. 8. Current–voltage characteristics for $b = 4$, $\gamma = 10^{-4}$, $a_E = \infty$ (heat flow perpendicular to the current, radial case), $(2\,R)/D = 10^{-3}$ and different a_L-parameters. The figures in brackets denote the relative temperature $T(r)/T_0$ in the axis of cylinder, the other values the relative mean temperature ϑ

$$\frac{j_z(r)}{(U_c/D)\,\sigma_0} = \frac{1}{2}\,\frac{U}{U_c}\,\frac{2\,R}{D}\,d_U\,e^{-2\left(\frac{b}{\vartheta} - \gamma\frac{U}{U_c}\right)}\,\frac{J_0\left(\frac{2\,R}{D}\,d_U\,\frac{r}{R}\right)}{J_1\left(\frac{2\,R}{D}\,d_U\right)}. \tag{4}$$

In (3, 4), R denotes the radius of the sample and j_z the current density. J_ν is the Bessel function of order ν with real argument. One can easily read from (3) that the heat flow is directed perpendicularly to the current. The electrical field is constant with respect to the coordinates in this case. At least the mean temperature must be estimated from the equation

$$-\frac{1}{\beta} + 2\left(1 + \frac{1}{\beta} - \vartheta\right)\frac{1}{\frac{2\,R}{D}\,d_U}\;\frac{J_1\left(\frac{2\,R}{D}\,d_U\right)}{J_0\left(\frac{2\,R}{D}\,d_U\right) - a_L\,d_U\,J_1\left(\frac{2\,R}{D}\,d_U\right)} = 0$$

in the radial case. In order to make the approximated solution (4) unequivocal it is necessary to demand that the current density (4) satisfies the inequality $j_z(r) > 0$ (for $j_z \leqq 0$ the condition I (8d) $\beta\,|\eta| \ll 1$ is no longer valid), i.e. the current and temperature distributions are restricted to regions with $(2\,R/D)\,d_U < \omega_0 \equiv 2.4048$ (ω_0 is the firth zero of J_0). This fact is demonstrated in Fig. 8. The current–voltage characteristic for $a_L = 0$ possesses such a part, in our model unstable (dashed line), and we must pass over to a better approximation (up to quadratic terms in η) in order to explain this unstable part of the characteristic. In general, the current–voltage curves anew show the S-shape behaviour of the z-dependent model, also different to Böer et al. [3].

In Fig. 9 the channel-like current density distribution is sketched. We see, contrary to [3], that the current density on the cylinder mantle always increases

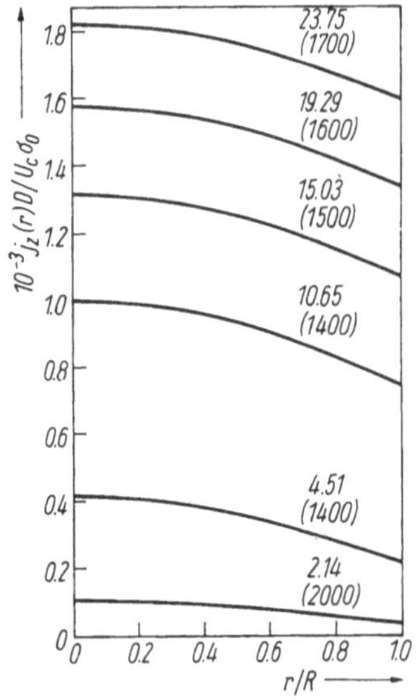

Fig. 9. Normalized current density distribution as function of the radial distance ($a_E = \infty$, $a_L = 10^{-2}$, $b = 4$, $\gamma = 10^{-4}$, $(2\,R)/D = 10^{-2}$). The values in brackets correspond to the applied voltage U/U_c, other figures denote the mean temperature ϑ

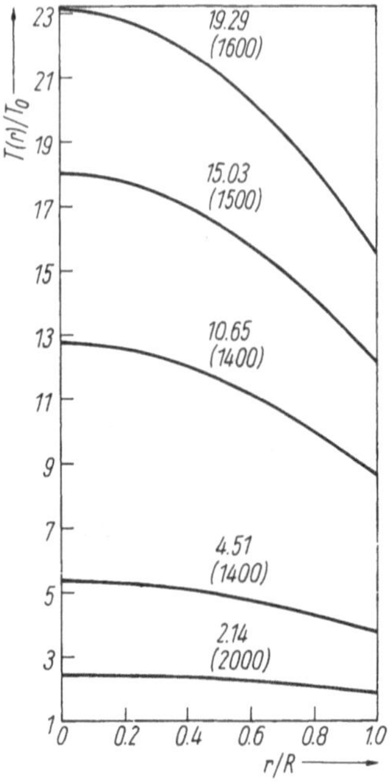

Fig. 10. Normalized temperature distribution as function of the radial distance ($a_E = \infty$, $a_L = 10^{-2}$, $b = 4$, $\gamma = 10^{-4}$, $(2\,R/)D = 10^{-2}$). Parameter: mean temperature ϑ

with increasing supplied power. Furthermore, first of all the current distribution has the tendency to become more and more inhomogeneous up to a certain mean temperature ϑ. Above this temperature the current distribution becomes more homogeneous again. The temperature distribution shows qualitatively a similar behaviour as the corresponding distribution in the axial case (Fig. 10). But the temperature as well as the current distribution does not possess an inflection point in this radial case. The existence of this point should be characteristic of the filament formation after [3, 5]. Finally, as it was pointed out already in [3, 4], the threshold voltage U_z increases with decreasing radius R (at a fixed thickness D) of the semiconducting cylinder (Fig. 11). The dependence of the

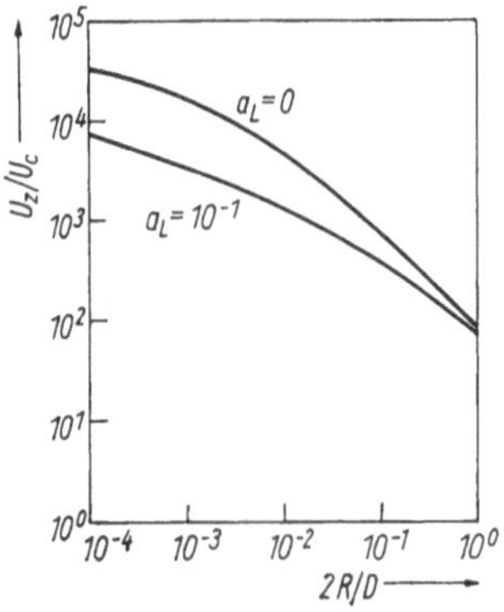

Fig. 11. Dependence of the normalized threshold voltage U_z/U_c on the ratio diameter to thickness $(2\,R)/D$ for different cooling parameters a_L ($a_E = \infty$, $\gamma = 10^{-4}$, $b = 4$)

threshold voltage on the parameters b and a_L is qualitatively the same as the dependence on b and a_E in the axial case (Fig. 3 and 4), likewise we obtain S-shape current–voltage characteristics only in the parameter range $b > 2$.

4. Discussion

We have discussed the dependence of the static current–voltage characteristic and the temperature, electric field, and current density distribution on several material and geometry parameters for a cylindrical semiconducting device on the basis of a temperature-dependent conductivity law I (2). Thereby it was possible to extend the results known in the literature to the whole temperature range with the help of the assumption that the temperature is quasi-homogeneous in the device. In the axial as well as in the radial case we have obtained always S-shape current–voltage characteristics in the parameter region $b > 2$, i.e. for conductivity–temperature curves steep enough. For that reason all semiconducting substances having such a conductivity behaviour and material parameters corresponding to $b > 2$ should show these switching characteristics if the thermal model is practicable. However, the accurate knowledge of the two cooling parameters a_E and a_L realized in experiment is necessary for determining the exact form of all characteristics. Therefore, it must be considered whether the parameter dependence of the switching characteristics calculated in Part I and II can be confirmed or not.

Acknowledgements

The authors are greatly indebted to Dr. H. Gengnagel, Dr. E. Steinbeiß, and Dr. H. J. Anklam for critical discussions and numerical calculations.

References

[1] W. Brodkorb and W. Haubenreisser, phys. stat. sol. (a) **11**, 755 (1972).
[2] K. W. Böer, phys. stat. sol. (a) **2**, 817 (1970).
[3] K. W. Böer, E. Jahne, and E. Nebauer, phys. stat. sol. **1**, 231 (1961).
[4] N. Klein, Thin Solid Films **7**, 149 (1971).
[5] T. Kaplan and D. Adler, Electrothermal Switching in Amorphous Semiconductors, Proc. 4th Internat. Conf. Amorphous and Liquid Semiconductors, August 9 to 13, 1971, Ann Arbor (Mich.).

(Received April 19, 1972)

phys. stat. sol. (a) **12**, 663 (1972)

Subject classification: 14.3.4 and 20.1; 22.1.2

Brown Boveri Research Center, Baden, Switzerland

Doping Inhomogeneities in Silicon Observed by a High-Resolution Electroreflectance Technique

By

R. SITTIG and W. ZIMMERMANN

An electroreflectance method has been used to reveal doping inhomogeneities in as-grown silicon. The local resolution was selected at 50 μm, but may be improved easily. The mean relative error of the measurement lies at about ± 2%. An area of 0.5 × 1.0 cm² has been investigated. The structure found by the electroreflectance measurements was confirmed by electrochemical striation etching. A quantitative interpretation of the measurements in terms of doping concentration was obtained by comparison with the four-point probe conductivity data.

Es wird eine Elektroreflexionsmethode zum Nachweis von Dotierungsinhomogenitäten im ursprünglich gewachsenem Silizium benutzt. Die örtliche Auflösung wurde zu 50 μm gewählt, kann aber leicht verbessert werden. Der mittlere relative Fehler der Messung liegt bei ungefähr ± 2%. Es wurde eine Fläche von 0,5 × 1,0 cm² untersucht. Die durch die Elektroreflexionsmessungen gefundene Struktur wurde durch elektrochemische Ätzung bestätigt. Es wurde eine quantitative Erklärung der Messungen mit der Dotierungskonzentration durch Vergleich mit Leitfähigkeitswerten der Vier-Punktmethode erhalten.

1. Introduction

Progress in large-scale integration and high-power devices demands narrow material tolerances in ever increasing sizes of silicon wafers, particularly with respect to local variations of doping. In spite of its importance the problem of controlling the factors which influence doping inhomogeneities during growth of silicon single crystals and further processing is poorly understood. To a large part, this is due to a general lack of adequate methods for measuring resistivity variations with sufficiently good local resolution. Up to now only the conventional four-point probe technique with about 1 to 2 mm resolution is fully developed. Recently, attempts to improve the well-known spreading resistance method have resulted in some progress [1, 2]. Spatial resolution of about 10 μm and good reproducibility is claimed. Both techniques do not leave a microscopically perfect surface after the measurement due to small scratches from the electrical contacts. They are destructive as far as further processing of the wafers towards devices is required. Non-destructive, contactless ways of probing the resistivity must resort to optical or microwave methods. The latter has so far not shown high resolution. IR-transmission topography, however, introduced by Sherman and Black [3, 4] and co-workers does show considerable promise but may be limited to materials with high doping levels.

In this paper we report on first results of our attempt to use electroreflection for detection and measurement of spatially varying doping levels.

2. Method

Electroreflectance spectroscopy is known as a very valuable way to obtain information on fine details of the electronic band structure of semiconductors [5]. All previous work was done using relatively large surface areas of the sample under study to meet the theoretical demands of nearly parallel incident light and the experimental difficulties of measuring the rather small electroreflectance signal. Therefore a spatial average of the properties investigated was recorded and the usually assumed constancy of doping may be quite true.

From the theoretical model developed to understand the phenomenon of electroreflection it follows that the signal contains information not only on the band structure as it does in its spectral variation, but that the magnitude of the signal depends also on the doping level of the specimen. The following consideration shows, how the variation of the doping level may be obtained from electroreflection measurement in a straightforward way.

An electric field applied perpendicular to the surface of a silicon slice produces changes of reflectance, ΔR, for light in the wavelength range 350 to 380 nm.

The relative change of reflectance at a fixed wavelength λ depends on the magnitude of the electric field, E, as

$$\left.\frac{\Delta R}{R}\right|_{\lambda} = r(E)\,, \tag{1}$$

where R is the reflectance at zero applied field.

One of the usual methods of generating an electric field in a semiconductor, is to construct a condenser-like arrangement with the specimen as one condenser plate. The field just below the surface now depends on applied voltage, difference of work functions between the specimen and the electrode, on surface charge density, and on doping concentration. The density of the fixed surface charges may be assumed to be constant over a surface of a single crystallographic orientation and identical pretreatment. If the distance between the sample and the second condenser plate is held constant, then only variations in the doping concentration N should cause the variation of the electric field strength near the surface

$$E = E(N)\,. \tag{2}$$

The electric field strength versus distance from the silicon surface is sketched in Fig. 1 for two doping concentrations. With the top potential for both cases

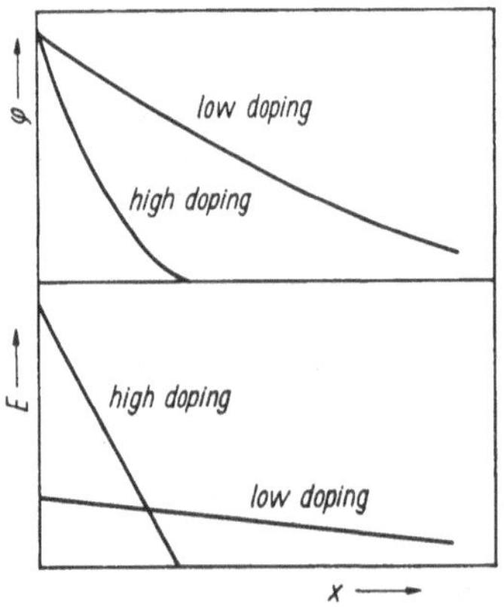

Fig. 1. Schematic dependence of potential and electric field inside a semiconductor for high and low doping concentrations

being set experimentally at the same level, the surface field strength is determined by the doping concentration.

From equations (1) and (2) electroreflectance signal and doping concentration are seen to be related as

$$r = f(N)\,. \tag{3}$$

The analytic form of this dependence is complicated, but can be calculated using some simplifying assumptions [5, 6]. For sufficiently small local deviations $\delta N(x, y) = N(x, y) - \bar{N}$ from the mean doping concentration $\bar{N}$ the function $f(N)$ may be linearized resulting in a proportionality between relative changes of electroreflectance signal and doping concentration:

$$\frac{r(x, y) - \bar{r}}{\bar{r}} = \frac{\delta r(x, y)}{\bar{r}} = K\,\frac{\delta N(x, y)}{\bar{N}}\,. \tag{4}$$

The factor K depends on $\bar{N}$ and on the band bending at the surface with and without applied voltage.

Relation (4) offers a new method for determining the variations of the doping concentration. The electroreflectance is measured using light of a fixed wavelength which is focused to a small spot on the specimen. The variations of the signal revealed by scanning the light spot over the surface are then proportional to the variations of the doping concentration, i.e. to the resistivity.

There are two main problems connected with this method. First of all the measurement of r needs to be related to N either by a theoretical analysis of formulae (4) or by a calibration procedure. We will return to this point later on but remark here that further work will be needed to resolve this problem satisfactorily.

The second point is entirely experimental. Because the electroreflectance signal is already inherently weak, reducing the light spot on the sample may reduce the signal-to-noise ratio to a very low level. Besides designing the optics properly, there are two possibilities to overcome these difficulties. To keep a high light flux we may enlarge the aperture angle of the incident light and, since we are not interested in the spectral dependence of the signal we may choose the wavelength of maximum signal-to-noise ratio.

3. Experimental

The originally lapped silicon slices are mechanically polished until they show a bright reflecting surface. Afterwards they are shortly treated with a chemical etch solution to remove surface damage. At a temperature of 390 °C an SnO_2 layer about 1000 Å thick is produced directly on the silicon. This layer shows high transmission in the near UV range and together with the silicon substrate it forms a heterojunction. A typical I–U characteristic of this is shown in Fig. 2. There is some variation in the characteristics of different samples but all show a definite blocking region. By varying the voltage in this region one generates an electric field at the silicon surface avoiding the trouble reported for the usual condenser-like arrangement.

The sample is mounted on a finely adjustable x–y support. This can be scanned in both directions perpendicular to the incident light beam so that a complete raster is obtained.

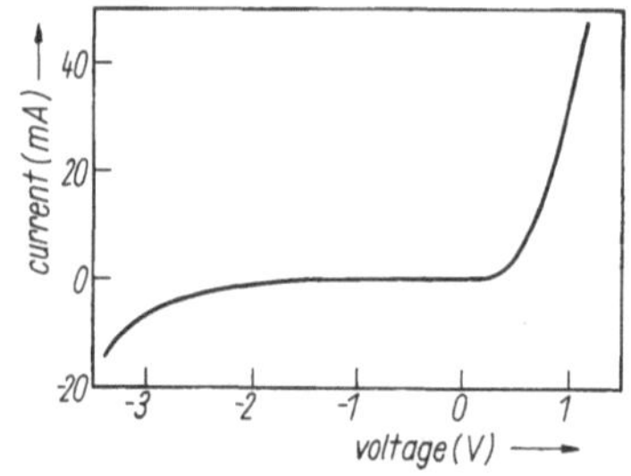

Fig. 2. Current–voltage characteristic of a heterojunction formed by a SnO_2 layer on a silicon slice. The silicon is n-type, 0.1 Ωcm, the active area 0.8 cm²

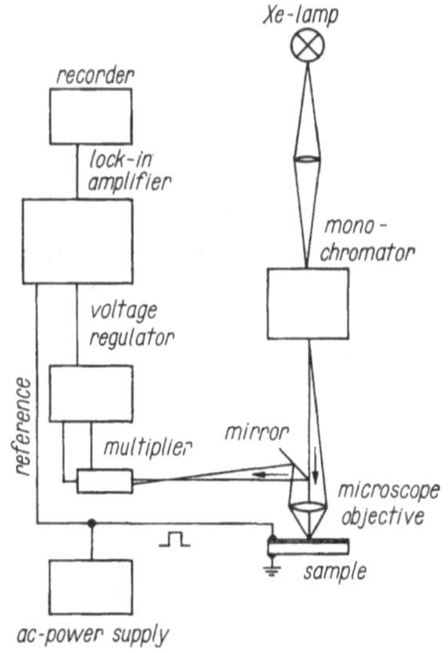

Fig. 3. Block diagram of the measuring apparatus

A schematic diagram of the apparatus is given in Fig. 3. We use a 450 W xenon-arc-lamp, which has a high intensity in the near UV region. The wavelength is selected by a monochromator. With the help of a microscope objective an image of the exit slit is formed on the silicon surface at one tenth of its actual size. Thus by varying the exit slit dimensions the form of the light spot may be adjusted.

The reflected light passes back through the objective and is then directed into a photomultiplier by a mirror.

To get rid of local variations of reflectance the multiplier voltage is controlled so as to give a constant direct current i. The relative change of reflectivity is then proportional to the ac component, Δi, which is measured by a lock-in amplifier and plotted by a recorder. All measurements are made at room temperature.

Some typical electroreflectance spectra at different applied voltages are shown in Fig. 4. As may be seen in this voltage region, the signal increase is nearly linear with increasing voltage. At still higher voltages saturation is observed and the spectra are shifted to longer wavelength.

For measuring the spatial variations of the electroreflectance signal we choose a voltage near the upper limit of the linear region. The scan velocity is chosen

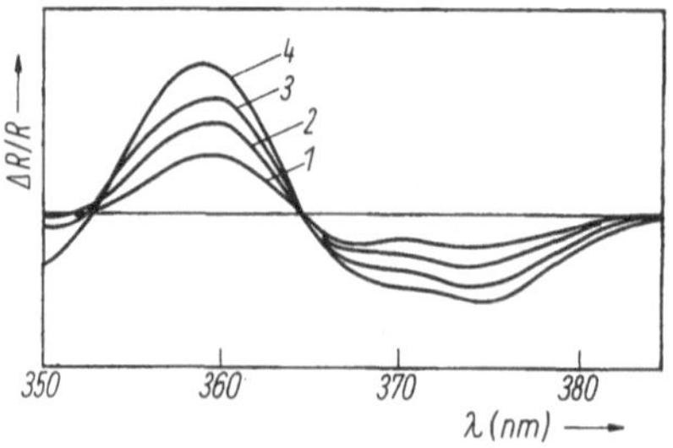

Fig. 4. Electroreflectance versus wavelength for different applied voltages. $\Delta R/R$ is given in arbitrary units. An ac square wave voltage is applied and centred at U_{dc} to give equal currents in the forward and reverse directions. The parameters are: (1) $U_{pp} = 1.8$ V, $U_{dc} = -0.6$ V; (2) $U_{pp} = 2.8$ V, $U_{dc} = -0.9$ V; (3) $U_{pp} = 3.8$ V, $U_{dc} = -1.2$ V; (4) $U_{pp} = 4.8$ V, $U_{dc} = -1.5$ V

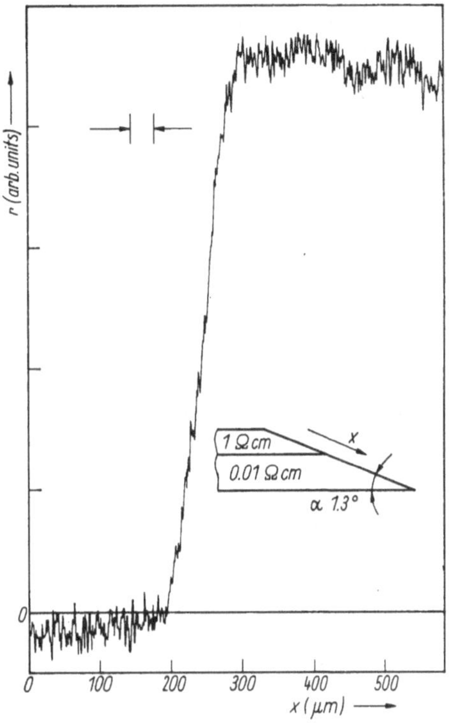

Fig. 5. Electroreflectance signal on a beveled n–n+ epitaxial layer. Spatial resolution is 35 μm, the integration time was chosen to be 0.3 s

to cover one light spot width in ten integration time constants. Setting the latter to 1 s and the spot width to 50 μm a run of 1 cm on the specimen surface takes 34 min. The mean error of the electroreflectance signal is then about ± 2%.

4. Results

In a first experiment an epitaxial n–n+ layer was investigated. On a substrate of a resistivity of 0.01 Ωcm a layer of 1 Ωcm was grown. The slice was beveled at an angle of 1.3° and measured in the region of the interface. As shown in Fig. 5, there was a large electroreflectance signal in the highly doped material but only a small negative signal on the lower doped side. This variation of the doping concentration by two orders of magnitude cannot be considered small, of course, so we cannot expect proportionality between electroreflectance and doping over the entire range. The drastic change in signal magnitude, however, showed the method was working correctly. Moreover, as the light spot width — 35 μm in this case — was smaller than the length over which the signal level changes, we could estimate a depth of the transition region to be about 1.4 μm in good agreement with penetration depth of the electric field in the lower doped material.

Supported by this rather crude experiment we now turn to the main problem, namely measuring only small variations of doping concentration. To try this the light spot was scanned along a diameter of a silicon slice. The result is

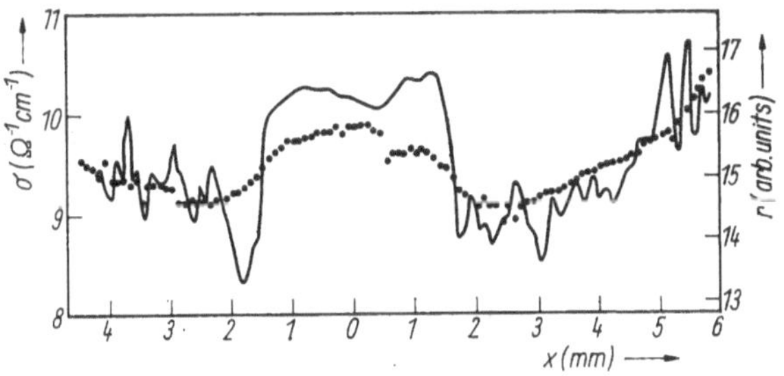

Fig. 6. Comparison between four-point probe conductivity measurement (●) and electroreflectance (–) along a diameter of an n-type silicon slice having a specified resistivity of 0.08 to 0.12 Ωcm. The light spot area was 50 × 100 μm², the wavelength 374 nm. The scale of the electroreflectance was chosen to fit the magnitude of the conductivity measurement. The zero levels of both scales coincide

shown in Fig. 6. For comparison the conductivity measurement by the four-point probe along the same diameter is plotted. The scale of the electroreflectance measurement was chosen to fit the magnitude of the conductivity curve. The zero levels of both scales are suppressed but they coincide.

As is known, the spatial resolution of the four-point probe amounts to 1 or 2 mm. Shorter variations of doping are smeared out by the measurement and therefore one expects a rather smooth curve for the conductivity versus position. Bearing this difference of the two methods in mind, the agreement between the two curves is excellent. From this we may also conclude that the proportionality factor K of equation (4) must be about 1[1]). So in this case the four-point probe measurement may serve as a calibration of the electroreflectance signal.

To get an impression of the variations of doping over an area of the slice, 20 curves of the electroreflectance measurement parallel to that of Fig. 6 were put together to form a three-dimensional model. A photograph of this is shown in Fig. 7.

A mesa-like structure is clearly seen at the centre of the slice — position (0, 0). It is surrounded by roughly annular minima and maxima, with the mean value of r increasing to the outside edge. At the right-hand side of the model ten of these minima and maxima are clearly distinguishable. The distance between two neighbouring maxima is about 500 μm near the centre and decreases to 300 μm near the edge. The variation of the signal between neighbouring maxima and minima lies between 4 and 8%. The maximum signal variation over the whole area amounts to ± 20% as is normal for the doping variation in as-grown silicon crystals.

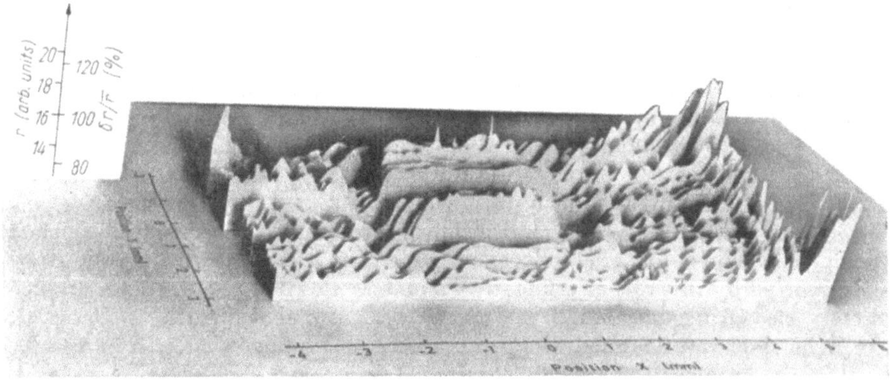

Fig. 7. 20 curves of the electroreflectance measurement as in Fig. 6, put together to form a model. Note that the zero level of the measurement lies 12 cm below the model plane. As the proportionality factor between electroreflectance and relative doping concentration is almost 1 in this case (see text), this represents a quantitative model of the doping across an area of 5 × 10 mm² of a silicon slice

---

1) A recent theory by Aspnes [7] shows that under special conditions, which he called surface barrier electroreflectance (SBER), r is proportional to the doping concentration N as well as to the ac voltage applied. This proportionality between r and the voltage can be seen in Fig. 4. The negligible variation of the spectral feature of the curves shown represents a further test confirming that we were working in the SBER region. Under these conditions K should always be equal to 1 as was determined from our measurements. Therefore these data also confirm the theory for silicon which Aspnes has tested on germanium.

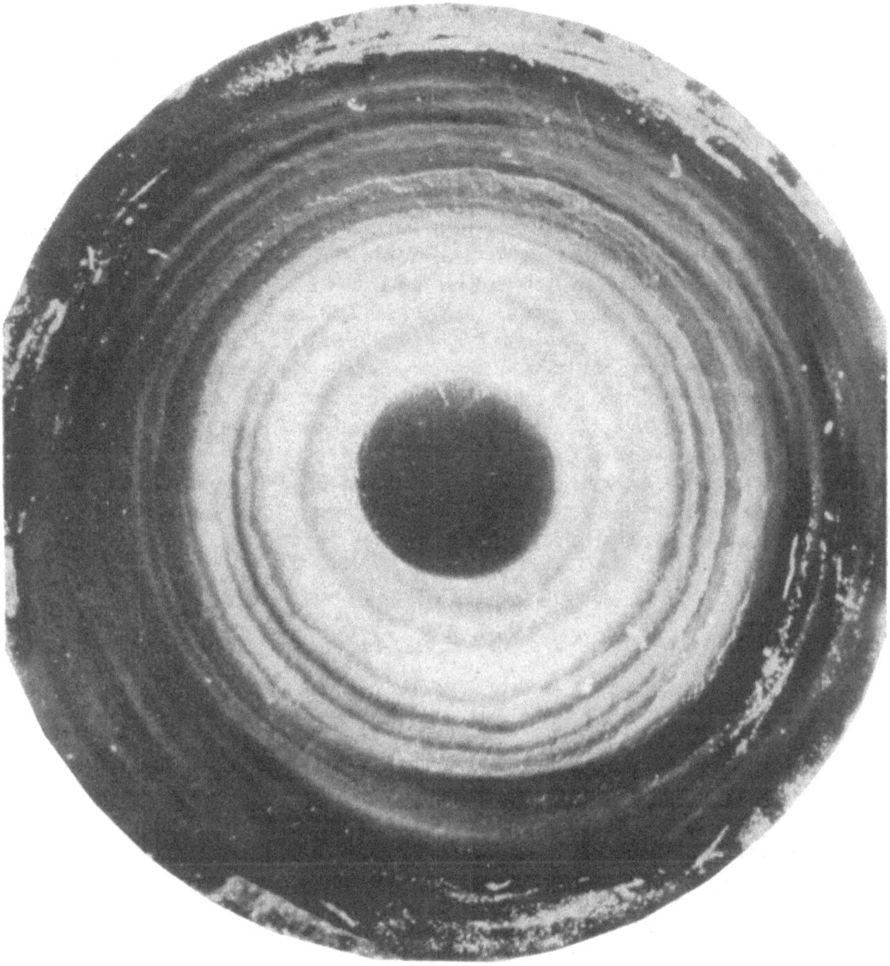

Fig. 8. Striation etch pattern of a silicon slice from the same rod as that measured in Fig. 7

To confirm this structure a striation etch was used. A picture of the etch pattern revealed for a neighbouring slice of the same rod is shown in Fig. 8. As in the model we see a more strongly etched circular area of 3 mm diameter, the centre of which is shifted by 0.1 mm from the centre of the slice. Especially remarkable is the spiral structure of the doping concentration. This can also be seen in the model. On the left-hand side the mesa edge has a small shoulder which forms a separate first maximum on the right and is increased to a large maximum after a full turn. Furtheron the increasing doping concentration towards the edge is confirmed by the etch picture as well as the mean distances between neighbouring maxima. Therefore we can say that the striation pattern revealed by the etching technique and that of the electroreflectance measurements are in excellent agreement.

5. Conclusions

It has been shown that by the electroreflectance method variations of the doping concentration can be detected with high spatial resolution in a rather simple and straightforward way. For the first time a nearly quantitative image of the doping variations across a large area of a silicon slice was obtained.

The quantitative connection between the electroreflectance signal and doping concentration at arbitrary doping levels has not yet been established, but we believe that the theoretical description, as well as calibration measurements will allow quantitative determination of doping by this method.

As the new method is non-destructive and shows a good accuracy, it already seems to be superior to the usual techniques. Moreover, it can be improved and simplified for example by using a spectral lamp and a narrow band filter instead of the xenon arc and the monochromator. Perhaps conditions of measurement are better defined by using a metal electrode in vacuum for generating the electric field. The focused light may be shone through a pinhole in the electrode onto the sample.

The results shown, as well as these arguments, allow us to hope that the electroreflectance method may become a powerful tool in investigations of variations of doping concentration.

Acknowledgement

It is a pleasure to thank Dr. C. Schüler for many stimulating and helpful discussions.

References

[1] R. G. MAZUR, J. Electrochem. Soc. **114**, 255 (1967).
[2] T. L. CHU and R. L. RAY, Solid State Technol. **14**, 37 (1971).
[3] B. SHERMAN and J. F. BLACK, Appl. Optics **9**, 802 (1970).
[4] D. C. GUPTA, B. SHERMAN, E. D. JUNGBLUTH, and J. F. BLACK, Solid State Technol. **14**, 44 (1971).
[5] M. CARDONA, Solid State Phys., Suppl. **11**, 165 (1969).
[6] A. MANY, Y. GOLDSTEIN, and N. B. GROVER, Semiconductor Surfaces, North-Holland Publ. Co., Amsterdam 1965.
[7] D. E. ASPNES, Phys. Rev. Letters **28**, 168 (1972).

(Received May 10, 1972)

phys. stat. sol. (a) **12**, 671 (1972)

Subject classification: 14.3.3; 16; 22.1.2

Brown Boveri Research Center, Baden, Switzerland

Measurement of Spatial Variations of the Carrier Lifetime in Silicon Power Devices

By

W. ZIMMERMANN

Lifetime measurements with spatial resolution of about three carrier diffusion lengths have been made in silicon high power diodes and thyristors. The method uses a pulsed 40 keV electron beam of 10 μm diameter to generate excess carriers the decay of which is measured by the built-in p–n junction. Lifetimes larger than 0.3 μs can be measured, but may easily be extended to smaller values. An example is given of a diode exhibiting lifetime variations within a factor of four in a range of a few millimeters.

An Silizium-Hochleistungsdioden und Thyristoren wurden Trägerlebensdauern mit einer Ortsauflösung von drei Diffusionslängen gemessen. Mit einem 40-keV-Elektronenstrahl von 10 μm Durchmesser werden Überschußträger erzeugt, deren zeitliches Abklingen dann gemessen wird. Die Werte der gemessenen Lebensdauern waren aus apparativen Gründen nach unten auf 0,3 μs begrenzt. Es wird ein Beispiel einer Diode angegeben, die in einem Bereich von wenigen Millimetern Lebensdauerschwankungen um einen Faktor vier aufwies.

1. Introduction

Lifetime of injected carriers in the high resistivity base region is an important parameter in high power devices [1, 2]. No information is yet available concerning the existence of possible spatial variations in lifetime. Those variations may exist as a consequence of built-in lateral inhomogeneities as in some types of thyristors, e.g. gate and emitter shortings. Another question is whether unknown and unwanted defects might act as centres of locally increased recombination. These can all be expected to have a strong influence on device characteristics. None of the well established lifetime measuring methods in diodes seem to be capable of giving quantitative information on local variations of lifetime. The usual recovery methods [3 to 8] for example always integrate over the whole volume filled with excess carriers. It would be possible, of course, to prepare a number of small diodes out of a large one, but local resolution is still far removed from its absolute limit, which in this case is the carrier diffusion length. With the conventional light spot technique [9 to 11], as we shall see later, inhomogeneities of diffusion length or lifetime may be revealed, but it is often impossible to obtain quantitative results. This is possible, however, using pulsed instead of steady state excitation and measuring the decay of the excess carriers. (As a probe the p–n junction always present in a diode is used.) It will be shown that spatial variations of lifetime in diode structures, at least in the low conductivity base region and some distance away from the p–n junction, can be quantitatively interpreted.

2. Principle of the Method

Excess carriers are generated near the surface of a diode structure by irradiation with an electron beam (Fig. 1a). The carriers diffuse into the bulk, and those reaching the junction are extracted and give rise to a photocurrent i_f. After a period long enough to establish steady state conditions, the beam is switched off. The relaxation time τ_r of the decaying photocurrent is, in the simplest case, identical with lifetime τ of excess carriers,

$$i_f(t) = i_{f0} \exp(-t/\tau_r) = i_{f0} \exp(-t/\tau) . \tag{1}$$

This is true if excitation takes place more than about two diffusion lengths away from the junction. Nearer by, the current probe, i.e. the junction, extracts too large a number of minority carriers thus disturbing recombination which results in a faster current decay. The correct $i_f(t)$ then depends not only on carrier recombination but in a complicated way on the boundary conditions provided by the p–n junction on one side and the ohmic n–n$^+$ contact on the other side of the diode.

In general, the function $i_f(t)$ can only be found using numerical methods, if realistic boundary conditions and other factors are considered. The full treatment of this problem is presently being studied in this laboratory.

For the time being, however, we concentrate on those cases where $i_f(t)$ is a simple exponential, i.e. where injection is some diffusion lengths away from the boundaries. We will show that under this condition the time constant τ_r is identical with lifetime τ by comparison with values obtained by the conventional light spot technique.

3. Experimental Details and Specimen Preparation

Diode structures were always used for the measurements, the diodes being cut and polished perpendicular to the p–n junction. A narrow electron beam of 10 μm diameter was allowed to impinge on this surface (Fig. 1a). The energy of the electrons was between 20 and 50 keV. Since τ_r deduced from the measurements was found to be independent of the energy of the primary electrons,

a

b

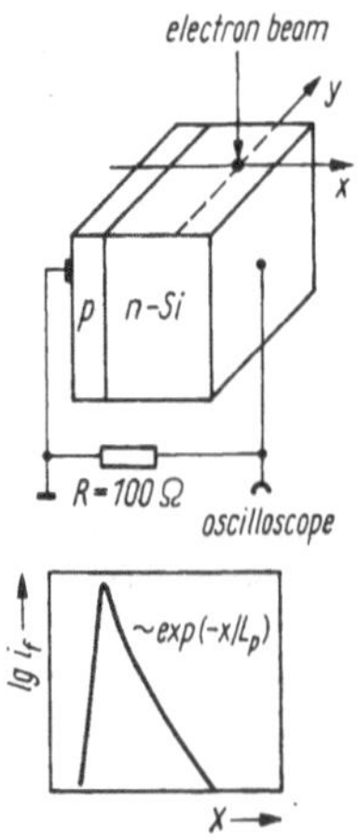

Fig. 1. a) Specimen arrangement during measurement. The diode can be moved in the x–y directions relative to the electron beam. b) Expected dependence of photocurrent i_f on position x

this energy was usually chosen as 40 keV. At this energy electrons are absorbed in silicon within a depth of about 10 μm [12], each incident electron generating about 1.1×10^4 electron–hole pairs. Beam currents could be varied from 0.1 to 100 nA. The induced photocurrent i_f was measured at the resistance R with a Tektronix 561 B oscilloscope with a 3A9 amplifier, then sampled and fed to a Hewlett Packard 7562 A Logarithmic Converter and finally plotted on a recorder. Two independent methods could be used: the conventional light spot technique, exhibiting a variation of the photocurrent with position x of the beam on the diode (Fig. 1b), and a measurement of the decay of the photocurrent, when the electron beam was pulsed but stationary at position x, revealing relaxation times of the current. For the pulsed photocurrent mode the electron beam was pulsed with a repetition frequency of 50 Hz. The duration of the pulses was long enough to establish a steady state photocurrent in the device, and the pulse rise and fall times were short enough to have no influence on the measured rise and fall times τ_r of the photocurrent i_f. The smallest measurable relaxation time was limited to $\tau_r \approx 0.3$ μs by the high frequency cut-off of the 3A9 amplifier. Because of this, lifetime measurements were possible only in the weakly doped n-base regions of the specimens, and not in the highly doped p-regions where they are expected to be smaller. In this case, 0.3 μs was always found only as an upper limit.

For measurements on thyristors, the silicon was removed from the emitter or anode side until the high resistivity n-base had been reached. Au (Sb) was then evaporated on to this new surface and alloyed at a temperature of 380 °C for several minutes in a N_2 atmosphere. Such contacts exhibited good injection qualities.

The surface to be irradiated by the electron beam was lapped and polished, finishing with 0.25 μm diamond paste. A final etch showed no influence on the results of the measurements and was not usually carried out. Most but not all of the specimens had been prepared from large area devices. Their dimensions were about 8×8 mm², with thicknesses between 200 and 800 μm. The p–n junctions lay at a depth of between 40 and 100 μm. All measurements were made at room temperature.

4. Results

Our first objective was to prove the usefulness and reliability of the new method. For this purpose we compare values for τ obtained from the time constants of the pulsed photocurrents to those resulting from determinations of the diffusion lengths by the conventional light spot technique.

It is convenient that the latter method can also be performed with the electron beam apparatus.

On varying the distance x between the p–n junction and the irradiated area, results of the type sketched in Fig. 1b are expected. The photocurrent i_f varies exponentially with distance up to at least a few diffusion lengths away from the junction,

$$i_f = i_{f0} \exp(-x/L_p) , \tag{2}$$

where L_p is the diffusion length of injected holes. Nearby this dependence is steeper, due to increasing removal of carriers by diffusion into the junction itself [13] and possibly surface recombination [14]. The number of holes removed

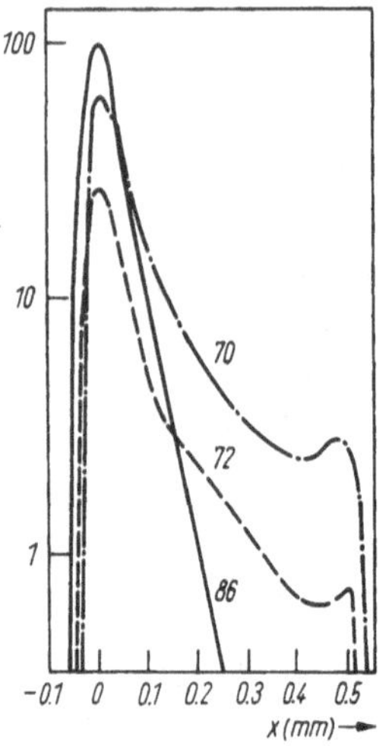

Fig. 2. Measured dependence of photocurrent i_f on position x for three diodes. The p–n junction is at $x = 0$ ($x \leqq 0$: p-Si, $x \geqq 0$: n-Si). All diodes are about 550 μm thick

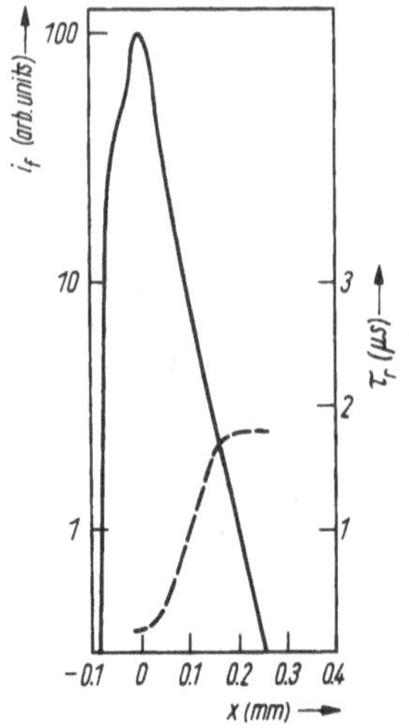

Fig. 3. Dependence of photocurrent i_f and relaxation time τ_r (dashed line) upon position x for diode 77. The diode thickness is about 550 μm. The diffusion length as determined from the diffusion mode signal is $L_p = 45$ μm, corresponding to $\tau = 1.65$ μs. The limiting value of the measured relaxation time at large x is $\tau = 1.8$ μs

by the junction is then no longer negligible compared to the number disappearing by recombination, which is a condition for the validity of equation (2).

Three examples of such measurements are given in Fig. 2. The expected exponential dependence is observed for specimen 86 over a large range of x. This behaviour is always found when the thickness d of the diode is large compared to the diffusion length L_p ($d \gg L_p$). This condition no longer holds for the other two specimens. An exponential dependence is observed over a short range in diode 72, but not at all in diode 70. However, it is just this diode 70 which is representative of properly designed devices, i.e. where diffusion length in the base region is not small compared with thickness.

The results summarized in Fig. 2 demonstrate quite clearly the principal shortcomings of the light spot method. The detailed analysis of van Roosbroeck [14] uses a few simplifying assumptions which break down in many practical cases where the light spot method is used. Quite frequently, the small signal case is not fulfilled because of large additional carrier concentrations. Furthermore, in diodes, the dimension perpendicular to the p–n junction is not always large compared to the carrier diffusion length. In other cases, excess carriers are generated near to or even within the back n–n^+ contact.

For the case of diodes 72 and 70, the increase of the photocurrent at the back n–n^+ contact is due to the built-in potential drop at this junction [15]. The lowly doped n-base material of our devices ($N_D \approx 10^{14}$ cm^{-3}) gives rise to a difference in energy levels of the conduction bands in the n-Si and the n^+-Si, $E_c(n) - E_c(n^+) \approx 0.3$ eV, which is capable of removing electrons generated nearby. This effect contributes to the overall photocurrent i_f measured in the external resistance R. In effect, only a very rough estimate of the diffusion length L_p in elements of this type is possible.

We will now turn to the pulsed photocurrent mode for determining the lifetime τ. As we shall see, this method is capable of yielding reliable results for cases where the previously treated diffusion mode fails.

On injecting electrons a distance x away from the p–n junction the photocurrent i_f is found to vary as

$$i_f(x, t) = i_{f0}(x) \exp\left(-t/\tau_r(x)\right) . \tag{3}$$

The measured dependence of the time constant τ_r on x is shown in Fig. 3. Only a few diffusion lengths away from the junction, $x \gtrapprox 3 L_p$, the relaxation time τ_r reaches a limiting value and becomes identical with the lifetime τ of injected carriers. The smaller $\tau_r(x)$ nearer to the junction is due to the increased proportion of carriers removed by diffusion into the junction and not by recombination alone [13], which is also a condition for the validity of equation (4) as it was for equation (2). Since rise and fall are governed by the same time constant [16], only the latter will be discussed further.

It will be shown by comparing measurements obtained using the diffusion mode and the time dependent mode, that the limiting value of the relaxation time is identical with carrier lifetime. This comparison was made on diodes similar to 86 and 72 in Fig. 2 which showed a definite exponential behaviour at least over a certain range of x, from which diffusion lengths could be calculated.

From diodes of the type 72 and 86, where it was possible to deduce diffusion lengths from the definite exponential decay of the diffusion mode measurements, carrier lifetimes were calculated from

$$\tau_p = \frac{L_p^2}{D_p}, \tag{4}$$

using $D_p = 12.9\ \text{cm}^2\,\text{s}^{-1}$, which corresponds to a hole mobility μ_p of $500\ \text{cm}^2\,(\text{Vs})^{-1}$. The mean deviation between τ_p and the limiting values τ of the pulsed photocurrent mode was $\pm 9\%$, thus giving identical results within the limits of error, which were estimated to be $\pm 10\%$ in each case.

Three specimens exhibited a strictly exponential behaviour, but also an increase of photocurrent near the n–n$^+$ contact. The relaxation time, nevertheless, is constant near this contact. We conclude from this that carrier lifetimes may also be measured as relaxation times of the photocurrent in cases where the exponential behaviour of the diffusion mode experiment is disturbed by the back contact, provided that the pulsed photocurrent mode gives constant relaxation times within a certain range of position. This conclusion was then extended to the more common cases, where in the diffusion mode no exponential behaviour at all was observed (e.g. Fig. 4). The diffusion length calculated from the measured lifetime, in this case,

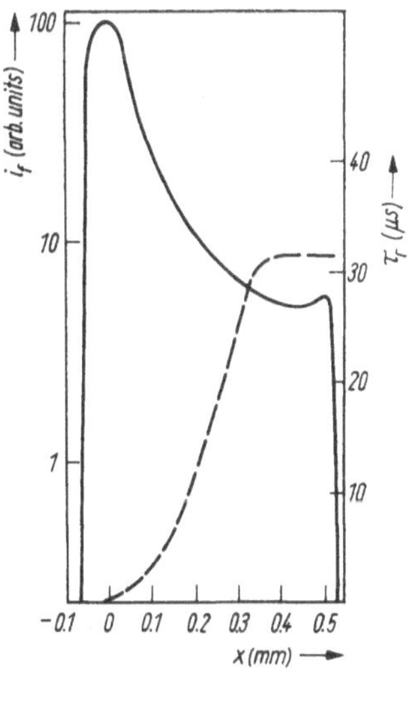

Fig. 4. Dependence of photocurrent i_f and relaxation time τ_r (dashed line) upon position x for diode 76. The limiting value of the measured relaxation time at large x is $\tau = 31.5\ \mu\text{s}$, corresponding to a diffusion length $L_p = 198\ \mu\text{m}$

was $L_p = 198$ μm, but the distance d between p–n and n–n$^+$ junctions was only $d = 500$ μm $= 2.5\, L_p$. It is clear then that the influence of the n–n$^+$ contact begins before a strictly exponential dependence is reached.

5. Spatial Variations of Lifetime

It was mentioned above that, with the pulsed light spot technique, it should be possible to study lifetime variations in a device. Several examples of such variations have been observed, and one will be given in the following. The usual behaviour of most of the specimens examined, however, was to show constant lifetime within the limits of error.

Both the diffusion mode and the time dependent pulsed mode have been used to reveal the lifetime inhomogeneities. Again, with the diffusion mode it was possible to get only qualitative information, at least in cases where the thickness d of the device compared to the diffusion length L_p was not large enough to establish clear exponential behaviour of the diffusion mode signal.

Results are given in Fig. 5 and 6 for the pulsed and diffusion mode, respectively. In Fig. 5 the position y of the beam was moved parallel to the junction over the total length of the specimen, the irradiated area at $x = 450$ μm being close to the back n–n$^+$ contact (see Fig. 1a). While over most of the specimen lifetime varies between $\tau_{max} = 28$ μs and $\tau \approx 20$ μs, there is a deep lifetime hole at $y = 8.3$ mm, where the measured value is $\tau_{min} = 8$ μs. The minimum value may be even smaller but could not be determined because at $x = 450$ μm, $y = 8.25$ mm, the signal became too small to be measured. Diffusion lengths corresponding to $\tau_{max} = 28$ μs and $\tau_{min} = 8$ μs are $L_{p\,max} = 193$ μm and $L_{p\,min} =$ $= 101$ μm, respectively. This difference is much more than expected and is too great to be tolerated in correctly functioning devices.

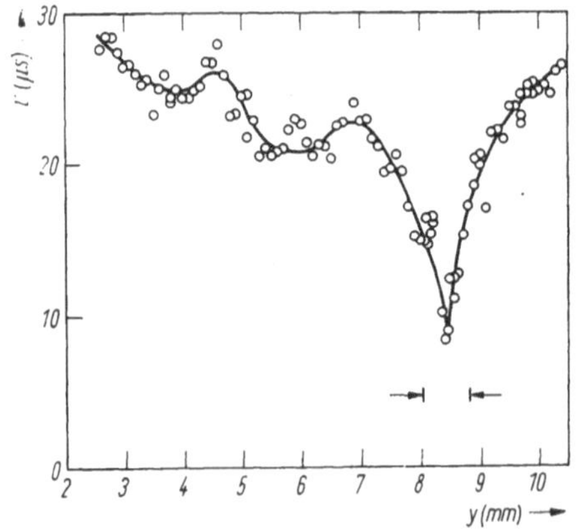

Fig. 5. Lifetime as a function of position y in diode 72. The arrows mark the region within which the curves of Fig. 6 have been measured

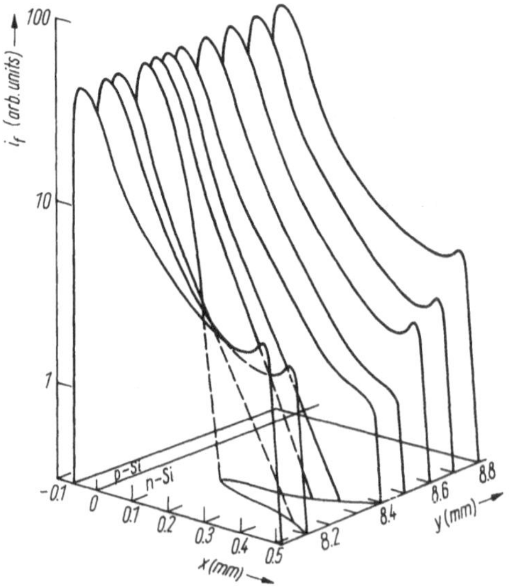

Fig. 6. Dependence of photocurrent i_f on position x for diode 72 at various positions y. The fall-off of the diffusion length near $y = 8.25$ mm is clearly seen. The different scales in x and y should be noted

Diffusion mode measurements have been carried out over the whole length of the diode, but do not change remarkably except within the region marked by the arrows in Fig. 5. Within this region, i.e. $8.0 \leqq y \leqq 8.8$ mm, a dramatic change is observed (Fig. 6). Interpretation of the result is straightforward: While diffusion length is large enough at the sides of the marked region to establish a figure comparable to crystal 70 in Fig. 2, it drops down to a minimum value of $L_{p\,min} = 20$ µm at $y = 8.25$ mm, corresponding to $\tau_{min} = 0.3$ µs. At $y = 8.15$ mm and $y = 8.35$ mm, $L_p = 84$ µm and $\tau = 5.5$ µs.

The cause for this large localized drop in lifetime is not clear at present. It is possible that a fault in the alloying process of the back $n-n^+$ contact is responsible. Possibly a structural defect, such as a micro-crack, could have acted as a nucleus for a larger Au precipitate which then forms a centre of increased recombination.

This specimen with its continuous variation of diffusion length over a large scale is also an illustrative example of the diffusion mode signal dependence upon diffusion length relative to diode thickness: strictly exponential behaviour is observed between $y = 8.15$ mm and $y = 8.35$ mm, where diffusion length is small enough, while outside this region this behaviour is already disturbed by the influence of the $n-n^+$ contact.

6. Conclusions

With the pulsed light spot technique described above, lifetime variations in silicon power diodes have been studied. While the well established steady state light spot technique, here referred to as diffusion mode, allows only qualitative interpretation in many cases, quantitative interpretation can always be made using the time dependent mode. One example from several similar specimens was given in which the lifetime varied by at least a factor of 4 over a range of a few millimeters.

The spatial resolution of the method is limited by the carrier diffusion length, the limits of error being less than $\pm$ 10%.

The technique proved also to be very suitable for studying the dependence of lifetime on temperature [17] and injection level, since the irradiation intensities used are low enough to avoid temperature changes in the excited volume. In addition, lifetime variations in high power thyristors due to an inhomogeneous lateral geometry (e.g. shorted emitter structure) may easily be detected. Studies on these subjects are in preparation.

Acknowledgements

The author wishes to thank J.-P. Roth for his skilful technical assistance, Dr. C. Schüler for many helpful suggestions, and his colleagues Dr. R. Sittig and Dr. M. J. Hill for numerous stimulating discussions.

References

[1] Y. C. Kao and J. R. Davies, IEEE Trans. Electron Devices **17**, 652 (1970).
[2] A. Herlet and K. Raithel, Solid State Electronics **9**, 1089 (1966).
[3] R. H. Kingston, Proc. IRE **42**, 829 (1954).
[4] B. Lax and S. F. Neustadter, J. appl. Phys. **25**, 1148 (1954).
[5] K. Schuster, Solid State Electronics **8**, 427 (1965).

[6] L. W. DAVIES, Proc. IEEE **51**, 1637 (1963).
[7] R. J. BASSETT, Solid State Electronics **12**, 385 (1969).
[8] S. C. CHOO and R. G. MAZUR, Solid State Electronics **13**, 553 (1970).
[9] G. JUNGK und H. MENNINGER, phys. stat. sol. **5**, 169 (1964).
[10] H. HIGUCHI and H. TAMURA, Japan. J. appl. Phys. **4**, 316 (1965).
[11] C. MUNAKATA, Microelectronics and Reliability **5**, 267 (1966).
[12] A. F. MAKHOV, Soviet Phys. — Solid State **2**, 1934 (1960).
[13] S. STRÄSSLER, private communication.
[14] W. VAN ROOSBROECK, J. appl. Phys. **26**, 380 (1955).
[15] W. RUPPEL, phys. stat. sol. **5**, 667 (1964).
[16] S. M. RYVKIN, Photoelektrische Erscheinungen in Halbleitern, Akademie-Verlag, Berlin 1965 (p. 360).
[17] W. ZIMMERMANN, phys. stat. sol. (a) **10**, K49 (1972).

(Received May 10, 1972)

phys. stat. sol. (a) **12**, 679 (1972)

Subject classification: 11; 2; 22.1.2

Physico-Technical Research Institute, Gorkii

On the Peculiarities of Silicon Amorphization at Ion Bombardment

By

A. I. Gerasimov, E. I. Zorin, P. V. Pavlov, and D. I. Tetelbaum

Pecularities of amorphous layer production in silicon under ion bombardment with middle energy ions were examined by electron diffraction. He^4, B^{11}, C^{12}, N^{14}, Ne^{20}, P^{31}, Ar^{40}, As^{75}, and Xe^{131} ion beams with energy of 50 keV were used. It was established that for light ions in a definite range of doses two amorphous layers are formed — an external layer and an internal one (sandwich structure). Doses necessary for amorphization and the thicknesses of the amorphous layers were determined. The results are discussed in terms of a model connected with the radiation defect accumulation. It is shown that for heavy ions the amorphization doses and the thicknesses of amorphous layers agree with the calculated distribution and concentration of radiation defects without taking account of their migration and annealing; for light ions under the given experimental conditions it is necessary to take account of the decreasing of a number of defects in the zone of displacements due to their migration and (or) annihilation.

Методом электронографии исследованы особенности формирования аморфных слоев в кремнии при бомбардировке ионами средних энергий. Использовались ионные пучки He^4, B^{11}, C^{12}, N^{14}, Ne^{20}, P^{31}, Ar^{40}, As^{75} и Xe^{131} с энергией 50 кэв. Установлено, что для легких ионов в определенном интервале доз образуются два аморфных слоя — наружный и внутренний („сэндвич"-структура). Определены дозы, необходимые для аморфизации, и толщины аморфных слоев. Результаты обсуждаются на основе модели, связанной с накоплением радиационных дефектов. Показано, что для тяжелых ионов дозы аморфизации и толщины аморфных слоев находятся в соответствии с расчетным распределением и концентрацией радиационных дефектов без учета миграции и отжига последних; для легких же ионов при условиях эксперимента необходимо учитывать убыль количества дефектов в зоне смещений вследствие их миграции и (или) аннигиляции.

1. Introduction

The problem of amorphization of solids under ion bombardment recently has become intensively discussed (see, for example, [1]). The suggested models of this phenomenon might be reduced to three basic ones: a) formation of locally disordered regions along ion tracks with subsequent overlapping of these regions [2 to 4]; b) formation of amorphous "nuclei" in the displacement spikes due to the drain of interstitial atoms [5]; c) destruction of the crystal lattice occurring after the radiation defect concentration has achieved a definite value [2, 6].

As it has been shown in [6] the data on amorphization of Ge and Si with phosphorus and of Ge with boron ions agree with the model c). Further experiments carried out for a wide range of ion masses have revealed some new regularities which make it necessary to return to this problem.

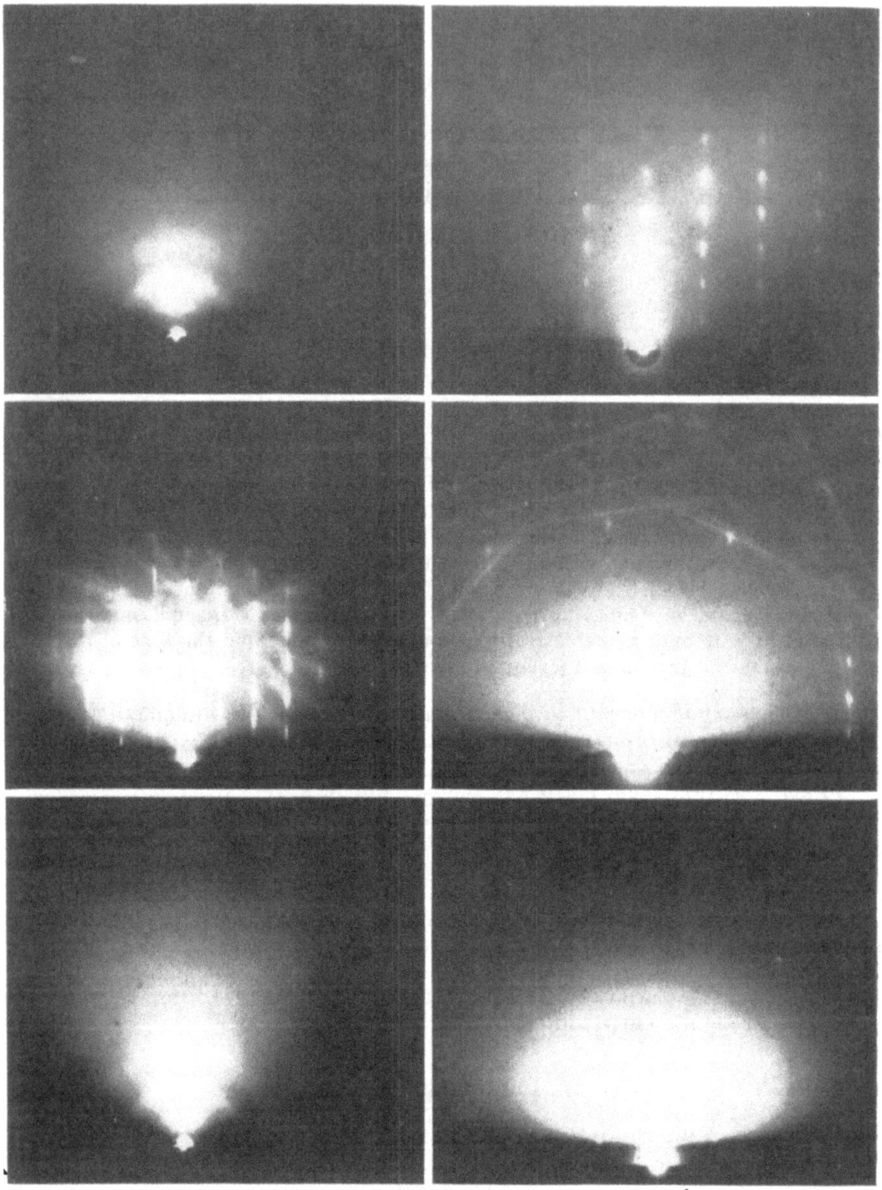

a d b e c f

Fig. 1. Electron diffraction pattern of silicon irradiated with boron ions ($l\lambda = 15.4$ mm Å). a) to c) $E = 50$ keV, $\Phi = 1.2 \times 10^{16}$ cm^{-2}, $j = 1.9 \times 10^{13}$ cm^{-2} s^{-1}; a) before etching, b) and c) after removal of the layers of 0.04 and 0.012 μm respectively. d) $E = 20$ keV, $\Phi = 1.9 \times 10^{16}$ cm^{-2}, $j = 1.2 \times 10^{13}$ cm^{-2} s^{-1}; a layer of 0.04 μm was removed. e) and f) $E = 50$ keV, $\Phi = 6.2 \times 10^{15}$ cm^{-2}; a layer of 0.18 μm was removed; e) $j = 6.2 \times 10^{12}$ cm^{-2} s^{-1}, f) $j = 6.2 \times 10^{13}$ cm^{-2} s^{-1}

2. Experimental Procedure

Investigations were carried out with 7.5 Ω cm n-Si and 1 Ω cm p-Si cut in the (111) plane. No special care to exact orientation relative to the ion beam had been undertaken. He^4, B^{11}, C^{12}, N^{14}, Ne^{20}, P^{31}, Ar^{40}, As^{75}, and Xe^{131} ions with an energy of 50 keV were used. The target temperature was 22 °C and the dose rate j in each series of experiments was held constant.

Investigation of structural changes over the depth was carried out by successive removal (using anodic oxidation) followed by taking reflection electron diffraction patterns. To reveal the structure of the layer nearest to the surface (up to several hundred Å of depth) more precise etching (5 min boiling in 60% HNO_3) was used; in this case (28 ± 5) Å were removed per cycle.

3. Sandwich Structure of a Damaged Layer

Investigation of the structure of the surface region over the depth after B^{11} bombardment showed that in a definite range of doses (5×10^{15} to 3×10^{16} ions cm^{-2} at $j = 2 \times 10^{13}$ ions $cm^{-2}\,s^{-1}$) this region represented a sandwich consisting of a thin external amorphous layer (EAL) directly adjacent to the surface, of an internal amorphous layer (IAL), and a monocrystal interlayer (MI) between them. Corresponding electron diffraction patterns are presented in Fig. 1 a, b, c. A similar sandwich structure was observed in the case of Ne^{20} ions, but it was absent for P^{31}, Ar^{40}, As^{75}, and Xe^{131}. Thus, under the given experimental conditions the formation of the sandwich structure took place for relatively light ions only. (For He^4 at an energy $E = 20$ keV and $j = 6 \times 10^{13}$ ions $cm^{-2}\,s^{-1}$ IAL was not observed in the examined dose range, i.e. up to 6.2×10^{16} ions cm^{-2}, but for the largest dose EAL was present.)

It is important to note that EAL represented true amorphous silicon but not a layer of contamination. Indeed, on the electron diffraction patterns from EAL the same diffuse rings were present as on those from IAL, the values of sin (θ/λ) for the regions with the highest intensity coinciding with the values of sin (θ/λ) for amorphous silicon [7]. Besides, if EAL occurred as a result of cracking processes and polymerization of organic vapour under ion bombardment, it would be expected that the dose at which this layer is formed would increase with ion mass (due to the increasing sputtering ratio), the other conditions being the same. In reality the reverse was true (see below).

The cases where a contamination film was present might be easily distinguished from EAL according to the electron-diffraction picture. After the film had been removed the electron diffraction pattern took the form shown in Fig. 1 a, and on further etching a sandwich structure with two layers of amorphous silicon was revealed.

The thickness of EAL (h_{ext}) increased linearly with dose Φ (Fig. 2). Extrapolation of the dependence of $h_{ext} = f(\Phi)$ to the intersection with the abscissa gives the threshold dose necessary for the beginning of EAL formation.

IAL also expanded with dose, but slower than according to a linear law. So, the thickness of IAL increased by 30% with the change of the Ne^+ dose from 9×10^{14} up to 2×10^{15} ions cm^{-2}. Expanding of IAL to the surface took place more rapidly than into the depth of the specimen.

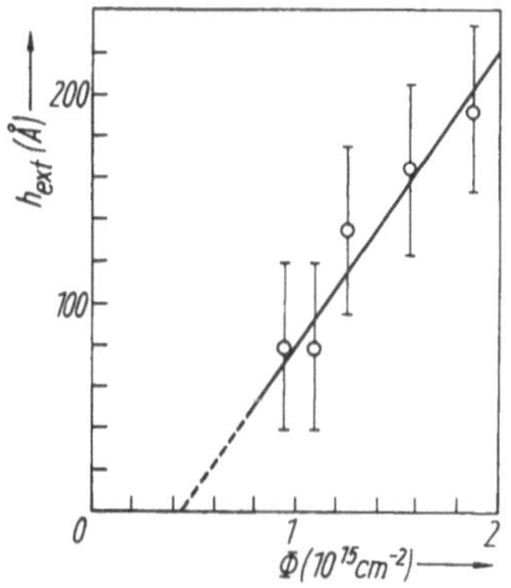

Fig. 2. Dependence of h_{ext} on dose at irradiation with boron ions ($j = 6.2 \times 10^{12}$ $cm^{-2}\,s^{-1}$)

The degree of MI perfection gradually decreases approaching IAL whereas the boundary between MI and IAL was rather sharp. Sometimes inclusions of polycrystalline silicon in MI were observed (Fig. 1d).

4. Amorphization Dose

The formation of the sandwich structure requires a specification of the conception "amorphization dose". In general three critical doses should be distinguished:

a) the dose of surface amorphization (DSA), i.e. the dose at which EAL is formed;

b) the dose of internal amorphization (DIA), i.e. the dose at which IAL is formed;

c) the dose of overall amorphization (DOA).

If EAL is not formed, DOA coincides with the dose at which IAL, when expanding, approaches the surface; if EAL formation does take place, DOA corresponds to the dose at which EAL joins with IAL.

In this connection it should be noted that in [6] for all the cases except of irradiation of Si with boron, DOA has been concerned, and in [8] the experimentally determined amorphization doses for Li^7 and B^{10} ions actually represent DSA what must be taken into account in the interpretation of these data. In Fig. 3 the dependence of DSA and DOA on ion mass, obtained by us at $j = 1.2 \times 10^{13}$ ions $cm^{-2}\,s^{-1}$, are presented. Both dependences have an expo-

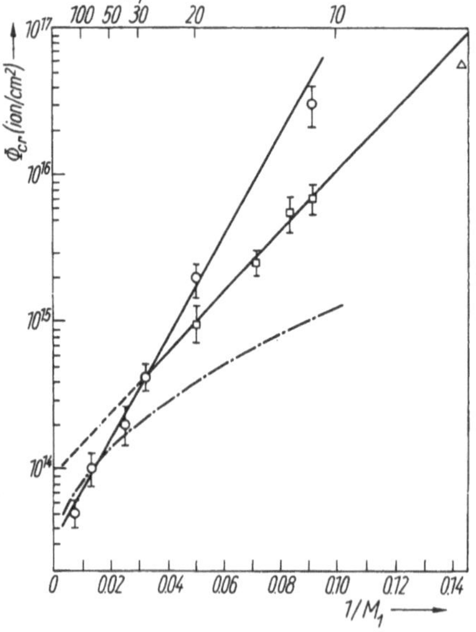

Fig. 3. Dependence of Φ_{cr} on ion mass. ○ Φ_{cr}^{ov}; □ Φ_{cr}^{s}; △ Φ_{cr}^{s} [8]

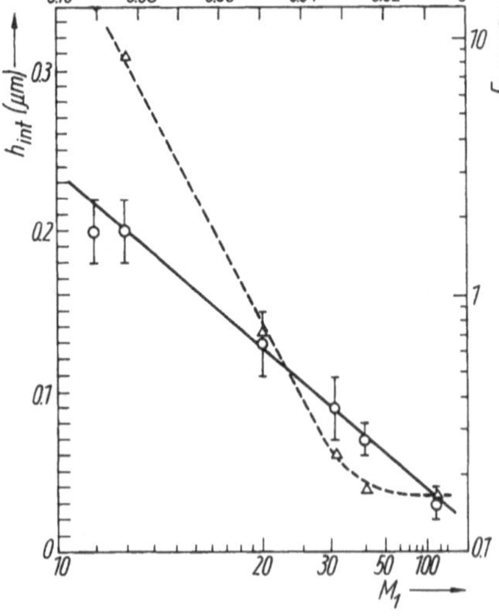

Fig. 4. Dependence of h_{int} and F_d on ion mass

nential form and may be approximated by the empirical relations

$$\Phi_{cr}^{ov} = 3.3 \times 10^{13} \exp\left(\frac{79}{M_1}\right) \text{ ions cm}^{-2} ,$$

$$\Phi_{cr}^{s} = 7.7 \times 10^{13} \exp\left(\frac{50}{M_1}\right) \text{ ions cm}^{-2} \text{ (at } M_I \gtrsim 30) ,$$

where Φ_{cr}^{ov} is DOA, Φ_{cr}^{s} is DSA.

If one supposes that radiation-defect annealing and the defect diffusion out of the irradiated layer does not take place at room temperature, then according to model c) (see Introduction) Φ_{cr}^{ov} must approximately correspond to a definite mean concentration $\bar{n}_d = N_d/N_0$ of displaced atoms (N_d is the mean number of displaced atoms per cm^3, N_0 the atomic density of Si). The dashed line in Fig. 3 corresponds to $\Phi_{0.3}$, doses at which $\bar{n}_d = 0.3$. The value of $\bar{n}_d$ has been calculated according to the Kinchin-Pease-Sigmund relation [9]; the threshold of displacement was accepted to be 22 eV [10], and as thickness of the irradiated layer the depth was accepted at which according to Monte-Carlo calculations [11] the concentration of displaced atoms was 0.2 of the concentration in the distribution maximum right of the latter. It is seen that for heavy ions Φ_{cr}^{ov} is close to $\Phi_{0.3}$, whereas for light ions $\Phi_{cr}^{ov} \gg \Phi_{0.3}$.

More exact information on the extent to which the amorphization dose is related to the calculated concentration of the displaced atoms may be obtained from the data on the depth (h_{int}) of the distant boundary of IAL. The values of h_{int} obtained in the dose range 6×10^{12} to 1.2×10^{13} ions cm^{-2} s^{-1} are presented in Fig. 4. The dependence of h_{int} on M_1 may be approximated by the relation

$$h_{int} = 0.02 + \frac{2.2}{M_1} \ \mu\text{m} .$$

The dashed line in Fig. 4 shows the concentrations of the displaced atoms F_d at depths h_{int}, $n_d(h_{int}) = N_d(h_{int})/N_0$, calculated by the Monte-Carlo method using the model presented in [11]. The calculation was carried out without taking account of the annealing of defects and their migration.[1]) It is seen that for heavy ions $n_d(h_{int})$ slightly changes with ion mass and is approximately 0.2 μm, but for light ions $n_d(h_{int})$ is much higher. The reasons for such a behaviour will be discussed in the next section.

It is important to note that for light ions the degree of disorder at constant dose and target temperature T_t close to room temperature depends on T_t [12, 13] and j [14]. The dependence upon j was also observed in our experiments (see Fig. 1e, f.

5. Discussion

Let us now discuss the data on the internal amorphous layer not concerning yet the reasons for formation of the external amorphous layer. The results presented here show that two cases are possible.

Case A: Amorphization doses correspond to a definite mean concentration of displaced atoms, and the thickness of the amorphous layer corresponds to the depth at which the concentration of the displaced atoms decreases up to a definite value.

---

[1]) The effect of saturation associated with the decrease of the number of the non-displaced atoms during bombardment was not taken into consideration in this calculation.

Case B: Amorphization takes place at far higher doses, and the amorphous-layer thickness does not correspond to the above presented depth.

At room temperature and moderate dose rates the case A is realized for heavy ions and the case B for light ions. As the dose rate increases or the target temperature is lowered, the boundary between ion masses M_1 for which the cases A or B are realized, is apparently shifted to lower masses.

The regularities pointed out may be explained basing upon the model c) (see Introduction), realized as follows: Let us consider the evolution of radiation defects formed in the displacement cascade. Generally speaking it is necessary to take into account the following processes:

1. diffusion of vacancies and interstitial atoms;
2. association of vacancies and interstitial atoms into complexes (compound defects);
3. recombination of vacancies with the interstitial atoms;
4. annealing of compound defects and their diffusion.

For simplicity let us suggest that at the irradiation temperature the processes 3 and 4 are absent. [2]) Besides, it may be suggested that due to a higher mobility in comparison with vacancies the interstitial atoms play no essential role for radiation-defect accumulation in the irradiated zone. Thus, at dose rates low enough when the evolution of radiation defects has time to complete before the next cascades would occur at the same place, defect accumulation in the displacement zone is determined mainly by the competition of the processes 1 and 2. Amorphization at a given point takes place when the vacancy concentration including the vacancies combined into complexes reaches a definite value. For heavy ions with a given energy the initial vacancy density in the cascade is high, hence the process 2 will dominate over the process 1 and no essential loss of vacancies from the displacement zone occurs. For light ions the reverse situation takes place that results in an increase of amorphization dose and deformation of the profile of the final distribution of defects. With increasing dose rate or decreasing target temperature (the latter results in an increase of effective time τ of radiation-defect evolution) the probability of superposition of displacement cascades during the time τ is increased and therefore causes a rising contribution of the process 2 and consequently a reduction of vacancy losses from the irradiation zone.

It should be noted that the increase in ion energy must lead (though to a less extent) to the same effect as the reduction of M_1. As far as a thermal spike accompanying the displacement cascade is concerned, it is not excluded that it playes an essential role in the case of the heaviest ions and (or) heavy targets for which the thermal energy is concentrated in a smaller volume. In this case a local melting is possible which should result in a deviation from the regularities of amorphization presented above.

Now we proceed to the discussion of the reasons for the formation of an external amorphous layer. Four mechanisms of EAL formation may be suggested[3]):

---

[2]) It is likely that recombination is inhibited by the energy barrier, the existence of which for the case of Ge was shown in [15] by extremely high mobility of interstitial atoms, and probably by spatial separation of vacancies and interstitial atoms during the development of the displacement cascade [16].

[3]) The first two mechanisms were proposed in [16] as an explanation of the amorphization phenomenon at all.

1. Migration of the displaced silicon atoms and their escape to the surface with subsequent growth of an amorphous phase. However, such a mechanism would result in a *proportional* but not in a simply *linear* dependence of h_{ext} on Φ as observed in the experiments.

2. Sputtering resulting in an accumulation of vacancies near the surface which are not compensated by the interstitial atoms. But it is hardly possible that such a mechanism causes the formation of an amorphous layer as thick as several hundred Å since according to the existing points of view the sputtered ions are knocked out from the depth of the order of only several interatomic distances [17]. However, it is not excluded that this mechanism initiates EAL origination.

3. Vacancy drain into the surface layer seems the most probable reason for EAL formation. It is known that the thin layer of a crystal adjacent to the surface is in a strained state and can serve as a potential well for vacancies (and as a potential hill for interstitial atoms). Such a drain will evidently play a greater role for light ions (case B) when the vacancy migration out of the maximum of their initial distribution in the cascade is important. By this may be explained the fact that EAL has not been observed for heavy ions.

4. It is possible that in a thin surface layer the threshold of displacement, E_d, is less or the amorphization of this layer takes place at lower critical concentrations of radiation defects than in the bulk. However, in both cases EAL would be equally formed for light and heavy atoms what does not take place.

References

[1] R. Kelly and N. Q. Lam, Radiat. Eff. **10**, 247 (1971).
[2] E. I. Zorin, D. I. Tetelbaum, Yu. S. Popov, and Z. K. Granitsina, Fiz. tverd. Tela **6**, 2017 (1964).
[3] J. R. Parsons, Phil. Mag. **12**, 1159 (1965).
[4] F. F. Morehead and B. L. Crowder, Radiat. Eff. **6**, 27 (1970).
[5] Yu. V. Martinenko, Fiz. tverd. Tela **11**, 1968 (1969).
[6] A. I. Gerasimov, E. I. Zorin, P. V. Pavlov, and D. I. Tetelbaum, Dokl. Akad. Nauk SSSR **192**, 324 (1970).
[7] H. Richter and G. Breitling, Z. Naturf. **13a**, 988 (1959).
[8] V. M. Gusev, Yu. V. Martinenko, and K. V. Starinin, Kristallografiya **14**, 1050 (1969).
[9] P. Sigmund, Radiat. Eff. **1**, 15 (1969).
[10] P. L. F. Hemment and P. R. C. Sevens, J. appl. Phys. **40**, 4893 (1969).
[11] P. V. Pavlov, D. I. Tetelbaum, E. I. Zorin, and V. I. Alekseev, Fiz. tverd. Tela **8**, 2679 (1966).
[12] J. E. Westmoreland, J. W. Mayer, F. H. Eisen, and B. Welch, Appl. Phys. Letters **15**, 308 (1969).
[13] S. T. Picraux, J. E. Westmoreland, J. W. Mayer, R. R. Hart, and O. J. Marsh, Appl. Phys. Letters **14**, 7 (1969).
[14] F. H. Eisen and B. Welch, Radiat. Eff. **7**, 143 (1971).
[15] V. V. Khainovskaya and L. S. Smirnov, Fiz. tverd. Tela **9**, 2043 (1967).
[16] N. B. Pridachin and L. S. Smirnov, Fiz. Tekh. Poluprov. **5**, 166 (1971).
[17] P. Sigmund, Phys. Rev. **184**, 383 (1969).

(Received May 12, 1972)

phys. stat. sol. (a) 12, K65 (1972)

Subject classification: 18; 21.1

Oregon Graduate Center, Beaverton

Magnetic Studies on $TbNi_2$ and $DyNi_2$

By

H. OESTERREICHER, J. STANLEY, and R. PITTS

The present study is part of a general attempt to clarify the considerably complex magnetic behavior of rare earths (R) in a matrix of host metals. Both $TbNi_2$ and $DyNi_2$ have been studied previously by Skrabek and Wallace (1) and by Farrell and Wallace (2) and were found to order ferromagnetically at 45 and 30 K respectively. The saturation magnetization at 4.2 K obtained, however, in fields up to about 20 kOe was low in comparison with the free tripositive rare earth values namely 7.8 and 9.2 μ_B. Since YNi_2, $LnNi_2$, and $CeNi_2$ are Pauli paramagnets (3) and $TbNi_2$ has been shown by neutron diffraction (4) to harbor magnetic ions only on Tb sites it is safe to assume that Ni does not exhibit a magnetic moment in these compounds.

Bleaney (5) has treated the relatively low saturation values of the series of RNi_2 compounds as a sign for crystal field effects being comparable or larger in magnitude than the exchange interaction It appears, however, that Bleaney's estimates of the magnitude of the crystal field splitting may have been somewhat high since no sizeable deviations from Curie-Weiss law were observed by Farrell and Wallace (2) down to very low temperatures (15 K in the case of $TmNi_2$ for instance). Moreover, Mössbauer spectroscopy on $DyNi_2$ by Ofer and Segal (6) has revealed 98% of the theoretical tripositive Dy moment ordered at the rare earth site so that the crystal field splitting in this case would appear to be rather small. $TbNi_2$ on the other hand has been shown by neutron diffraction (4) to contain besides a ferromagnetic component with 7.2 μ_B an antiferromagnetic G type structure which naturally would account for relatively low saturation moments.

The situation in pseudobinary systems of the form RT_2-RAl_2 (R: rare earth, T: transition metal) on the other hand is considerably complex (7 to 11) comprising stepwise increases in magnetization with field, linear increases of magnetization even at the highest fields and inversions with temperature, pointing at the presence of sizeable disordered components, anisotropy, domain wall effects and possibly minor effects from the splitting of crystal field states.

It was of interest therefore to study the magnetization mechanism of the single binaries $TbNi_2$ and $DyNi_2$ in more detail, down to lower temperatures and with higher fields, in order to compare it with materials where we do not as yet have the additional information as it is available in the present case.

$TbNi_2$ and $DyNi_2$ were prepared by induction melting under argon in MgO crucibles. Debye-Scherrer diagrams confirmed the presence of pure $MgCu_2$ type crystal structure with a = 7.142 (7.160) (12) and 7.135 (7.155)Å (12) respectively. Magnetic measurements were carried out on a Foner vibrating sample magnetometer. Applied fields were corrected for demagnetizing fields according to the formula $H_{int} = H_{appl} - n\sigma d$, where H_{int} and H_{appl} are internal and applied fields respectively, N was taken $\frac{4\pi}{3}$ for the nearly spherical specimens, σ is the magnetization, and d the density. More details on experimental procedures have been given elsewhere (10).

Magnetization curves for $TbNi_2$ and $DyNi_2$ are shown in Fig. 1 and 2 respectively. Both compounds show signs of saturation in fields of about 10 kOe but exhibit rising magnetizations even at the highest fields (applied fields up to 46 kOe). Saturation values were obtained by extrapolating the magnetization to zero reciprocal internal field a procedure which is somewhat questionable, however, due to some curvature in the extrapolation curves which in turn is connected with the almost linear increases in magnetization. Saturation values are 6.55 (1.6 K) and 6.32 (4.2 K) for $TbNi_2$ and 9.02 (1.6 K) and 8.88 (4.2 K) for $DyNi_2$.

The case of $DyNi_2$ seems to be an especially interesting one. Here we have almost linear increases in magnetization typical for a large variety of intermetallics containing lanthanides in a substance that according to Mössbauer evidence harbors the full tripositive rare earth moment. In this case therefore the linear increases in magnetization clearly should be connected with strong anisotropy rather than being

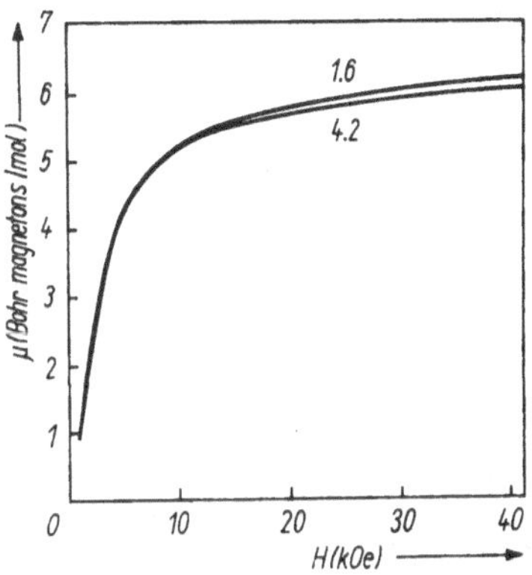

Fig. 1

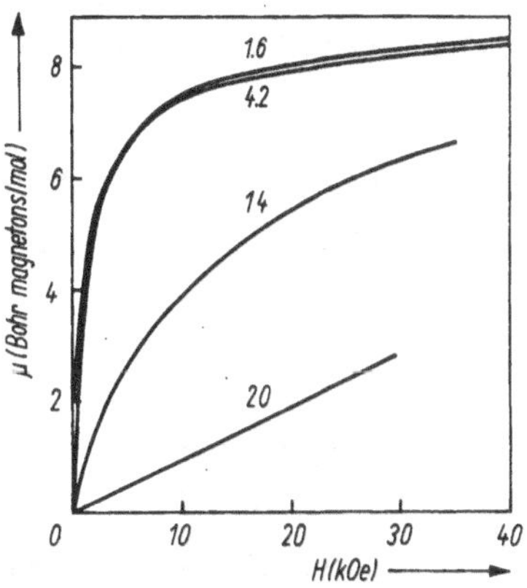

Fig. 2

Fig. 1. Magnetic moment μ vs. internal field H at different temperatures for $TbNi_2$

Fig. 2. Magnetic moment μ vs. internal field H at different temperatures for $DyNi_2$

a result of purification of the spectroscopic ground state by the field due to large crystal field splittings. The low saturation moments therefore should have a natural explanation.

The relatively low saturation values on $TbNi_2$ on the other hand seem to be connected with the presence of antiferromagnetic components observed with neutrons and the saturation behavior can be taken as a characteristic example for such a situation. Here it is not clear whether the pronounced increases in magnetization are a result entirely of anisotropy or partly stem from an ordering into the field direction of the antiferromagnetic contributions. Unfortunately, there is little help for a diagnosis of the two cases from the saturation behavior only since both are so similar.

The findings on $DyNi_2$ point out that strong anisotropy can be a major cause of the many cases reported so far of relatively low ordered moments of rare earths in intermetallic hosts, even in crystallographically completely ordered substances.

References

(1) E.A. SKRABEK and W.E. WALLACE, J. appl. Phys. 34, 1356 (1963).

(2) J. FARRELL and W.E. WALLACE, Inorg. Chem. 5, 105 (1966).

(3) M. MANNSMANN and W.E. WALLACE, J. chem. Phys. 40, 1167 (1964).

(4) G.P. FELCHER, L.M. CORLISS, and J.M. HASTINGS, J. appl. Phys. 36, 1001 (1965).

(5) B. BLEANEY, Proc. Roy. Soc. A276, 28 (1963).

(6) S. OFER and E. SEGAL, Phys. Rev. 141, 448 (1966).

(7) H. OESTERREICHER, L.M. CORLISS, and J.M. HASTINGS, J. appl. Phys. 41, 2326 (1970).

(8) H. OESTERREICHER, Solid State Commun. 10, 525 (1972).

(9) H. OESTERREICHER, J. Phys. Chem. Solids 32, 1403 (1971).

(10) H. OESTERREICHER, J. appl. Phys. 42, 5137 (1971).

(11) H. OESTERREICHER, phys. stat. sol. (a) 7, K55 (1971).

(12) M.V. NEVITT, in: Electronic Structure and Alloy Chemistry of Transition Metals, Ed. P.A. BECK, Wiley, 1964.

(Received June 28, 1972)

phys. stat. sol. (a) 12, K69 (1972)
Subject classification: 14.3.1; 22.2.3

Physics Department, Indian Institute of Technology, New Delhi

Transport Phenomena in Thin Films of n-InSb

By
D.P. TEWARI and J. KAMAL[1)]

In earlier communications (1, 2) we have reported the results of an analytical investigation of size dependence of transport properties of thin films of metals and semimetals. The band structure of carriers in these films was taken to be parabolic, though the shape of the constant energy surfaces was properly accounted for. These analyses, though made for metals and semimetals, could be applied to semiconductors in which flat band conditions exist (i.e. zero surface potential). But in many of the semiconductors, even under this assumption, the simple model of band structure (1) does not fit well and the transport phenomena are largely dependent on the actual band structure.

In this communication, we have studied the transport phenomena in thin films of a nonparabolic semiconductor. To specify the problem we have chosen a n-InSb film, since its band structure is very well known; the effect of the sample size on the band structure has, however, been neglected here. The analysis has been given for two temperature regions in which a relaxation time can be defined:
i) low temperatures 20 to 100 $^{\circ}$K (ionized impurity scattering),
ii) high temperatures 300 to 500 $^{\circ}$K (polar optical scattering).
Since the Debye temperature, Θ_D, for InSb is 280 $^{\circ}$K and a relaxation time for polar optical scattering is definable only for $T \gg \Theta_D$, the analysis could not be made for the temperature range between 100 and 300 $^{\circ}$K. We have, nevertheless, extrapolated our results of the two regions to the intermediate temperature range to have a qualitative appreciation of the variation of the mobility with thickness over the entire temperature range.

We assume the geometry of the arrangement such that the surface is parallel to the current flow. The scattering at the surface has been considered to be diffuse,

1) Work supported by CSIR (India).

neglecting the presence of any potential at the surface; the latter assumption appears to be an oversimplification of facts but its effect on the mobility as shown by Juhasz (3) and Anderson (4), is quite small for low values of surface potential.

In particular we have studied the size dependence of mobility for various values of temperatures taking the energy dependence of the carrier mass and relaxation time into account; this has not been taken into account by any worker (5 to 11) so far. Solving the Boltzmann transport equation by the method adopted in our earlier communications (1, 2), we get the following expressions corresponding to the two analytic forms of the distribution function required to satisfy the boundary conditions for diffuse scattering (12, 1):

$$f_1^+ = \frac{eE}{\hbar}\tau\frac{\partial f_o}{\partial k_x}\left[1 - \exp\left(-\frac{z\,m^*}{\tau\hbar k_z}\right)\right], \quad v_z > 0\,; \qquad (1a)$$

$$f_1^- = \frac{eE}{\hbar}\tau\frac{\partial f_o}{\partial k_x}\left[1 - \exp\left(\frac{2d - z}{\tau\hbar k_z}\right)\right], \quad v_z < 0\,. \qquad (1b)$$

τ in equations (1a) and (1b) for the two regions of temperatures may be written as (13, 14)

$$\text{(i)} \quad \tau_i = \frac{2^{1/2}\,\varepsilon^{3/2}\,\varkappa^2\,m^{*1/2}}{\pi\,e^4\,N_i\,g}, \qquad (2a)$$

$$\text{(ii)} \quad \tau_{po} = \frac{(2m^*\varepsilon)^{1/2}}{e\,E_o\,(2N_q + 1)}, \qquad (2b)$$

where m^* is the energy dependent mass given by (15)

$$m^* = L/(2\varepsilon + \varepsilon_G)\,; \qquad (3)$$

the rest of the symbols used in equations (1), (2), and (3) are the same as used in references (2, 13 to 15).

Using equations (1) the current density may be obtained as

$$I = -e\left[\int_{-\infty}^{\infty}\int_{-\infty}^{\infty}\int_{-\infty}^{0} v_x f_1^+ dk_x dk_y dk_z + \int_{-\infty}^{\infty}\int_{-\infty}^{\infty}\int_{0}^{\infty} v_x f_1^- dk_x dk_y dk_z\right] . \qquad (4)$$

Equation (4) gives the current distribution across the thickness of the film. The average current density for a film of thickness 2d adjacent to the z = 0 surface is given by

$$\bar{I} = \frac{1}{2d}\int_0^{2d} I\, dz . \qquad (5)$$

In terms of the current density, the mobility may now be written as

$$\mu = \frac{1}{n} I_x/E_x ,$$

where n is the carrier concentration given by

$$n = \frac{1}{\pi}\int f_o k^2 dk . \qquad (6)$$

Since an analytical evaluation of equations (5) and (6) is not possible, we have solved thêm numerically.

The plot of the ratio of the bulk mobility to the mobility of the film, μ_B/μ, in Fig. 1, reveals that at low temperatures it increases with temperature while at higher temperature it decreases. The behaviour of the curves can be explained as follows:

At low temperatures, ionized impurity scattering is dominant, where the carrier mean free path increases with increasing temperature. As a result the effective surface scattering is more important and the ratio of μ_B/μ in accordance with equations (1a) and

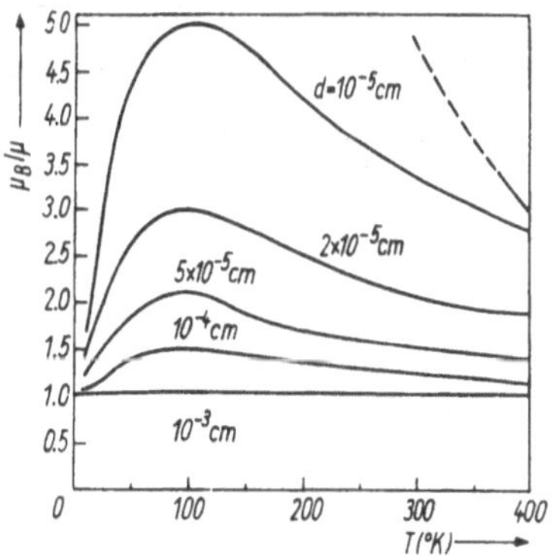

Fig. 1. Variation of μ_B/μ with temperature in InSb for various film thicknesses. The broken curve is due to the experimental results of Berchtold and Huber (16)

(1b) increases. Exactly, the reverse is true at high temperatures where the polar optical mode of scattering is dominant.

Experimental results of Berchtold and Huber (16) have been shown by a broken line in Fig. 1. At high temperature the qualitative agreement between our results and those of Berchtold and Huber seems to be fairly good. At somewhat lower temperatures the discrepancy in results increases which could be attributed to neglect of surface potential and some of the scattering mechanisms.

Thanks are due to Prof. M.S. Sodha and Dr. V.K. Tripathi for many helpful discussions.

References

(1) J. KAMAL and D.P. TEWARI, phys. stat. sol. (a) 2, 643 (1970).
(2) J. KAMAL and D.P. TEWARI, phys. stat.sol. (b) 43, 357 (1971).
(3) C. JUHASZ, Ph.D. Thesis, London University, 1968.
(4) J.C. ANDERSON, Adv. Phys. 19, 311 (1971).
(5) A. AMITH, J. Phys. Chem. Solids 14, 142 (1960).
(6) J.N. ZEMEL, Phys. Rev. 112, 762 (1958).
(7) J.R. SCHRIEFFER, Phys. Rev. 97, 641 (1955).
(8) R.H. KINGSTON and S.T. NEUSTADTER, J. appl. Phys. 26, 718 (1955).
(9) R.F. GREENE, J. Phys. Chem. Solids 14, 291 (1960).
(10) J.N. ZEMEL and R.L. PETRITZ, Phys.Rev. 110, 1263 (1958).
(11) R.F. GREENE, Phys. Rev. 97, 641 (1955).
(12) E.H. SONDHEIMER, Adv. Phys. 1, 1 (1952).
(13) M. KINCH, Proc. Phys. Soc. 88, 1257 (1966).
(14) E.M. CONWELL, Solid State Phys. Suppl. 9, Chap. 3 (1967).
(15) E.O. KANE, J. Phys. Chem. Solids 1, 249 (1957).
(16) K. BERCHTOLD and D. HUBER, phys. stat. sol. 33, 425 (1969).

(Received May 23, 1972)

phys. stat. sol. (a) 12, K73 (1972)
Subject classification: 20.2; 20.3; 22.8

Institute of Crystallography, Academy of Sciences of the USSR, Moscow (a), and Mendeleev Institute of Chemical Technology, Moscow (b)

Stimulated Emission from the New $LiGd(MoO_4)_2$:Nd^{3+} Crystal Laser

By

A.A. KAMINSKII (a), A.A. MAYER (b), N.S. NIKONOVA (b), M.V. PROVOTOROV (b), and S.E. SARKISOV (a)

It is known that TR^{3+} (rare-earth) ions in crystals of $A^I A^{III}(B^{VI}O_4)_2$ type (where A^I: Li^+, Na^+, K^+, Rb^+; A^{III}: La^{3+}, Y^{3+}, Bi^{3+}, TR^{3+}, and B^{VI}: Mo^{6+}, W^{6+}) with scheelite structure form the multitude of activator centres (1 to 3). This is reflected by the spectral properties of these crystals as their absorption and luminescence spectra result from a superposition of spectra belonging to different types of centres. The variety of activator centres is reflected in processes that occur in the stimulated emission conditions.

In this communication we present preliminary results of investigations of a new laser based on an $LiGd(MoO_4)_2$ crystal activated with Nd^{3+} ions. This material crystallizes in tetragonal syngony and has the space group $C_{4h}^6 - I4_1/a$ with parameters $a_o = 5.20$ and $c_o = 11.32$ Å. Its crystal structure was determined in (4). The $LiGd(MoO_4)_2$ crystal have also the following characteristics: melting point 1035 ^{o}C (5); hardness (Knoop scale) 700 kp/mm^2; thermal expansion coefficient (200 to 600 ^{o}C) $\alpha_\perp = 8.5 \times 10^{-6}$ and $\alpha_\parallel = 10.7 \times 10^{-6}$ deg^{-1}; refractive index $n_\perp \approx n_\parallel \approx 1.95$; density 5.44 g/cm^3 (hydr.) and 5.25 g/cm^3 (X-rays); transparency range 0.4 to 5.2 µm.

Stimulated emission at (10599 $\pm$ 1)Å ($\approx$ 9435 cm^{-1}) due to a $^4F_{3/2} \rightarrow {}^4I_{11/2}$ transition has been observed at room temperature.

The Nd^{3+}-doped $LiGd(MoO_4)_2$ crystals used in our work were grown by the Czochralski method. The concentration of the activator varied from 0.1 to 10 wt% (to the initial charge). The experimental results obtained from the study of absorption ($^4I_{9/2} \rightarrow {}^4F_{3/2}$), luminescence (Fig. 1), and laser spectra were used to establish the sequence of Stark levels of the $^4F_{3/2}$ and $^4I_{11/2}$ manifolds at 77 ^{o}K between which induced transition takes place. These sequences are shown in Fig. 2.

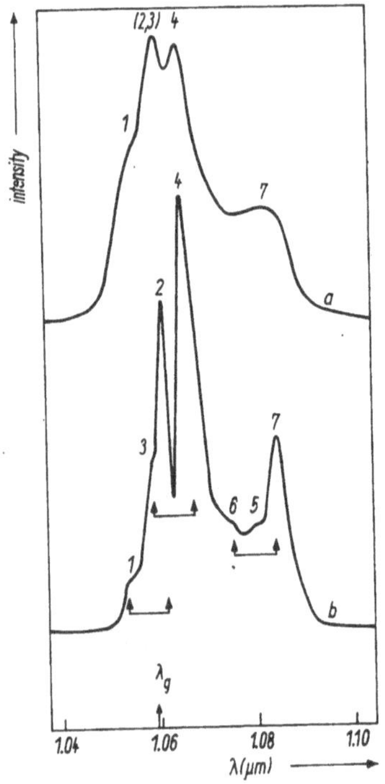

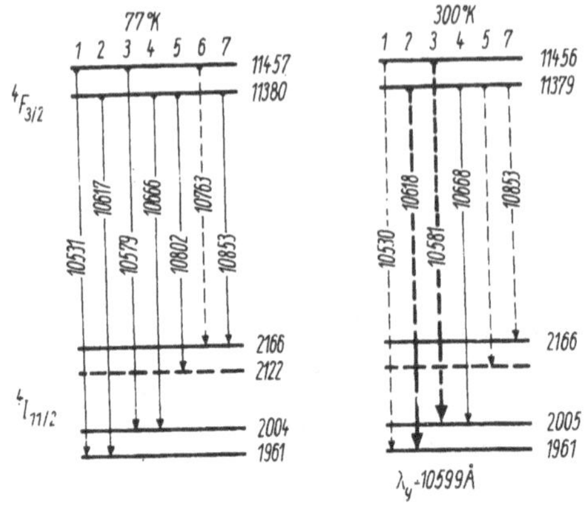

Fig. 2. Crystal-field splitting of the $^4F_{3/2}$ and $^4I_{11/2}$ manifolds of Nd^{3+} ions in $LiGd(MoO_4)_2$ crystals at 77 and 300 °K. Level positions are in cm^{-1} and the transitions in Å. The laser transition is denoted by a heavy arrow

Fig. 1. Luminescence spectra of $LiGd(MoO_4)_2:Nd^{3+}$ crystal (≈ 2 wt%, $\vec{F} \perp \vec{c}$), transition $^4F_{3/2} \rightarrow {}^4I_{11/2}$; (a) for 300 and (b) 77 °K

The corresponding numeration used to denote the lines in the luminescence spectra and the transition has been adopted to facilitate their comparison. Levels of the $^4I_{9/2}$ manifold at 77 °K are located at 0.96, 165, 239, and 433 cm^{-1}. For $LiGd(MoO_4)_2:Nd^{3+}$ (to ≈ 2 wt%) crystals the lifetime of the excited state $^4F_{3/2}$ at 300 °K is ≈ 140 μs.

A cylindrical laser rod 4 mm in diameter, ≈ 20 mm long was fabricated from a crystal of $LiGd(MoO_4)_2$ containing ≈ 2 wt% Nd^{3+} ions. The rod axis (F) was perpendicular to a crystallographic axis (c). For the pulsed laser we used an illuminator of elliptical cross section with an ISP-1000 type Xe lamp. The optical resonator consisted of external spherical mirrors with multilayer dielectric coatings in a confocal configuration (R ≈ 576 mm). The transmission of the mirrors at 1.06 μm was about 0.7%. At 300 °K the laser begins generating a line A with λ_g = 10599 Å when the electrical energy supplied to the pump reaches ≈ 3 J.

References

(1) A.M. MOROZOV, M.N. TOLSTOI, P.P. FEOFILOV, and V.N. SHAPOVALOV, Optika i Spektroskopiya 22, 414 (1967).

(2) A.M. MOROZOV, L.G. MOROZOVA, and P.P. FEOFILOV, in: Spektroskopiya kristallov, Nauka, 1970 (p. 153).

(3) G.M. ZVEREV and G.YA. KOLODNY, Zh. eksper. teor. Fiz. 52, 337 (1967).

(4) P.V. KLEVTSOV and L.P. KOZEEVA, Izv. Akad. Nauk SSSR, Ser. neorg. Mater. 5, 1842 (1969).

(5) KH.S. BAGDASAROV, G.A. BOGOMOLOVA, A.A. KAMINSKII, A.M. PROKHOROV, and T.M. PROKHORTSEVA, Dokl.Akad. Nauk SSSR 197, 557 (1971).

(Received June 28, 1972)

phys. stat. sol. (a) 12, K77 (1972)
Subject classification: 10.1 and 17; 21.1

Uraler Polytechnisches Institut, Sverdlovsk

Der funktionelle Zusammenhang zwischen der Exoelektronenemissionskinetik und den Verfestigungs- und Entfestigungsprozessen von Metallen bei plastischer Verformung

Von

R.I. MINTS und V.P. MELEKHIN

Die Exoelektronenemission ist durch die Gesamtheit der Elementarprozesse bedingt. Ihre Rolle wird durch äußere und innere Faktoren bestimmt. Es ist deshalb trotz der zahlreichen Arbeiten (1 bis 4), die der Untersuchung des Einflusses von Defekten auf die Elektronenemission gewidmet sind, immer noch offen, ob eine Korrelation zwischen den Emissionsparametern und dem plastischen Deformationsprozeß existiert. Es ist zweckmäßig, die Untersuchung der Exoemission auf das Studium ihres Zusammenhanges mit den sie prinzipiell bedingenden Faktoren zurückzuführen.

Das Ziel der vorliegenden Arbeit ist die Untersuchung des Zusammenhanges zwischen der Exoelektronenemission und den Prozessen der plastischen Verformung von Metallen - der Verfestigung und Entfestigung.

Bei einachsiger Dehnung mit einer konstanten Geschwindigkeit im Vakuum von 5×10^{-6} Torr wurden gleichzeitig die Exoemissionsintensität (I) und Verfestigung (σ) als Funktion des Deformationsgrades (ε) registriert. Ein Sekundärelektronenvervielfacher diente als Exoelektronendetektor. Eine ausführliche Beschreibung der Apparatur ist in (5) enthalten.

Als Untersuchungsobjekt dienten polierte und im Vakuum von 5×10^{-5} Torr getemperte Cu-Al Legierungen mit unterschiedlichem Aluminiumgehalt (2,3; 4,5; 7,4 Gew.%). Die Korngröße betrug nach dem Tempern in allen Proben ≈ 200 µm. Die Stapelfehlerenergie (γ) ändert sich für diese Legierungen zwischen 65 erg/cm^2 für Cu bis 2 erg/cm^2 für Cu + 7,4% Al (6, 7).

Es ist bekannt, daß die plastische Verformung von Polykristallen von Verfestigung und Entfestigung begleitet ist. Wie es aus (8 bis 10) folgt, wird die Entfestigung durch den dynamischen Erholungsprozeß verursacht. Dieser Prozeß wird von der Umverteilung, Annihilation und dem Austritt der Kristallgitterfehler auf die Metall-

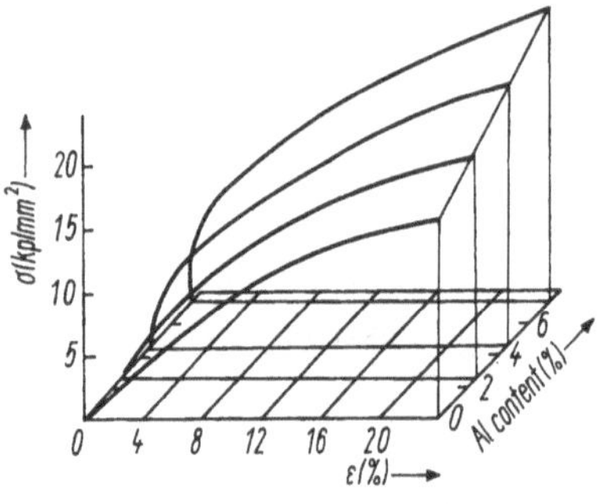

Fig. 1

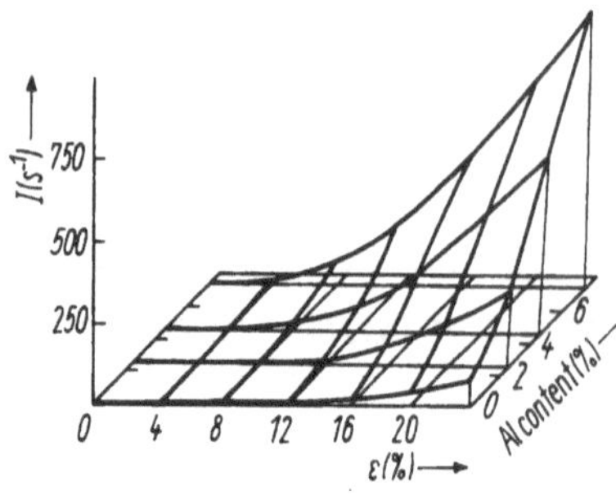

Fig. 2

Fig. 1. Der Einfluß des Aluminiumgehaltes auf die Verfestigungskinetik der Cu-Al Legierungen

Fig. 2. Der Einfluß des Deformationsgrades (ε) auf die Exoelektronenemissionsintensität (I)

oberfläche gekennzeichnet. Dabei ist die Intensität der dynamischen Erholung bei gleichen homologen Temperaturen und Deformationsgeschwindigkeiten für das Metall mit größerer Stapelfehlerenergie größer (11).

Für Polykristalle ist der Übergang vom Verfestigungs- zum Entfestigungsstadium nicht so deutlich ausgeprägt wie für Einkristalle. Es ist deshalb zweckmäßig, für die Bestimmung des Deformationsgrades, bei dem dieser Übergang festzustellen ist, statt der $\sigma(\varepsilon)$-Abhängigkeit die $\lg\sigma(\lg\varepsilon)$-Abhängigkeit aufzutragen, die eine Gerade mit einem Knick im Punkt ε_0 ergibt. Das Gebiet $\varepsilon < \varepsilon_0$ entspricht dem Verfestigungsstadium und $\varepsilon > \varepsilon_0$ dem Entfestigungsstadium (11).

Die experimentallen $\sigma = f(\varepsilon)$-Kurven sind in Fig. 1 im üblichen Maßstab für Cu mit unterschiedlichem Aluminiumgehalt aufgetragen. In der Tabelle sind die ε_0-Werte zusammengestellt, die sich bei Darstellung derselben Kurven in $\lg\sigma$ - $\lg\varepsilon$ Koordinaten ergaben.

Die Abhängigkeit der Exoelektronenemissionsintensität (I) vom Deformationsgrad für die untersuchten Legierungen ist in Fig. 2 angegeben, aus der folgt, daß sich der Einsatz der Exoelektronenemission mit Vergrößerung des Aluminiumgehaltes in der Legierung stark ins Gebiet kleinerer Deformationen verschiebt. Dabei fällt der Deformationsgrad (ε_1), bei dem sich ein merklicher Emissionsstrom re-

Fig. 3. Der Einfluß der Stapelfehlerenergie (γ) auf den Exponent α

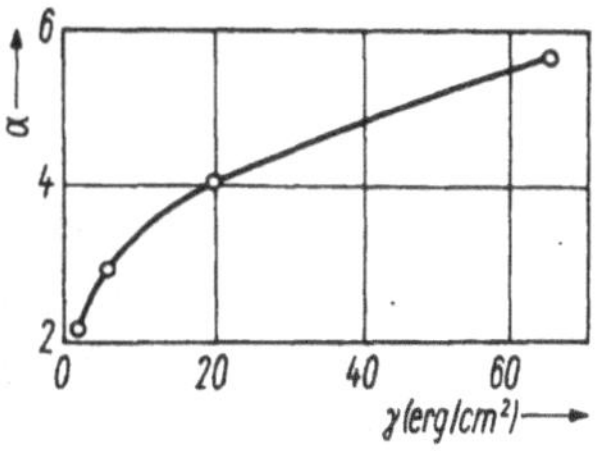

gistrieren läßt, gut mit dem Wert von ε_o zusammen.

Man kann also feststellen, daß im Deformationsgebiet, in dem eine dynamische Erholung auftritt, eine Elektronenemission festzustellen ist.

Die in logarithmischem Maßstab aufgetragenen Kurven $I = f(\varepsilon)$ werden gut durch Geraden approximiert. Die Steigungen dieser Geraden sind für die untersuchten Legierungen in der Tabelle angegeben.

Tabelle 1

Metall	γ (erg/cm^2)	ε_o (%)	ε_1 (%)	α
Cu	65	10,0	11,0	5,8
Cu + 2,3% Al	20	5,5	7,0	4,1
Cu + 4,5% Al	5	0,6	2,0	2,9
Cu + 7,4% Al	2	0,4	1,0	2,2

Die Analyse der erhaltenen Daten zeigt, daß die $I(\varepsilon)$-Abhängigkeit bei $\varepsilon > \varepsilon_o$ folgende Form hat:

$$I \approx \left(\frac{\varepsilon}{\varepsilon_o}\right)^{\alpha} ,$$

ε_o und α sind Parameter, die von der Stapelfehlerenergie abhängen.

Der Abnahme der Stapelfehlerenergie entspricht eine Verringerung des Exponenten (Fig. 3). Mit Vergrößerung des Legierungsgrades (bei Herabsetzung der Energie für Stapelfehlerbildung) verringert sich α, was auf eine schwächere Abhängigkeit der Exoelektronenemissionsintensität vom Deformationsgrad hinweist.

Die erhaltenen Daten weisen auf die Existenz eines funktionellen Zusammenhanges zwischen der Exoelektronenemission und den Grundcharakteristiken des plastischen Verformungsprozesses hin.

Literatur

(1) R. CLAYTOR und F. BROTZEN, J. appl. Phys. 36, 3543 (1965).

(2) VON VOOS und F. BROTZEN, J. appl. Phys. 30, 1639 (1959).

(3) J. LANGENECHER und D. RAY, J. appl. Phys. 35, 2586 (1964).

(4) B. SUJAK, Acta phys. Polon. 20, 889 (1961).

(5) V.P. MELEKHIN, V.S. KORTOV und R.I. MINTS, Zavodskaya Laboratoriya 35, 996 (1969).

(6) H. OTTE, J. appl. Phys. 38, 217 (1967).

(7) A. HOVY, Pryamoe nablyudenie nesovershenstv v kristallakh, Metallurgiya, Moskva 1964 (S. 198).

(8) M. POLANY und E. SCHMID, Naturwissenschaften 17, 301 (1929).

(9) L.D. SOKOLOV, Soprotivlenie metallov plasticheskoy deformatsii, Metallurgizdat, Moskva 1963.

(10) V.F. TERENTEV, Dokl. Akad. Nauk SSSR 185, 83 (1969).

(11) L.D. SOKOLOV und Y.P. DMITRIEV, Izv. Akad. Nauk SSSR, Ser. met. 4, 154 (1971).

(Received June 8, 1972)

phys. stat. sol. (a) 12, K81 (1972)

Subject classification: 13.4 and 14.3.4; 22.1.1

Solid State Electronics Laboratory,
Department of Electrical Engineering, University of Cincinnati, Ohio

Double-Injection Negative Differential Resistance in Compensated Gold-Doped Germanium

By

N. L. VANVARI and H. THURMAN HENDERSON

A simple model of a semiconductor may be constructed based upon the analysis of Lampert and Rose (1), and Lampert (2), where a deep acceptor level E_a (trap) is completely compensated by a shallow donor level E_d, as shown in Fig. 1, and the hole trapping cross-section σ_p exceeds the electron trapping cross-section σ_n. This model predicts, on a log current versus log voltage scale, a unity slope (Ohm's law) giving way to a slope two, which in turn gives way to a negative differential resistance as the deep trap is depopulated by hole injection and the hole lifetime increases. (Ashley and Milnes (3) have analyzed the case of partial compensation.)

Here, an attempt has been made to prepare a diode which at room temperature would closely approximate the Lampert semiconductor model. This effort resulted

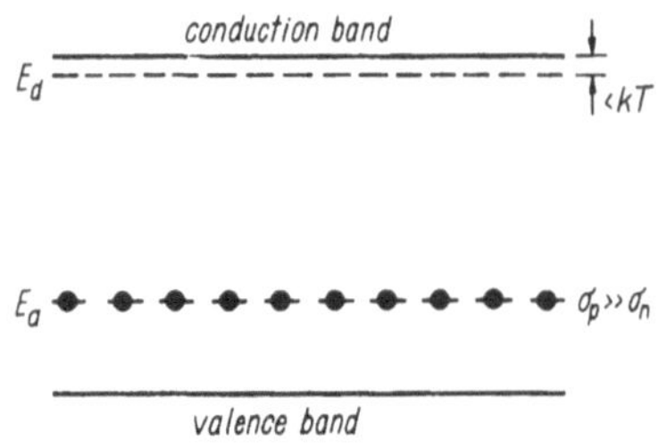

Fig. 1

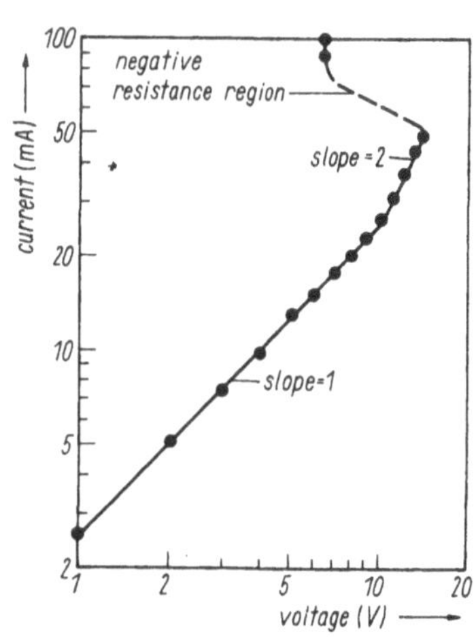

Fig. 2

Fig. 1. Simplified energy band model

Fig. 2. Observed characteristic curve at room temperature

in a germanium crystal doped with gold to 1.2×10^{15} cm^{-3} compensated with a shallow donor with an ionization energy less than kT (where k is Boltzmann's constant and T is the absolute temperature) to 0.7×10^{15} cm^{-3}. These values, determined by Hall measurements, also showed the deep level to be the normal first ionization level of gold at 0.15 eV above the valence band. Electrodes consisted of evaporated and sintered gold alloy: 3% antimony for the cathode and 5% gallium for the anode. The planar diode was 635 μm (25 mils) thick, and the resistivity of the material was 6.5 Ωcm.

Fig. 2 shows the resulting volt-ampere characteristic, which gives precise conformity with the Lampert theory in power law dependence and onset of negative resistance. A more detailed analysis is now being made to check agreement with the predicted multiplicative constant, which involves a lifetime term.

Finally, the authors wish to acknowledge support for this project provided by the National Aeronautics and Space Administration (NASA) through the University of Cincinnati Institute of Space Sciences, Grant NGR-36-004-014.

References

(1) M.A. LAMPERT and A. ROSE, Phys.Rev. 121, 26 (1961).

(2) M.A. LAMPERT, Phys. Rev. 125, 126 (1962).

(3) K.L. ASHLEY and A.G. MILNES, J. appl. Phys. 35, 369 (1964).

(Received June 12, 1972)

phys. stat. sol. (a) 12, K83 (1972)
Subject classification: 10.1; 21.6

Nuclear Research Center, Negev, Beer-Sheva

Interaction of Positrons with Defects in Plastically Deformed Metals

By
J. BARAM, M. BLAU, and M. ROSEN

The positron annihilation lineshape obtained with a Ge(Li) spectrometer may be approximated analytically in the same manner as an angular correlation curve (1), as the sum of an inverted parabola and a Gaussian, attributed to conduction electrons and core electrons, respectively. When annihilations take place in a crystal lattice with vacancies or dislocations, positrons are trapped by the defects, following a trapping model reported in (2). Vacancies and dislocations act as attractive centers for the positrons. A larger fraction of positrons annihilate with low-momentum conduction electrons, causing relatively little Doppler broadening. The resultant narrowing of the annihilation line has been observed with high resolution Ge(Li) spectrometer (3).

In (3), the annihilation lineshape was used to measure the percentage of positrons trapped by dislocations in plastically deformed Al and Cu. Both experimental results (4) and the theoretical trapping model (2), account for the saturation of the trapping process.

In a perfectly annealed metal, there is no significant amount of annihilations in dislocations (or vacancies). In contrast, in a heavily deformed metal, all the annihilations take place in defect sites (4). In (3), a lineshape parameter S was obtained, which is in fact a measure of the contribution to annihilations in the parabolic part of the lineshape. S can be expressed as

$$S = \frac{S_p (N_d - N_o) + S_s N_d}{N_o} \tag{1}$$

which is equivalent to

$$\frac{N_d}{N_o} = \frac{S - S_p}{S_d - S_p} , \tag{2}$$

where N_d is the number of positrons annihilating in defects, N_o is the total number of annihilating positrons, S_p is the value of S in a perfect lattice, S_d is the value of S in a heavily deformed lattice, or, more precisely, the value of S when saturation of trapping is obtained.

But, from the differential equations governing the trapping model (when escape is assumed to be negligible)

$$q = \lambda_p \frac{N_d/N_o}{1 - N_d/N_o} , \qquad (3)$$

where q (s^{-1}) is the trapping rate of positrons by defects and λ_p is the annihilation rate in perfect lattice. This leads to

$$q = \lambda_p \frac{S - S_p}{S_d - S} . \qquad (4)$$

The trapping rate depends on the defect concentration ϱ, the average velocity of thermalized positrons $\bar{v}$ and the trapping cross section σ:

$$q\,(s^{-1}) = \varrho\,(cm^{-3})\,\bar{v}\,(cm\,s^{-1})\,\sigma\,(cm^2) . \qquad (5)$$

By means of equations (4) and (5), it is therefore possible to calculate the trapping cross-section of positrons by defects, using the experimentally determined values of the S parameters, as obtained from the lineshape narrowing.

Experimental values of S, S_p, and S_d, can be obtained in a similar manner to that reported in (3). Each stage of deformation yields a value S, starting from S_p, for the non-deformed lattice, up to S_d, for the saturation value. To each stage of deformation corresponds a defect density ϱ, which can be deduced from strain-hardening theory (5).

This theoretical model was applied to a 99.999% purity Ag polycrystal deformed by rolling, up to 80% thickness reduction. The lineshapes were recorded in a Ge(Li) spectrometer, whose resolution for the 514 keV line of ^{85}Sr was 1.15 keV FWHM. Saturation was obtained after 30% reduction. The annihilation lineshapes for the perfectly annealed and the 30% deformed samples are given in Fig. 1. Assuming a parabolic stress-strain relationship in that range of deformation, we obtain the

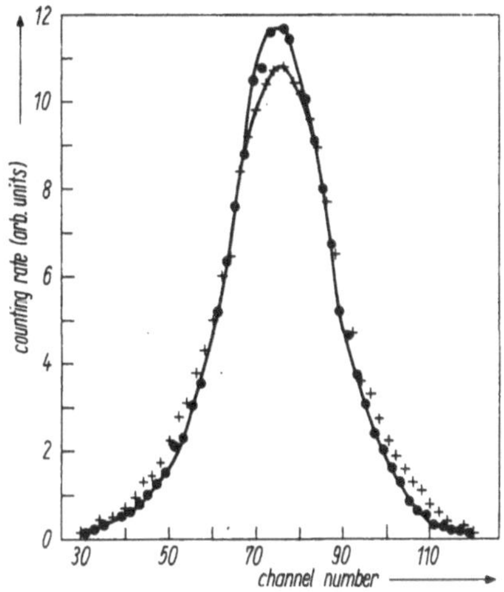

Fig. 1. Annihilation lineshapes - Ag (5N purity) - normalized to unit area;
● perfectly annealed sample,
+ 30% deformed sample

value of 5×10^{-15} cm^2 for the trapping cross-section of positrons by dislocations in Ag.

This relatively simple model seems to be useful for trapping cross-section determination. Using appropriate quenching experiments, one should be able to separate trapping in vacancies from trapping in dislocations, and decide whether positrons behave differently in point defects compared to line defects. More elaborate experiments are in progress.

References

(1) I.G. VARLASHKIN and E.F. ZGANGAR, Nucl. Phys. A130, 182 (1969).

(2) B. BERGERSEN and M.J. STOTT, Solid State Commun. 7, 1203 (1969).

(3) I.K. MACKENZIE, J.A. EADY, and R.R. GINGERICH, Phys. Letters (Netherlands) 33A, 279 (1970).

(4) P. HAUTOJÄRVI, A. TAMMINEN, and P. JAUHO, Phys. Rev. Letters 24, 459 (1970).

(5) D. KUHLMANN-WILSDORF, Trans. MS AIME 224, 1047 (1962).

(Received June 15, 1972).

phys. stat. sol. (a) 12, K87 (1972)

Subject classification: 10.1; 21; 21.1; 21.1.1

Department of Physics, University of Guelph, Canada

The Specific Responses of Trapped Positrons to Solid State Defects

By

I.K. MACKENZIE

The object of this note is to present experimental evidence that, when positrons annihilate in traps, they do so in a sufficiently distinctive manner to reveal qualitative features of the traps. This raises the possibility that positron trapping measurements may serve to eliminate some of the uncertainties which remain in the interpretation of annealing phenomena in metals (1). We are also interested in the possibility that positron trapping may be sufficiently specific to determine non-destructively the size distribution of voids introduced by fast neutron irradiation (2).

In discussing phenomena associated with positron trapping in solids the usual classification in terms of insulators, semiconductors and metals is not very meaningful. It is more useful to think of only two classes of solids, namely those which permit the formation of positronium and those which do not. We shall be concerned here only with materials of the latter class and therefore ignore any experimental techniques related specifically to positronium. We must extract conclusions from measurements of positron lifetime spectra and centre-of-mass momentum distributions.

Measurement of the momentum spectra can be made by the high resolution technique of angular correlation or by the high efficiency technique of Ge (Li) spectroscopy (3). Using the latter method we have measured the Doppler broadened lineshape in a sample of five nines purity copper from room temperature to the melting point. The lineshape is described quantitatively by a parameter, S, which gives the fraction of all events lying within 1.12 keV of M_oC^2. It is clear from Fig. 1 that this parameter changes by 5.1% when an annealed specimen is heated to a sufficiently high temperature to cause "saturation" with vacancies. The inset figure shows that saturation is well established since no detectable lineshape change occurred above 1200 K in spite of the exponentially increasing vacancy density. When the same material was deformed 50% by rolling at room temperature a change of 8.6% resulted.

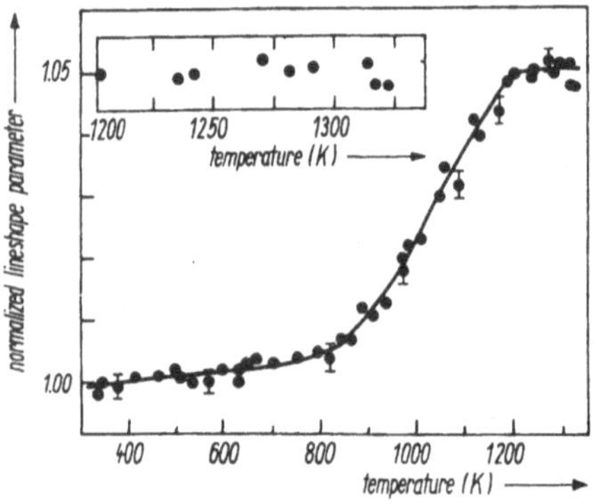

Fig. 1

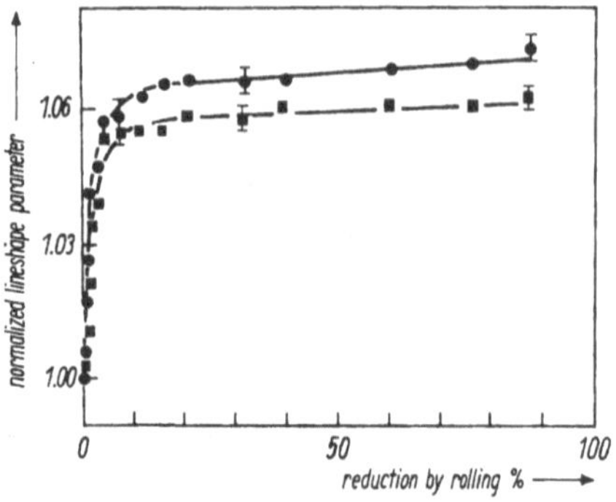

Fig. 2

Fig. 1. The temperature dependence of the annihilation lineshape from room temperature to melting. The inset figure has the same vertical scale as the main figure but the horizontal scale is expanded to emphasize the saturation. The material is copper of five nines purity

Fig. 2. The dependence of annihilation lineshape in "Ferrovac E" iron of 99.97% purity on the degree of deformation and on the time elapsed after deformation. The upper set of data applies to freshly deformed material and the lower set after 36 h of recovery

This may be regarded as the lineshape parameter characteristic of trapping by edge dislocation in copper since any point defects introduced by the deformation migrate rapidly at room temperature.

The positron apparently distinguishes quite clearly between vacancies and edge dislocations in copper. Fig. 2 shows data on trapping in "Ferrovac E" iron of 99.97% purity which implies, rather less convincingly, that a similar discrimination applies in that material. The upper curve gives the dependence of lineshape on deformation when the measurement was carried out over the period from 1000 to 3000 s after deformation. The lower curve applies to the same specimens after they had aged for 36 h at room temperature. The lineshape change certainly reflects some kind of recovery process; further data shows that the time constant is about 800 s.

We reason that since vacancies are immobile in iron at room temperature, the positrons are distributed between vacancy and dislocation traps, the latter proportion increasing with deformation. This would account for the much more rapid approach to saturation than that observed in Cu and Al where vacancies play no role at room temperature (3). The recovery is attributed primarily to the migration of carbon interstitials which must move preferentially to neutralize the dislocation traps. This implies that the bulk of the positrons in an aged specimen annihilate in vacancies which is in keeping with the general lowering of the lineshape parameter and the flattening of the plateau region. Snead et al. (4) observed effects of the migration of C interstitials at a higher temperature but, because of the much longer time involved in their angular correlation measurements, may not have been sensitive to the fast migration component.

The most persuasive evidence of sensitivity to defect properties is provided by positron lifetime measurements in alkali halides reported by LeBlanc and Davis (5). A particular specimen of KCl showed a decay spectrum with two easily resolved components of similar intensity with lifetimes of 200 and 660 ps, the latter associated with a trap of unspecified type. The transition rate of the trapped component varied linearly with pressure up to 55 kbar by which point the lifetime was reduced to 475 ps. This dramatic change occurs in a defect much smaller than those of greatest interest in the study of voids. However, in the absence of contrary evidence, we can hope that the positron retains its size sensitivity in the range of large defects.

In summary the positron shows promising signs of specific response to traps and, considering that it shows no reaction to interstitials, it should play a very useful role in annealing studies. The situation with respect to void studies is more obscure because of a lack of data on large defects but the extrapolation from existing data is encouraging.

It is a pleasure to acknowledge the research support provided by the National Research Council of Canada and to thank my present hosts, the Technical University of Denmark, for facilities and stimulation.

References

(1) J.W. CORBETT, in: Vacancies and Interstitials in Metals, Eds. A. SEEGER, D. SCHUMACHER, W. SCHILLING, and J. DIEHL, North-Holland Publ. Co., Amsterdam 1970 (pp. 977 to 997).

(2) S.F. PUGH, M.H. LORETTO, and D.I.R. MORRIS (Eds.), Voids Formed by Irradiation of Reactor Materials, British Nuclear Energy Society, 1969.

(3) I.K. MACKENZIE, J.A. EADY, and R.R. GINGERICH, Phys. Letters (Netherlands) 33A, 279 (1970).

(4) C.L. SNEAD, Jr., A.N. GOLAND, J.H. KUSMISS, H.C. HUANG, and R. MEADE, Phys. Rev. B3, 275 (1971).

(5) R. LEBLANC and D. DAVIS, in: Positron Annihilation, Second International Conference, Kingston, Canada, Vol. 2, Eds. B.G. HOGG and A.T. STEWART, 1971 (p. 28).

(Received May 24, 1972)

phys. stat. sol. (a) 12, K91 (1972)

Subject classification: 20.2; 22.4.4

Institut für angewandte Physik der Universität Karlsruhe

Tunable Laser Emission of CdS_xSe_{1-x} Graded Band Gap Crystals by Two-Photon Excitation

By

J. BILLE, H. LIEBING, P. MENGEL, and G. SCHEIBER

Laser emission has been obtained from CdS_xSe_{1-x} graded band gap crystals using two-photon excitation by a Q switch ruby laser. The wavelength could be varied by moving the crystal across the laser beam in the grading direction. The onset of laser emission was deduced from an abrupt increase of the emission intensity and from line narrowing at the threshold.

Recently (1) we have reported on continuously tunable laser emission in graded band gap CdS_xSe_{1-x} crystals excited with 215 keV electrons. Using two-photon excitation, volume excitation of the crystals takes place because the penetration depth of the exciting radiation is about 20 times larger than that of 215 keV electrons. Therefore, we now got a greater output power of the tunable semiconductor laser, which may be interesting for technical applications.

The single crystals used in the experiment were grown by a modified Piper-Polich method as described in (2). The samples were cut in the form of rectangular parallelepipeds with a pair of carefully treated plane-parallel faces, forming a planar resonator. The distance between the plane-parallel faces was about 2 mm. The light from the ruby laser was focused by two cylindrical lenses onto one of the sample faces perpendicular to the resonator axis and the grading direction (see Fig. 1).

The energy of the exciting pulse was 0.75 J, its duration 15 ns and its repetition rate one pulse per minute. The pumping level was determined by means of neutral

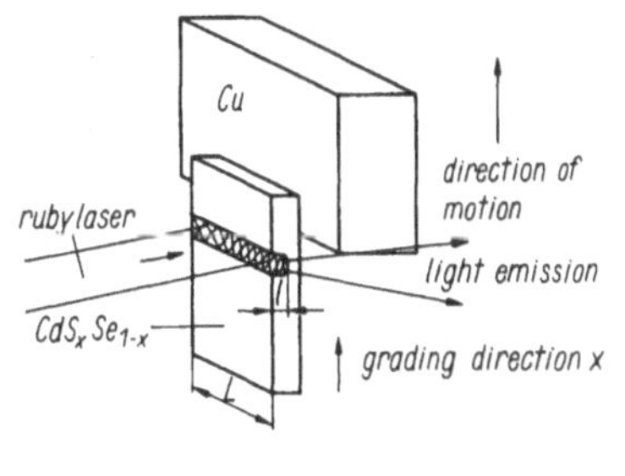

Fig. 1. Experimental arrangement; L cavity length, l penetration depth

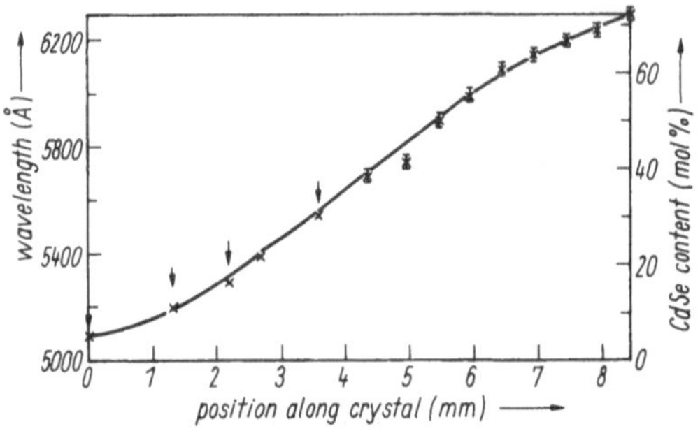

Fig. 2. Emitted laser wavelength as a function of position along crystal. Selenide concentration after data of Hurwitz (3)

optical calibrated filters. The radiation from one of the resonator faces was directed into a double-monochromator (Zeiss MM12) and was detected with a photomultiplier (Valvo XP 1002).

Fig. 2 shows the dependence of the emission wavelength on the position along the sample. The corresponding crystal composition (mol% CdSe) was deduced from data of Hurwitz (3). The band gap shift across the entire crystal of about 8.5 mm length was 0.4 eV corresponding to 1200 Å.

The laser emission was, however, limited to a region of 3.5 mm, corresponding to wavelengths from 5100 to 5550 Å. Outside of this region no laser emission was obtained due to additional losses in the crystal cavity with increasing selenide concentration.

The onset of laser emission was deduced for each point in Fig. 2 by plotting the emission intensity I_K as a function of the incident ruby laser intensity I_R. Four typical characteristics are shown in Fig. 3. First, below the threshold the emission intensity increases as the square of pump intensity I_R. Then I_K sharply grows by a few orders of magnitude which corresponds to the self-excitation threshold.

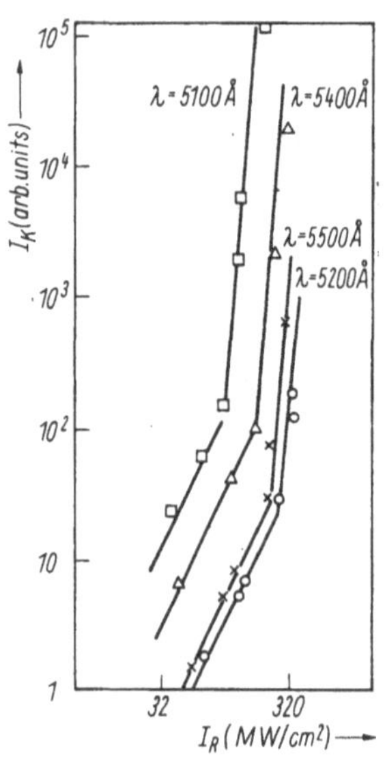

The intensity of the crystal laser was measured with a photodiode (RCA 935) and suitable optical filters and was found to be of the order of a few kilowatts.

Fig. 3. Emitted light intensity (I_K) as a function of incident ruby laser intensity (I_R) at wavelengths of 5100, 5200, 5400, and 5550 Å

References

(1) J. BILLE, B.M. KRAMER, P. REIMERS, W. RUPPEL, and R. STILLE, phys. stat. sol. 36, K71 (1969).

(2) P. REIMERS, phys. stat. sol. 35, 707 (1969).

(3) C.E. HURWITZ, Appl. Phys. Letters 8, 243 (1966).

(Received July 4, 1972)

phys. stat. sol. (a) 12, K95 (1972)

Subject classification: 14.3.3; 20.3; 22.2.1

VEB Kombinat Halbleiterwerk Frankfurt/Oder, Werkteil Stahnsdorf

Injection Dependence of Phase Shifts of Recombination Emission of Sinusoidal-Modulated Electroluminescence Diodes

By

R. DOSS

The recombination emission of Fabry-Perot-type electroluminescence diodes shows phase shifts in the active region, which are dependent on the minority carrier injection and graduation (doping concentration, acceptor doping gradient) of the p-n junction (1). In agreement with the model (1) the phase curve of the recombination region is determined by sequential continuous regions free from charge (low-level injection). At high-level injection, the current induced by the field is not negligible outside the space charge because the potential acts not only on the transition region.

The diode junction shall be excited by a variable forward voltage which is superimposed by a small alternating voltage of frequency ω with constant amplitude. The GaAs diodes prepared by Zn diffusion with different graduation were measured by a point-by-point scanning of the light emitting area of the near field whereas the detector only detects radiation emitted approximately vertical to the scanning area (2).

Fig. 1 and 2 show the local dependence of the phase shift and near field intensity curves at different injection levels. The phase dependence according to

$$\Delta\varphi = \frac{\omega}{4}\,\frac{\tau}{L}\,(n_{ini}\,,\,p_{ini})\,\Delta x \qquad (1)$$

shows the local dependence of injected minority carrier concentration p_{ini}, n_{ini} at given doping. Regions with constant phase are related to the exponential dependence of the recombination rate R(x). At maximum intensity R(x) is only exponential in differential ranges, thus $\Delta\varphi \rightarrow 0$. The concentration of minority carriers increases in the transit region with increasing injection level, hence deviations from the exponential behaviour arise (3, 4).

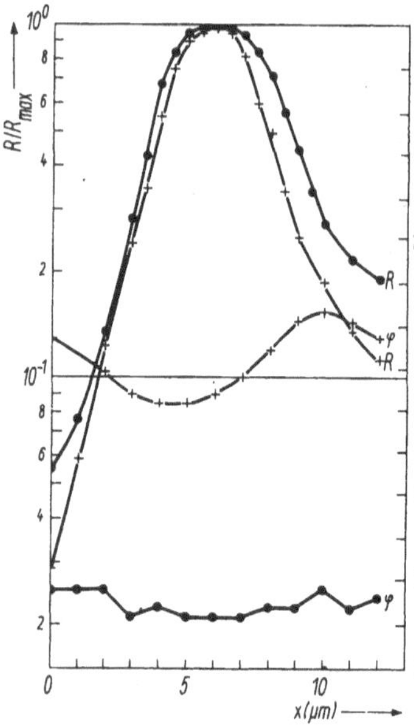

Fig. 1

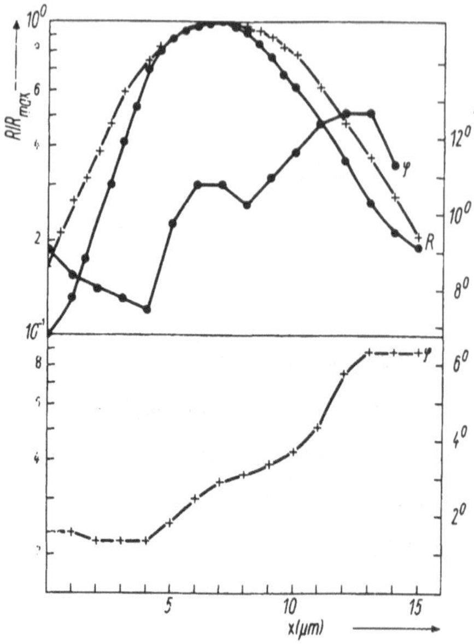

Fig. 2

Fig. 1. Local dependence of the phase shift and near field intensity curves at different injection levels; T = 300 °K, f = 30 MHz, $\Delta N_A/\Delta x \approx 8\times10^{22}\,cm^{-4}$, $N_D = 2.5\times10^{18}\,cm^{-3}$. + $2\times10^2\,A\,cm^{-2}$; ● $2\times10^3\,A\,cm^{-2}$

Fig. 2. Local dependence of the phase shift and near field intensity curves at different injection levels; T = 300 °K, f = 30 MHz, $\Delta N_A/\Delta x \approx 5\times10^{22}\,cm^{-4}$, $N_D = 5\times10^{17}\,cm^{-3}$. ● $2\times10^2\,A\,cm^{-2}$, + $2\times10^3\,A\,cm^{-2}$

Phase differences decrease with increasing injection level, therefore phase homogeneity of recombination emission improves. Simultaneously with local delay time differences the absolute delay time between electrical and optical signals changes.

Experimental results are in agreement with theory (1), so that the influence of phase shifts by optical absorption is negligible within the width of quarter intensity of the emitting area.

With our measuring arrangement (2) it is possible to determine on the basis of the theoretical formulae minority carrier lifetime τ, diffusion length L, and local

injection course of luminescence and laser diodes respectively when the doping profile is known.

The author thanks Dr. K. Unger, J. Schnabel, and F. Kugler for helpful discussions.

References

(1) R. DOSS, phys. stat. sol. (a) 9, K61 (1972).

(2) H. WENDT, W. SCHUBERT, and R. DOSS, DDR WPG 01n/157281 (1971).

(3) C.J. HWANG, J. appl. Phys. 42, 2757 (1971).

(4) A.K. JONSCHER and M.H. BOYLE, Symposium on GaAs 1966, Paper 12A.

(Received June 16, 1972)

phys. stat. sol. (a) 12, K99 (1972)
Subject classification: 14.3.1; 12; 22.1.3

Sektion Physik der Humboldt-Universität zu Berlin, Bereich Kristallographie

Zur Deutung der Temperaturabhängigkeit der elektrischen Leitfähigkeit von Tellurschichten auf Glas

Von

C. ALBERS

Tellur gehört der Raumgruppe D_3 an und weist deshalb einen deutlichen Piezoeffekt auf. Die Änderung der spezifischen Leitfähigkeit als Folge eines uniaxialen bzw. hydrostatischen Druckes wird als Piezowiderstandseffekt bezeichnet und ist an Tellureinkristallen mehrfach nachgewiesen worden (1 bis 5). Bezieht man die Widerstandsänderung nicht auf den Druck bzw. Zug, sondern auf die elastischen Eigenschaften, so wird im allgemeinen von einem Elastowiderstandseffekt gesprochen (4). An dünnen Tellurschichten wurde der Elastowiderstandseffekt bereits nachgewiesen (6, 7). Untersuchungen der elektrischen Transporterscheinungen an dünnen Tellurschichten auf unterschiedlichen Substraten durch verschiedene Autoren wie auch eigene Untersuchungen zeigen recht uneinheitliche Ergebnisse und starke Abhängigkeiten vom gewählten Substratmaterial. Wir vermuten deshalb, daß der Einfluß des Substrates, insbesondere der zum Tellur unterschiedliche thermische Ausdehnungskoeffizient, bisher zu wenig beachtet wurde. Bei einer Temperaturänderung übt bei unterschiedlichen Ausdehnungskoeffizienten das Substrat bekanntlich erhebliche mechanische Kräfte auf eine dünne Schicht aus.

Wir wollen im folgenden an Hand eines sehr einfachen Modells zeigen, daß im Temperaturbereich unterhalb Zimmertemperatur (Ausläufergebiet) die Leitfähigkeitsänderung infolge des Piezowiderstandseffektes merklich werden kann. Hierzu machen wir die folgenden stark vereinfachenden Annahmen:

1. Das Substrat sei bezüglich seiner thermischen Ausdehnung isotrop (z.B. Glas).
2. Die Tellurschicht sei polykristallin, d.h. im räumlichen Mittel seien die Transporteigenschaften und der thermische Ausdehnungskoeffizient der Schicht ebenfalls isotrop.
3. Sowohl der lineare thermische Ausdehnungskoeffizient α_S des Substrates als auch der entsprechende Koeffizient α_{Te} des Tellurs seien in erster Näherung temperaturunabhängig.

4. Die Tellurschicht hafte fest auf dem Substrat, so daß sie einer Temperaturänderung bzw. Kontraktion folgt. Eine Rückwirkung der Schicht auf das Substrat wird vernachlässigt.

Wir beziehen unsere Überlegungen auf einen Temperaturbereich unter 20 ^{o}C, in dem in den Proben reine Störleitung vorliegt. Eine Temperaturänderung wirkt dann nicht auf die Ladungsträgerkonzentration. Wir interessieren uns für die Widerstandsänderung $\Delta \varrho$ der Tellurschicht bei einer auf das System Schicht - Substrat wirkenden Temperaturänderung ΔT, also für die Größe $\Delta\varrho/\Delta T$. Diese Größe kann formal wie folgt geschrieben werden:

$$\frac{\Delta \varrho}{\Delta T} = \frac{\Delta \varrho}{\Delta p} \frac{\Delta p}{\Delta l} \frac{\Delta l}{\Delta T} . \tag{1}$$

Dabei beschreibt $\Delta\varrho/\Delta p$ den Einfluß des Druckes auf den Schichtwiderstand, $\Delta p/\Delta l$ die Kompressibilität und $\Delta l/\Delta T$ die lineare thermische Ausdehnung. Dann würde die relative Widerstandsänderung unter Berücksichtigung der Tatsache, daß nur die Differenz der Ausdehnungskoeffizienten von Schicht und Substrat auf die Schicht wirksam wird, folgende Form haben:

$$\left(\frac{\Delta \varrho}{\varrho \Delta T}\right)_{Te} = \left(\frac{\Delta \varrho}{\varrho \Delta p}\right)_{Te} \left(\frac{l \Delta p}{\Delta l}\right)_{Te} \left[\left(\frac{\Delta l}{l \Delta T}\right)_{S} - \left(\frac{\Delta l}{l \Delta T}\right)_{Te}\right] . \tag{2}$$

Der erste Faktor auf der rechten Seite von (2) ist der Piezowiderstandskoeffizient Π, der zweite Faktor die elastische Konstante c. Die eckige Klammer enthält die Differenz der thermischen Ausdehnungskoeffizienten, so daß wir (2) auch in der folgenden Form schreiben können:

$$\left(\frac{\Delta \varrho}{\varrho \Delta T}\right)_{Te} = \Pi_{Te} c_{Te} \left[\alpha_S \alpha_{Te}\right] . \tag{3}$$

Aus der Literatur entnehmen wir die folgenden Werte:
$\Pi_{Te} = 35 \times 10^{-12}$ cm^{2}dyn^{-1} (nach Zitat (3)), $c_{Te} = 3{,}8$ bis 7×10^{11} dyn cm^{-2} (nach (8)).
Außerdem wollen wir für die Differenz der Ausdehnungskoeffizienten einen plau-

siblen Wert von $10x10^{-6}$ grd^{-1} annehmen. Dann ergibt sich die relative Widerstandsänderung zu $\Delta\varrho/(\varrho \Delta T) = 2x10^{-4}$ grd^{-1}. Experimentelle Untersuchungen an Tellurschichten auf Glas ergeben einen Wert, der etwa eine Größenordnung über dem berechneten liegt. Das ist verständlich, da einerseits das Modell stark vereinfacht wurde, andererseits aber auch verschiedene bekannte Einflüsse nicht berücksichtigt worden sind. Das Modell gibt aber z.B. die Abhängigkeit vom Substratmaterial richtig wieder. So entsteht kein Widerspruch zu den von uns kürzlich veröffentlichten Ergebnissen der elektrischen Transporteigenschaften von Tellurschichten auf Glimmer (9). Die mittleren Ausdehnungskoeffizienten der polykristallinen Tellurschicht und des Glimmers unterscheiden sich nur wenig, so daß eine solche Schicht wesentlich geringeren Verspannungen unterworfen ist.

Literatur

(1) A.N. GOLDOBIN, L.V. LESHENKO und E.T. SHARNOPOLSKAYA, Fiz. tverd. Tela 3, 3247 (1961).

(2) A.N. GOLDOBIN und L.V. LESHENKO, Selen, tellur i ikh primenenie, Sbornik Statei, Baku 1965.

(3) K.H. HERRMANN und G. OELGART, phys. stat. sol. 24, 289 (1967).

(4) K.H. HERRMANN und J. PUHLMANN, phys. stat. sol. 29, 193 (1968).

(5) W. BECKER, W. FUHS und J. STUKE, phys. stat. sol. (b) 44, 147 (1971).

(6) R.L. PARKER und A. KRINSKY, J. appl. Phys. 34, 2700 (1963).

(7) M. GRANVEAUD und Y. PETROFF, phys. stat. sol. (a) 3, 629 (1970).

(8) P. GROSSE, Der Halbleiter Tellur, Springer Tracts on Modern Physics, Bd. 48.

(9) C. ALBERS und R. LINK, phys. stat. sol. (a) 12, 89 (1972).

(Received June 28, 1972)

phys. stat. sol. (a) 12, K103 (1972)
Subject classification: 15; 22

Institute of Inorganic and Physical Chemistry,
Academy of Sciences of the Azerbaidzhan SSR, Baku

Effect of Partial Substitution of Lead by Gallium, Indium, and Thallium on Thermoelectric Properties of PbTe

By

P.G. RUSTAMOV, M.A. ALIDZHANOV, and CH.I. ABILOV

Based on measurements of electric and physical properties, it was supposed (1, 2) that PbTe is a semiconductor with complex band structure. The valence band consists of two subbands, separated from each other by an energy gap of the order $\Delta = 0.05$ to 0.10 eV (3) and the width of forbidden zone $\Delta E = 0.30$ eV. It was shown in paper (4) that the proposed complex band structure of lead telluride is applicable also to solid solutions based on it. It is known that lead telluride and solid solutions based on it are promising materials to be used in thermoelectric devices and infrared technique (5 to 9). Therefore it is of interest to trace the influence of tellurides of group III elements gallium, indium, and thallium on the thermoelectric properties and band structure of lead telluride.

Procedure Earlier we found that the solubility of gallium, indium, and thallium tellurides in PbTe is higher than 5% (7).

For our studies the solid solutions based on PbTe were synthesized from elements of 99.999% purity by direct fusing in a temperature oven, in quartz ampoules evacuated to 10^{-5} Torr at a temperature of about ≈ 900 °C.

X-ray powder patterns of the specimens were taken using FeK_{α} radiation with an exposure of 40 h. The homogeneity of specimens was checked by x-ray diffraction analysis, microstructure analysis, and measurement of the microhardness. For electric measurements the ingots were annealed at 600 °C for 450 h and parallelepipeds were cut from them. Thermal electromotoric force, thermal conduction, and Hall constant were measured by a compensation method. The Hall effect was measured in a magnetic field of 8 kOe. Measurements were made with two directions of magnetic field and current.

Experimental results As was shown by X-ray analysis (Fig. 1) the formation of solid solutions based on PbTe is accompanied by a slight change of the lattice

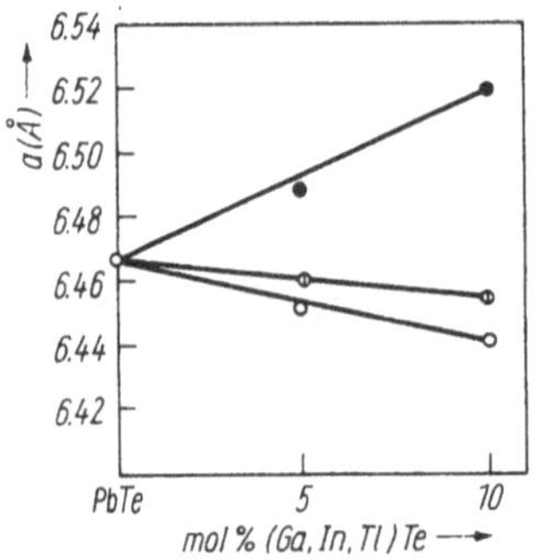

Fig. 1

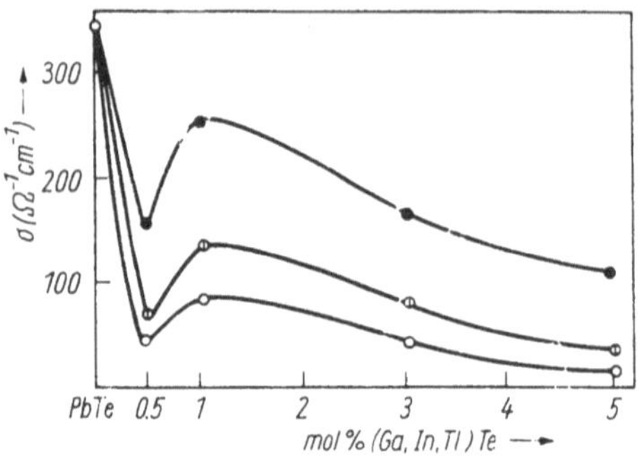

Fig. 2

Fig. 1. Variation of lattice parameter of solid solutions based on PbTe
o GaTe, ⊕ InTe, ● TlTe

Fig. 2. Conductivity variation of alloys based on lead telluride
o GaTe, ⊕ InTe, ● TlTe

constant. The lattice parameter increases linearly with the amount of TlTe, and decreases slightly with that of GaTe and InTe. The dependence of conductivity on Ga, In, Tl concentration in the alloys has a minimum at $x = 0.005$ and is slightly decreasing with higher content of the substituting elements above $x = 0.01$ (Fig. 2).

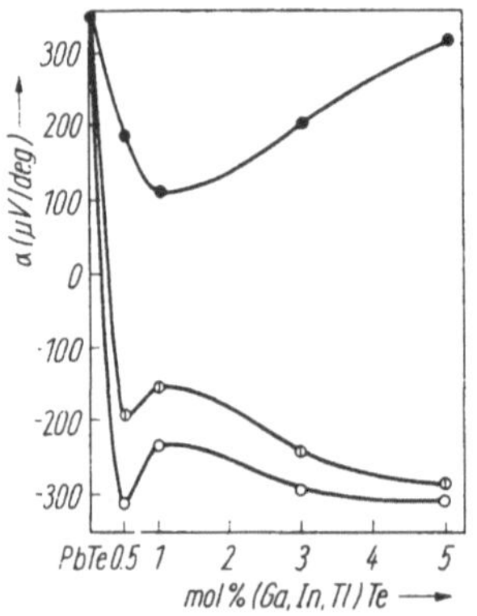

Fig. 3

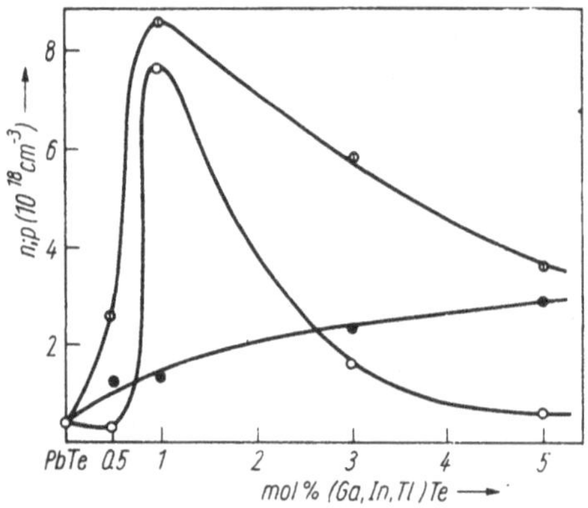

Fig. 4

Fig. 3. Variation of thermal electromotoric force of alloys based on lead telluride
o GaTe, ⊕ InTe, ● TlTe

Fig. 4. Carrier concentration as a function of composition in solid solutions based on PbTe
o n-GaTe, ⊕ n-InTe, ● p-TlTe

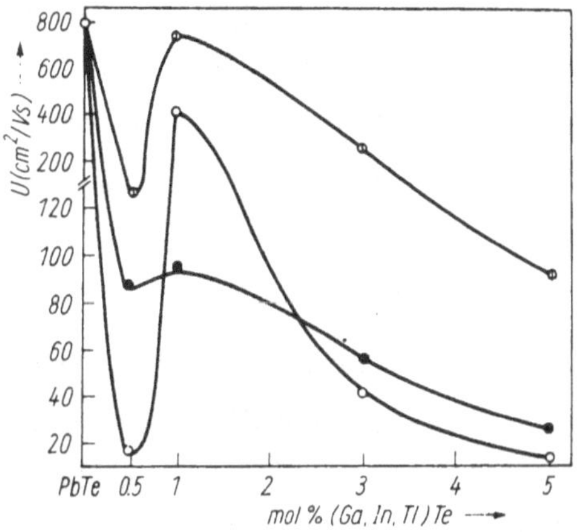

Fig. 5

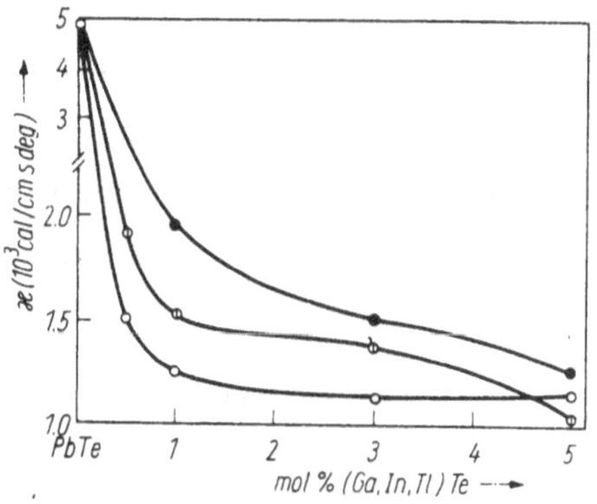

Fig. 6

Fig. 5. Mobility variations of current carriers in alloys based on PbTe
○ GaTe, ⦶ InTe, ● TlTe

Fig. 6. Variation of thermal conductivity of alloys based on PbTe
○ GaTe, ⦶ InTe, ● TlTe

The dependence of the thermal e.m.f. on composition at room temperature for solid solutions Pb_{1-x} (Ga, In, Tl)$_x$Te shows a minimum at X = 0.005; in the case of GaTe and InTe an inversion of sign takes place, the absolute value of the thermal e.m.f. rising with increasing content of Ga, In, and Tl tellurides in the alloys (Fig. 3). Fig. 4 and 5 show the dependence of concentration and mobility of current carriers on composition in solid solutions based on PbTe. The concentration of current carriers shows a maximum at about 1 mol% of (Ga,In)Te in the alloy, while a slight increase is observed in the case TlTe. The mobility of current carriers in the case of lead substituted by Ga, In, and Tl up to 0.5 mol% varies just as the conductivity; for higher concentrations of Ga and In larger changes of mobility are observed than in the case of Tl. As seen from Fig. 6, there is a first sharp decrease of thermal conductivity of solid solutions based on PbTe as compared with that of PbTe up to 1 mol% of (Ga, In, Tl)Tl, and then a slight decrease.

Discussion In solid solutions of $Pb_{1-x}Tl_xTe$ thallium atoms (r = 0.95 Å) substitute lead atoms (r = 0.84 Å), placed within octahedral interstitials indirectly extending the periods of α-solid solutions. However, in solid solutions of Pb_{1-x}(Ga,

In)$_x$ Te the lattice parameters would slightly decrease due to the smaller atomic radii of gallium (r = 0.62 Å) and indium (r = 0.812 Å) as compared to that of lead. In papers (9 to 11) Ga and In are shown to be donors in lead telluride, while Tl forms acceptors. It is suggested that lead vacancies being atomic point defects may serve as donors or acceptors depending on the substituting element.

Based on the thermal e.m.f. it was determined that the conductivity of PbTe is p-type, but when GaTe and InTe are added the conductivity of the alloys is n-type.

Addition of TlTe resulted in no change of the conductivity type, i.e. remained p-type. Therefore, in lead telluride (Ga, In)Te form donors and TlTe acceptors. The deviation from stoichiometry in PbTe is known to have a strong effect on concentration and type of current carriers. Lead excess over stoichiometric composition reduces and tellurium excess increases the carrier concentration. An excess of lead is dissolved in PbTe and occupies vacancies, resulting in a smaller number of defects, while an excess of tellurium, on the contrary, results in new lead vacancies, i.e. new lattice defects (11). Besides it was revealed that admixtures at Pb lattice sites reduces the mobility of electrons and holes more than those at Te sites (6). The above holds also for the variation of concentration and mobility of current carriers in solid solutions Pb_{1-x} (Ga, In, Tl)$_x$ Te.

The variations of conductivity and thermal e.m.f. with composition can be explained by assuming the defect nature of lead telluride. It is assumed that in solid solutions based on PbTe at x = 0.01 the Ga, In, and Tl atoms fill the vacancies in lead telluride. This results in a smaller number of point defects acting as acceptors or donors and therefore, in a higher concentration of current carriers. The further increase of the content of the substituting component, when the possible number of vacancies has been already filled, leads probably to the beginning of solid solution formation. The results of the study have shown that the thermal lattice conductivity is reduced much stronger than the mobility in alloys containing 1 mol% of (Ga, In)Te. Accordingly the ratio of mobility to thermal conductivity is 3.3×10^5 and 5.3×10^5 (it is 1.8×10^5 for p-PbTe) and may serve to increase the thermoelectric properties of materials.

References

(1) A.A. ANDREEV, Fiz. tverd. Tela 8, 2818 (1966).

(2) R.N. TAYBER and J.B. CADOFF, J. appl. Phys. 38, 3714 (1967).

(3) L.M. ROGERS, Brit. J. appl. Phys. 1, 845 (1968).

(4) A.J. STRAUSS, Phys. Rev. 157, 608 (1967).

(5) A.J. ROSENBERG and E. WOLOV, J. Phys. Chem. Solids 28, 1079 (1969).

(6) A.J. ROSENBERG, R. CRIEVSON, J. WOLLEY, and P. NIKOLIC, Trans. MS AIME 230, 342 (1964).

(7) P.G. RUSTAMOV, M.A. ALIDZHANOV, and CH.I. ABILOV, Theses of Rep. Conf. Thermoelectric Materials and Methods of Their Studies, Acad. Sci. SSSR, Kishinev 1971.

(8) A.J. ROSENBERG, J. Electrochem. Soc. 107, 2690 (1960).

(9) B.B. FRITS, in: Thermoelectric Materials and Transformers, Mir, Moscow 1964 (p. 153).

(10) A.F. GILSON, Proc. Phys. Soc. B65, 378 (1952).

(11) N.V. KOLOMIETS, E.YA. LEV, and L.M. SYSOEVA, Fiz. tverd. Tela 5, 10 (1963).

(Received March 8, 1972)

phys. stat. sol. (a) 12, K109 (1972)
Subject classification: 18.2 and 18.4; 21

Oregon Graduate Center, Beaverton

Metamagnetism in ErNiAl and TmNiAl[1)]

By

H. OESTERREICHER

This work is a preliminary note on the magnetic properties of two of the compounds RTAl where R : rare earth and T: Ni and Cu. A more detailed account of this family of compounds shall appear elsewhere (1). With this study we have been interested to explore the type and strength of cooperative effects in an intermediate situation between the electron concentration (e.c.) of the well studied binaries of rare earths with transition metals and Al.

Both compounds have been found to crystallize with the Fe_2P type structure (2). The series RNiAl with exception of TmNiAl has been investigated previously by Leon and Wallace (3) although not always at the exact stoichiometry. This study showed ErNiAl to order ferromagnetically with a Curie temperature of 16 K and a saturation moment of 7.4 μ_B extrapolated from measurements in fields up to 20 kOe.

Both materials have been prepared by melting together metal pieces in MgO crucibles under argon with induction techniques. Debye-Scherrer diagrams confirmed a pure Fe_2P structure. Magnetic studies were carried out on powdered specimens using a Foner magnetometer. Demagnetization corrections were applied according to the relation $H_i = H_a - \frac{4\pi}{3} \sigma d$, where H_i and H_a are internal and applied fields respectively, σ the magnetization, and d the density of the specimen.

A thermomagnetic analysis run at 20 kOe shows ErNiAl and TmNiAl to order ferromagnetically at 15 and 12 K respectively. In the paramagnetic region Curie-Weiss law is obeyed with Weiss temperatures of -1 and -11 K and effective magnetic moments of 9.8 and 7.8 μ_B for the Er and Tm compound respectively. The magnitude of the effective moments corresponds to the free tripositive rare earths and shows that Ni is nonmagnetic in these compounds presumably due to electron transfer from the electropositive ligands. Fig. 1 shows the magnetization behavior of ErNiAl and TmNiAl at various temperatures. In both cases S type saturation behavior develops at the lowest temperatures. On decreasing the field the magnetiza-

---

1) Part of this work has been supported by a grant from the Office of Naval Research.

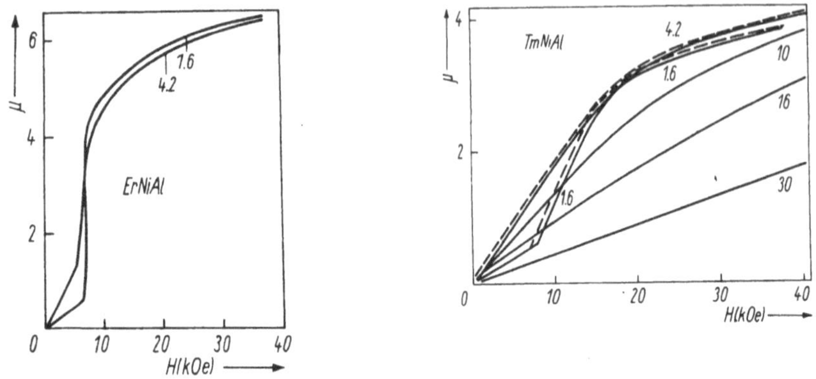

Fig. 1. Magnetic moment μ (Bohr magnetons per formula unit) vs. internal field H (kOe) at various temperatures (K) for ErNiAl and TmNiAl. The dashed curves represent measurements with decreasing fields

tion follows the initial magnetization curve with some minor hysteresis so that no remanence is left at zero field. This is the typical behavior of a metamagnet.

It is especially interesting to note that the S type saturation behavior relaxes at temperatures just below the ordering temperature to approach the behavior of a pure ferromagnet. Therefore, it is suggested that a delicate balance between ferromagnetic and antiferromagnetic exchange interactions exists and that at the lower temperatures the later predominate. An antiferromagnetic exchange interaction indeed appears already indicated in the negative Weiss constants. The relatively low critical fields (below 10 kOe) are somewhat unusual for a metamagnet.

Generally it has been observed that metamagnetism often is connected with somewhat buckled and distorted structures. This is also the case with the Fe_2P structure as shall be outlined in more detail in the subsequent paper.

References

(1) H. OESTERREICHER, J. less-common Metals, in the press.

(2) A.E. DWIGHT, M.H. MULLER, R.A. CONNER, J.W. DOWNEY, and H. KNOTT, Trans. MS AIME 242, 2075 (1968).

(3) B. LEON and W.E. WALLACE, J. less-common Metals 22, 1 (1970).

(Received June 26, 1972)

phys. stat. sol. (a) 12, K111 (1972)

Subject classification: 14.3.3 and 20.3; 22.2.1; 22.2.4

Sektionen Chemie (a) und Physik (b) der Karl-Marx-Universität Leipzig, Arbeitsgemeinschaft $A^{III}B^{V}$-Halbleiter

Local Scanning of the Emission Delay Time of Epitaxial (Ga,Al)As/GaAs Luminescence Diodes

By

E. BUTTER (a), B. JACOBS (a), K. JACOBS (a), K. UNGER (b), and A. ZEHE (b)

GaAs luminescence diodes, driven by current pulses or by a modulating ac, show a time delay between electrical and optical signal. This delay time is not constant over the light emitting region (1, 2 cf. also 3).

Doss recently reported the local dependence of near field intensity and phase shift for GaAs luminescence diodes produced by Zn diffusion (2), excited by a high-frequency small signal ac, superimposed on a forward biased dc.

This note is concerned with results of similar measurements on epitaxial (Ga,Al)As/GaAs diodes. The influence of annealing after the crystal growth is studied, too.

The diodes were fabricated by growing Zn-doped (Ga,Al)As layers on Sn-doped GaAs substrates from a non-stoichiometric melt. Zn diffusion before and during the epitaxial growth yields the following typical diode structure: n-GaAs (80 μm)/ p-GaAs (2.4 to 3.8 μm)/p^+-(Ga,Al)As (20 μm). The doping concentrations in the GaAs substrate and in the epitaxial layer and in the vicinity of the p-n junction were determined by electrical and radioanalytical methods, respectively (4). Typical values are $N_D = 1x10^{18}$ cm^{-3}, $(N_A - N_D) = 2x10^{19}$ cm^{-3} in the bulk of the (Ga,Al)As layer. The diodes operated under the following conditions: j = 200 A cm^{-2}, f = 30 MHz, T = 300 K.

In Fig. 1 the distribution of the near field intensity and phase shift is shown on typical examples of our diodes in a plane perpendicular to the p-n junction. The measurements were performed by local scanning with the apparatus described in (5).

Two parts of the GaAs/(Ga,Al)As wafer from which the unannealed samples were fabricated, were heat-treated under different conditions. This resulted in other $\Delta\varphi$ (x) curves which are shown in Fig. 1, too.

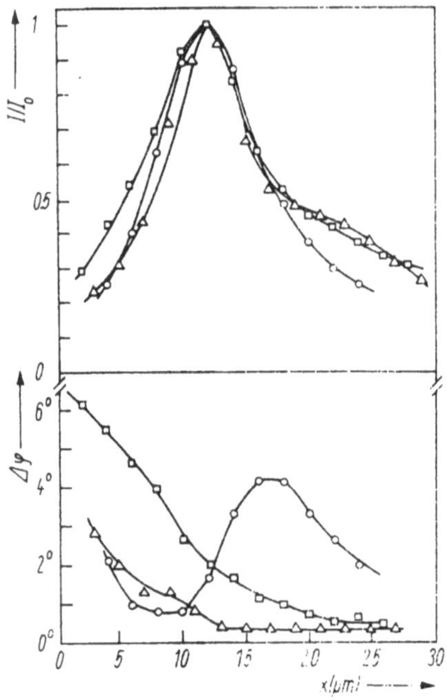

Fig. 1. Observed local dependence of phase shift and near field intensity for different annealing conditions. o Unannealed ; Δ 90 min at 935 °C; □ 90 min at 900 °C

The $\Delta\varphi$ (x) curve of our unannealed diode is in qualitative agreement with that measured by Doss (2). During the annealing Zn diffuses somewhat deeper into the substrate and the p-n-junction is therefore in greater distance from the Al containing layer. The recombination region (p-GaAs) becomes wider and the doping concentration profile in this region becomes flatter.

Besides, this annealing results in a higher degree of compensation of the Sn-doped substrates by migration of this amphoteric dopant from Ga sites onto As sites. A third effect of annealing is the removal of inhomogeneities in the dopant distribution, possibly generated during the growth process. All these effects influence the $\Delta\varphi$ (x) curves, as shown in Fig. 1.

Fig. 2 shows calculated φ (x) curves and near field intensity distributions. The φ (x) curves were computed under the assumption that the phase shift is a result of a charge storage and that the propagation of the injected carriers is diffusion controlled. The influence of a locally different absorption coefficient and refractive index was considered for the calculations, too. The agreement between theory

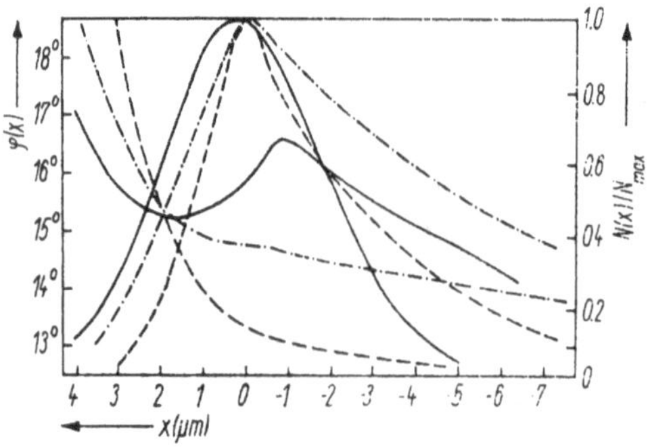

Fig. 2. Calculated curves for phase shift and near field intensity, taking into account different absorption profiles around the recombination region

and experiment is good. The calculations will be discussed in more detail in a following paper (6).

With definite doping concentrations and annealing conditions it is possible to produce diodes with only very small variations in the phase shift ($\leqq 1^{o}$) over the light emitting region, what is of importance for some practical applications.

The authors acknowledge valuable discussions and good cooperation with R. Doss.

References

(1) H. WENDT, private communication, 1970.

(2) R. DOSS, phys. stat. sol. (a) 9, K61 (1972).

(3) V.N. DERYAGIN and L.E. MARASIN, Fiz. Tekh. Poluprov. 5, 1981 (1971).

(4) K. JACOBS, B. JACOBS, and E. BUTTER, to be published.

(5) H. WENDT, H. SCHUBERT, and R. DOSS, DDR WPG 01n/157281 (1971).

(6) A. ZEHE, to be published.

(Received July 13, 1972)

phys. stat. sol. (a) 12, K115 (1972)

Subject classification: 14.3.3; 22

Institutsteil für magnetische Werkstoffe Jena des Zentralinstitutes für Festkörperphysik und Werkstofforschung der Deutschen Akademie der Wissenschaften zu Berlin

Calculation of a Thermally Produced Current-Voltage Characteristic of N-Type

By

C. MATTHECK

Certain semiconducting materials have a dc conductivity which varies rapidly at a critical temperature T_c (1 to 3). If the conductivity is homogeneous, its temperature dependence may be described approximately by

$$\sigma(T) = \begin{cases} \sigma_1 & \text{if } T < T_c, \\ \sigma_2 & \text{if } T > T_c, \end{cases} \qquad \text{where} \quad \sigma_1 > \sigma_2 .$$

To calculate the current-voltage characteristic (I(U) curve) we consider the geometry sketched in Fig. 1. It is assumed that the insulator has the same thermal properties as the semiconducting material. As well-known from the thermal model of current instabilities (4) below a critical applied dc voltage U_c^1 the semiconducting material is in the conducting state σ_1. At U_c^1 a low conducting filament along the axis x = 0 will be produced. For another critical voltage U_c^2 the whole semiconducting material is in the conducting state σ_2. In the thermal model the I(U) curve can be calculated from the thickness a of the filament which is determined by solving the heat transport equation

$$\frac{d^2T}{dx^2} = - \frac{\sigma U^2}{\varkappa d^2}$$

in the regions with the solutions

$$T_1(x) = - \frac{\sigma_2 U^2}{2\varkappa d^2} x^2 + \frac{U^2}{2\varkappa d^2} a^2(\sigma_1 - \sigma_2) + \frac{\sigma_1 U^2}{\varkappa d^2} lL - \frac{U^2}{\varkappa d^2} aL(\sigma_1 - \sigma_2) - \frac{\sigma_1 U^2}{2\varkappa d^2} l^2 + T_o$$

for $0 < x < a$,

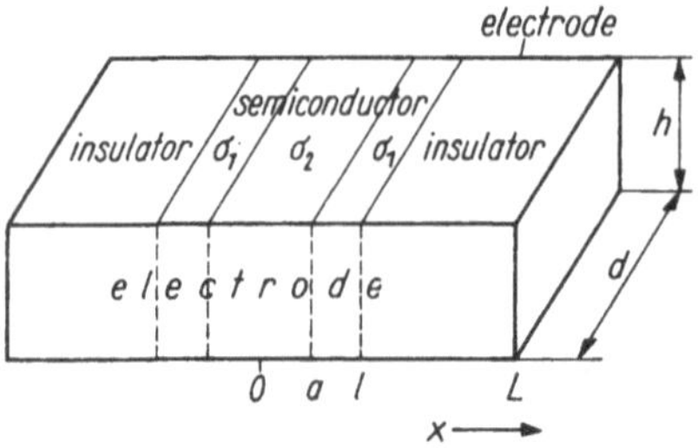

Fig. 1. Sketch of the used geometry

$$T_2(x) = -\frac{\sigma_1 U^2}{2\varkappa d^2}x^2 + \frac{U^2}{\varkappa d^2}a(\sigma_1 - \sigma_2)x + \frac{\sigma_1 U^2}{\varkappa d^2}lL - \frac{U^2}{\varkappa d^2}aL(\sigma_1 - \sigma_2) - \frac{\sigma_1 U^2}{2\varkappa d^2}l^2 + T_o$$

for $a < x < l$,

$$T_3(x) = \left[-\frac{\sigma_1 U^2}{\varkappa d^2}l + \frac{U^2}{\varkappa d^2}a(\sigma_1 - \sigma_2)\right]x + \frac{\sigma_1 U^2}{\varkappa d^2}lL - \frac{U^2}{\varkappa d^2}aL(\sigma_1 - \sigma_2) + T_o$$

for $l < x < L$,

satisfying the boundary conditions

$$T_1(a) = T_2(a); \quad \left.\frac{dT_1}{dx}\right|_{x=a} = \left.\frac{dT_2}{dx}\right|_{x=a}; \quad T_2(l) = T_3(l); \quad \left.\frac{dT_2}{dx}\right|_{x=l} = \left.\frac{dT_3}{dx}\right|_{x=l};$$

$$T_3(L) = T_o = \mathrm{const}; \quad \left.\frac{dT_1}{dx}\right|_{x=0} = 0 .$$

$\varkappa$ is the thermal conductivity and d, h, l, L, are explained in Fig. 1. In this model the dc voltage, U, is connected with the thickness of the filament, a, by the relation $T_1(x=a) = T_c$ which gives

$$U = \left[\frac{\varkappa d^2(T_c - T_o)}{\sigma_1\left(\frac{a^2}{2} - \frac{l^2}{2} - La + lL + qaL - qa^2\right)}\right]^{1/2}$$

with $q = \sigma_2/\sigma_1$. $a = 0$ gives U_c^1 and $a = l$ gives U_c^2. The dc current I can be calculated by integrating the heat transport equation and is given by

$$I = \frac{\left[\varkappa d^2(T_c - T_o)\right]^{1/2}\left\{\sigma_1\left[2(l-a)h + 2qah\right]\right\}}{d\left[\sigma_1\left(\frac{a^2}{2} - \frac{l^2}{2} - La + lL + qaL - qa^2\right)\right]^{1/2}} .$$

The I-U characteristic is plotted in Fig. 2. To demonstrate the type of the I-U curve we used the special values h = d = l = 5 μm; $\sigma_1 = 10^3\ \Omega^{-1}\mathrm{m}^{-1}$; $L = 10^4$ μm; $T_c - T_o = 50$ deg; $\varkappa = 80\ \mathrm{kg\ m^2\ s^{-3}\ deg^{-1}}$ with q as a parameter.

We obtain an N-type I(U) characteristic where the region of negative differential resistance is determined by the value of q. For q-values above the critical value

$$q_c = \frac{1 - l/L}{1 - l/2L}$$

derived from the relation I(a = 0) = I(a = 1) no N-type appears. In the region of negative differential resistance the device can work as an oscillator, because an increasing voltage lowers the current and the temperature. Therefore the resistance is decreased and the current increases again. Higher values of the current give a higher temperature and a higher resistance follows. This will repeat with a frequency which is determined by the geometry as well as the kinetics of the filament formation.

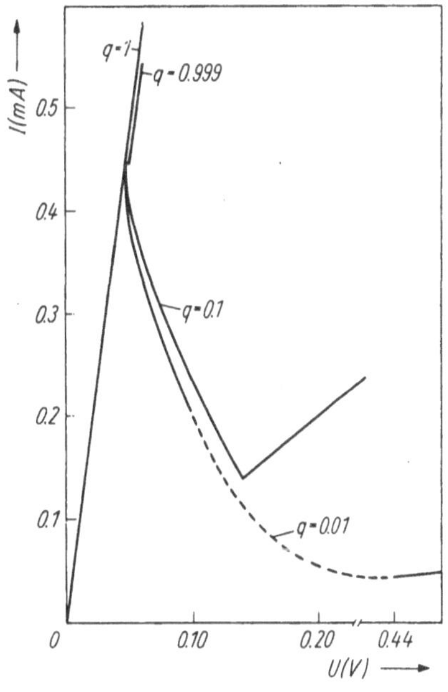

Fig. 2. Current-voltage characteristic for a fixed geometry with q as a parameter

Acknowledgements

I would like to thank Dr. W. Haubenreißer, Dr. H. Gengnagel, and Dr. U. Röpke for stimulating discussions.

References

(1) M. KAHN, Amer. Ceram. Soc. Bull. 50, 676 (1971).

(2) NGUYEN VAN DONG, C.R. Acad. Sci. (France) 272, 646 (1971).

(3) W. HEYWANG, J. Mater. Sci. 6, 1214 (1971).

(4) E. NEBAUER and E. JAHNE, phys. stat. sol. 8, 881 (1965).

(Received July 20, 1972)

phys. stat. sol. (a) 12, K119 (1972)
Subject classification: 16; 2; 13.4; 22.3.1

A. F. Ioffe Physico-Technical Institute, Academy of Sciences of the USSR, Leningrad

Photoactivation in Vitreous $Tl_2Se-As_2Te_3$

By

B. T. KOLOMIETS, T. N. MAMONTOVA, G. I. STEPANOV, and E. A. SMORGONSKAYA

In this work the kinetics of photoconductivity of vitreous $Tl_2Se-As_2Te_3$ at low temperatures was investigated. When illuminating the samples with light in the range of the fundamental absorption a stepwise (S-type) increase of photoresponse was observed if the samples were previously desactivated, i.e. the state of the samples was close to thermodynamic equilibrium. Such form of the increase of photoresponse had been earlier observed for some crystalline semiconductors (1 to 3) and is connected with the phenomenon of photoactivation.

The samples were plates of 150 to 200 μm thickness cut from bulk material synthesized by usual melting of the components in vacuum at 700 $^{\circ}$C.

For the desactivation before the measurements the samples were kept in the dark for a long time at room temperature and then cooled to 77 $^{\circ}$K. During the measurements the sample was illuminated with light in the range of the fundamental absorption edge with photon energy 0.8 to 0.9 eV. The intensity of the exciting light did not exceed 2×10^{13} photons/cm^2 s. The duration of the pulses light was 5×10^{-3} s. The method of the measurements was described in (4).

The experimental results are presented in Fig. 1. The relaxation curves are plotted both in normal and logarithmic scales. As one can see from the inset in Fig. 1, when the light is switched on a small leap of conductivity (σ), changing to a plateau is observed, then the conductivity increases to saturation (σ_{st}). When switching off the light the photoconductivity at first sharply drops and then the rate of the drop decreases. The transition to the equilibrium state occurs during a very long time so the sample can be considered to be in quasi-equilibrium. Under the second photoexcitation of such a sample the plateau is not observed and the photoresponse increases rather sharply.

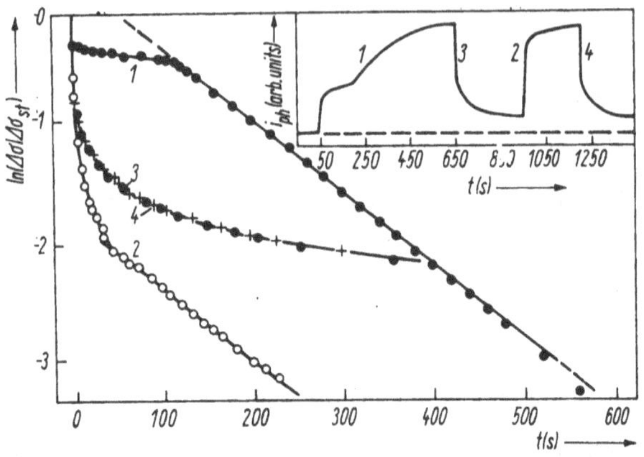

Fig. 1. Increase (1, 2) and decay (3, 4) of the photoresponse of vitreous $Tl_2Se-As_2Te_3$ previously desactivated and activated with light, respectively

The character of the relaxation curves obtained suggests that in the gap of $Tl_2Se-As_2Te_3$ there exist two groups of local levels acting as β-type traps and recombination levels under the given conditions. The energy position of these groups of levels in vitreous $Tl_2Se-As_2Te_3$ was determined earlier using the spectral distribution of optical quenching and induced photoconductivity (5) and thermostimulated conductivity (6) and corresponds to 0.45 eV and 0.51 to 0.56 eV below the conduction band.

The photoactivation phenomen can be explained in the following way. Let the capture probability of the majority carriers for the traps be much larger than that for the recombination levels and the probability of thermal emptying of the traps be small (deep traps, low temperatures). Under illumination at first all non-equilibrium carriers are trapped, so a considerable time interval can pass before during excitation the quasi-Fermi level is displaced enough to provide the increase of free carrier concentration.

The possible scheme of electronic transitions leading to the photoactivation phenomenon is shown in Fig. 2. If the lifetime of minority carriers with respect to recombination levels is assumed to be much less than the times of capture and recombination for majority carriers, in our case of band-to-band absorption all transitions of interest occur in the way analogous to the transitions when the excitation takes place from the recombination levels to the band. Then the model suggested in (7) for level-to-band excitation can be used, and the parameters of

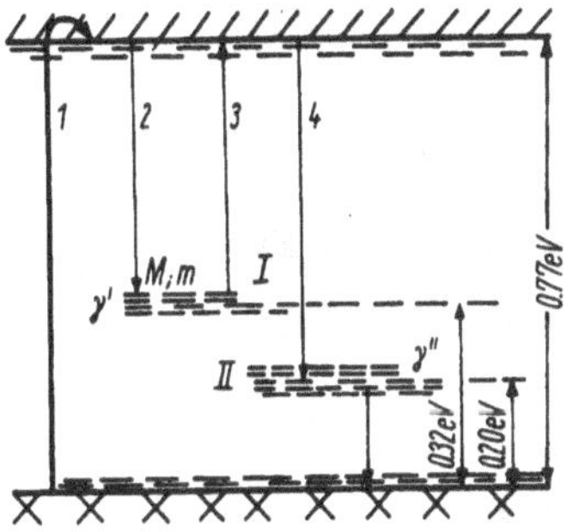

Fig. 2. Possible scheme of the electronic transitions explaining photoactivation of vitreous $Tl_2Se-As_2Te_3$.
1: F; 2: $\gamma' n(M-m)$; 3: Pm; 4: $\gamma'' n(n+m)$

the traps can be determined from the following kinetic equations:

$$\frac{dn}{dt} = F - \gamma' n(M-m) + Pm - \gamma'' n(n+m) \;;$$

$$\frac{dm}{dt} = \gamma' n(M-m) - Pm \;.$$

When t = 0, n = m = 0. Here n, m are the concentrations of free and trapped carriers, respectively, M is the concentration of traps, γ' and γ'' are the probabilities of capture of the carriers by traps and ionized recombination centres, respectively, F is the excitation rate, and P the probability of thermal emptying of the traps.

For a small time interval ($t \ll 1/\gamma''$) this system of equations is obviously simplified. If the interval during which only the traps capture the majority carriers is large enough the situation arises that all carriers are trapped and the concentration of free carriers is constant with time, i.e.

$$\frac{dn}{dt} = 0 \;;$$

$$\frac{dm}{dt} = F \;.$$

This situation corresponds to the plateau on the curve of the photoresponse increase. The plateau begins at $t_1 \approx 1/(\gamma' M)$ and ends at $t_2 \approx M/F$. The values t_1 and t_2 correspond to the lifetime of the carriers with respect to the traps and the time of complete filling up all the traps. The height of the plateau is defined by the additional concentration of carriers during the photoactivation process

$$\Delta n = \frac{F}{\gamma' M} .$$

Using the above relations for t_1, t_2, and Δn it is easy to determine the concentration of the traps M, the capture probability γ', and the capture cross section $s = \gamma'/v$ for the traps (v is the thermal velocity of the carriers):

$$M = \frac{t_2}{t_1} \Delta n; \qquad \gamma' = \frac{1}{M t_1} .$$

The values of t_1, t_2, and Δn were determined from the experimental data to 1 s, 120 s, and 1.8×10^{8} cm^{-3}, respectively. So $M = 2.16 \times 10^{10}$ cm^{-3}, $\gamma' = 4.6 \times 10^{-11}$ cm^{-3} s^{-1}, and $s = 8.5 \times 10^{-18}$ cm^{2}.

Thus the photoactivation phenomenon like a number of other effects indicating the complication of the energy spectrum of local states in vitreous semiconductors, can be used for the determination of the parameters of the corresponding levels. In this connection the temperature dependence of the photoactivation phenomena is of particular interest because it can help to clear up the nature of the local centres in vitreous semiconductors.

References

(1) R.H. BUBE, Phys. Rev. 99, 1105 (1955).

(2) A. ROSE, Phys. Rev. 97, 322 (1955).

(3) R.H. BUBE, Proc. IRE 43, 1836 (1953).

(4) B.T. KOLOMIETS, T.M. MAMONTOVA, and G.I. STEPANOV, Fiz. tverd. Tela 7, 1630 (1965).

(5) G.I. STEPANOV, Autoref. Cand. Dissertation, Leningrad 1968.

(6) A.M. ANDRIESH, Autoref. Cand. Dissertation, Leningrad 1964.

(7) K.W. BÖER and H. VOGEL, Ann. Phys. (Germany) 76, 1869 (1949).

(Received July 25,1972)

Erratum

To the Short Note by J.E.L. BISHOP

Dynamic Ferromagnetic Domain Wall Response to Incremental AF Fields and Becker Wall Area Measurement Technique

phys. stat. sol. (a) 11, K75 (1972)

the author requests the following rectifications:

Just below the centre of page K76 read correctly $W = 4\omega M_s^2 d^3/(\pi^3 \varrho \gamma)$ instead of $W = 4\omega M_s^2 d^2/(\pi^3 \varrho \gamma)$ and on page K77, tenth line from the bottom, read correctly $\varrho \approx 47\,\mu\Omega$ cm instead of $\varrho \approx 0.47\,\mu\Omega$ cm.

Pre-printed Titles

Papers to be published

in "physica status solidi (a) - applied research"

Vol. 13, No. 1

Review Article

H.A. BUCKMASTER, R. CHATTERJEE, and Y.H. SHING, Alberta

The Application of Tensor Operators in the Analysis of EPR and ANDOR Spectra

Original Papers

J. KALADĖ, E. MONTRIMAS, and A. PAŽĖRA, Vilnius

Kinetics of Drift Current of Small Charge and Determination of Lifetime of Charge Carriers

J. STRINGER, J. HILL, and A.S. HUGLIN, Liverpool

Hall Effect in Cadmium-Mercury Alloys

V.A. BURAVIKHIN, V.V. LITVINTSEV, V.G. KAZAKOV, A.I. USHAKOV, and V.A. POMAZKIN, Irkutsk

Phase Composition Effect on the Magnetic Properties of Thin Fe-Ni Films

K. THIESSEN and LE TRONG TUYEN, Berlin and Karl-Marx-Stadt

Application of Nematic Liquid Crystals for the Investigation of p-n Junctions and Isolating Layers

L. SCHULTZ and H. FREYHARDT, Göttingen

Anisotropic Critical Currents in Heterogeneous Lead-Sodium Alloys

B. J. MULDER, Eindhoven

Optical Properties of Crystals of Cuprous Sulphides (Chalcosite, Djurleite, $Cu_{1.9}S$, and Digenite)

A.A. GUZEV, G.L. KURYSHEV, and S.P. SINITSA, Novosibirsk

Scattering Mechanisms in Inversion Channels of MIS Structures on Silicon

B.W. JAMES, Salford, Lancashire

Comments on the "Paper Ultrasonic Wave Propagation in Tetragonal Crystals and Measurements of the Elastic Constants of Strontium Molybdate"

K. NAUKKARINEN, Helsinki

Berg-Barrett X-Ray Topographic and Etch Pit Studies of Dislocations in Tellurium Single Crystals

C. PAORICI, C. PELOSI, and G. ZUCCALLI, Parma

Epitaxial Growth of CdTe on CdS Substrates

P. GÖRNERT, Jena

Magnetocrystalline Anisotropy of Co^{2+}-Substituted Mg-Ferrite Single Crystals with Different Inversion Degrees

G.C. BHAR and R.C. SMITH, Southampton

Optical Properties of II-IV-V_2 and I-III-VI_2 Crystals with Particular Reference to Transmission Limits

G. BAUER and F. KUCHAR, Wien

Impact Ionization in Heavily Doped n-InAs and n-InSb

I. KUNZE and P. MÜLLER, Berlin

ITC Investigations of CdF_2

C. RAU, München

Galvanomagnetic Measurements of the Annealing Kinetics in Quenched Aluminium

H. EBEL and M.F. EBEL, Wien

Photoelectron Spectrometric Investigations of Thin Gold Films Evaporated onto Glass

M.J. AUBIN and A.T. TRUONG, Quebec

Scattering in High Mobility $Cd_{2.8}Zn_{0.2}As_2$

R.G. MAEV and E.F. TOKAREV, Moscow

The Features of the Thermo-Electric Skin Effect in the Presence of Shallow Levels

S. ISOMURA and K. MASUMOTO, Tokyo

Preparation and Some Optical Properties of $ZnGeP_2$ and $CdSiP_2$

M.N. CHANDRASEKHARAIAH and S. RANGANATHAN, Varanasi, India

Interpretation of Field-Ion Images from Ordered Ni_4Mo

T. MOOKHERJI, Huntsville, Alabama

Electron Spin Resonance of Ultraviolet Radiation Induced Defects in ZnO Thermal Control Coating Pigment

R. VANSELOW and R.D. HELWIG, Wisconsin-Milwaukee

Destruction and Microcratering of Field Electron Microscope Tips by Microparticles

S.I. KOVALENKO, E.I. INDAN, and A.A. KHUDOTEPLAYA, Kharkov

An Electron Diffraction Study of Thin Films of the Binary Mixtures of Rare Gases

W. BÜRGER, Hermsdorf

Lorentz Electron Microscopic Studies on the Motion Threshold Field of Cross-Tie Walls and on the Behaviour of Cross-Tie Walls at Inhomogeneities in Thin Magnetic Fields

W.J. TROPF and D. KUHLMANN-WILSDORF, Charlottesville, Virginia

The Migration of Supersaturated Constituents to Surface Sinks

T.J. MAGEE, R.H. MORRISS, and C.D. MELVIN, New Orleans, Lousiana

Diffraction Contrast from Small Spherical Voids in Silicon

R.P. HUEBENER and R.T. KAMPWIRTH, Argonne, Illinois

Intermediate State Structure in Superconducting Films of Indium

J. STOEMENOS, S. KOKKOU, and N.A. ECONOMOU, Thessaloniki

On the Superstructure of Germanium Telluride Single Crystalline Films

E. HARTMANN and W. WINDSCH, Leipzig

Changes of the Domain Structure in Vanadyle-Doped TGS Observed by EPR

W. GERLACH, H. SCHLANGENOTTO, and H. MAEDER, Frankfurt/Main

On the Radiative Recombination Rate in Silicon

D. CHATAIN, C. LACABANNE et M. MAITROT, Toulouse et Villeurbanne
Étude des relaxations dans le polyhexamethylene adipamide (nylon 6.6) par la méthode des thermocourants de depolarisation

H. KATTO and C.T. SAH, Urbana, Illinois
Frequency Response of the Surface State Admittance in Weakly Inverted Thin SiO_2-Si MOS Capacitors

P. GERTHSEN, K.H. HÄRDTL, and A. CSILLAG, Aachen
Mobility Determinations from Weight Measurements in Solid Solutions of $(Ba,Sr)TiO_3$

M.R. BENNETT and J.G. WRIGHT, Norvich
Amorphous Films of the Transition Elements

K. YONEMITSU, Tokyo
Variation of the Electrical Resistivity and the Hall Coefficient of Al-Ag Alloys during Pre-precipitation

J. SWORAKOWSKI, Wrocław
Photoemission from Molecular Crystals

L.M. VALIEV, I.G. KERIMOV, S.KH. BABAEV, and Z.M. NAMAZOV, Baku
Galvanomagnetic Properties of $CuCr_2Se_4$

Papers to be published
in "physica status solidi (b) - basic research"
Vol. 53, No. 1

Review Article

W. HAUBENREISSER, W. BRODKORB, A. CORCIOVEI, and G. COSTACHE, Jena and Bucharest
A Status Report of Green's Function Theory in Uniaxial Ferromagnetic Unpinned Thin Films

Original Papers

YA.A. IOSILEVSKII, Moscow
The Debye Frequency Tensor of Single Crystals

D.L. RODE, Murray Hill, New Jersey
Electron Mobility in Ge, Si, and GaP

R. BRUNO and H.J. JODL, München
Luminescence Side Band and Raman Spectra

J.A. SUSSMANN and Y. WEISSMANN, Yavne and Tel-Aviv
Application of the Quantum Theory of Diffusion to H and D in Niobium

H.E. SCHAEFER and G. LAMPERT, Stuttgart
The Ferromagnetic After-Effect at 55 K in Electron-Irradiated Nickel

P. KRYGEL, J. LIPPOLD, and S. SCHWABE, Berlin
Evidence of Anomalous Z_1 Dependence of Interaction Potential by Multiple Scattering Experiments

L.E. VOROBEV, V.I. STAFEEV, and A.U. USHAKOV, Leningrad
Infra-Red Birefringence on Hot Electrons n-Type Ge

S. GRANDE, V.A. SHUTITOR und L.N. FERSHTAT, Leipzig und Leningrad
Akustische Kernresonanz an Na^{23} im NaF-Einkristall

KWOK LEUNG YIP and W. BEALL FOWLER, Bethlehem, Pa.
The Electronic Structure of NaCl:Cu

A. HUBERT, Stuttgart
Attractive Interactions between Flux Lines as Derived from a Generalized Neumann-Tewordt-Functional

J. DUNDURS, Evanston, Illinois
Dislocation and Disclination Contents of Line Defects

A.S. SALEH and H.Y. FAN, Lafayette, Indiana
Optical and Magnetooptical Absorption in Pure HgTe

P.S. GROVER and R. CHAUDHRY, Delhi
Electronic Structure of Binary Disordered Alloys (Al-Mg)

V.P. SOLDATOV, V.I. STARTSEV, T.I. VAINBLAT, and L.A. DANILENKO, Kharkov
On Creep Strain Delay at the Superconducting Transition

R.G. SELSBY, Rio Piedras, Puerto Rico
The Applicability of Diffusion Theory to the Study of Luminescent Emission from a Doped Organic Crystal

W. SCHILZ, Hamburg
Propagation of Surface Spin Waves in the Inhomogeneous Field of Rectangular YIG Samples

G. MUNSCHY, Strasbourg
On the Exciton Binding to Neutral Impurities in Semiconductors

H. BAESSLER and H. KILLESREITER, Marburg
Hot Carrier Injection into Molecular Crystals and Its Relevance to the Field Dependence of Photocurrents

H. KILLESREITER and H. BAESSLER, Marburg
Field Dependence of Exciton-Induced, Contact-Limited Photocurrents in Molecular Crystals

R.R. DOGONADZE and A.A. KORNYSHEV, Moscow
Phenomenological Theory of Polar Systems

H. DEN HARTOG, D.O. WELCH, and B.S.H. ROYCE, Princeton, New Jersey
The Diffusion of Calcium, Phosphors, and OD^- Ions in Fluorapatite

N. CONELL, Cambridge, Massachusetts
The Temperature Dependence of the Absorption Edge in Some Amorphous Semiconductors

S.C. BHARGAVA and P.K. IYENGAR Trombay, Bombay
Spin Fluctuations of Fe^{3+} Ions in Cobalt-Zinc Ferrites Using the Mössbauer Effect

R.C. CHOPRA and P.W. PANDON, Bombay
Hyperfine Magnetic Fields at As in Fe, Co, and Ni

K. CHOUNTAS, W. DÖNITZ, K. PATHANASSOPOULOS, and G. VOGL, München
Influence of Neutron Irradiation at Low Temperatures on Internal Friction and Bulk Modulus of Silver

M. GLOGAROVA, C. FRENZEL, and E. HEGENBARTH, Prague and Dresden
The Behaviour of $(NH_4)_2Cd_2(SO_4)_3$ under Pressure

R. HARRIS, M. PLISCHKE, and M.J. ZUCKERMANN, Montreal
Coherent-Potential Approximation for Alloys with Random Off-Diagonal Elements: Heisenberg Ferromagnet

B. PIETRASS, Dresden
Selfconistent Lattice Dynamical Theory of Structural Phase Transitions in Perovskite-Type Crystals

W. JOHN, G. LEHMANN, and P. ZIESCHE, Dresden
The Korringa-Kohn-Rostoker and Augmented Plane Wave Method for Non-Spherically Symmetric Muffin-Tin Potentials

U. KNOHL, München
Zur Ultrarot-Absorption in Ionenkristallen

K.L. NGAI, Washington
Comments on Field Emission in Solids: Multiple Particle Tunneling and Singularities

G. HEGER and R. GELLER, Tübingen

Neutron and Mössbauer Studies of the Planar Antiferromagnet $KFeF_4$ in the Critical Region

H. STOLZ and R. ZIMMERMANN, Berlin

Exchange Interactions and Local Field Effects in Exciton Theory

M. CARDONA, C.M. PENCHINA, W. GUDAT, E.E. KOCH, and P.Y. YU, Stuttgart and Hamburg

Optical and Photoelectric Properties of the Lead Chalcogenides

B. ALBANI, F. GOEDSCHE, G. HEBER, and G. RÖPKE, Dresden

Functional Integral Approach for an Extended Hubbard Model: Two-Centre Coupling

O. GOEDE and A. BÜCHTEMANN, Berlin

Piezoemission Study of Free Excitons in CdS

J. BILLE, H. LIEBING, and P. MENGEL, Karlsruhe

Exciton Interaction in CdSe

O.P. GUPTA and P.R. WALLACE, Montreal

Effect of Trigonal Warping on Landau Levels of Graphite

19. Magnetic Resonance
20. Optical Properties
    - 20.1 Spectra. Optical Constants (X-Ray Spectra See 20)
    - 20.2 Lasers
    - 20.3 Luminescence (X-Ray Spectra See 20)
21. Metals. Alloys
    - 21.1 Transition Metals (Group Sc to Zn) and Their Alloys
        - 21.1.1 Fe and Fe Alloys
    - 21.2 Alkali Metals
    - 21.3 Alkali-Earth Metals
    - 21.4 Rare-Earth Metals
    - 21.5 Actinides
    - 21.6 Noble Metals
    - 21.7 Semi-Metals (Semiconducting Alloys See 22.7)
22. Semiconductors and Ionic Crystals
    - 22.1 Elements
        - 22.1.1 Germanium
        - 22.1.2 Silicon
        - 22.1.3 Group VI Elements
    - 22.2 III–V Compounds
        - 22.2.1 Arsenides
        - 22.2.2 Phosphides
        - 22.2.3 Antimonides
        - 22.2.4 Mixed Crystals
    - 22.3 IV–IV, II–V, and III–VI Compounds
        - 22.3.1 Mixed Crystals
    - 22.4 II–VI Compounds
        - 22.4.1 Sulphides
        - 22.4.2 Selenides
        - 22.4.3 Tellurides
        - 22.4.4 Mixed Crystals
    - 22.5 Halides
        - 22.5.1 Silver Halides
        - 22.5.2 Alkali Halides
        - 22.5.3 Alkali-Earth Halides
        - 22.5.4 Mixed Crystals
    - 22.6 Simple Oxides (CdO, ZnO See 22.4)
        - 22.6.1 Mixed Crystals
    - 22.7 Semiconducting Intermetallic Compounds
    - 22.8 Tertiary and Higher Compounds
        - 22.8.1 Salts of Oxy-Acids (Carbonates, Phosphates, Silicates Including Mica etc.)
        - 22.8.2 Spinels, Garnets, Ferrites
    - 22.9 Organic Semiconductors
23. Solidified Gases

Attention

The preceding classification scheme is closely akin to the scheme used by Physics Abstracts to classify the various fields of solid state physics. A paper will either be classified under one or more of the subsections or, if none of these are relevant, it will just appear under the general section heading (i. e. sections and subsections are equivalent in this sense).

The *subject classification* given at the head of each paper begins with the main subject matter of the paper, and this is followed, if necessary, by figures referring to subsidiary subject matter and the type of solids investigated.

The papers published in this issue are listed in the order of their subject classification at the end of the table of contents.

32914

Manuscripts and letters for physica status solidi (b) — basic research and physica status solidi (a) — applied research

should be sent to

DDR, 102 Berlin, Neue Schönhauser Str. 20

or

Internationaler Buchversand GmbH, 1 Berlin 19, Eschenallee 21.

The object of "physica status solidi (b)" and "physica status solidi (a)" is to publish rapidly important and topical results concerning fundamental research in the field of solid state physics and aspects of application and experimental techniques of fundamental solid state research, respectively. The issues of "physica status solidi (b)" and "physica status solidi (a)" will be published on the 1st and the 16th day of each month, respectively. Either series will contain the following categories:

1. Review articles written by arrangement with the editors. They will be remunerated.
2. Original papers selected from those which have not been submitted for publication elsewhere, after being reviewed. (The latest date for acceptance of contributions is the 8th and the 23rd day of each month for "physica status solidi (b)" and "physica status solidi (a)", respectively, 50 days before the delivery of the journal.)
3. Short notes published in the off-set part of the journal. They are, if necessary after being reviewed, selected by the editorial staff. (The latest date for acceptance of manuscripts for this part is the 12th and the 27th day of each month for "physica status solidi (b)" and "physica status solidi (a)", respectively, 16 days before the delivery of the journal.)
4. Pre-printed titles of papers to be published in the next issue of both series.

Editorial Information

Manuscripts should be submitted preferentially in English. Besides, papers in German, French, and Russian are admitted. Authors are asked to observe the following rules:

1. Manuscripts have to be written as concisely as possible. As most, 20 and 3 pages (28 lines per page; abstracts, tables, references, captions of figures included) are admitted for original papers and short notes, respectively. All manuscripts should be type-written on one side only. Two copies are required. Badly arranged manuscripts are returned.
2. The author's name, initials, and institution should be given.
3. An abstract of about ten lines in English and a second one in German, French, or Russian are required for original papers.
4. Formulae should not be written too small and with pencil. Separate lines for formulae are desirable. Indistinct writing will lead to a return of the manuscript.
5. Drawings should be enclosed on separate sheets of paper; photographs should be submitted, on white paper, only for review articles and original papers. Duplicates are required.
6. Captions for the figures should be submitted on separate sheets of paper, and special instructions with regard to the size of illustrations should be given on these.
7. References should be numbered in order of appearance and should be given, on a separate sheet, in the form used in physica status solidi.
8. Units, symbols, etc. should be used in accordance with the recommendations given by the IUPAP.
9. Each paper should contain a proposal for the subject classification according to the classification scheme appearing on the second and third cover of each issue.

The authors will be sent galley proofs of review articles and original papers. Proof reading of short notes and pre-printed titles will be done by the editorial staff.

For each original paper and short note a total of 75 reprints is available free of charge. The number of reprints free of charge per review article is 25. Requests for additional reprints (against payment) should be given together with the submitted manuscript. **The deadline for orders of additional reprints is the 15th and the 30th day of each month** for "physica status solidi (b)" and "physica status solidi (a)", respectively, two weeks before the delivery of the journal.

FSC
www.fsc.org
MIX
Papier | Fördert gute Waldnutzung
FSC® C083411

Kolibri 360 GmbH
Ferdinand-Jühlke-Straße 7
99095 Erfurt, Deutschland
produktsicherheit@kolibri360.de